Africa South of the Sahara

TEXTS IN REGIONAL GEOGRAPHY
– A Guilford Series –
Edited by James L. Newman, Syracuse University

THE EUROPEANS: A GEOGRAPHY OF PEOPLE, CULTURE, AND ENVIRONMENT
Robert C. Ostergren and John G. Rice

AFRICA SOUTH OF THE SAHARA: A GEOGRAPHICAL INTERPRETATION
SECOND EDITION
Robert Stock

Africa South of the Sahara

A Geographical Interpretation

– *Second Edition* –

ROBERT STOCK

THE GUILFORD PRESS
New York London

© 2004 The Guilford Press
A Division of Guilford Publications, Inc.
72 Spring Street, New York, NY 10012
www.guilford.com

Printed in the United States of America

This book is printed on acid-free paper.

Last digit is print number: 9   8   7   6   5   4   3   2   1

Library of Congress Cataloging-in-Publication Data

Stock, Robert F.
   Africa south of the Sahara : a geographical interpretation / Robert Stock.— 2nd ed.
      p.   cm. — (Texts in regional geography)
   Includes bibliographical references and index.
   1-57230-868-0 (pbk.)
   1. Africa, Sub-Saharan—Geography.   I. Title.   II. Series.
DT351.9.S76 2004
916.7—dc22

                                    2004002113

For Evelyn, Matthew, and Rachel

Preface

Africa South of the Sahara: A Geographical Interpretation, Second Edition, attempts to take a fresh look at the geography of the continent. In studying regions such as Africa south of the Sahara, geographers have moved beyond their traditional emphasis on "where" and "what" to new perspectives examining the political, social, economic, and environmental dynamics of how regions function. This book focuses on a range of contemporary issues that challenge Africans, but always in the context of their historical genesis. It seeks to challenge long-established myths and conventional truisms about Africa and its people. It will be, I hope, for many but an initial step in a longer process of learning about Africa, and finding ways of supporting, even from afar, the African quest for self-determination and development.

Over the past decade, Africa has experienced many changes. Some of these have been for the worse, such as the rapid growth of the HIV/AIDS pandemic and political anarchy in countries such as the Democratic Republic of the Congo and Liberia. Yet there have also been significant accomplishments—democratization and the increased role of civil society to give but one example—that receive far too little recognition in the rest of the world. New contributions to the academic literature have challenged several taken-for-granted ideas about Africa.

While this second edition retains the same basic structure as the first, there have been substantial revisions to the text. The discussion of physical geography has been increased from one to three chapters. There are several new chapters on topics such as children, water resources, and Western perceptions of the Continent. Significant changes have been made to the content of most other chapters. Each chapter now contains a list of pertinent websites, reflecting the massive growth of the Internet as a resource for research on African issues.

The maps and diagrams reflect the care and attention to detail of two fine cartographers. Keith Bigelow prepared the new and revised figures, while other figures reproduced from the first edition are the work of Ross Hough. Many of the photographs were obtained from the Canadian International Development Agency photo library. Thanks to Pierre Vachon for his assistance in obtaining these photographs. It is also a great pleasure to have been able to include several photographs taken by a number of former students of mine—Deborah Alsen, Keith Child, Roy Maconachie, and Kevin Rondi.

I am very grateful for the insightful comments on the initial draft of the manuscript by Jim Newman, Greg Myers, and another anonymous reviewer. Comments from many users of the first edition—from students as well as from faculty—provided insights into what they liked and what they hoped to see the next time. I acknowledge with thanks the encouragement of former colleagues at Queen's University, particularly Bruce Berman, Jonathan Crush, John Holmes, Alan Jeeves, Jayant Lele, Colin Leys, David Macdonald, Barry Riddell, and Bob Shenton. Finally, to my wife, Evelyn Peters, my heartfelt thanks for her encouragement and unwavering support.

Contents

Africa South of the Sahara

Introduction

Africa in the New Millennium

Roots of Crisis

During the 1950s, Africa was a continent awakening to the prospects of independence. Ghana's independence in 1957 received extensive publicity; not only was it the first black African state to regain its independence, but it also had in Kwame Nkrumah a charismatic leader who gave voice to the aspirations of people throughout the continent.

Africa in the 1960s could be characterized as a continent in transition. During the decade, 31 countries in Africa became independent, 17 of them in 1960 alone. It was a time of great optimism, because independence implied that Africans would again control their own destinies. Governments could pursue policies that would be in the national interest, rather than for the benefit of colonial overlords. Development theorists such as Walter Rostow claimed that modernization was imminent, provided that appropriate policies were put in place to encourage investment. Volunteers from the U.S. Peace Corps and from other countries arrived by the thousands, optimistic that their training, energy, and enthusiasm could speed the development process.

By the early 1970s, much of the initial optimism had faded; Africa was a continent in limbo. Development had proved much more complex and difficult to achieve than modernization theorists or political leaders had envisaged. Although the first decade of independence had brought some notable progress— large increases in school enrollment, for example—the benefits were very unevenly distributed. Regional, ethnic, and political tensions were unleashed in some countries, resulting in violent clashes and coups. As the decade proceeded, the economies of more and more countries faltered in the face of world petroleum price increases, coupled with stagnating demand and low prices for other primary products. The 1970s also witnessed the Sahelian drought, and with it, the widespread recognition of an African environmental crisis. Whereas some analysts saw the growth of economies in a few countries (particularly Kenya and Côte d'Ivoire) as signs of hope, many others viewed the future with growing pessimism.

During the 1980s, it became clear that Africa was a continent in decline. The news from Africa was almost always bleak: starving children in Ethiopia and Somalia; civil wars in Sudan, Angola, and other countries; and the

3

slaughter by poachers of elephants and other wildlife species. Countries such as Zambia and Tanzania had to abandon their African socialist models of development because of reduced export revenues and growing debts. The growing burden of debt affected all countries, even those such as Côte d'Ivoire that had seemed most prosperous. Few new development projects could be undertaken. Existing infrastructure— roads, for example—deteriorated because there was not enough money for maintenance.

By the 1990s, the prevailing view was that Africa was a continent in crisis. The rapid spread of HIV/AIDS was beginning to pose an ominous threat to economies, societies, and health care systems. In addition, the economic situation continued to deteriorate. Even in countries like Ghana, where interventions by the International Monetary Fund (IMF) and the World Bank succeeded in creating a semblance of stability, the situation was better characterized as stable poverty than as stable prosperity. In Liberia, Somalia, Rwanda, and the Democratic Republic of the Congo (formerly Zaire), bloody civil wars brought about the collapse of the state, and with it increasing anarchy and suffering. In the absence of new, innovative ideas on how real progress toward development might be achieved, Africa was largely left on its own, without committed assistance from abroad.

During this first decade of the new millennium, Africa is a continent at the crossroads. Most African countries achieved relatively little economic and social progress during the 1990s. Africa has largely been passed by, as the rising tide of global economic growth has transformed many other parts of the Third World. HIV/ AIDS has had a devastating impact on all aspects of the economy and society of many African countries, and continues to grow explosively in areas that were previously relatively less affected. Recognizing that Africa is at a crossroads in its development, Western leaders have begun to speak about providing substantial support for African development, in the form of debt relief and new aid programs. Many African leaders have also recognized the need for renewal, and have addressed this challenge by creating a new pan-African organization, the African Union, to replace the increasingly ineffective Organization of African Unity. Much depends on the translation of these good intentions into sustained action over the next several years.

Seeds of Hope

Although the overall situation in Africa south of the Sahara has deteriorated, there were several positive political developments during the 1990s, the most dramatic being the dismantling of apartheid in South Africa. The release of Nelson Mandela in 1990, after 27 years of imprisonment, set the stage for negotiations on a constitution to give all South Africans— regardless of race—rights of participation in the political process, as well as rights of residence and property ownership anywhere in the country. In a referendum in March 1992, white South Africans voted strongly to endorse the ongoing reform process. Prolonged negotiations produced an agreement for constitutional reform among groups representing most South Africans, leading to the country's first fully democratic, nonracial elections in April 1994, and a successful transition to black majority rule. Contrary to some predictions, the transition to black majority rule brought neither political chaos nor economic ruin.

These developments in South Africa have brought about important changes elsewhere in the region. After many years of pressure to leave Namibia, South Africa withdrew in 1990 and handed over power to an elected government led by the South West African People's Organization (SWAPO), the movement that had led the anticolonial struggle for three decades. South Africa also ended almost two decades of efforts to destabilize Angola and Mozambique. South Africa's withdrawal of support for insurgent groups has made possible political accommodation and reconstruction in these countries for the first time since they achieved independence in 1975. In the longer term, the end of apartheid has created opportunities for increased trade and regional cooperation between South Africa and neighboring countries. South Africa has played a constructive role in African relations—for example, acting as mediator in a number of disputes.

Since 1990, there has been a rapidly growing movement toward political pluralism. In response to increasing public pressure, the majority of military dictatorships and one-party states have moved toward multiparty political systems. Democratic changes in government have taken place peacefully in several countries, including Zambia, Mali, and Benin. Although it remains to be seen how thorough and successful the democratization movement will be, any development that makes autocratic leaders more accountable to the people is cause for hope.

The end of the Cold War has brought some benefit to Africa as a whole, even though the reduced flow of aid seems to have been one of its immediate consequences. Several wars and insurrections in the past were fueled largely by superpower rivalries. Millions of Africans have died in Ethiopia, Sudan, Angola, and Mozambique (to name only the most important examples) in wars sustained by arms shipments from the superpowers, whose primary interest in the conflicts was to counter each other. A promise of renewed development assistance made by the world's most powerful countries, the G8, in 2002 offers hope that global responsibility to help support African development has finally been recognized in more prosperous nations. This increase in aid has yet to materialize, however—and, judging by past performance, it may fall far short of expectations.

In any case, future development in will depend primarily on the mobilization of African resources and know-how. If greater African self-reliance means that more attention is paid to indigenous resources, significant benefits may be gained. In fields such as ethnomedicine and peasant agriculture, researchers are developing a greater appreciation for the inherent logic and value of African ways of doing things. The possibility of supporting and ultimately building upon what ordinary Africans already know and do represents an exciting alternative to development strategies that rely exclusively on the state and that ignore what people themselves can contribute to the development process.

Development is not only about government initiatives or international aid. It is about recognizing and building upon the accumulated experiences, talents, and energy of ordinary Africans. Across Africa, thousands of national and community-based nongovernmental organizations (NGOs) are engaged in efforts to improve the quality of life in their communities. The activities of NGOs range from reforestation projects, to support for persons with AIDS, to political agitation on issues of local importance. Grassroots initiatives are not new; cooperative efforts in support of development have been ongoing for countless generations. For example, women in many African societies have long practiced forms of revolving credit; typically members of a group contribute a set amount each week, and take turns collecting the entire sum. Revolving-credit schemes help women to raise capital for business or personal needs, but also serve as social support groups for those who are members.

Organizing Themes in the Geography of Africa South of the Sahara

This book is intended to be an introduction to contemporary Africa south of the Sahara. It examines the sociocultural, political, economic, and environmental processes that help to explain the patterns of human utilization of the continent and its resources, as well as the dynamics of change in Africa's geography.

Because the focus is on Africa south of the Sahara, several African nations are not discussed in this book. The five countries of North Africa—Egypt, Libya, Tunisia, Algeria, and Morocco—have stronger cultural and historical ties to the Mediterranean and southwestern Asia, and thus they have been excluded. The Indian Ocean island nations of Mauritius and Seychelles, although members of the African Union, have likewise been excluded. Not only are these states demographically and culturally distinct, but also their socioeconomic paths have increasingly diverged from those of continental African states. Other island nations—the Comoros, Madagascar, Cape Verde, Equatorial Guinea, and São Tomé e Príncipe—have been included because of their closer proximity, both spatially and socioeconomically, to continental Africa. South Africa has also been included, despite its unique political and social history and its greater modern development. With majority rule in place, South Africa has become the most influential member of the family of African nations.

The book is divided into ten units of three chapters each. Individual chapters explore different aspects of broad topic areas such as physical environment, demography, natural resources, and urban economies and societies; each of these represents a distinct area of scholarship with its own questions and debates. Although individual chapters explore particular topics, a number of broad themes draw the chapters together. These represent some of the more important points of debate in African studies, and as such are themes around which further reading or term assignments could be structured. Several of these themes are introduced below.

Unity in Diversity

The question of whether Africa south of the Sahara is better characterized by its unity or by its diversity has had a long history of debate within African studies. This book acknowledges the importance of both perspectives, but gives precedence to the theme of unity; that is, it emphasizes "unity in diversity."

Unity and diversity are both evident in Africa's physical geography. The ancient Precambrian rocks of Africa have been warped, faulted, weathered, and eroded over hundreds of millions of years, giving rise to a landscape characterized by vast plateaus and plains with comparatively few complicating physical features. Yet closer inspection reveals considerable topographic diversity, ranging from snow-capped volcanoes to deep rift valleys. The continent may be divided into vast regions sharing a similar climate and ecology: desert, semidesert, savanna, and tropical forest. Nevertheless, when these ecological regions are compared, the differences between them are immense.

Unity and diversity are also evident in the cultural geography of Africa south of the Sahara. There are thousands of distinct ethnic groups, each possessing a particular set of cultural attributes; several individual countries are home to between 100 and 300 ethnic groups. Despite the evident complexity of the African cultural map, we can also point to elements of emerging cultural unity in the form of stronger national and pan-African identities.

Unity and diversity are evident in the economic geography as well. African countries face significant problems of underdevelopment, including inadequate infrastructure, a weak industrial sector, and heavy reliance on raw-material exports. They are characterized by dual economies, consisting of a large indigenous sector of small-scale producers and a small modern sector. Not all African economies are identical, however. Some depend on mineral exports, whereas others rely on agriculture. Africa's economic geography also reflects attempts at different times and in different countries to implement a wide range of economic models: socialist, capitalist, and mixed.

The thematic structure of the book serves to emphasize the striking similarities in culture, historical evolution, and contemporary political and economic development, rather than the inevitable differences found among the 46 political units that constitute Africa south of the Sahara. Two or three vignettes are included in each chapter as a reminder of the complexity and diversity of the unique geographies of specific localities. The vignettes contain case studies that help to bring Africa's stories to life, as well as to illuminate a number of key development debates.

Underdevelopment and Development

Africa south of the Sahara is the poorest of the world's megaregions. In the 1990s, the number of African countries with per capita annual incomes less than $500 increased from 28 to 33. Of the 48 countries with per capita annual incomes of $500 or less, 33 are located in Africa south of the Sahara. In most of these countries, economic growth failed to keep pace with the rate of population increase between 1975 and 2000.

Except for Liberia, all parts of Africa south of the Sahara have experienced European colonial rule. Structures established during the colonial era have contributed significantly to African underdevelopment. The colonial legacy includes national boundaries that pay little regard to physical and cultural realities, as well as externally oriented economies based on the export of raw materials. The cultural legacy of colonialism is evident in religion (Christianity), education, language, and values. These imprints have proved to be very resilient indeed.

Although the colonial era has shaped contemporary Africa in many ways, the historical legacy of precolonial Africa is also important, especially in a symbolic sense. For inspiration and a sense of identity, African intellectuals look back to the accomplishments of their ancestors: the art of Ife and Benin; the architecture of Zimbabwe and Lalibela; the intellectual, political, and economic accomplishments of ancient Ghana and Mali.

The history of comparatively recent times has often had a more tangible significance. In countries such as Guinea–Bissau, Namibia, and Zimbabwe, prolonged struggles for independence have created a strong sense of national identity. Several countries, including Ethiopia, Uganda, Sudan, and Angola, have had lengthy periods of civil war and unrest that have set back prospects for development. Then there are the legacies of particular rulers that some countries have inherited: For example, Julius Nyerere (Tanzania) was characterized by many as a visionary, whereas Idi Amin (Uganda) was widely regarded as a brutal despot.

Probably the most important constraint on development is Africa's vulnerability within the global political economy. World markets for African raw-material exports have been characterized by large price fluctuations and a general downward trend, making sound longer-term planning precarious at best. To retain access to international credit, African governments have had to accept the intervention of the IMF and the World Bank, which some have characterized as a new form of colonialism. Moreover, although it has more need for development assistance than any other continent, aid to Africa has stagnated, especially following the decline in international tensions during the 1990s. Since 2001, the preoccupation with ter-

rorism has deflected attention from Africa to other parts of the world and brought about increased isolationism in the West.

Debates concerning African development, and the best strategy for achieving it, have raged continually since the advent of independence. Prospects at first seemed very good; modernization theorists promised that an age of prosperity could be achieved through savings, investment, and specialization. When the results of such policies proved to be disappointing, there came into focus dependency theories emphasizing that Africa had been underdeveloped rather than developed as a result of the European presence.

During the late 1970s and early 1980s, the underdevelopment-versus-development debate continued, albeit with an altered focus. Development theories shifted from an emphasis on the top-down, macro-scale approaches of modernization to bottom-up approaches emphasizing small-scale development projects benefiting local communities. These projects were often designed to enhance the provision of basic needs such as safe water, basic health services, and adequate nutrition.

During the early 1980s, the locus of development debates shifted again. The IMF and World Bank imposed structural adjustment policies emphasizing currency devaluation, less government involvement in the economy, and other measures designed to stimulate economic growth. They argued that inappropriate policies and inefficiency accounted for the disappointing pace of development. Underdevelopment theorists, conversely, attributed the deepening crisis to exploitative trade relationships, the massive burden of debt, and the heavy-handed intervention of the IMF and the World Bank.

The official discourse about development has given undue attention to macro-scale change, focused at the level of the nation-state. Most of the data collected by the United Nations and other agencies are aggregated at the national level, and as such hide regional, class, and gender-based disparities. Macro-scale models pay little attention to the differential impacts of development initiatives, and even less to the diverse priorities and coping strategies practiced within African societies at a local scale. Conventional studies of development have also failed to recognize the importance of complex patterns of exchange and interdependence—flows of people, goods, money, and information—that link different places.

Development is a virtually universal goal. Individual families struggle to make ends meet and to find some means of bettering the lot of their children. Groups of men and/or women come together and pool their resources in cooperative societies, hoping for mutual gain. Every community and region hopes for investment to spur local development; many agitate long and hard to gain official favor. At a national level, macro-scale economic and social policies are designed and implemented with the objective of achieving progress. Likewise, the international discourse about aid and the global economy always focuses on development as an ultimate goal.

Among the themes explored in this book are the context and significance of development decision making at different levels of resolution, ranging from the individual farmer's choice of what crops to produce to the decisions that governments make about development objectives and strategies. This is not to imply that there is necessarily a wide range of choice; individuals and governments may be equally constrained in their choices because of factors beyond their control. Moreover, members of a single family or community are bound to have diverse priorities for development, as well as different opportunities and constraints that affect what they can accomplish. In short, development—whether in theory or in practice—is a very complex and patchy process.

Aspects of the continuing debates about development and underdevelopment recur throughout this book and are of great importance in the interpretation of the condition of African societies and economies. Varying perspectives are presented on key development–underdevelopment debates. It is important that you think carefully about the issues involved, do further research on the subject, and make up your own mind. (Both the "Further Reading" and "Internet Sources" lists at the ends of chapters will assist you in doing this research. See

Vignette I.1 for some suggestions about using the Internet in research on Africa.) Nevertheless, it should be noted that my own views generally conform to an underdevelopment perspective that identifies various external influences, starting with the establishment of the slave trade and continuing to the present day, as a primary reason for the underdevelopment of Africa south of the Sahara.

African Societies in Cultural Context

African cultures reflect certain core values, such as the importance of cooperation and continuity. Social units such as the extended family, the community, and the ethnic group assume great importance; individual rights and individual accomplishments tend to have less importance in African tradition. The techniques of production are often specific to a culture and depend on knowledge passed from one generation to another. So too are rituals and traditions widely undertaken in conjunction with planting, harvesting, and other productive activities. Although the role of cultural inheritance may be less obvious in the daily lives of urban Africans, especially those who have substantial Western education, it does continue to influence values and behaviors. Examples are to be found in the continuing strength of extended families in the city and the maintenance of strong economic and social ties between urban-based and rural-based family members.

Nevertheless, cultural geographers are less inclined now to accept this unitary view of culture (i.e., as a set of attributes common to a group of people). When culture is treated unitarily, there is danger that the importance of differences *within* society may not receive adequate recognition. This argument has particular relevance for the study of African societies—where there tends to be clear differentiation between the cultural and economic lives of men and women, as well as between the lives of the rich and poor, and urban and rural dwellers. The relevance of gender is particularly evident in the division of labor. Women are responsible not only for rearing children and doing housework, but also, in most societies, for producing food and obtaining water

VIGNETTE I.1. Using the Internet to Study Africa

Only a few years ago, there were very few websites representing organizations based in Africa, and relatively few websites anywhere offering detailed information on Africa. This situation has changed, and it has resulted in a rich resource base for scholars and students interested in the continent. These web-based sources may be used to supplement printed materials—books, journals, and periodicals—for research about Africa.

In addition to a "Futher Reading" list, this introductory chapter contains a list called "Internet Sources," giving a range of comprehensive websites especially useful for research purposes. Each subsequent chapter also includes a list of several pertinent websites. Students are encouraged to become familiar with the Internet as a source of information about Africa, both as a resource for completing assignments and as a way of gaining an enriched understanding of current debates and developments.

Here are some suggested strategies for using the Internet as a source of information about Africa (the URLs for the identified resources are listed at the end of the chapter):

1. Use Internet-based research catalogues to identify recent articles and other key sources for research on your topic of choice. *Eldis*, a Development Studies database maintained by Sussex University, is an excellent resource for this purpose.
2. For access to recent articles in numerous (often obscure) academic journals based in Africa, or focused on Africa, see *African Journals OnLine*. Your university will have subscriptions to only a fraction of these journals.
3. Consult websites maintained by global organizations such as the World Bank, United Nations Development Program, and Economic Commission for Africa, for current data about social and economic conditions in African nations.
4. Stanford University Library offers an invaluable gateway to Internet-based resources about Africa. Identify websites for numerous organizations, including many NGOs based in Africa, and examine these websites to let African organizations tell their own stories. Other sites are maintained by groups such as Human Rights Watch that have active programs related to Africa.
5. Use one of several excellent current affairs sites to find out what is being said, on a day-by-day basis, in hundreds of newspapers in every African country; *allAfrica.com* is a comprehensive and easily used point of entry to survey the African print media.

Using the African Internet offers constant reminders of the paucity of resources that restricts the activities of African governments and NGOs. Many governments (especially at the state or regional level) have no website, and many others provide only basic information. Many African universities—perhaps the majority—are also absent from the Internet.

and fuelwood for household use. Past development efforts have often failed because of the assumption that in order to succeed, programs needed only to target men as the heads of households. The belated recognition of the importance of gender differences has given rise to development initiatives focusing specifically on the needs of African women and to a growing literature that explores many aspects of African women's lives.

Moreover, simply distinguishing between men and women, or between rural and urban residents, is not enough. The key to understanding African culture is to recognize the flexibility and fluidity of cultural categories. The meaning of constructs such as ethnicity may vary according to situational factors such as time and place. For example, individuals may express and utilize their ethnicity quite differently in urban and rural settings. Moreover, ethnicity is only one of several overlapping constructs that are used by individuals to define their identity; religion, age, education, and occupation are some others.

The Environment as the Material Basis for Development

African economies and societies depend fundamentally on the environment as a sustenance base. As such, environmental health is inevitably interwoven with economic and societal health. A significant majority of Africans are of primary producers—farmers, herders, fishers, and hunters—who depend directly on environmental resources for their sustenance. For farmers to obtain an adequate crop, they need access to reasonably fertile soil and favorable weather conditions. Not only profits, but also the survival of the family and community, are put in jeopardy when crops fail. African primary producers are conscious of the importance of the environmental resources upon which they depend, as is evident in the traditional strategies of resource protection that are used in many communities.

The prosperity of other Africans is also linked to the environment as a sustenance base. Most export earnings are derived from primary products, principally agricultural goods and minerals. Many urban-based Africans are employed in the large-scale processing of primary products, ranging from oil seeds to wood and minerals. Moreover, the cost of living is greatly affected by the availability of food, which in turn reflects environmental and other factors bearing on productivity in the countryside.

With the increasing global awareness of environmental issues, the question of development's sustainability has assumed great importance. *Sustainability* implies that strategies to meet the needs of current generations must also ensure that the ability of future generations to do likewise is not compromised. The publication in 1987 of *Our Common Future,* the report of the World Commission on Environment and Development, drew attention to these issues. A series of major international conferences on the environment—especially the Rio Earth Summit in 1992; the 1997 Kyoto meetings on global climate change, and the second Earth Summit, held in Johannesburg in 2002—resulted in treaties that seek to move sustainable development from theory to reality.

The sustainability of development takes on particular significance in the African context. The depletion of natural resources, especially soil, forests, water supplies, and wildlife, is of growing concern. Recurring episodes of drought and crop failure point to the seriousness of these environmental crises. Media representations of these issues have tended to portray Africans as either hapless victims or careless perpetrators of environmental crises. However, through research we are learning more about the subtle ways in which African societies manage the environment not only to guard against damage to the resource base, but also to enhance its productivity. It is time to rethink many of the taken-for-granted ideas about Africans and the environment that have had major impacts on development strategies for more than a century.

What happens to the environment, as society's sustenance base, cannot be separated from larger-scale development decisions—ranging from official directives to cultivate soil-depleting crops like cotton, to the construction of large dams that may force farmers and herders onto less productive and more vulnerable lands. In decisions of this sort, the prospects of immediate returns often have taken precedence over questions of sustainability and dangers of long-term effects on the environment.

It is hardly new that ignorance of the needs and structures of African societies and economies has given rise to inappropriate development decisions. People of European heritage have often referred to Africa as "the dark continent," implying not only that its people were dark-skinned, but also that African societies were mysterious and primitive. This book attempts to provide a broad and balanced introduction to the geography of Africa south of the Sahara, thus contributing to a greater knowledge of the continent, its peoples, its problems, and its prospects.

Further Reading

Africanists make use of a large number of national and international bibliographic sources that include the following:

African Historical Dictionary (a series of source-books for individual countries)

Cambridge Encyclopedia of Africa
A Current Bibliography on African Affairs
International African Bibliography

There are several journals of African studies with a wide variety of articles on topics of historical and contemporary interest. Some of the more important of these journals are the following:

Africa
African Affairs
African Studies Review
Canadian Journal of African Studies
Journal of African History
Journal of Modern African Studies
Journal of Southern African Studies
Review of African Political Economy
Rural Africana
Third World Quarterly

Current events and recent trends can be followed through these popular periodicals:

Africa
Africa Report
African Business
New African
New Internationalist
The East African
West Africa

For a list of organizations with a primary interest in some aspect of African development, see:
Review of African Political Economy, no. 54 (1992), pp. 126–132.

Internet Sources

Several extremely valuable websites, maintained by major universities, provide comprehensive listings of printed and online sources. In particular, see the following:

Columbia University. *African Studies Resources.* www.columbia.edu/cu/lweb/indiv/africa

Eldis. *Welcome to the Eldis Gateway to Development Information.* www.eldis.org

Harvard University. *African Studies Links.* www.fas.harvard.edu/~cafrica/links.shtml

Institute of Development Studies, University of Sussex. *Institute of Development Studies Info Services.* www.ids.ac.uk/info

Michigan State University. *MSU Global Access: Africa.* www.globalmichigan.com/geo/africa

Stanford University. *Africa South of the Sahara: Selected Internet Resources.* http://gill.stanford.edu/africa

University of Pennsylvania Library. *African Studies.* www.library.upenn.edu

The following is a comprehensive resource on African academic journals:

International Network for the Availability of Scientific Publications (INASP). *African Journals OnLine.* www.inasp.org.uk/ajol

The following sites provide excellent coverage, on a day-to-day basis, of news from Africa and about particular countries:

Africa on Line. www.africaonline.com
allAfrica.com www.allafrica.com
BBC World Service: Africa. http://news.bbc.co.uk/2/hi/africa
Channel Africa online. www.channelafrica.org
OneWorld Africa. www.oneworld.net/africa
Panapress. www.panapress.com
USAfrica Online. www.usafricaonline.com

In addition to these general sources, look for country-specific sources, such as comprehensive sites on specific countries (e.g., *Nigeria.com*), newspaper websites, and government websites. The sources above will be helpful in accessing these sites.

For current statistical information, see the following:

United Nations Development Programme (UNDP). www.undp.org
United Nations Economic Commission for Africa (UNECA). www.uneca.org
The World Bank. www.worldbank.org

Overview

The three chapters in this section serve as a brief introduction to Africa south of the Sahara and set the stage for the sections that follow.

Chapter 1 introduces the map of Africa. As in any new field of study, many of the names will be unfamiliar at first. Time spent now becoming better acquainted with the countries of Africa, their locations relative to other countries, and some of their basic characteristics will pay dividends later. "Knowing the map" is essential for grasping the geographical context of situations and events, past or present.

Chapter 2 examines the images of Africa that have dominated Western perceptions of the continent. It is argued that these mostly negative stereotypes—for example, in the media and in films—have helped to perpetuate Africa's underdevelopment. Learning about Africa requires us to recognize and reconsider received wisdom about the continent and its peoples. Chapter 2 also points to the growth of alternate views—for example, on indigenous knowledge—that offer a quite different perspective on Africa.

In Chapter 3, the focus turns to the people of Africa. This chapter looks at the changing interpretation of culture in geography. Some of the characteristics that are typical of African cultures are discussed, as is the question of the extent to which African culture represents a unified whole. Chapter 3 also discusses the various cleavages that divide Africans as a whole, as well as Africans who are members of a particular cultural group. These cleavages include ethnicity, religion, social class, and gender. Thus I emphasize the importance of not only the unity but also the diversity of African peoples.

1

The Map of Africa

The English satirist Jonathan Swift, writing in the early 18th century, commented on the prevalent ignorance about Africa and the African people:

> So geographers in Afric maps
> With savage pictures fill their gaps,
> And o'er uninhabitable downs
> Place elephants for want of towns.

A century later, Africa remained, in the public mind, "darkest Africa"—a mysterious and virtually unknown continent (see Vignette 1.1 and Figure 1.1). Even now, in the 21st century, Africa remains the least-known continent. The names of African countries are often in the news, but people generally know too little about these countries to give meaning to what they read and hear. Where is Mali? Is Malawi a different place? Is it Ghana or Guyana that is in Africa? What was the former name of Burkina Faso? Is Equatorial Guinea a part of Guinea? Simple questions such as these are difficult even for college-educated Westerners.

Africa covers a vast territory. At its widest from west to east and at its longest from north to south, the distance is almost the same: approximately 7,500 km. To put this into context, the distance from Los Angeles to New York is 4,470 km. Or, if you prefer, with a surface area of 24.6 million km^2, Africa south of the Sahara is about three times the size of the continental United States (see Figure 1.2).

Just as Africa has occupied a relatively small part of the consciousness of most Westerners, it has also been portrayed on world maps in a way that makes it seem smaller than it actually is. The widespread use of the Mercator and other scale-distorting projections has contributed to misperceptions about the relative sizes of landmasses. Because distortion within a Mercator projection increases markedly with distance from the equator, places at higher latitudes, such as Greenland and Canada, appear much larger than places of comparable size nearer the equator. For example, although Greenland appears to be roughly the same size as Africa on a Mercator projection, it is actually only 7.3% as large, or slightly smaller than Sudan or the Democratic Republic of the Congo.

The Political Map

The contemporary political map of Africa south of the Sahara (Figure 1.3) bears little resemblance to that of a hundred years ago, when the scramble

15

VIGNETTE 1.1. A Continent Not Yet "Discovered": A British Map of Africa, 1807

In 1807, when the highly regarded British cartographer Arrowsmith published a map of Africa (Figure 1.1), European explorers and slave merchants had been visiting coastal regions of Africa south of the Sahara for more than 300 years. The map testifies to how little they had learned about Africa's geography. Vast areas of the continent remain completely blank, lacking even the "elephants for lack of towns" to which Jonathan Swift had alluded.

In West Africa, the interior just beyond the "Coast of Guinea" remained totally unknown. Note, however, the comparatively great detail in the savanna regions south of the Sahara, providing evidence of the considerable significance of the trade linkages from this region to North Africa and beyond. The Niger River appears on the map, but it is shown flowing into Lake Chad—although the map's publication coincided almost exactly with Park's expedition, which demonstrated that the Niger flows into the Atlantic.

Especially intriguing is the range of mountains shown extending the full breadth of the continent, named the Mountains of Kong in West Africa and the Mountains of the Moon in Central Africa. The presence on this map of these mountains that in reality do not exist attests to the power of "received wisdom." These mountain ranges appear on virtually every African

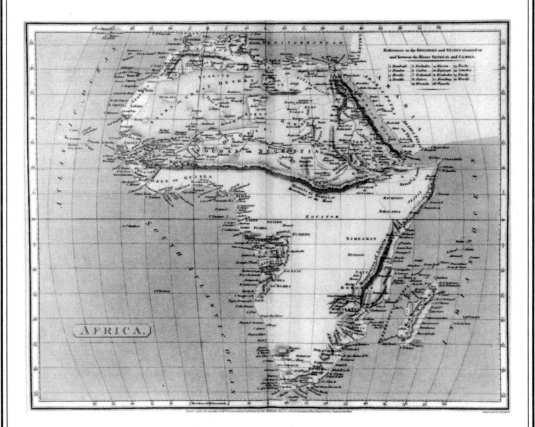

FIGURE 1.1. The Arrowsmith map of Africa, 1807—an indication of Europe's ignorance of Africa other than the coast at the time. Note especially the fictitious mountain range stretching the width of the continent.

VIGNETTE 1.1. *(cont.)*

---

map from the 15th to the 19th century. Cartographers relied primarily on earlier maps for the information they needed; since every map showed the Mountains of Kong, they surely existed! Indeed, some maps published as late as the 1870s show these phantom ranges, even though several European travelers had by then visited the region and failed to find any mountains.

For the most part, Arrowsmith stuck to the geographical "facts" (as he knew them), but he did editorialize in a couple of places, such as his reference to "Wild Hottentots" in South Africa.

Maps such as this, especially when two or more are viewed comparatively, provide fascinating insights into how Africa was perceived and how these perceptions gradually changed over time. They tell us more about Europe than they do about Africa.

---

to carve up the continent among European imperialist powers was in full swing. The details of how the colonial division of Africa was accomplished are discussed in Chapter 8. What is important at this point is to recognize that African borders are recent and often unrelated to either cultural/political realities or natural features. Sometimes this artificiality leads to international disputes, as indicated by the examples of Vignette 1.2 (and illustrated in Figure 1.4).

Another unfortunate colonial legacy is the extreme fragmentation of the political map. In all, there are 46 independent states, some of

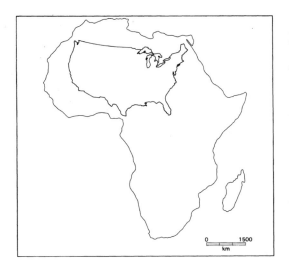

FIGURE 1.2. Africa and the continental United States: Relative sizes.

which are too small to be considered economically viable. The small sizes and national populations of the majority of African states (Table 1.1) are continuing constraints on development. Certain states have shapes that are unusual and unhelpful. Gambia is the most extreme example; it extends 325 km along the Gambia River and is no more than 30 km wide. In addition, except for a short coastline, Gambia is completely surrounded by Senegal.

Fifteen African states are landlocked (Figure 1.5). Most of these states share a common legacy of colonial indifference and neglect. Countries such as Mali, Niger, and Chad served as labor reserve areas from which workers were recruited for the plantations and mines of more prosperous colonies. The (relative) exceptions to this pattern of colonial neglect are Uganda, once described by Winston Churchill as the "pearl of Africa," and Southern and Northern Rhodesia (now Zimbabwe and Zambia, respectively), which were prosperous centers of mining and commercial agriculture. Ethiopia joined the ranks of Africa's landlocked states in 1993 after its coastal province of Eritrea, annexed in 1954, succeeded in gaining its independence after three decades of armed struggle.

With the notable exception of Botswana, Africa's landlocked states continue to be very poor and undeveloped. Only four of them (Zimbabwe, Botswana, Swaziland, and Lesotho) have per capita incomes greater than $400. They also tend to have small populations; only Ethiopia

FIGURE 1.3. The countries of Africa.

and Uganda have more than 13 million people. However, their greatest source of vulnerability results from a perpetual dependence on neighboring states for an outlet to the sea. This problem is exacerbated by the frequent absence of reasonable transportation linkages. Five of the landlocked states have no railroads, and even where linkages exist, political tensions between neighbors or within neighboring states may preclude the use of these railroads.

The experience of Zambia provides an example of the vulnerability of landlocked states. When an international embargo on trade with Rhodesia (the former Southern Rhodesia, now Zimbabwe) was imposed following that coun-

try's unilateral declaration of independence in 1965, the route by which Zambia had exported its copper and imported its oil and other goods was cut off. The alternate route via Zaire (now the Democratic Republic of the Congo) and Angola could not be used because of the state of insurgency in Angola. Mozambique could not be used because of the insurgency and the lack of an established route. To overcome the blockade, Zambia arranged to have a railroad and pipeline built to the port of Dar es Salaam in Tanzania. The Tazara railway was an engineering success, but because of the great distances involved, port congestion, and low volumes of freight, it has proved to be of limited useful-

VIGNETTE 1.2. Disputed Borders

On several international borders, disputes about territory have occurred as part of the legacy of the arbitrary political division of Africa under colonialism. These disputes have been the most serious in places where colonial borders have divided a particular ethnic group between two countries, or where it has been felt that historical–political affiliations have not been recognized. The postage stamps shown in Figure 1.4 illustrate how certain governments have sought to correct what they have perceived as long-term injustices.

The first stamp is from Mauritania and celebrates the annexation of the southern part of Spanish (now Western) Sahara. In 1974, Spain decided to abandon its colony, in large part because of the growing threat of a Moroccan invasion. Ignoring a judgment from the International Court of Justice that rejected the claims of Morocco and Mauritania, Spain signed a treaty in November 1975 with those two countries, under which Spanish Sahara was to be divided between Morocco and Mauritania. The Saharan people were not consulted about this agreement, and the Saharan liberation organization called POLISARIO has pursued a fierce guerrilla war of independence. Although the total population of Western Sahara was only about 100,000, POLISARIO managed to force the Mauritanians to withdraw and renounce their territorial claims in 1978. The struggle against Moroccan occupation continues.

The second stamp is from Somalia and shows the neighboring Ogaden region of Ethiopia as part of Somalia. This region has a predominantly Somali population and has long been claimed as part of "Greater Somalia." In 1977, Somalia invaded Ethiopia in an unsuccessful attempt to take over the Ogaden. Although Somalia's current state of turmoil means that there is no immediate threat of another invasion, the Somali people have not abandoned their belief that the Ogaden is rightfully theirs.

During the 1990s, the dispute over Somalia's borders turned inward. As the country descended into civil war and the central government effectively ceased to exist, the northwestern part of Somalia seceded and declared itself the Republic of Somaliland. Later, a second secessionist state, known as Puntland, was formed in northwestern Somalia. Somaliland has functioned as a de facto separate state for more than a decade, despite the lack of diplomatic recognition by any other country. Ethnicity did not influence the decision of Somaliland to attempt secession from Somalia, but colonial history was a factor: The self-declared Republic of Somaliland corresponds to the former British Somaliland, which was amalgamated with the Italian colony of Somalia at independence. Despite the initial success of Somaliland's secession, the absence of international diplomatic recognition means that longer-term prospects for its attempt to "turn back the map" appear bleak.

a   b

FIGURE 1.4. "Lay claim to thy neighbor." (a) Mauritania, 1976. This stamp celebrates Mauritania's ill-fated attempt to annex part of Western Sahara. (b) Somalia, 1964. This stamp shows parts of neighboring countries as Somali territory.

TABLE 1.1. Countries of Africa South of the Sahara

	Capital	Area (1,000 km²)	Population (1,000)	Per capita income ($)	Human development index
Angola	Luanda	1,246	13,134	240	0.403
Benin	Porto Novo	113	6,272	380	0.420
Botswana	Gaborone	600	1,541	3,300	0.572
Burkina Faso	Ouagadougou	274	11,535	230	0.325
Burundi	Bujumbura	28	6,356	110	0.313
Cameroon	Yaoundé	475	14,876	570	0.512
Cape Verde	Praia	4	427	1,330	0.715
Central African Republic	Bangui	623	3,717	290	0.375
Chad	N'Djamena	1,284	7,885	200	0.365
Comoros	Moroni	2	706	380	0.511
Congo (Dem. Rep.)	Kinshasa	2,345	50,948	—	0.431
Congo (Republic)	Brazzaville	342	3,018	630	0.512
Côte d'Ivoire	Yamoussoukro	322	16,013	660	0.428
Djibouti	Djibouti	22	632	840	0.445
Equatorial Guinea	Malabo	28	457	—	0.679
Eritrea	Asmara	94	3,659	170	0.421
Ethiopia	Addis Ababa	1,130	62,908	100	0.327
Gabon	Libreville	267	1,230	3,180	0.637
Gambia	Banjul	11	1,303	330	0.405
Ghana	Accra	239	19,306	350	0.548
Guinea	Conakry	246	8,154	450	0.414
Guinea–Bissau	Bissau	36	1,199	180	0.349
Kenya	Nairobi	583	30,669	360	0.513
Lesotho	Maseru	30	2,035	540	0.535
Liberia	Monrovia	111	2,913	—	—
Madagascar	Antananarivo	587	15,970	260	0.469
Malawi	Lilongwe	118	11,308	170	0.400
Mali	Bamako	1,240	11,351	240	0.386
Mauritania	Nouakchott	1,031	2,665	370	0.438
Mozambique	Maputo	802	18,292	210	0.322
Namibia	Windhoek	824	1,757	2,050	0.610
Niger	Niamey	1,267	10,832	180	0.277
Nigeria	Abuja	924	113,862	260	0.462
Rwanda	Kigali	26	7,609	230	0.403
São Tomé e Princípe	São Tomé	1	138	290	0.632
Senegal	Dakar	196	9,421	500	0.431
Sierra Leone	Freetown	72	4,405	130	0.275
Somalia	Mogadishu	638	8,778	—	—
South Africa	Pretoria	1,221	42,800	3,020	0.695
Sudan	Khartoum	2,506	31,095	320	0.499
Swaziland	Mbabane	17	925	1,290	0.577
Tanzania	Dodoma	945	35,119	280	0.440
Togo	Lomé	57	4,527	300	0.493
Uganda	Kampala	236	23,300	310	0.444
Zambia	Lusaka	753	10,421	300	0.433
Zimbabwe	Harare	391	12,627	480	0.551

Data sources: United Nations Development Programme (UNDP). *Human Development Report 2002.* New York: Oxford University Press, 2002. World Bank. *World Development Report 2002.* New York: Oxford University Press, 2002.

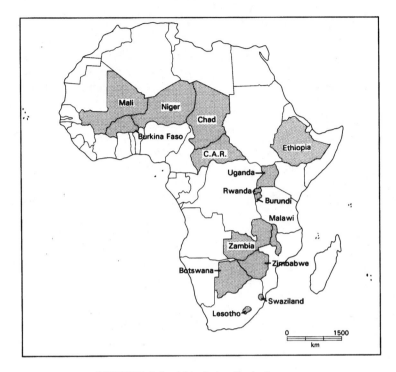

FIGURE 1.5. Africa's landlocked states.

ness. Although Zambia has become less vulnerable since the independence of Zimbabwe and the end of apartheid in South Africa, its long-term prospects remain constrained by its landlocked status.

One of the ongoing tasks for those involved in African studies is to "relearn" the map periodically as changes are made to place names or administrative structures (see Table 1.2). In several cases, name changes at independence or after independence represent a decision to replace colonial names with ones more historically and culturally relevant.

Several countries have also relocated their capital cities. In each case, the change has been justified as a means of bringing government closer to the people by abandoning colonial seats of government for smaller, more centrally located places.

Levels of Development

Maps are a powerful tool for displaying and analyzing spatial distributions, such as varia-

tions in wealth and the quality of life. The maps in Figures 1.6 and 1.7 illustrate contrasting approaches to the definition of development in the continent. Although Africa south of the Sahara is very poor as a whole, extreme variations of wealth and development exist across the continent. There are significant differences between the most and least developed countries in income, economic diversity, and quality of life. There are also large differences within each country—between urban and rural areas, and between the rich and the poor.

There is no universally accepted measure of development, in part because development is multidimensional and in part because there are disagreements about what development entails. The most widely used measure is per capita gross national income (GNI). In only six countries is per capita GNI above $1,000, while in seven countries it is less than $200 (see Table 1.1 and Figure 1.6). The $3,300 GNI per capita of Botswana is 33 times that of Ethiopia, which at $100 has the continent's lowest per capita GNI. However, while variations in per capita GNI are certainly important, aggregate

TABLE 1.2. Some Important Postindependence Changes to the Map of Africa

Countries renamed at time of independence

New name	Former name
Botswana	Bechuanaland
Djibouti	French Somaliland
Ghana	Gold Coast
Lesotho	Basutoland
Malawi	Nyasaland
Zambia	Northern Rhodesia
Zimbabwe	Rhodesia

Countries renamed since independence

New name	Former name
Benin	Dahomey
Burkina Faso	Upper Volta
Zaire (1971–1997)	Congo-Kinshasa
Democratic Republic of the Congo (1997–　)	Zaire
Tanzania	Tanganyika and Zanzibar

Name changes to capital cities

New name	Former name	Country
Banjul	Bathurst	Gambia
Harare	Salisbury	Zimbabwe
Kinshasa	Léopoldville	Dem. Rep. Congo
N'Djamena	Fort Lamy	Chad
Maputo	Lorenço Marques	Mozambique

New capital city established

New capital	Old capital	Country
Abuja	Lagos	Nigeria
Dodoma	Dar es Salaam	Tanzania
Lilongwe	Zomba	Malawi
Yamoussoukro	Abidjan	Côte d'Ivoire

national income data do not show how wealth is distributed in a society, or whether available wealth has been used to improve productivity or the quality of life.

Another measure, which is being used with increasing frequency, is the human development index (HDI) (see Table 1.1 and Figure 1.7). In the *Human Development Report,* published under the auspices of the United Nations Development Programme (UNDP), the HDI is described as an index of the range and quality of options available to people to shape their own destinies. The index is calculated annually by using measures of life expectancy, education, and per capita income, which are combined according to a methodology described in the report. The HDI scores for 2002 emphasize the underdevelopment of Africa, compared even to the most disadvantaged countries in other parts of the world. There are 27 African countries with lower scores than Haiti and Bangladesh, which have the lowest HDI ratings outside of Africa. Of 44 countries that were given an HDI rating in 2002, 38 are among the bottom 50 countries. The lowest HDI, 0.275, was assigned to Sierra Leone.

The HDI is a serious attempt to move beyond the limitations of per capita GNI. However, with another mix of variables or with a different weighting of variables, somewhat different results would emerge. Thus care should be exercised in drawing conclusions based on the proportional size of HDI scores for different countries or on the ranking of countries when their HDI scores are fairly similar. Nevertheless, this index is useful for focusing attention on broad differences in levels of national development and for identifying countries whose people are the most disadvantaged.

Measures such as GNI and HDI have another major weakness—namely, that they provide only national aggregate measures of development. National scores may be quite misleading for African countries where there are very large differences in income and human welfare between regions or social groups within the country. We need to keep in mind several weaknesses in national statistics. For example, economic data generally ignore or underestimate the value of women's work and of subsistence production. Some statistical measures of development also reflect Western cultural and economic biases.

The use of national aggregate measures also helps to perpetuate a vision of development as occurring naturally within the bounded territories of nation-states. When development is conceived at a national scale, the rich diversity of resources and of development responses at the local level does not receive appropriate attention. Nor are the many ways in which development is facilitated through connections that link diverse places—urban and rural, North and South—given due consideration. Robinson's *Development and Displacement* (see "Further Reading") expands upon these important ideas.

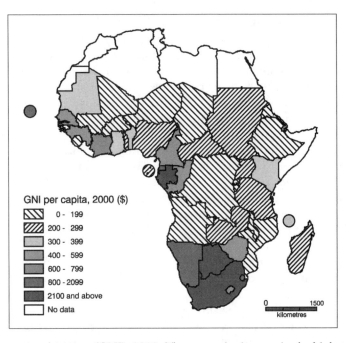

FIGURE 1.6. Gross national income (GNI), 2000. The per capita income in the highest income countries is about 30 times as large as that of the poorest countries. Data source: United Nations Development Programme (UNDP). *Human Development Report 2002*. New York: Oxford University Press, 2002.

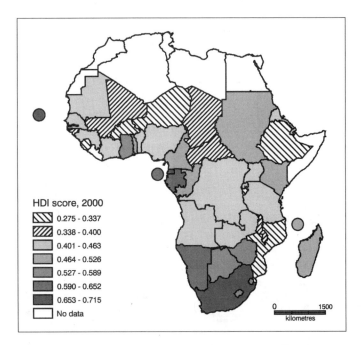

FIGURE 1.7. Human development index (HDI), 2002. When Figures 1.6 and 1.7 are compared, there is a broad correlation between GNI per capita and the HDI score. Data source: UNDP. *Human Development Report 2002*. New York: Oxford University Press, 2002.

Regional and Political Groupings

One approach to the definition of groupings of countries is membership in regional economic and political organizations based on shared culture and history. The most important political organization linking African states is the African Union, founded in 2002 as the successor to the Organization of African Unity. All African states, with the exception of Western Sahara, are members. Among the continent's 11 regional political–economic organizations, two stand out: the Economic Community of West African States (ECOWAS), which links 16 countries in West Africa, and the Southern African Development Community (SADC), composed of 12 states in southern Africa (see Figure 1.8a) plus Mauritius and Seychelles. The SADC states came together initially with the objective of reducing their dependence on South Africa. Following the abolition of apartheid, South Africa joined and became "first among equals" in SADC.

Several African nations are members of the Commonwealth of Nations or of La Francophonie (see Figure 1.8b). The former organization brings together states from all continents that were formerly British colonies; Cameroon

and Mozambique have also been accepted as members. La Francophonie brings together the former French and Belgian colonies of Africa with other French-speaking nations. During the 1990s, four countries that had formerly been under Portuguese or Spanish control also joined the organization.

In addition to groups defined by membership in an organization, regional groupings are often defined on the basis of geographical proximity and perceived similarity. Figure 1.9 shows some of the commonly used informal regional groupings of countries in Africa south of the Sahara. Note that there is no single defining characteristic, and also that there is less than complete agreement on which countries should be included in each group.

The term *West Africa* commonly refers to countries west of the Cameroon–Nigeria border, an important physical and cultural dividing line in the continent. *The Sahel* countries form a significant subregion within West Africa characterized by desert-margin environments and (especially in recent years) recurring drought. *East Africa* consists of Kenya, Tanzania, and Uganda, the members of the East African Community (EAC; see Chapter 30), but Rwanda and Burundi are usually included be-

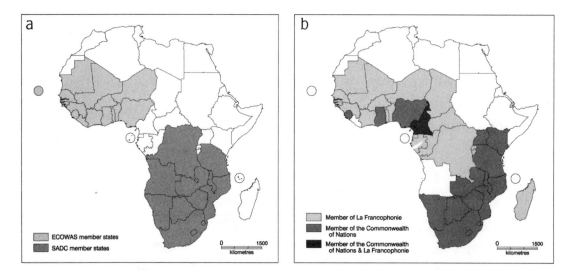

FIGURE 1.8. Examples of major political and economic groupings: (a) Economic Community of West African States (ECOWAS) and Southern African Development Community (SADC) states. (b) La Francophonie and the Commonwealth of Nations.

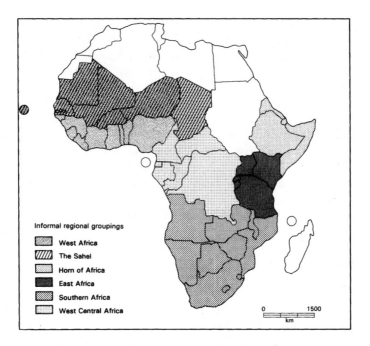

FIGURE 1.9. Informal regional groupings of countries.

FIGURE 1.10. Major rivers, lakes, and coastal waters.

cause of their strong economic ties with the others. Since the end of apartheid, *southern Africa* has become a meaningful grouping for the first time. Previously, there had been a group referred to as the *Frontline States*, defined by their proximity to South Africa and opposition to apartheid. Other informal regional groupings include the four countries of the *Horn of Africa* (Ethiopia, Eritrea, Somalia, and Djibouti; Sudan is sometimes included in this group) and the states of *west central Africa* (anchored by Cameroon to the north and the Democratic Republic of the Congo to the south).

The Physical Map

At first glance, the physical map of Africa looks rather uninteresting. The coastline of the continent is often straight and uncomplicated; there are only a few identifiable seas, gulfs, and other adjoining bodies of water. Topographically, the vast, gently undulating plateaus create an impression of uniformity, especially when there are no great mountain ranges such as the Himalayas or Rockies to catch one's eye.

A closer inspection, however, reveals considerable variety in Africa's topography. For example, there are spectacular escarpments up to 2,000 m fringing the southern African coast; the escarpment known as the Drakensberg Mountains in South Africa is especially spectacular. Then there is the world's largest rift valley system, extending from southern Mozambique through eastern Africa to the Red Sea and beyond. And to this list can be added the magnificent volcanic peaks, notably Mounts Kenya, Kilimanjaro, Elgon, and Cameroon, which rise to between 4,000 m and almost 6,000 m above sea level.

A half-dozen major river systems together drain some four-fifths of Africa south of the Sahara (Figure 1.10). Four rivers stand out: the Nile, the Congo, the Niger, and the Zambezi. Others of regional note are the Orange, the Limpopo, the Kasai, the Ubangi, the Benue, the Volta, and the Senegal.

Because the coastline is regular and has very few substantial indentations, Africa has few good natural harbors. The scarcity of harbors, along with the presence of escarpments and major rapids near the mouths of many rivers, impeded early European attempts to explore and exploit the continent.

Becoming familiar with the locations of prominent physical features as well as other elements of the African map is not a particularly important end in itself. However, familiarity does provide a basis for interpreting specific issues and situations, each of which occurs in a particular context that is spatial, environmental, social, political, and economic. Thus the maps in this chapter serve to establish the spatial context for our study of the geography of Africa south of the Sahara.

Further Reading

Thematic and regional atlases address diverse aspects of the geography of Africa. The first Griffiths volume is a useful general source, while the others listed below are more specialized in nature.

Griffiths, I. L. L. *An Atlas of African Affairs*, 2nd ed. London: Routledge, 1994.

There are several excellent historical atlases of Africa:

Ajayi, J. F. A., and M. Crowder. *Historical Atlas of Africa*. London: Longman, 1985.

Fage, J. D. *An Atlas of African History*. New York: Africana, 1980.

Freeman-Grenville, G. S. P. *New Atlas of African History*. Englewood Cliffs, NJ: Prentice-Hall, 1991.

Griffiths, I. L. L. *Africa on Maps Dating from the Twelfth to the Eighteenth Century*. Leipzig, Germany: Editions Leipzig, 1968.

McEvedy, C. *Penguin Atlas of African History*. New York: Penguin USA, 1996.

For a fascinating study of fact and fiction in early maps of Africa, see the following source:

Bassett, T. J., and P. W. Porter. "From the best authorities: The Mountains of Kong in the cartography of West Africa." *Journal of African History*, vol. 32 (1991), pp. 367–414.

Here are two useful series of atlases, each volume of which pertains to a particular country:

Barbour, K. M., J. O. C. Oguntoyinbo, J. O. C. Onyemelukwe, and J. C. Nwafor. *Nigeria in Maps*. New York: Africana, 1982. (Other volumes in the series deal with Sierra Leone [1972], Malawi [1972], Tanzania [1971], Zambia [1971], and Liberia [1972].)

Les Atlas Jeune Afrique: République Centrafricaine. Paris: Editions Jeune Afrique, 1984. (The Jeune Afrique atlas series also includes volumes on Africa [1973], Congo [1977], Niger [1980], and Senegal [1980].)

The following are examples of thematic atlases on Africa:

Christopher, A. J. *The Atlas of Changing South Africa*, 2nd ed. London: Routledge, 2000.

Diesfeld, H.J., and H.J. Hecklau. *Kenya: A Geomedical Monograph*. Heidelberg, Germany: Springer, 1978.

Food and Agricultural Organization of the United Nations (FAO). *Atlas of African Agriculture*. Rome: FAO, 1986.

Murray, J. *Cultural Atlas of Africa*, revised edition. New York: Checkmark Books, 1998.

Norwich, I. *Norwich's Maps of Africa: An Illustrated and Annotated Carto-Bibliography*, 2nd ed. Norwich, VT: Terra Nova Press, 1997.

For ideas on African underdevelopment in global perspective, see the following sources:

Robinson, J. *Development and Displacement*. Oxford: Oxford University Press, 2002.

Smith, D. *The State of the World Atlas*, 6th ed. London: Penguin, 1999.

Internet Sources

The following sites provide access to national maps for African countries:

Northwestern University. *Africa Base Map*. www. library.northwestern.edu/africana/map/

WorldAtlas.com www.worldatlas.com/webimage/ countrys/af.htm

Yunlong Xia. *Digital Map Page: Africa*. http://informatics.icipe.org/databank/

Several university libraries have websites devoted to their Africa collections:

Caruso, J. S. Columbia University. *Maps and Power in Modern African History*. www.columbia. edu/cu/lweb/indiv/africa/mappower

University of Texas Library On-line. *Perry Castaneda Library Map Collection: Africa Maps*. www.lib.utexas.edu/maps/africa.html

2

Imagining Africa: Roots of Western Perceptions of the Continent

This chapter focuses on the power of ideas—ideas about Africa that have become deeply embedded in Western society. These ideas range from widely held assumptions about how Africans are despoiling their environment through backward agriculture and rapid population growth, to the endlessly repeated tropes about tribalism, violence, and decay. Other ingrained images emphasize the exotic: vast herds of wildlife, and colorfully dressed people practicing traditional rituals.

Where do such ideas come from, and why do they persist? Much of this chapter looks at the roles of various Western groups—such as travelers, missionaries, scholars, and the mass media—in disseminating these notions. And why is it important to think about the power of "received wisdom"? Not only do deeply embedded assumptions and stereotypes about Africa stand in the way of effective learning about the continent, but also they have too often been the basis for ill-conceived academic research and development policies. Learning often starts with "unlearning."

Orientalism

The publication in 1978 of Edward Said's book *Orientalism* (see "Futher Reading") had a profound influence on thinking about the cultural and political relationship between the Western world (the so-called "Occident") and the East/South (the "Orient"). *Orientalism* focused on the writing of Western scholars, travelers, and officials about the colonized world and its peoples. This discourse typically portrayed the colonized subject as exotic, mysterious, deviant, and often dangerous. According to Said, writing that asserted the superiority of the West became instrumental in the domination of colonized peoples. The colonized world was objectified as essentially a thing for study, display, and control.

Said's thesis extended earlier works by anticolonialist writers such as Frantz Fanon (*Wretched of the Earth; Black Skin, White Masks*). Fanon had written about the role of ideas in the maintenance of European control over its colonies, and the inculcation of a sense of inferiority among subjects who were taught to aspire to be Europeans but could not be accepted as such.

Said's writing has been criticized for focusing heavily on culture while ignoring the economic dimensions, and emphasizing the role of discourse while paying too little attention to "concrete" geographies of underdevelopment. However, the ideas espoused in *Orientalism* provide a valuable framework for looking at ideas about Africa.

The Traveler's Gaze

Prior to the 19th century, European knowledge of Africa was virtually confined to the coast. The discoveries of the 19th century explorers—Livingstone, Barth, Speke, and Stanley, among others—who ventured into the interior of the continent were reported in great detail through the publications of geographical societies and newspapers to a European and North American public that could not get enough. The first wave of explorers was followed by the arrival of natural scientists, geographers, and anthropologists, who came to collect specimens for museums, and to describe in detail the land and its people.

The Rise of Tourism

Shortly after the colonial annexation of Africa at the turn of the 20th century, the first tourists began to arrive. The massive game populations in the eastern and southern parts of the continent proved irresistible to the aristocrats and capitalist tycoons from North America and Europe. Although some came only to view and photograph the fauna, many came to collect trophies. These early safaris were lavish affairs, befitting the expensive tastes of their clientele.

Among the most famous of the early visitors was Theodore Roosevelt, who arrived with one of his sons in 1909, shortly after the end of his term as president of the United States. Traveling with 250 porters and guides, Roosevelt made his way across British East Africa, entered the Belgian Congo, and then followed the Nile to Khartoum. Roosevelt and his son killed 512 animals and collected over 1,000 specimens for the Smithsonian Institution. Every adventure was eagerly reported in the North American and European press, and pictured on hundreds of different postcards (Figure 2.1).

In recent years, African countries have done more to promote historical sites such as Great Zimbabwe and the churches of Lalibela, as well as the rich cultural heritage of the continent.

FIGURE 2.1. Former U.S. President Theodore Roosevelt poses with a hunting trophy, Kenya, 1909. Roosevelt's widely publicized trip helped to popularize the African safari. Photo: Phelps Publishing Company.

Nevertheless, what is promoted in contemporary tourist literature is often a caricature of African culture. Consider these excerpts from a tourist brochure:

> This tour is designed for viewing the African people . . . visits are made to primitive areas where one must expect inconvenience. . . . The most fascinating of the West African countries is the Cameroons where native behavior has existed without interruption since the beginning of time . . . two-thirds of the population still practices the ancient pagan religion of ritual figures, magic men and evil witches. (Travcoa Africa, 1979, p. 26)

However, for most Europeans and North Americans, tourism in Africa is synonymous with the wildlife safari. African society fades into obscurity; the safari experience offers not just Africa-as-wildlife, but also luxurious comfort in the wild. In another tourist brochure, the safari is described as follows:

> Shamwari is an African dream. A game reserve in which a multitude of plant, animal, and birdlife unfold the very soul of an untamed Continent, along with the most luxurious means of experiencing your safari holiday. Shamwari's highly trained game rangers, with skilled service staff will ensure a memorable adventure, personalized to your needs. (www.places.co.za/html/5636. html)

National Geographic *Pictures Africa*

The majority of North Americans and Europeans are unlikely to take a trip to Africa for a holiday. Nevertheless, there is widespread fascination with the continent and its attractions. For more than a century, since the age of European exploration, *National Geographic* magazine has played a major role in bringing pictures of Africa to the world, and has been instrumental in shaping public perceptions about the continent.

In their book *Reading National Geographic* (see "Further Reading") Lutz and Collins characterize *National Geographic* as "America's lens on the world." It possesses a subscriber base of about 10 million, but is also widely used in schools and displayed in the waiting rooms of many professional offices. The magazine's pop-

ularity has rested in large part on its success in replicating and legitimizing the world view of white, upper-middle-class America. The publishers of *National Geographic* have managed to situate it both as a readily accessible scientific journal and as a magazine providing exotic pleasure to its readers.

The main attraction of *National Geographic* is its lavish use of spectacular photographs that relegate the written text to a secondary role. Photographs have the aura of objectivity, of showing the world as it is. However, this appearance of objectivity is misleading. Every aspect of presentation in *National Geographic*—from the selection and modification of photographs, to their arrangement and the writing of captions—is undertaken with utmost care to convey a desired "message." The images reproduce and subtly reinforce traditional gender divisions of labor and racial hierarchies of dominance.

For most of the past century, Africa in *National Geographic* has been the mysterious, often intimidating, and barely tamed "dark continent." Numerous colonial-era articles featured the adventures of intrepid white travelers encountering exotic peoples in remote places. Wildlife stories were featured prominently, and Africans were repeatedly portrayed in demeaning ways. The dominant images began to shift in the 1950s and 1960s; stories and the photographs illustrating them gave greater prominence to the promise of progress and modernization in newly independent African countries. Wildlife stories remained a staple, but exotic portrayals of African culture became less common.

As the 1990s progressed, however, the coverage of Africa in *National Geographic* increasingly resembled that of a half-century earlier. Modern Africa was hardly to be seen. A number of articles featured colorful ceremonies of "traditional" peoples, and photos of bare-breasted women appeared with greater frequency. A lengthy three-part series in 2000–2001 described a "megatransect" of the Congo rain forest in a style reminiscent of the colonial-era explorers' journals—indeed, accentuated by the availability of modern graphic and photographic techniques as means of sensationalization (Vignette 2.1).

VIGNETTE 2.1. Megatransect: *National Geographic's* Africa, 2000–2001

This vignette discusses a series of articles published a few years ago in *National Geographic* (see the list in the footnote below). It is argued that the troubling picture of Africa in these articles owes much to the magazine's editorial decisions. You are strongly urged not just to read the vignette, but also to examine the series of articles in *National Geographic.*

In September 1999, conservationist J. Michael Fay embarked on a 15-month journey through 1,900 km of dense forest from the northern Congo (Democratic Republic) to the coast in Gabon. He traveled by foot, accompanied by a small group of African porters and guides. The stated purpose of the trip was to record for the ecology of this large, almost uninhabited, and little-explored region of forest that is increasingly threatened by logging. *National Geographic,* along with the Wildlife Conservation Society, sponsored the journey.

Faye's journey was reported in astonishing detail, in three stories with a total length of 96 pages. What is most remarkable about these articles, however, is not their length, but rather the portrayal of this part of Africa and its people, and of Faye's obsessive determination to conquer the fearsome jungle on his own terms.

Bold graphics are used throughout the three articles. Several of these bold headings accentuate the sense that danger is everywhere in the African forest. Here are some examples:

- "There are dire diseases, armed poachers, and other sorts of threats" (Part 1, pp. 10–11).
- "After dark there comes a weird, violent, whooshing noise that rises mystifyingly" (Part 1, pp. 16–17).
- "Where does Ebola lurk between outbreaks? What species in the forest—a small mammal? an insect?—serves as its reservoir host?" (Part 2, p. 20).

Many of the photographs in the articles are dark and shadowy, and several are poorly focused. These photographs help to convey the image of the forest as a place where danger lurks everywhere. Several of the photos show the region's people unsympathetically. Note, for example, a photo entitled "Subsistence hunting' (Part 1, pp. 24–25) that shows a man almost completely enveloped in darkness holding his bloody (well-illuminated) prey.

Faye himself is portrayed in the manner of the great white explorer. Not only does the success of the trip depend on his drive and conviction, but also, like the explorers of times past, he repeatedly loses patience with his African helpers and harangues them. The following bold headings are meant to show Faye's unwavering zeal, but perhaps the zeal is simply disturbing:

- "It looked like a daunting endeavor—far too arduous and demented to tempt an ordinary ecologist, let alone a normal human being" (Part 2, p. 17).
- "I need to ship these boys home . . . they are haggard, totally worn out. No matter how good they were they are just going to go down one by one" (Part 2, p. 20).

The overall message of the articles is clear: There is little hope for the forest, and implicitly for Africa itself. In its selective and manipulative portrayal of Africa, *National Geographic* both *reflects* and *molds* current sentiments about Africa in the West.

Sources for quotes: D. Quammen. "Megatransect." *National Geographic,* October 2000, pp. 2–29. D. Quammen. "Green abyss: Megatransect Part 2." *National Geographic,* March 2001, pp. 2–37. See also D. Quammen. "End of the line: Megatransect Part 3." *National Geographic,* August 2001, pp. 74–103.

The White Man's Burden

Europe's View of Its Role in Africa

Europe's colonization of Africa grew from its sense of historical destiny. Europeans' long-standing views that their continent's societies were more progressive and rational than those of people in other parts of the world were espoused with increasing conviction during the Renaissance. Other societies not only were seen as inferior, but were considered by many as sub-human. Thus European domination of other societies was seen as natural and inevitable.

The industrial revolution increased demand for raw materials, and for secure markets where manufactured goods could be sold. Despite the obvious commercial imperative that was driving colonial expansion, Europeans preferred to characterize their conquests as noble, selfless deeds: It was the "white man's burden" to spread European civilization to all regions (see Figure 2.2). This benevolent view of colonialism as having provided a foundation for African development has proven to be very resilient, even many decades after the end of colonial rule and despite copious evidence of colonialism's darker legacy.

Missionaries as Bearers of the White Man's Burden

Christianity, conceived as a European religion, helped to provide a spiritual, conceptual frame of reference for Europe's sense of destiny. Europe's progress was seen as God-given; by implication, other societies following other religions were inferior. European expansion and conquest provided access to new regions and peoples who could be converted to Christianity. In turn, Christian missionaries assisted in the colonial project by encouraging converts to accept their earthly fate and to look instead toward a heavenly reward. David Livingstone, the most famous of the 19th-century missionaries in Africa, asserted in 1857 that Christianity and commerce were inseparable vectors of civilization.

Although some missionaries sought to protect Africans from the worst excesses of colonial rule, many others turned a blind eye. During King Leopold II of Belgium's reign of terror in the Congo at the turn of the 20th century, most missionaries worked closely with colonial officials and helped to implement forced labor. Many of the children attending the early mission schools in the Congo were taken forcibly

FIGURE 2.2. "Carrying Light into the 'Dark Continent.'" Texaco plays on the image of the colonial "civilizing mission" in this 1930s postcard. Photo: unknown.

from their families and sent for training that would prepare them to serve the colonial authority as soldiers and petty officials.

Even though missionaries in Africa lived frugally, they needed to raise substantial amounts of money in Europe and North America to carry on their work. Like fund raisers today, they sought to create both a sense of the urgency of their work and the success they were achieving through their efforts. Picture postcards, countless hundreds of which were published by the missionary societies to publicize their work, showed an Africa being transformed through missionary activity (see Figure 2.3). Many cards showed Africans learning trades or learning the general discipline of work. Others used the adoption of Western clothing as a metaphor for civilization; the change from unclothed to

clothed was equated with a transition from savagery to civilization. Some cards used images of Africans in traditional dress or performing traditional rituals to warn patrons of the dire consequences of failing to spread the gospel faster.

Christian missionaries have played a major role in shaping North American and European attitudes about Africans—and about themselves as supporters of the "civilizing mission." Colonial-era missionary postcards and other documents typically showed Africans in an unfavorable, patronizing way. Indeed, the contemporary fund-raising strategies of certain Christian charities—for example, television infomercials that seek funds for famine relief and child sponsorship—have been criticized for perpetuating an image of Africans as passive, tragic victims.

a   b

Missions de Scheut : Congo. Au garde-à-vous.

FIGURE 2.3. Early 20th century postcards were widely used by missions to advertise their work in Africa. (a) Congolese mission schoolchildren "on guard for thee." (b) Written message seeks financial support: "If we had started our missionary effort earlier perhaps I might have sent a different picture." Photos: (a) Missions de Scheut. (b) *Missionary Review of the World.*

Hollywood's Africa

Hollywood has contributed to the shaping of Western images of Africa. Africa was merely the backdrop for the series of movies based on the Edgar Rice Burroughs *Tarzan* books, and for the famous 1951 film *The African Queen*. Although more recent films set in Africa have given greater prominence to Africans, they have continued to offer white outsiders' perspectives on the continent. The widely acclaimed 1985 film *Out of Africa* focused on the struggles of a white woman in colonial Kenya to retain control of her farm and to protect her African farm workers. Africans also take a subsidiary role in *Gorillas in the Mist*, the 1988 film about the crusade of Dian Fossey to save Rwanda's mountain gorillas.

The 1987 movie *Cry Freedom* tells the story of the death of Steve Biko, a prominent leader in the struggle against apartheid in South Africa, but does so through the eyes of a white newspaper editor. In giving prominence to the white character, so that the film would be more attractive for white audiences in North America and Europe, Hollywood provided a distorted image of the antiapartheid struggle. Similarly, the 2002 film *Black Hawk Down* fails to provide the historical context of the crisis in Somalia.

Several recent films have been criticized for unsympathetic, if not racist, portrayals of Africans. In the case of *Dogs of War*, a 1981 film about white mercenaries, the African characters are uniformly crude and cowardly. The 1994 Disney movie *The Lion King* has also been criticized for the racialization of the features and accents of "good" and "bad" characters.

The Western Media's Tragic Continent

The Western news media provide less coverage of events in Africa than in any other part of the world. The coverage of Africa, especially outside such major newspapers as *The New York Times* and *The Guardian*, tends to be not only sporadic but also sensationalized and stereotypical. Africa has not been well served by the Western press.

During the apartheid era, the South African government put considerable effort into its attempts to obtain sympathetic media coverage in the West. Politicians and journalists thought to be sympathetic to apartheid were flown to South Africa, given royal treatment, and taken on carefully staged guided tours, with the expectation that they would write stories supporting the apartheid regime. The State Department of Information published copious amounts of propaganda material for circulation worldwide.

Certain types of stories have dominated the news from Africa. Famine has been a recurring theme, typified by the extensive coverage given to the Ethiopian famine of the mid-1980s. Another recurring theme has been political, ethnic, and religious violence, such as that reported in Somalia and Rwanda during the early 1990s. Underlying these stories is the recurring suggestion that Africa is a tragic continent, beset with problems of an unforgiving environment, endemic tribal rivalries, and corrupt and incompetent officials. Africa is portrayed as a continent unable to help itself and unable to put outside aid to good use to bring about progressive change.

The image of Africa as a tragic continent has been underpinned by the use of dramatic visual images of suffering—for example, desperate, gaunt famine victims in Ethiopia; piles of bones from the victims of Rwanda's genocide; and gun-toting child rebels in Sierra Leone. Certain specific images have been repeated so frequently that they have come to define the crisis in the minds of viewers and readers. Indeed, the availability of dramatic photographs is a good predictor of the extent of coverage a given story receives. It is not necessarily the most serious famines or crises that receive the most international attention.

The involvement of whites in a particular news story greatly increases the likelihood of international coverage, and it also tends to distort the coverage. The seizure of white farms in Zimbabwe received extensive media coverage and was used to account for the widespread food shortages in that country in 2002. As a result, equally serious food shortages in every other country in the southern part of Africa, and natural causes of shortages (such as prolonged, severe drought), faded into the background in the news stories.

Whereas stories of African tragedy have been commonplace, Africa's success stories have generated little media interest. There has been much less coverage of Uganda's economic recovery since the mid-1980s than there was of either the terror under Idi Amin, or the continuing guerrilla campaigns of the Lord's Resistance Army in northern Uganda. Where successes have been recognized, they have often been treated as exceptional—South African reconciliation after the fall of apartheid as the individual legacy of President Mandela, for example, rather than as a national achievement.

The Western media's negative, stereotyped reporting of African events has been instrumental in convincing the Western public, as well as politicians, that Africa is a hopeless case. Africans appear to be passive victims; images of self-reliance and innovation are rarely shown. The negative imagery has made it easy for people in the West to absolve themselves of any responsibility for Africa's problems, while also claiming credit for saving the continent through humanitarian interventions.

Alternate Images: New Perspectives in Historical–Cultural Studies and the Arts

Although many negative stereotypes about Africa have been quite resilient, newer scholarly writing and artistic works have increasingly challenged the old, negative images and have provided a different point of view—an African perspective.

Rewriting African History

The discipline of history has tended to focus mostly on important people and major events—wars and territorial expansion, political regimes and their legislative legacy, and the like. Moreover, history has usually reflected the perspective of the conquering power, rather than that of conquered peoples. The rise of postcolonial and poststructural approaches has given rise to an alternate, African-focused history of the continent.

Postcolonial scholars reverse the gaze, looking at African history from an African rather than a European perspective, and focusing on dynamics at the grassroots rather than in centers of power. Oral history has proven to be a powerful tool to capture the stories of lived experience of ordinary Africans under colonialism. Oral history has also provided a means of recovering vanishing traces of knowledge about precolonial times.

Postcolonial research has uncovered the varied ways in which local communities resisted colonial rule—not only through strikes and armed struggle, but also through daily acts of resistance and noncompliance. For example, the songs of workers engaged in forced-labor projects were not always what they seemed; it was common for these songs to subtly mock detested officials or hated policies. Often women played an important role in organizing protests at the grassroots.

The increased interest in local histories has underscored the danger of overgeneralizing about a continent as vast and diverse as Africa. Whereas the colonial view strongly reflected the interests and biases of the center—of London, Paris, Brussels, and Lisbon—the view from the periphery reflected the vast diversity of ways in which ordinary Africans experienced and resisted colonial rule. Differences of culture, gender, ecology, and economy all shaped the experience of colonialism, as did external factors such as location relative to centers of colonial development (e.g., mines, plantations, and ports) and the unevenly distributed impress of missionary activity. For poststructuralists, the emphasis on diversity at the grassroots has been accompanied by a strong resistance to what they consider to be "totalizing" discourses—overgeneralization—about Africa.

Afrocentrism

Afrocentrism is a term often applied to the school of historical–cultural studies that portrays Africa as the true source of human civilization. Afrocentrists typically emphasize the cultural unity of all Africa; thus ancient Egyptian civilizations and diverse black African civilizations are treated as part of a pan-African legacy. African civilization is portrayed idealistically, placing an emphasis on the strength, inventiveness, peacefulness, and spirituality of the continent's people. Bernal's *Black Athena* and Asante's *Afrocentricity* (see "Further Reading") are examples of this school of thought.

Afrocentrist writing has helped to instill a heightened sense of pride, particularly among African Americans and Africans in their cultural heritage. However, Afrocentrism as historiography has been criticized for its overgeneralization and idealization of the African cultural–historical legacy. It is essentially an inversion of the Eurocentrism that dominates conventional historical writing. Most historians argue that the antidote to Eurocentric biases and blind spots is a wealth of carefully crafted historical research to set the record straight—not the sweeping, utopian narratives that have been characteristic of much Afrocentric scholarship.

Africa and the Arts

African creative genius is contributing to a more informed understanding of the continent and its people. Artistic creativity is hardly a new phenomenon; African music, dance, painting, sculpture, and other art forms have evolved over countless generations. (Nor have African works of art gone unrepresented in Western museums; indeed, passionate struggles are now being waged over the rightful ownership of traditional African artifacts currently exhibited in non-African institutions, as Vignette 2.2 describes.) The focus in this brief section is on contemporary writers, artists, and musicians, whose work has helped to develop a pan-African identity and to educate non-Africans about the continent.

The volume of, and critical acclaim for, African creative writing have continued to grow since independence. Although a disproportionate number of successful novelists, poets, and playwrights have come from a few countries, including Nigeria, South Africa, Kenya, and Senegal, most African countries have produced well-known, successful writers. Countless students in Africa and in other parts of the world have been introduced to works by writers such as Chinua Achebe, Wole Soyinka, and Ngugi wa Thiong'o, and in the process have come to know more about Africa.

African popular music represents a fusion of diverse influences, principally African, European, and African-American. African popular musicians have been very eclectic and innovative in synthesizing these various musical influences to create distinctly African music. There are many regional and stylistic musical styles. Most of the popular performers have gained attention for the rhythmic and instrumental qualities of their music. Fela Anikulapo-Kuti gained worldwide fame, and some notoriety, for his unique styles of music that celebrated African genius and mercilessly mocked the Nigerian state.

Whereas African music has a large international following, African cinema has made slower inroads abroad. Because many of Africa's most accomplished filmmakers are from Francophone countries (especially Burkina Faso, Mali, and Senegal), African film has remained virtually unknown to the North American public. With the exception of South Africa, the film industry in English-speaking countries has received less critical acclaim. Since 1972, the best new productions have been shown every two years at the African Film Festival in Ouagadougou, Burkina Faso.

Alternate Images: New Perspectives in Social Science and on Development

Challenging "Received Wisdom"

A glance at many introductory geography textbooks reveals a range of widely assumed knowledge about Africa. For example, African examples are commonly used to discuss the environmental implications of overpopulation. Such "received wisdom" has exerted a powerful influence not only on popular perceptions but also on social science research, since new research projects are strongly shaped by the existing academic literature. The selection and design of most development projects are likewise shaped by conventional "received wisdom."

Recent research has looked anew at the relationships linking population, land use methods, and environmental degradation. The findings of many studies in all parts of Africa have challenged views that have long been taken for

VIGNETTE 2.2. African History and Culture in Western Museums

The portrayal of African history and culture in non-African museums is a contentious issue that has at times pitted African intellectuals and political leaders, and often African Americans as well, against European and North American archeologists, museum curators, and academics interested in African culture. These disputes form part of a larger struggle by aboriginal and colonized peoples to reclaim ownership of their cultural heritage.

Many of the finest works of African art are to be found in museums and private collections outside of Africa. Many of these works were seized as war booty during the colonization of Africa. Others have been stolen—and continue to be stolen—from religious shrines, palaces, and museums in Africa, eventually finding their way into Western art markets. The attempts by African governments to have these artifacts returned have generally been rebuffed, largely because of the fear that a repatriation of antiquities long held by Western museums would set a precedent leading to an eventual decimation of their collections. Africans argue that many of the objects of art held by Western museums are very significant cultural and religious symbols, and as such belong "at home" rather than in a museum display.

African concerns about Western museums extend beyond the ownership of antiquities and sacred objects to the way in which Africans are portrayed. Consider, for example, the controversy that erupted just prior to the 1992 Barcelona Olympic Games over a display in the museum at Banyoles, Spain of the stuffed remains of an African, believed to have been stolen from a grave about 1830 by two French naturalists. "The Negro," as the exhibit is known, has been on display since 1916. A local African cultural center referred to the exhibit as a disgrace, a horror, and a serious mistake. The International Olympic Committee, at the request of several African nations, asked that the exhibit be withdrawn, but the town council refused and rejected claims that the exhibit was racist.

A rather more complex controversy erupted at the Royal Ontario Museum in Toronto in 1990 over an exhibit entitled "Into the Heart of Africa." Through the display of photographs Canadian missionaries had taken and artifacts they had collected, the exhibit was intended to explore the cultural arrogance of these missionaries who worked in Africa in the 19th and early 20th centuries. However, opponents denounced the display because it contained negative and stereotyped images of Africans; they argued that the display could misinform rather than educate impressionable young viewers. The display was withdrawn, but not before there had been heated debate about the messages conveyed by museum displays and the rights of particular communities to be involved in deciding how their own histories and cultures should be interpreted.

Although progress toward more sensitive museum policies has been slow, there have been victories. In 2002, France returned the skeleton and bottled organs of Saartje Baartman to South African officials, almost two centuries after she had been transported to Europe and displayed naked as though she were a circus freak under the name "Hottentot Venus" in Paris and London. After her death, a plaster cast of her body was prepared and remained on display, along with her skeleton, in the Museum of Mankind in Paris until 1974. The return of her remains took place several years after a formal request from President Mandela to the President of France. Saarje Baartman's state funeral and burial in South Africa were deeply emotional experiences for many, especially for her own Khoisan people, who had waged a spirited fight for the return of her remains. .

granted. For example, the image of relentlessly advancing deserts appears to have been greatly exaggerated. When new research succeeds in challenging accepted views, social scientists and students are challenged to look with fresh eyes at other widely accepted assertions. Governments and development agencies are challenged to revisit their policies, and are reminded of the importance of local surveys as a prelude to decision making about development.

Research has also challenged conventional views about HIV/AIDS in Africa. Initial publications on the apparent African origin of the disease attributed its emergence to a variety of customs characterized as exotic and deviant, such as initiation ceremonies and sexual practices. Many of these assertions were based on extremely flimsy evidence, such as early-20th-century writings of missionaries and colonial officials. There is now a vast literature that has thoroughly refuted the myths of deviant African behavior, and instead has provided a nuanced understanding of many aspects of HIV/AIDS in Africa.

Indigenous Knowledge

The study of indigenous knowledge has been a major focus for social scientists working in Africa, especially since the late 1970s. These studies have looked not only at what is known, but also at the uses made of this knowledge, its conceptual underpinnings, and its transmission from generation to generation. Such studies require prolonged observation and detailed interviews of both leading practitioners (the prime guardians of indigenous knowledge) and ordinary people.

Research on African farming systems has demonstrated the economic and ecological value of practices such as intercropping (growing more than one crop together), agroforestry (integrating useful trees into farming systems), and shifting cultivation, all of which were assumed until fairly recently to have little or no value. This has changed most agricultural development from a focus on *replacing* indigenous farming systems to approaches that *build upon* these systems. The former belief that African farmers were resistant to change has been successfully challenged by research on the diffusion of new seeds and methods, and on the role of experimentation in indigenous agriculture.

Although studies of indigenous knowledge have done much to change perceptions of African culture, concerns have been raised about ethical issues. These include the ownership of indigenous knowledge and the rights of individuals and societies for recognition and compensation when knowledge developed over many generations is used to create modern drugs or other products with commercial value.

Participatory Development

Participatory approaches have become the accepted "best practices" in development work. In the past, professional experts appeared, sometimes without notice, to implement development projects conceived in a country's capital city or even abroad. There was little if any allowance for consultation with those directly affected; project managers were usually outsiders who often did not speak the local language.

Participatory development is meant to reverse the old power dynamics by insisting that development be directed from the local level. Local communities are to be empowered to identify their own needs and priorities. The role of the expert is redefined as that of a facilitator—helping the community to analyze its own situation, to set goals for development, and to develop an implementation plan. The effective use of local knowledge, together with the mobilization of local resources, is critical to successful participatory development.

Participatory development is not a panacea. Not only do communities very seldom speak with a single voice, but also power relations between the rich and the poor, and between men and women, are reflected in the process. Development professionals and the state often manipulate the process to further their own agendas.

Further Reading

The historical construction of Africa as an "idea" is explored in this book:

Mudimbe, V. Y. *The Idea of Africa*. Bloomington: Indiana University Press, 1994.

There is a large literature on constructions of Africa in the news media. See the following, for example:

Fair, J. "War, famine, and poverty: Race in the construction of Africa's media image." *Journal of Communications Inquiry*, vol. 17 (1996), pp. 5–22.

Hawk, B. G., ed. *Africa's Media Image*. New York: Praeger, 1992.

Myers, G., T. Klak, and T. Koehl. "The inscription of difference: News coverage of the conflicts in Rwanda and Bosnia." *Political Geography*, vol. 15 (1996), pp. 21–46.

Phelan, J. M. *Apartheid Media: Disinformation and Dissent in South Africa*. Chicago: Chicago Review Press, 1987.

Walsh, G. *The Media in Africa and Africa in the Media: An Annotated Bibliography*. Lochcarron, Scotland: Hans Zell, 1996.

For a riveting account of the media's coverage of the Ethiopian famine, see the following video:

Consuming Hunger (Part I: Getting the Story; Part II: Shaping the Image). Maryknoll, NY: Maryknoll World Productions, 1988.

Representations of Africa in geography textbooks are discussed in this article:

Myers, G. "Introductory human geography textbook representations of Africa." *Professional Geographer*, vol. 53 (2001), pp. 522–532.

The following sources look at the portrayal of Africa and Africans in Western popular culture, with particular reference to films and *National Geographic* magazine:

Cameron, K. M. *Africa on Film: Beyond Black and White*. New York: Continuum, 1994.

Davis, P. *In Darkest Hollywood: Exploring the Jungles of Cinema's South Africa*. Athens: Ohio University Press, 1996. (Video version: *In Darkest Hollywood*. Villon Films, 1994.)

Lutz, C., and J. L. Collins. *Reading National Geographic*. Chicago: University of Chicago Press, 1993.

Mayer, R. *Artificial Africas: Colonial Images in the Times of Globalization*. Hanover, NH: University Press of New England, 2002.

Pieterse, J. N. *White on Black: Images of Africa and Blacks in Western Popular Culture*. New Haven, CT: Yale University Press, 1992.

There is a growing literature on colonial travel writing. See these works, for example:

Blunt, A. *Travel, Gender, and Imperialism: Mary Kingsley and West Africa*. New York: Guilford Press, 1994.

Kearns, G. "The imperial subject: Geography and travel in the work of Mary Kingsley and Halford Mackinder." *Transactions of the Institute of British Geographers*, vol. 22 (1997), pp. 450–472.

The following sources offer a range of postcolonial perspectives:

Blaut, J. M. *The Colonizer's Model of the World: Geographic Diffusionism and Eurocentric History*. New York: Guilford Press, 1993.

Crush, J., ed. *Power of Development*. London: Routledge, 1995.

Godlewska, A., and N. Smith. *Geography and Empire*. Oxford: Blackwell, 1995.

Said, E. *Orientalism*. New York: Pantheon, 1978.

Said, E. *Culture and Imperialism*. New York: Knopf, 1993.

An Afrocentric interpretation of civilization is offered in the following sources:

Asante, M. J. *Afrocentricity*. Trenton, NJ: Africa World Press, 1988.

Bernal, M. *Black Athena: The Afroasiatic Roots of Classical Civilization* (2 vols.). New Brunswick, NJ: Rutgers University Press, 1987 and 1991.

There are numerous sources on indigenous knowledge, participatory development, and related topics. See the following, for example:

Leach, M., and R. Mearns, eds. *The Lie of the Land: Challenging Received Wisdom about the African Environment*. Oxford: James Currey, 1996.

Internet Sources

There are several interesting sites that focus on different aspects of African contemporary popular culture:

Festival Panafricain du Cinéma et de la Télévision de Ouagadougou. **www.fespaco.bf**

Heinemann Publishers. *African Writers Series*. **www.africanwriters.com**

Innis Library, University of Toronto. *African Cinema*. **www.utoronto.ca/innis/library/africanfilm.html**

Motherland Nigeria: Music and Samples. **www.motherlandnigeria.com/music.html**

Revue Noire. **www.revuenoir.com**

3

Culture and Society

Africa is a diverse continent, containing many hundreds of linguistic and ethnic groups. African societies are also divided by differences of gender, social class, and religion. While acknowledging the importance of the tremendous social and cultural diversity of the continent, we must also ask whether there are attributes that are common to all Africans, or at least to broad groups of African peoples. Cultural geographers continue to struggle with the challenges of understanding African culture and describing the dynamic processes of cultural change that continue to reshape it.

Identities are defined not only by a group's self-image, but also by the group's image in the eyes of others. As Chapter 2 has shown, non-Africans possess a limited number of often stereotypical images of Africa and Africans, garnered from such diverse sources as the news media, film, missionaries, and politicians' speeches. Many Africans are well aware of these external images, and their sense of identity is partly shaped in response to them.

Countless Cultures, or One?

Traditional Scholarly Views

Numerous studies by cultural anthropologists and geographers of the past have documented the cultural diversity of Africa, providing us with rich detail about many societies at specific points in time. These cultural studies describe the social structure, livelihoods, religious beliefs, and other characteristics of a community or tribe. They tend to emphasize the uniqueness of cultural characteristics, which are often seen as representing functional adaptations to the local environment. These traditional studies conceptualize culture as a set of recognizable attributes—the beliefs and practices held in common by a group of people living in a particular region. Culture is transmitted from one generation to another through the socialization and education of children within the family and community.

Traditional anthropological studies contributed to the Western perception of African cultures as exotic, mysterious, and tradition-bound. Although some things were seen as admirable—the harmonious connections of Africans to nature, for example—many of these studies have also implied that development would be difficult to achieve and likely to undermine social and ecological balances. Africa's cultural fragmentation, as well as the perceived enmity of different groups, served to justify a colonial rule that (it was argued) would protect African societies from each other.

FIGURE 3.1. African popular wisdom, written as graffiti on a wall, Zanzibar. In Swahili, *Sali kabla hujasaliwa* ("Pray before you are prayed for"). Photo: D. Alsen.

There is another view of African culture reflected in the older academic literature and in the Western imagination. This view, exemplified by the writings of the anthropologist Jacques Maquet (see "Further Reading"), depicted Africa as a single cultural entity characterized by a broad uniformity. Maquet argued that Africans from south of the Sahara share a common identity, which he called *Africanity*. The shared characteristics, shared experience, and shared world view that define Maquet's Africanity were thought to prevail despite the vast size of the continent and the great diversity of languages, ethnic identities, and religions. According to Maquet, several characteristics give rise to Africanity:

- A black skin, although not all black people are African.
- Belief systems that recognize the presence of spiritual forces capable of affecting humans on a daily basis, everywhere in the environment.
- A "chain of life" that links past and future generations through the honor given to the spirits of ancestors and the nurturing of the next generation.
- A dependence on the land (African societies

have traditionally depended on hunting, herding, and farming for their sustenance).
- A respect for the wise and judicious authority exercised by revered elders and traditional leaders.
- The use of artistic expression, particularly storytelling, sculpture, music, and dance, for many purposes (e.g., veneration of ancestors, education of youth). (See Figure 3.2.)
- A deep sense of pride about the outstanding accomplishments of Africans, both past and present.
- A shared history of humiliation and oppression under colonialism.

Neither of the above-described traditional views of African culture is viewed with enthusiasm by Africans or by most contemporary scholars. Both approaches emphasize the adherence to tradition as a fundamental characteristic of African cultures. In both cases, the primary focus is on rural cultures little affected by change. The view that Africans share a common cultural heritage has been criticized for failing to pay due attention to the great cultural diversity of the continent. In short, both are seen as alien frames of reference that reinforce stereotypical views of Africa.

FIGURE 3.2. Calabash carvers, Jigawa State, Nigeria. Much African art involves the decoration of everyday objects with designs that are often intricate and imbued with cultural meaning. Photo: author.

A New Cultural Geography

Scholars now think of culture in new ways. In the new cultural geography and related fields, culture is viewed as a fluid, dynamic phenomenon (see Vignette 3.1 for an example). Because culture changes constantly and unpredictably, views of "culture as tradition"—as a given "package" of attributes—are rejected. Culture is also seen as *contested*, meaning that various factions within a society have different values and aspirations. Thus it is normal to find tension about cultural issues between young and old, men and women, "innovators" and "traditionalists." Identities are individual and complex; all people are characterized by multiple identities related to their age, gender, religion, social class, and so on. Finally, culture is seen to be a socially constructed phenomenon. As such, it reflects subjective, selective, and often self-serving identities, rather than a set of objectively defined characteristics. Culture is socially constructed from within—built, for example, on a community's own memories of triumph and tragedy. It is also socially constructed from without—reflecting the definitions formulated by outsiders, especially conquerors or others in positions of dominance.

What this means is that there is not a single Maasai culture (to use this one East African pastoral society as an example), but many. The traditions guarded by male elders do not necessarily reflect the aspirations of younger Maasai who have moved to the city and are employed in the public service. The cultural identities of men and women are not identical. Whereas some Maasai have embraced Islam, others have become Christian. The cultural influences of the West (propagated through the school curriculum, through television and movies, and through interaction with tourists and missionaries) have not had a uniform impact on Maasai communities. In short, the stereotypical view of the Maasai as spear-carrying pastoralists subsisting on milk and blood taken from their cattle does not help us to understand who they are today.

Indeed, appearances may be deceiving. Although the Maasai may appear from afar to be a group that epitomizes African ethnic identity in its traditional sense, their history tells a different story. The Maasai did not emerge as a distinct group until the 18th and 19th centuries, following many centuries of intermingling of diverse groups that had migrated into the rift valley region of Eastern Africa. Even their pastoral way of life is quite recent; it was only in the 18th century that the ancestors of the Maasai ceased to farm and began to rely solely on pastoralism.

VIGNETTE 3.1. Continuity and Change in the Hausa Cult of *Bori*

The word *tradition* usually evokes an image of reverence for the ideas and rituals of the past and perhaps of resistance to change. This case study illustrates that so-called "traditional" religious–cosmological belief systems may in reality be extremely fluid, reflecting and responding to changes that are taking place in the broader social milieu. It also provides an example of a cultural institution that has come to be interpreted and used very differently by men and women.

Bori is a spirit possession cult widely practiced in segments of Hausa society in northern Nigeria and Niger. *Bori* is based on a belief in an invisible spirit world that parallels the visible world of humans. There are spirits of every social origin and status, and of every disposition. They are deeply involved in the daily lives of humans. Some spirits seek to establish a friendly relationship with humans and may bring them prosperity. Others are antisocial and cause various types of misfortune. *Bori* adepts act as intermediaries between the spirit and human worlds, and they are especially involved in the treatment of diseases believed to be caused by spirit attack.

The origins of *bori* are to be found in the pre-Islamic religion of the Hausa people, now observed only in a few isolated rural areas where people have resisted conversion to Islam. This religion involves several types of rituals to mediate the relationship between the spirit world and humanity. For example, heads of households make sacrifices to the spirits prior to the agricultural season to ensure that a good crop is obtained. The details of these sacrifices are carefully specified—for example, the exact coloring of the animal to be killed, as well as the location, timing, and method of sacrifice.

Where Islam is firmly established, most of the rituals associated with traditional Hausa religion have long since been discontinued. However, *bori* continues to be practiced widely, even in the most thoroughly Islamic urban areas. The majority of *bori* members in the city consist of women. Most men condemn *bori* as un-Islamic and claim that its adepts are usually prostitutes or, in the case of male participants, homosexuals. For women, whose social role has been regulated by the implementation of Islamic traditions such as seclusion (purdah), *bori* provides a medium of social interaction that is relatively free of male authority. The ceremonies of spirit possession become metaphors for the concerns of women, ranging from the frequent death of children (whose diseases are attributed to spirits) to rivalry among cowives in polygamous households. Women are often able to use the requirements of *bori* to negotiate concessions from their husbands, who fear the consequences of antagonizing the spirits.

During the colonial era, the British attempted to ban *bori*, fearing that it could become a medium for political resistance. However, it survived and experienced a strong resurgence in the 1950s. *Bori* has become an important force in political life in Hausaland, not only because it facilitates women's political participation, but also because it promises politicians a way of enlisting supernatural support for their campaigns.

Far from being merely a repository of "tradition," *bori* is a mirror of social change. Although formerly important concerns, such as smallpox, have receded owing to modern vaccination, new social problems, such as the urban breakdown of traditional values and authority, have come to the fore in Hausaland. These new forms of social malaise are attributed by many Hausa to changes in the spirit world, such as the arrival of previously unknown spirits. The ability of the *bori* cult to reflect and respond to the changing times accounts more than anything else for its survival and continuing relevance.

An Emerging African Identity?

What, then, of the concept of a single African identity, such as that proposed by Maquet? There is little support at present for the concept of a primordial African world view—an identity that has existed since time immemorial and is based on an assumed common relationship to the land and a shared history. The colonial contribution to a sense of shared identity has also been questioned; colonial rulers did not seek to cultivate a common African identity. Even within the artificial political units they created, they used divide-and-rule tactics that often accentuated supposed cultural differences.

In some countries and at particular points in time since independence, citizens have felt a heightened sense of national pride. Ghana under Nkrumah and South Africa following the demise of apartheid provide excellent examples. Nevertheless, the growth of national cultural identities has been at best sporadic and uneven.

Four decades after the end of the colonial era in most countries, a stronger sense of pan-African identity is emerging, especially in the larger cities. In large part, this reflects the impact of Western education and of the modern media. Modern Africans are reasonably well informed about events in other parts of the continent. During the fight against apartheid, they identified closely and personally with the struggle of their South African brothers and sisters. When Kenyans win medals in Olympic track competitions, or when soccer teams from such countries as Cameroon, Nigeria, and Senegal perform well in the World Cup, their triumphs are celebrated throughout the continent as victories for Africa.

The values, attitudes, and lifestyles of young Africans are also shaped increasingly by their exposure to global popular culture. Many of the visual markers of that culture—the music, the soft drinks, and the fashions—are readily seen not only in large cities, but in many rural areas as well. This is not to say that young Africans have forgotten or rejected their local cultural heritage. Rather, it points to the multifaceted and flexible nature of cultural identities in the contemporary world: Kikuyu, Kenyan, and African, but also village-born, male, and Christian, among others.

Societal Diversity

Africa has a rich diversity of cultures, and this diversity has had marked effects on political life. During the 1960s, the term *pluralism* was widely used to refer to the complex patterns of ethnicity, language, and religion in African nations. The forging of a national identity from often seemingly incompatible groups was considered to be a major challenge for newly independent states. Unfortunately, studies of pluralism were mostly descriptive and provided little guidance about policies for facilitating nation building. Moreover, they failed to reflect the fluidity and flexibility that are characteristic of ethnic and other identities in contemporary Africa.

The following discussion looks at several key dimensions of sociocultural diversity, notably ethnicity, language, religion, and social class. Gender relations in African societies are discussed at length elsewhere (see Chapter 19).

Ethnicity

Ethnicity has been defined as affiliation with or loyalty to a group sharing a common sense of origin, real or artificially constructed. Members of most ethnic groups share the same language, culture, and political and economic institutions. *Ethnicity* has replaced the pejorative terms *tribe* and *tribalism,* with their connotations of primitive feuding between hostile rivals. Ethnicity is a deceptive construct, however, since ethnic divisions are often quite imprecise. Certain African groups share a common ethnic identity that overrides evident cultural differences, whereas other groups with no significant cultural differences claim separate ethnic identities. Nor do ethnic groups usually occupy discrete and exclusive territories (contrary to common Western belief), as Figure 3.3 illustrates in regard to Sierra Leone. It is because identities are socially constructed, rather than objectively defined, that such apparent contradictions occur.

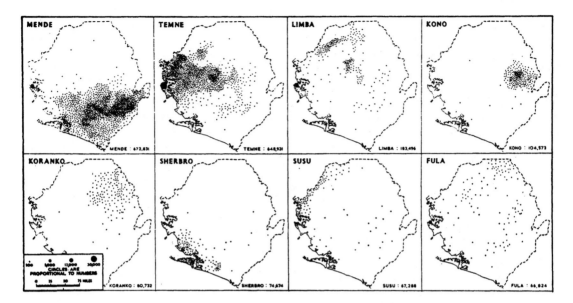

FIGURE 3.3. Sierra Leone: Distribution of major ethnic groups. Note that although groups tend to occupy relatively easily identified core regions, ethnic territories are seldom discrete and exclusive to a single group. Source: J. I. Clarke. *Sierra Leone in Maps.* New York: Africana, 1972, p. 37. © 1969 by J. I. Clarke, S. J. A. Nelson, and K. Swindell. Reprinted by permission.

Some tribes were essentially colonial creations. Ethnic identities were heightened by strengthening the power of chiefs as nominal rulers and by establishing new ethnically defined administrative units. Certain loosely affiliated groups were for the first time given chiefs and other traditional institutions. For example, the Igbo of southeastern Nigeria had no chiefs and only a loose sense of common identity prior to the colonial era.

Colonial tribalization was instrumental in legitimating not only the colonial conquest that allegedly brought peace to a feuding continent, but also policies of separate development applied to different groups. Ethnic identities were defined in relation to characteristics (e.g., supposed cultural sophistication) ascribed to various groups by colonial rulers. Certain groups were considered to be dependable workers, while others were deemed recalcitrant and lazy. Such labeling not only helped to shape and to justify policies that created gross regional inequities in development, but also sanctioned the forceful recruitment of migrant labor from certain regions and ethnic groups.

Regardless of how feelings of ethnic identity

originated, they did not wither after independence. African nations, however, vary greatly in the extent of their ethnic diversity. A few nations, notably Somalia, Lesotho, Swaziland, and Botswana, are relatively homogeneous and contain no substantial ethnic minorities. At the other extreme are highly diverse nations such as Tanzania, Cameroon, and Nigeria, which contain hundreds of ethnic groups. In Tanzania and Cameroon, no ethnic group is large enough to be politically dominant. However, Nigeria has three very large and powerful ethnic groups that dominate some 300 smaller groups in the country. The Hausa, Yoruba, and Igbo, whose current populations are about 20–25 million each, are more appropriately considered ethnic *nations.* Minority groups such as the Nupe, Ijaw, Tiv, and Kanuri, with 1–4 million members, are comparatively small in the context of Nigeria but are hardly insignificant. The national populations of 15 African countries are less than the 3.2 million population of the Tiv in Nigeria, yet the Tiv are only the eighth largest Nigerian ethnic group.

Political conflicts with ethnic dimensions have always received considerable attention in the Western media. The 1966 riots and the subsequent civil war in Nigeria, and the recurrent Hutu–Tutsi conflicts in Rwanda and Burundi (Vignette 3.2), are typical examples of crises that the Western media have chosen to interpret as conflicts about ethnicity. Unfortunately, contrary evidence has not received the attention it deserves. Not only have ethnic rivalries not been a major problem in many countries in Africa, but also the intensity of ethnic rivalries has receded in many nations. By giving excessive coverage to ethnic disputes, the media has helped to reinforce the colonial image of black African nations as hopelessly divided and unviable.

Several recent case studies help to underscore that ethnic identity in reality has always been fluid and socially constructed. This is especially the case in urban settings. *Pastimes and Politics*, Fair's study of cultural identities in Zanzibar during the first half of the 20th century (see "Further Reading"), provides an example. Leisure activities, popular culture, and modes of dress became vehicles through which former slaves sought to redefine their status in society. Women who had formerly been slaves adopted the veiled dress of the local ruling class, and men joined soccer teams that helped to forge long-lasting social and political connections.

Language

Africa has great linguistic diversity; there are close to 2,000 distinct African languages. Linguists continue to disagree about the best way of classifying African languages. Even where languages are considered to be related, the processes underlying the relationship usually remain a mystery.

Students of African languages have identified over 100 groups of languages—that is, two or more closely related languages or a single language unrelated to any other—belonging to four major language superfamilies, of which Niger–Congo is the largest. It includes both the many languages of West Africa and the approximately 800–1,000 Bantu languages that constitute a subfamily within Niger–Congo. Numerous languages of the Afro-Asiatic group are found in the area adjoining the Sahara, both in West Africa and in the Horn of Africa. Hausa and Amharic are its most important members. The Nilo-Saharan languages of north central Africa and the Khoisan languages of southwestern Africa constitute the other two distinct language superfamilies. Both are spoken by far fewer people than either the Niger–Congo or the Afro-Asiatic group.

The distribution and importance of individual African languages have gradually but continuoually evolved. Many of the more obscure languages, spoken by a few dozen to a few thousand people, are dying out and are being replaced by expanding lingua francas, pidgins, and official European languages. Many Africans are multilingual, typically speaking at least one lingua franca and a European language, in addition to their mother tongue and perhaps other local languages. Multilingualism enables people to function effectively in a diverse society; it also helps to break down societal divisions.

In many regions, people from different cultural backgrounds must use a lingua franca to converse (Figure 3.4). Swahili, a Bantu language with many Arabic loan words, developed along the East African coast and is now spoken throughout East Africa. It is also increasingly prevalent in adjacent parts of central Africa. Hausa, spoken by at least 50 million people, is the most important lingua franca in West Africa, particularly in Nigeria and Niger. Arabic is widely spoken in Islamic societies, especially in Sudan. A few other languages, including Malinke in Senegal and neighboring countries and Bemba in south central Africa, are used as regional lingua francas. Creole and pidgin languages, which are derived from English and other European tongues combined with local African languages, serve as lingua francas in certain coastal areas of West Africa that have had a long history of African–European contact.

African lingua francas possess various advantages that permit them to expand at the expense of small local languages. They are widely used in the printed and spoken media and have their own literatures. The use of lingua francas as languages of instruction in the school systems of countries where they are spoken has facilitated their spread. The growth of lingua francas has contributed significantly to the less-

Vignette 3.2. "Ethnicity" and Political Violence in Rwanda

The curse of ethnic conflict has been a recurring theme in coverage by the Western media of events in Africa, while other dimensions of conflicts have received little attention. The result has been to solidify in the minds of the Western public the idea that so-called "tribalism" is an *African* problem.

The background to the violence in Rwanda during 1994, resulting in the deaths of between 500,000 and 800,000 people, illustrates the importance of looking beyond stereotypes of tribal conflict. The violence in Rwanda was mostly the result of ethnic differences, but not exclusively so. Moreover, the ethnic dimension of the crisis reflects a complex history of relations between different segments of Rwandan society, extending over many centuries. Of special importance in this history is the manipulation of ethnic relations that occurred under colonialism.

The people of Rwanda, the Banyarwanda, embrace three subgroups: the Hutu (about 85% of the population), the Tutsi (about 15%), and the Twa (1%). The groups commonly intermarry, speak the same language, and share a common culture. The Tutsi were the ruling class in the Kingdom of Rwanda before the colonial conquest, maintaining a "feudal" domination of the Hutu and Twa. Status and wealth, vested in their ownership of cattle, were associated with being Tutsi. Tutsi who lost their cattle and became poor came to be identified as Hutu, and Hutu who became wealthy could become Tutsi. The system was more one of class than of ethnicity.

The rift between the groups deepened under German, and later (after World War I) Belgian, colonialism. People were classified formally as Hutu or Tutsi on the basis of cow ownership; anyone with fewer than 10 cows was deemed to be Hutu. In communities that had Hutu chiefs, the chiefs were replaced by Tutsi. Chiefs were made to administer a brutal regime of forced labor and taxation that made life unbearable for the Hutu and for the poorer Tutsi. As independence approached, the Belgians ended their policy of using the Tutsi as administrators, and supported the emergence of an all-Hutu political party. Communal violence broke out in 1959, resulting in the deaths of many chiefs and citizens. The Belgians seized the opportunity to replace many Tutsi chiefs with Hutu. In January 1961, with Belgian support and in defiance of the United Nations, the Tutsi monarchy was abolished and Rwanda was declared a republic.

The years following independence saw recurring violence, directed primarily by the governing Hutu majority against the Tutsi. Many thousands were killed. By 1964, an estimated 150,000 Banyarwandan refugees had fled to neighboring countries. The Rwanda government continued to play on ethnic fears among those who remained; for example, people were required to carry identity cards that named the group to which they belonged. Nevertheless, Hutu and Tutsi continued to live and work side by side, usually without undue tension.

Banyarwanda refugees who had fled to countries such as Uganda became essentially stateless. Rwanda refused to allow repatriation, claiming that the country was overpopulated and that many Banyarwanda wishing to enter Rwanda had for generations been citizens of another country. Banyarwanda living in Uganda were viewed as outsiders and, because of their economic success, were viewed with hostility. The sense of alienation lay behind the decision of some Banyarwanda to form the Rwanda Patriotic Front (RPF) and to invade Rwanda in 1989.

A study undertaken by Malkki in Tanzania among Hutu refugees from Burundi living in Tanzania sheds light on the psychological impact of exile. Hutu refugees, especially those living in camps, created an elaborate mythical history of Hutu–Tutsi relations and their envisaged return home. Their sense of identity as Hutu was profoundly affected by their experience of displacement.

Between 1989 and 1994, there were intermittent battles and several failed attempts to ne-

(cont.)

VIGNETTE 3.2. *(cont.)*

gotiate a peaceful end to the conflict between the RPF and the Rwandan government. A wave of genocidal violence, perpetrated primarily by government paramilitary groups, commenced in April 1994 following the death of the president in a plane crash. Although most of the victims were Tutsi, many Hutu distrusted by the governing faction were also killed. There were also many acts of retaliation by the RPF and its supporters. As violence escalated, the RPF began to move on the capital, Kigali. The government stepped up its campaign of fear by warning that "the lords" were trying to return. In this climate of terror and fear, it is hardly surprising that after the defeat of government forces by the RPF, nearly 2 million Rwandans fled to Zaire (later the Democratic Republic of the Congo) and other neighboring countries, fearing for their lives.

Although Rwanda has been relatively peaceful and stable during the decade following the genocide, the same cannot be said for the adjoining eastern Congo region. The presence of huge numbers of Rwandan refugees, controlled by the exiled former government and its militia, served to destabilize the region. Fighting broke out in 1996, involving the exiled Rwandans, the armed forces of Zaire, and a local resistance movement called the Alliance of Democratic Forces for the Liberation of Congo–Zaire (ADFL). Within weeks of the outbreak of fighting, the refugee camps had been shut down and most of their residents returned to Rwanda. Ultimately, the ADFL defeated the Zaire government forces, and the ADFL leader, Laurent Kabila, became president in 1997. However, fighting in the eastern Congo continued unabated for several years, resulting in many deaths and widespread destruction. The intervention of troops from Rwanda and Uganda (opposed to Kabila's regime), and from Zimbabwe and other African countries (supporting Kabila), transformed an internal, regional conflict into an international war.

Based on A. Des Forges. *Leave None to Tell the Story: Genocide in Rwanda.* New York: Human Rights Watch, 1999. L. Malkki *Purity and Exile: Violence, Memory, and National Cosmology among Hutu Refugees in Tanzania.* Chicago: University of Chicago Press, 1995. C. Watson (V. Hamilton, ed.). *Exile from Rwanda: Background to an Invasion.* Washington, DC: U.S. Committee for Refugees, 1991.

---

ening of ethnic rivalries and the emergence of stronger national identities. In Kenya, Tanzania, and Uganda, Swahili is strong enough to be recognized as an official national language.

The colonial languages—English, French, Portuguese, and Spanish—have continued to spread because of their role in education. The continued use of English and French is virtually assured because of their importance as international languages of science and technology. European languages will retain their position as official national languages because most countries have no indigenous language suitable for this purpose. Even where other languages have the potential to become national languages, European languages have become increasingly entrenched and widely spoken.

Religion

Christianity, Islam, and traditional religions are widely practiced. In a very general sense, Christianity is dominant in southern, central, and west central Africa, whereas Islam is prevalent in semiarid and savanna regions from the Atlantic to the Red Sea and along the Indian Ocean coast from the Comoros northward. However, the pattern of religious observance is far more complex. Many countries, including Nigeria and Côte d'Ivoire, have significant regional differences in religious affiliation. In many regions, religious *syncretism* is common; that is, many people who are nominally Christian or Muslim continue to practice elements of a traditional religion as well. Because religious affiliation for many Africans is fluid and com-

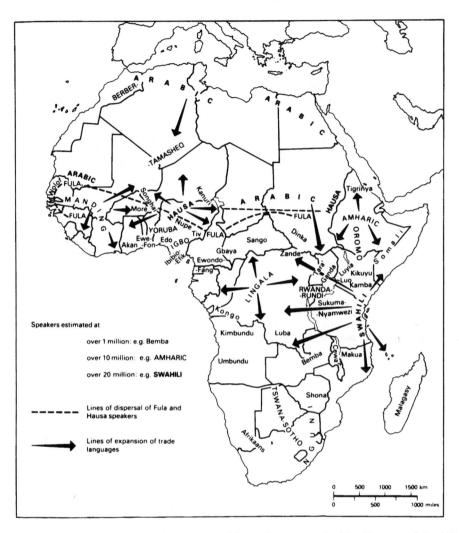

FIGURE 3.4. Lingua francas and other important African languages. Arabic, Hausa, and Swahili are the most important and fastest-growing of Africa's lingua francas. Source: D. Dalby. "African languages." *Africa South of the Sahara, 1991.* London: Europa, 1990, p. 96. © 1990 by Europa Publications Limited. Reprinted by permission.

plex, religious data such as those in Figure 3.5 must always be viewed with caution.

Traditional religion encompasses all the belief systems and religious practices that are specific to a culture. Various kinship and occupational groups typically depend on specific deities for protection and prosperity, and take primary responsibility for performing the rituals needed to ensure continuing good fortune. The deities have a variety of characteristics and are commonly associated with particular elements in the natural environment. The persis-

tence of traditional religions varies greatly, reflecting both the duration and strength of Christian and Islamic missionary activity and the resilience of individual cultures. Certain groups, such as the Mossi of Burkina Faso, have resisted conversion. The boundaries between religions are blurred by the retention of elements of traditional belief systems by African Christians and Muslims. As Vignette 3.1 has shown, even where Islam or Christianity has become very firmly established, beliefs and practices based on indigenous religion may continue to

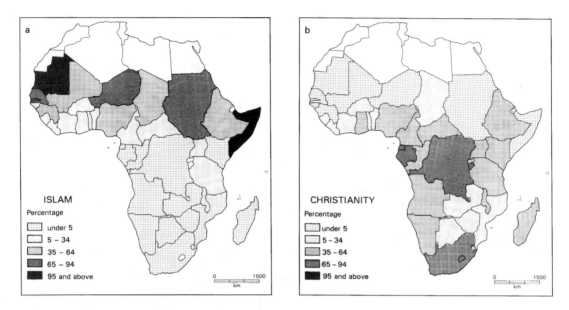

FIGURE 3.5. The distribution of (a) Islam and (b) Christianity. Islam predominates in the states adjacent to the southern edge of the Sahara, Christianity further south. Nigeria and Ethiopia are the only states with large proportions of both.

be important, although they may be expressed in new ways.

African Christianity dates from about A.D. 400, when two Christian kingdoms emerged: Nubia, in what is now Sudan, and Axum, in the region that is now Ethiopia. The Ethiopian Orthodox Church, the state church of the Ethiopian Empire, has maintained continuing ties to major centers of Orthodox Christendom from its formation until the present.

Except for sporadic missionary efforts during the era of slavery, European interest in converting Africans to Christianity only commenced in the 19th century. The continent was then subdivided into religious spheres of influence dominated by different Protestant and Catholic missions, each promoting its own version of Christianity. Missionaries were actively involved in providing health care and education, which were used as inducements to conversion. The missions were often the only sources of such services in areas neglected by the colonial states.

As a response to the uncompromising attitude of most missionaries toward traditional customs and religions, numerous independent African churches emerged. These independent churches combine Christian teachings and ceremonies with elements of African culture and traditional religion. Examples of independent churches include the Zionists of South Africa, the Cherubim and Seraphim of Nigeria, and the Kabunga of the Democratic Republic of the Congo. Although these churches generally remained aloof from anticolonial struggles, their original rejection of European religious dominance had real political undercurrents. The African independent churches remain a distinct and very important force in the societies of many African countries.

During the late 20th century, evangelical movements began to exert a growing influence on African Christendom, in large part through the global reach of U.S. television evangelism. Evangelical churches are the fastest-growing denominations in many countries. In parts of northern Nigeria, rivalries between Muslims and evangelical Christians have intensified in recent years.

The geography of Islam reflects long-standing patterns of interaction between Africa and the Islamic societies of Arabia and North Africa. Initially, Islam was mainly the religion of the ruling classes; it began to take hold south of the Sahara when the ruling class of the ancient kingdom of Ghana converted in the 12th century. Islam became established at an early date elsewhere, especially in other parts of the West African savanna, in the Swahili states along the eastern coast, and in northern Sudan. Islam grew remarkably during the 19th century as a result of militant Islamic reform movements centered in present-day Sudan, Nigeria, and Mali. In each case, charismatic leaders launched *jihads* (holy wars) and established orthodox Islamic states. These states revitalized Islam within their borders, and extended its reach through the conquest of adjacent regions.

The militant tradition of Islam provided the impetus for determined resistance to colonial expansion and, later, passive resistance to colonial rule. Under colonialism, Muslims rapidly expanded their influence, both numerically and spatially, as a result of their own proselytizing activities. Muslims often describe Islam as a religion for black people and, by contrast, characterize Christianity as a white religion. New converts are attracted by the apparent unity and discipline of Islamic communities (Figure 3.6). They may also be attracted by the relative ease of becoming a Muslim; the process commences with a simple profession of faith and typically proceeds gradually over many years and even generations. Islam has tended to be more patient than Christianity about the reluctance of converts to abandon traditional cultural practices. On the other hand, there has been determined resistance to conversion in certain countries, such as in the southern Sudan.

There has been a tendency in the West to see Islam as a monolithic whole, and—especially since the events of September 11, 2001—to

a   b

FIGURE 3.6. Islam in Africa. (a) Mosque, Niamey, Niger. The mosque is the focal point for religious, and often political, life in Islamic communities. (b) Islamic scholar at prayer, Niger. Photos: CIDA (R. Lemoyne).

demonize Islam as a fundamentalist faith linked to support for terrorism. In reality, Islam is a very diverse faith, with several distinct brotherhoods characterized by different styles of worship and different stances on social and political issues. Mahmoud Mohamed Taha, a Sudanese religious leader who has called for liberal democratic reforms and increased opportunities for women, represents one of several reformist traditions within Islam. Senegalese Islam is dominated by the influence of Sufi (mystical) brotherhoods such as the Mourides, a group organized in 1886 to resist colonial rule. During the colonial era, the Mourides created many new settlements where members grew groundnuts (peanuts). They gradually expanded into other facets of the economy, and became a potent economic force. The Sufi brotherhoods also served as a social welfare organizations, offering material support to poorer members. Since independence, the brotherhoods have continued to be a powerful political force in Senegalese society, and have contributed positively to the development of the country's stable democracy.

Islamic fundamentalism has posed a continuing challenge to the governments of certain states, particularly Sudan and Nigeria. Many Muslims are worried about the implications of increasing Western influences. Although most Muslims seek a pragmatic accommodation of the two, a determined minority continues to advocate the establishment of orthodox Islamic states purged of corrupting Western influences. Fundamentalist organizations in several countries have received financial support from religious organizations in Saudi Arabia, and to a lesser extent from Libya. In Nigeria, Islamic fundamentalism has become an increasingly powerful political force. The status of *sharia*, the Islamic legal code, has become the focus of heated disputes between Muslims and non-Muslims, as well as between fundamentalist and liberal Muslims.

Social Class

Social classes reflect the unequal distribution of political and economic power in a society. Ruling classes are in a position to exploit the weak and consolidate their own privileged position,

using tools ranging from alliances, both internal and international, to ideology. Class divisions cut across ethnic, linguistic, and religious categories.

The study of social formations has remained mostly the preserve of Marxist scholars. Liberal and conservative writers have tended to focus on horizontal cleavages such as ethnicity and religion, or to use variables such as income or occupation as indicators of status. There is no attempt in the following discussion either to define the various social classes precisely or to adhere to Marxist categories. Rather, the objectives are to describe in general terms the changing character of vertical cleavages in African societies, and to emphasize that the dynamics of class interaction are at least as important as those of ethnicity and religion.

Class composition and the dynamics of class relations evolve in response to changing economic, political, and sociocultural circumstances. In precolonial times, many African societies were structured hierarchically, with chiefs and royal families at the top. The ruling classes consolidated their power by using their control over access to land, customary obligations and taxes payable to them, and their ability to wage war. Other precolonial societies were more egalitarian, with no tradition of chieftaincy.

Patterns of stratification were radically different in the colonial state. Europeans occupied the top positions in the social hierarchy. There emerged a new class of indigenous elites, with some Western education and access to certain official positions. These small "assimilated" classes had legal status that conferred limited political rights in the Portuguese, Belgian, and French colonies. The power of traditional chiefs was most often eroded under colonialism; although the chiefs retained their positions, they had very little meaningful power. However, chiefs who in some areas had previously wielded limited power rose to prominence during the colonial era. Although most Africans continued to rely on agriculture for their sustenance, farmers became more differentiated. Some achieved relative prosperity through commercial cropping and trade, but many disadvantaged farmers were forced to leave home to seek seasonal employment to earn money for their taxes.

The class structure shifted once again after independence. Europeans either left or assumed less visible roles. The dominant class was a *state bourgeoisie,* consisting of politicians, civil servants, teachers and other professionals, and senior military officers. Not only were areas of power formerly reserved for colonial officials opened to Africans, but also the entire state apparatus was precipitously expanded and diversified. Entry and promotion depended on educational qualifications and connections to those in power. However, opportunities become progressively more limited for those lacking political connections. Meanwhile, the formal power of traditional rulers continued to recede.

The turn to neoliberal economic policies, initiated through the structural adjustment programs imposed by the International Monetary Fund (IMF) during the 1990s, has diminished the role of the public sector and increased opportunities for the accumulation of wealth by well-placed, innovative entrepreneurs. The result has been increased disparities in wealth between the richest members of society and all others, including not only the poor but middle-class, salaried workers as well. Policy measures such as reducing or ending subsidized food imports, reducing public-sector employment, and imposing fees for health care and education have greatly increased the cost of living.

In most countries, the attitudes and behavior of the dominant classes have generated much cynicism from the masses. The Wa Benzi, as they are called in East Africa, are blamed for mismanagement and the deepening economic crises. Although relatively few of them own Mercedes-Benz cars, take shopping trips in Europe, live in palatial estates, and wear luxurious clothing, these supposed symbols of success are popularly viewed as indicators of corruption and betrayal.

In addition to the ruling classes, several groups with varying levels of wealth and power are to be found in the city:

- Often relatively prosperous petty capitalist traders, transporters, and artisans.

- Lower-status employees of larger businesses and governments.
- Poor people who make ends meet in informal-sector jobs they have often created for themselves.
- The unemployed, who are mostly dependent on others for their sustenance.

With the rapid growth of cities, distinctly urban social classes have emerged and made their mark. Although the urban masses have little political influence in most countries, states have tacitly recognized their potential for protest by implementing certain policies benefiting the urban poor. For example, many countries in the 1960s and 1970s subsidized urban consumers by importing staple foods for sale below market prices. These policies were often reversed during the 1980s as a result of the implementation of structural adjustment packages.

With few exceptions, most rural people have fared poorly since independence. Although more services are now provided in rural areas, the rural–urban gap in service provision continues to grow. Government policies have depressed rural incomes and stimulated urban-ward migration. Benefits from rural development projects have gone mostly to large farmers and entrepreneurs, and have accelerated the growth of a landless class that survives by selling its labor. Despite their numerical dominance, the rural masses have even less political clout than their urban counterparts.

The flow of people and money between rural and urban areas contributes to class formation in both settings. For example, rural–urban migration has been the primary source of growth for urban classes, especially the urban poor. Migrants to the city commonly maintain ties to their ancestral homes through gifts, investments in property, and involvement in hometown improvement associations. The "big men" in rural society are often urban-based, although they will not necessarily be "big men" in the city.

Civil Society

Political scientists have paid considerable attention in recent years to the role of civil society in

African social change. *Civil society* refers to organized groups that exist between, and help mediate relations between, the state and society as a whole. Many of these groups are based on economic interests (e.g., business associations, cooperatives, and trade unions); others represent particular regional, ethnic, or religious perspectives (e.g., church organizations and the Muslim brotherhoods); and still others are nascent political opposition groups (e.g., the Forum for the Restoration of Democracy [FORD] in Kenya). Many civil society groups have also been organized to represent the interests of women.

Given the weakness of the state and of democratic institutions in most African countries, civil society organizations have often played a critical role in representing the interests of segments of society excluded from power. They have formed a core around which opposition to the government could coalesce. In other cases, civil society organizations have worked very closely with the state. We shall return in Chapter 27 to the subject of civil society and its significance for local development.

Exploring the Diversity of African Society

Africa's writers have provided a rich and diverse body of literature, in which many of the themes introduced in this chapter are explored through the eyes of ordinary Africans who are forced to grapple with the challenges and dilemmas of societal change in the course of simply living their lives. Works by African writers can be an invaluable resource for geographers wishing to develop a better understanding of the dynamics of tradition and change in African societies.

There has been much notable writing by authors communicating in French and Portuguese, but the largest and most accessible body of African fiction is that written in English. Several classic works originally written in French have been translated into English, including novels by Ousmane Sembene (Senegal) and Ferdinand Oyono (Cameroon). More and more novels are also being written in African languages such as Swahili, Kikuyu, Yoruba, and Igbo in some parts of the continent.

Nigeria has produced many writers of note, including Wole Soyinka (who won the 1986 Nobel Prize for Literature), Chinua Achebe, Cyprian Ekwensi, Festus Iyayi, and Buchi Emecheta. Novels written by Ghana's Ayi Kwei Armah and Kofi Awoonor have also been widely acclaimed. East Africa has a thriving literary tradition, in which the Kenyan novelist Ngugi wa Thiong'o has achieved prominence. In South Africa, the violence of apartheid has been the focus for provocative works not only by members of the oppressed communities, but also by white liberal writers such as Alan Paton and Nadine Gordimer.

Further Reading

The following sources are recommended as introductions to the cultural identity of Africa:

Amadiume, I. *Re-Inventing Africa: Matriarchy, Religion, and Culture.* London: Zed Books, 1998.

Appiah, K.A. *In My Father's House: Africa in the Philosophy of Culture.* New York: Oxford University Press, 1992.

Clark, L. *Through African Eyes: Cultures in Change.* New York: Praeger, 1969.

Maquet, J. *Africanity: The Cultural Unity of Black Africa.* New York: Oxford University Press, 1972.

Mazrui, A. *The Africans: A Triple Heritage.* Boston: Little, Brown, 1986. (This volume was written to accompany an excellent PBS documentary series with the same title.)

Mudimbe, V.Y. *The Invention of Africa: Gnosis, Philosophy, and the Order of Knowledge.* Bloomington: Indiana University Press, 1988.

There are many sources on ethnic issues in specific countries. For general surveys on ethnicity and language, see the following sources:

Greenberg, J. *The Languages of Africa.* Bloomington: Indiana University Press, 1966.

Fair, L. *Pastimes and Politics: Culture, Community, and Identity in Post-Abolition Urban Zanzibar, 1890–1945.* Athens: Ohio University Press, 2002.

Kuper, L., and M. G. Smith, eds. *Pluralism in Africa.* Berkeley: University of California Press, 1969.

LeVine, L. "Conceptualizing 'ethnicity' and 'ethnic conflict': A controversy revisited." *Studies in Comparative International Development*, vol. 32 (1996), pp. 45–75.

Shaw, T. "Ethnicity as the resilient paradigm for Africa: From the 1960s to the 1980s." *Development and Change;* vol. 17 (1986), pp. 587–605.

Vail, L., ed. *The Creation of Tribalism in Southern Africa.* London: James Currey, 1989.

For contrasting perspectives on the meaning of Maasai ethnicity, see the following:

Beckwith, C. *Maasai.* New York: H.N. Abrams, 1980.

Spear, T., and R. Waller, eds. *Being Maasai: Ethnicity and Identity in East Africa.* Athens: Ohio University Press, 1993.

An extensive literature on African religion includes the following sources:

Blakely, T. D., et al., eds. *Religion in Africa: Experience and Expression.* Portsmouth, NH: Heinemann, 1994.

Brenner, L., ed. *Muslim Identity and Social Change in Sub-Saharan Africa.* Bloomington: Indiana University Press, 1993.

Isichei, E. *A History of Christianity in Africa: From Antiquity to the Present.* Grand Rapids, MI: Eerdmans, 1995.

Sanneh, L. O. *West African Christianity: The Religious Import.* Maryknoll, NY: Orbis, 1983.

Simone, A.M. *In Whose Image?: Political Islam and Urban Practice in Sudan.* Chicago: University of Chicago Press, 1994.

Taha, M. M. *The Second Message of Islam.* Syracuse, NY: Syracuse University Press, 1987.

Villalón, L. A. *Islamic Society and State Power in Senegal: Disciples and Citizens in Fatick, Senegal.* Cambridge, UK: Cambridge University Press, 1995.

Issues related to social class in Africa are discussed in the following sources:

Lubeck, P., ed. *The African Bourgeoisie.* Boulder, CO: Lynne Rienner, 1987.

Sklar, R. "Patterns of social conflict: State, class, and ethnicity." *Daedalus,* vol. 111 (spring 1982), pp. 71–98.

Stichter, S., and J. Parpart, eds. *Patriarchy and Class: African Women in the Home and Workforce.* Boulder, CO: Westview Press, 1988.

Zeilig, L., ed. *Class Struggle and Resistance in Africa.* Cheltenham, UK: New Clarion Press, 2002.

Studies of African civil society are found in the following sources:

Harbeson, J. W., D. Rothchild, and N. Chazan, eds. *Civil Society and the State in Africa.* Boulder, CO: Lynne Rienner, 1994.

Makumbe, J. "Is there a civil society in Africa?" *International Affairs*, vol. 74 (1998), pp. 205–217.

For an introduction to themes explored in African novels, read the following source:

Riddell, J. B. "Let there be light: The voices of West African novels." *Journal of Modern African Studies,* vol. 28 (1990), pp. 473–486.

Internet Sources

There are many fascinating sites about African culture, some of them general and others specific to particular regions or groups. Check out the following:

Arthur, G., and R. Rowe. *Akan Cultural Symbols Project.* www.marshall.edu/akanart/

Ethnologue Country Index: Languages of Africa. www.ethnologue.com/country_index. asp?place=Africa

Fung, K., Stanford University. *Africa South of the Sahara: Culture and Society.* www-sul.stanford. edu/depts/ssrg/africa/culture.html

Isizoh, C.D. *African Traditional Religion.* www. afrikaworld.net/afrel

Natural History Museum of Los Angeles County. *Africa: One Continent, Many Worlds.* www.nhm. org/africa

School of Art and Art History, University of Iowa. *Art and Life in Africa Project.* www.uiowa.edu/ ~africart

Smithsonian Institution. *Smithsonian National Museum of African Art.* www.nmafa.si.edu

United Nations Educational, Scientific and Cultural Organization (UNESCO). *Culture and UNESCO: Africa.* www.unesco.org/culture/ww/ africa

The Physical Environment

The chapters that follow describe the physical environment of Africa and outline the physical geographical processes shaping it. Although this book is about the human geography of Africa south of the Sahara, the fundamental importance of the environment for human activity cannot be denied. This is particularly the case in Africa, where the majority of people are primary producers relying directly on the environment for their sustenance, and where so many concerns have been raised in recent years about the nature of environmental degradation and its significance for humans.

Chapter 4 surveys the geomorphology and geology of Africa. To understand the broad physical uniformity of the African continent, it is important to understand the dynamics of plate tectonics and the associated processes of folding, faulting, and volcanism. The processes that created Africa as we know it continue to alter the continent very gradually, both on the macro scale (e.g., the fracturing of the continent along its rift valley systems) and on the micro scale (e.g., processes of weathering, erosion, transportation, and deposition by water and wind).

In Chapter 5, the climate is described. Africa is bisected by the equator, and its Northern and Southern Hemisphere climates are essentially mirror images of each other. Major variations in the amount, duration, timing, and reliability of precipitation are important determinants of agricultural potential. Climate change has been a major concern in recent years, especially because of recurring droughts in several regions. Nevertheless, climate change is hardly a new phenomenon in Africa.

Biogeography and ecology are examined in Chapter 6. Africa's vegetation is diverse, ranging from tropical rain forest to desert. Its vegetation zones broadly mirror its climate regions. Much has been *written about threats to the African environment from human development. Deforestation and desertification are complex issues that have too often been analyzed in a simplistic manner. However, new research is offering quite different perspectives on the impacts African societies have had on their environment.*

4

Geology and Geomorphology of Africa

The African continent has an exceptional degree of physiographic uniformity. This is particularly evident in the vastness of its plains and high plateaus, the long escarpments abruptly separating one physiographic unit from another, and the infrequently indented coastline. The apparent regularity partly reflects the widespread occurrence of Precambrian bedrock; most exposed materials consist of Precambrian outcrops, the weathered remnants of Precambrian formations, or sedimentary deposits originally derived from Precambrian rocks. The fact that particular tectonic and weathering processes have operated at a grand scale, both spatially and temporally, has also contributed to the broad uniformity of the physical landscapes that we see today.

Although it is appropriate to recognize the broad uniformity of Africa's geology and physiography, it is also important to acknowledge the tremendous diversity of physical landscapes. Each region has its own particular geological history that needs to be documented. The geological histories of particular regions have continuing significance for human society; for example, these are linked to the distribution of valuable mineral and fossil fuel deposits, as well as to the qualities of the soil (which in turn are linked to opportunities for agriculture).

Relief Features

Geographers often distinguish between *Low Africa* and *High Africa*, separated by a line running from northern Angola to northwestern Ethiopia (Figure 4.1). Low Africa, located northwest of this line, is characterized by sedimentary basins and low plains usually under 500 m above sea level. Low Africa also contains several isolated upland regions with altitudes of 1,000–4,000 m, the most important of which are (1) the Tibesti and Aïr Massifs and the Ahaggar Mountains of the Sahara, and (2) the Guinea and Adamawa Highlands further south. Plateaus and plains 1,000–2,000 m above sea level dominate High Africa, which forms the southeastern portion of the continent. The plateaus are bounded by often spectacular escarpments, including the Great Escarpment, which parallels the coastline from Angola to southern Mozambique. There are also several prominent

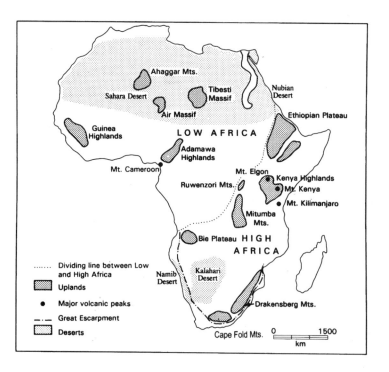

FIGURE 4.1. Physiography: Major physical features.

high plateaus and mountain systems, most notably the Drakensberg Mountains (Lesotho and South Africa), the Mitumba Mountains (Democratic Republic of the Congo, Rwanda, and Burundi), the Kenya Highlands, and the Ethiopian Plateau.

Africa's surface has been gently warped to form several large drainage basins ringed by areas of higher elevation (Figure 4.2). Among the most important basins are the Sudan (Nile), Congo, Chad, El-Djouf (Niger), and Kalahari. Some contain large river systems (e.g., the Niger and Congo) that reach the ocean, while others (e.g., the Chad and Kalahari) have rivers that drain toward the centers of their respective basins. The basins owe their formation to complex patterns of uplift and warping of the continent. This uplift has contributed to downcutting by major rivers, in some cases (such as the Congo and Zambezi) creating deep gorges, spectacular waterfalls, and rapids.

The increased erosion that came with continental uplift has gradually worn down the uplifted basin margins and has increased sedimentation within the basins and along the coasts.

Repeated cycles of crustal uplift have been linked to accelerated erosion and deposition, and to the resultant formation of a number of quite well-defined erosion surfaces occurring at different elevations. Thus prolonged, intensive erosion and deposition are key to understanding the vast expanses of plains and plateaus that characterize Africa's topography.

Geological Evolution and Structure

Africa is structured around three major zones that have remained very stable since the Precambrian era, which ended more than 590 million years ago. These stable *cratons* were centers of mountain-building activity some 1.5 billion years ago. Subsequent weathering and erosion gradually removed the mountains; only their inner cores remain. The northwest African craton is located in the western Sahara, the Congo craton in west central Africa, and the Kalahari craton in the southern part of the continent.

The areas between these large formations are

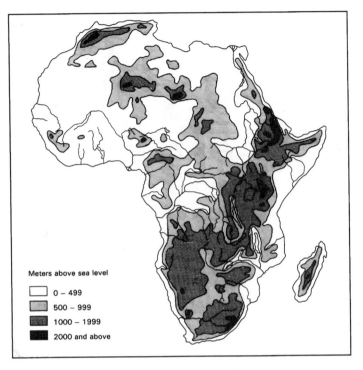

Meters above sea level

☐ 0 – 499

▨ 500 – 999

▨ 1000 – 1999

■ 2000 and above

FIGURE 4.2. Physiography: Relief and drainage.

formed of rocks that are, in comparison, younger and geologically more complex. They consist of rocks formed through volcanic processes (igneous rocks); through the deposition of weathered, eroded, and transported sediments (sedimentary rocks); and through the alteration of existing bedrock under conditions of intense heat and pressure (metamorphic rocks). Some of these geological formations are more than 1 billion years old, but are still younger than rocks of the cratons. Other formations are much younger; the processes that have shaped Africa continue to the present day.

Large areas of northern and western Africa contain deposits of sedimentary rocks, formed some 400–500 million years ago during the Paleozoic era, that lie on top of older Precambrian rocks. Much of southern Africa, from the Congo basin to the Cape region of South Africa, was blanketed with sediments some 250–350 million years ago. These rocks, known as the Karoo deposits, consist of sediments derived from till laid down by massive glaciers, eroded sediments carried from the newly created Cape fold mountains, and basalts of volcanic origin.

Sedimentary deposits have continued to be laid down over the past 65 million years in coastal areas and in the large, shallow drainage basins that occupy much of Africa's interior. Sediments, loosened by weathering and erosion from higher elevations, are deposited layer upon layer by rivers and, in arid regions, by wind. These processes of weathering, erosion, and deposition were intensive as a result of the increased uplift and warping of the African landmass following the demise of the megacontinent of Gondwanaland.

Rocks of volcanic origin have also continued to be produced. Accompanying the formation of the Red Sea, massive layers of volcanic basalt were deposited on the Ethiopian Plateau during the Tertiary period (Figure 4.3). More localized younger volcanic formations have been created (and are still being created) in several major volcanic zones, among them Mount Cameroon and the Ruwenzori Mountains.

The ongoing processes that continue to reshape Africa's landmass include both tectonic activities (faulting, folding, and volcanic activity) and the processes of weathering that

FIGURE 4.3. Volcanic plateau, near Lalibela, Ethiopia. Several layers of basalt are visible on the mountainside. Photo: E. Peters.

weaken and alter rocks on and near the earth's surface. Weathering is a prelude to other processes—erosion, transportation, and deposition—that redistribute sediments, bringing about the gradual wearing down of higher-elevation areas and the slow infilling of lower-elevation sedimentary basins.

Plate Tectonics and the Formation of Africa

The physiography of Africa reflects tectonic movements that have occurred over hundreds of millions of years. The very existence of the continent is the result of tectonic forces that caused the breakup of the ancient megacontinent of Pangaea. Currents of magma rise through the Earth's mantle and diverge as they encounter the solid lithosphere, subjecting the lithosphere to tremendous tensional forces. Eventually, the lithosphere fractures and opens a new rift. Volcanic activity often occurs simultaneously, creating lava flows and possibly volcanic peaks. As a result of the formation of new crust, together with the constant tension from the upwelling currents of magma, the rift gradually widens and in time forms separate landmasses divided by water. The Mid-Atlantic Ridge, where the African and South American tectonic plates

meet, is an example of a zone where these processes may be observed; magma continues to upwell and form new crust along the line of the ridge.

The evolution from Pangaea to Africa occurred in several stages, commencing some 250 million years ago. The first stage saw Pangaea divided into two megacontinents known as Laurasia and Gondwanaland, each of which was destined to fragment further. For Gondwanaland, the second stage involved the separation of the Indian subcontinent, Australia, Antarctica, and Madagascar. Next, South America separated from Africa. Africa assumed what was essentially its present shape about 15 million years ago, with the separation of the Arabian Peninsula and creation of the Red Sea. The great rift valley systems of eastern Africa are the results of continuing continental fragmentation owing to tectonic forces. In short, the processes that gave rise to Africa are continuing to reshape it. Although these processes are hardly discernible in relation to the scope of human history, they are quite recent and fast-developing when viewed in geological time.

The evidence that Pangaea and then Gondwanaland existed is varied and convincing. The shapes of Gondwanaland's constituent parts, when placed side by side like pieces in a jigsaw puzzle, fit quite closely. The adjoining parts of the reassembled megacontinent have similar rock types, identical fossil glacial deposits, and otherwise unique fossil life forms. The configuration of ocean-bottom geological formations also points to the progressive separation of landmasses that were formerly a single unit.

Tectonic Landforms

Tectonic landforms are those that result from the deformation and reshaping of the earth's crust, through intense compressional and tensional forces linked to subsurface flows of magma. Folded mountains, rift valleys, and volcanoes are the major features created by these massive forces.

Folded Mountains

Folded mountains occupy only a small area in Africa south of the Sahara, and they are relatively low compared to the folded mountains found on other continents. The Cape Fold Mountains consist of a series of ranges, interspersed with broad valleys, aligned parallel to the southern coast of South Africa. The highest of these ranges are 2,150 m above sea level. Their formation took place approximately 250–300 million years ago, during the Permian period.

The only other folded mountains on the African continent occur on its northern margins in Morocco, Algeria, and Tunisia. The formation of the Atlas ranges took place in several stages that were concurrent with the formation of the Alps and Pyrenees in Europe. They were formed through the convergence of the African and Eurasian tectonic plates, which compressed, folded, and uplifted marine sediments from the Tethys Sea. The Atlas Mountains cover a much larger area and are higher (up to 4,160 m) than the Cape Fold ranges.

Faulting

Strong and widespread uplift and deformation of the continent resulted in the fracturing of bedrock along lines of weakness. As a result, rift valleys formed along these fault lines. Africa's most spectacular rift valleys occur in eastern and central Africa, extending some 5,600 km from the Red Sea coast of Djibouti to southern Mozambique. These are divided into two major sections, the Eastern and Western Rifts. Vignette 4.1 provides further details about these spectacular landscape features, particularly about the diverse structure and appearance of different portions of the rift valleys.

Rift systems are also found in other parts of Africa. One extends parallel to the Nigeria–Cameroon border, on a line passing through Lake Chad, Mount Cameroon, and the offshore island of Fernando Po. Many geologists consider it to be an initial sign that western Africa will eventually separate from the rest of the continent, in much the same way that Madagascar and the Arabian Peninsula broke away previously. The Benue River and a portion of the middle Niger River also follow major fault lines.

Volcanism

Africa has widespread evidence of volcanic activity, extending from the Precambrian era to the present. As previously noted, increased volcanism has often occurred in conjunction with the sequential breakup of Gondwanaland. The Karoo deposits of southern Africa and the massive basalt flows that covered the Ethiopian Plateau are the two most important examples of this phenomenon.

Associated with the rift valleys in eastern Africa are several massive volcanic peaks, rising from 4,000 to almost 6,000 m in elevation. Most of these mountains, including Mount Kilimanjaro, Mount Kenya, Mount Elgon, and Mount Meru, are dormant. Several peaks along the Congo border with Uganda and Rwanda are active volcanoes. Eruptions on occasion pose a hazard for humans, as in 2002 when lava from an eruption of Mount Nyiragongo buried about one-quarter of the city of Goma in the Congo. Nevertheless, volcanism also brings benefits. East Africa's spectacular peaks attract many tourists, and dense populations are often supported on fertile volcanic soils. The weathering of soils of volcanic origin has formed productive soils that support very intensive agriculture and

VIGNETTE 4.1. The Rift Valleys of Africa

The rift valleys of eastern Africa (Figure 4.4) are among the world's most impressive physiographic features. They extend from the Jordan River valley, through the Red Sea and East Africa, to central Mozambique. This represents a distance of over 7,000 km. Africa has two major rift valley systems. The Eastern Rift extends from the Afar Depression on the Red Sea coast through central Ethiopia, and thence through central Kenya (Figure 4.5). The Western Rift runs from north of Lake Victoria through Lake Tanganyika and Lake Malawi to the southern Mozambique coast.

Different parts of these rift valleys developed in different ways and have distinct geological characteristics. The Afar Depression, located in the vicinity of Djibouti, continues to widen at a rate of several centimeters per year. The region experiences constant earth tremors and much volcanic activity, both signs of the geological instability of this region. From southern Ethiopia to central Kenya, spectacular escarpments rise 1,000–1,500 m above the linear valley floor. In the Lake Tanganyika basin, vertical displacements of the geological strata of up to 6,000 m have been found. The faulted landscapes of the southern portions of both rift systems tend to be rather less impressive.

The difference between the apparent depth of the rifts and the actual displacement of strata is attributable to the erosion of valley sides and the accumulation of eroded sediments and often of volcanic lavas in the valley bottom. In the vicinity of Lake Naivasha in Kenya, some 1,800 m of volcanic lavas have been deposited on top of the original valley floor. Volcanic activity associated with the rift valleys has also created several massive volcanic peaks, the largest of which is Mount Kilimanjaro. Although some sectors have much volcanic activity, volcanism is absent from other areas, such as in the Lake Tanganyika sector. The form of faulting also varies, with simple tensional faulting and vertical displacement creating steep-sided escarpments in some areas, and complex patterns of parallel faulting creating stepped escarpments elsewhere.

East Africa's large lakes, with the notable exception of Lake Victoria, are located within the rifts. Lake Tanganyika, Lake Malawi, and Lake Turkana are the largest of these rift valley

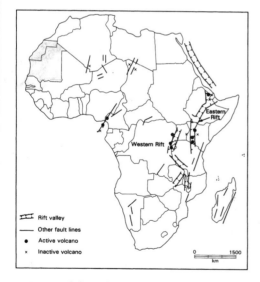

FIGURE 4.4. Rift valleys and volcanic peaks.

FIGURE 4.5. Escarpment along the edge of the Great Rift Valley near Naivasha, Kenya. Photo: author.

high population densities in Rwanda and a number of other regions.

Several areas of volcanic activity are associated with volcanic *hot spots*—namely, locations where intensive plumes of magma rise toward the surface. Areas located above hot spots are subjected to local heating and typically have frequent volcanic eruptions. The active volcanoes of Mount Cameroon and western Uganda, as well as the island volcanoes of the Comoro Islands and Réunion in the Indian Ocean and St. Helena and Ascension in the South Atlantic Ocean, all sit astride hot spots. So too do the volcanically active Tibesti Massif and Ahaggar Mountains of the central Sahara.

Climatic Influences: Weathering and Erosion

Climate is a fundamental determinant of landscape formation. Water facilitates chemical weathering in moist climates, and as it flows over the surface provides the energy needed to erode soil and weathered rock and to redistribute the sediments thus formed through processes of transportation and deposition. Many are surprised to learn that glaciation has also helped to shape Africa (see Vignette 4.2). Quite different forms of weathering, erosion, transportation, and deposition that depend much less on water occur in arid regions.

Landforms and Processes in Humid Tropical Regions

In more humid regions (over 1,000 mm of rainfall), deep chemical weathering occurs—especially where there is a combination of high temperatures, open-jointed Precambrian rocks, and the presence of decaying vegetative matter that can increase the acidity of infiltrating rainwater. The mineralogy and other characteristics of the bedrock determine the rate of weathering in particular areas. Chemical changes to certain constituent minerals cause the bedrock to become weaker and eventually to disintegrate. Chemical weathering commonly affects the top 15 m or more of bedrock in humid tropical regions. Because of the seasonality of rainfall in savanna regions, chemical weathering tends to be less effective, and thus results in a smaller average depth of weathered material.

In relatively flat areas subjected to prolonged chemical weathering, well-defined layers of mineral accumulation often develop below the surface. These layers, known collectively as *duricrusts*, may have different compositions. In humid tropical areas, the leaching of mineral matter downward through the soil has created brick-red layers, called *ferricrete*, that have high concentrations of iron and aluminum. Silica-rich layers known as *silcrete* occur in several parts of southern Africa. In somewhat drier regions, limestone layers—*calcrete*—were formed

VIGNETTE 4.2. Glaciers in Africa?

Few realize that the tropical continent of Africa has been shaped significantly by glaciers, or indeed that glaciers are still to be found in Africa. Massive continental glaciers covered large parts of Africa on several occasions. Evidence of the extent of these glacial periods is provided by deposits of *tillites*—glacially deposited sediments long since transformed into solid rock. These deposits are found in diverse locations, including southern Africa, the Congo basin, and the northwestern Sahara.

The greatest known extent of glaciation in Africa occurred during the Carboniferous era, some 300 million years ago. The Dwyka glaciation covered not only large parts of southern Africa, but also much of what are now western Australia, Antarctica, and southern South America. The striking similarity of the tillites in found these widely separated continents has provided strong support for the theory of plate tectonics and the common origin of these continents as part of the single megacontinent of Gondwanaland.

Today glaciers are found at the summits of a few of Africa's tallest peaks—notably Mount Kilimanjaro (Figure 4.6), Mount Kenya, and the Ruwenzori range, despite the fact that they are all located very close to the equator. During the last ice ages in Europe and North America, these glaciers were much larger than today, and some other highland areas such as the Ethiopian Plateau were covered in ice. The extent of moraine deposits suggests that ice covered up to 800 km² of East Africa's mountain peaks, compared to only 10 km² today.

Africa's remnants of glaciers have fared badly in recent decades, due to global climate change and an increase in atmospheric pollution. Mount Kilimanjaro's glaciers declined in size by about 80% during the 20th century, while Mount Kenya's have shrunk by 40% since 1963. Glaciologists estimate that the last of Africa's glaciers could disappear as early as 2010–2020 (*IAI Newsletter*, March–June 2001, pp. 14–17). Perhaps more important than the minor environmental impacts that will accompany the melting of Africa's last glaciers is the strong symbolic message from this imminent event about the threat posed to environmental systems worldwide by human-induced climate change.

FIGURE 4.6. Mount Kilimanjaro, Tanzania. A small icecap crowns the summit of this magnificent volcanic peak. At 5,895 m above sea level, it is the highest mountain in Africa. Photo: CIDA.

through the subsurface accumulation of calcium carbonate.

When exposed by the eroding of the overlying soils, these mineral-rich layers dry out and become rock-like. This process may also occur as a result of climate change (i.e., as conditions become drier). Many duricrusts are relict features that are the results of chemical weathering and mineral accumulation many thousands or even millions of years earlier under different climate conditions. However, intense chemical weathering also continues to occur today. Farming is almost impossible on duricrust hardpans that have been exposed through erosion. In some areas, outcrops of duricrust forming caps over softer underlying materials form flat-topped hills known as *mesas*, or define the break of slope of an escarpment. Deposits of *bauxite*—duricrusts with exceptionally high concentrations of aluminum—are mined in several countries, and some ferricretes are exploited as a source of iron ore. Duricrust deposits are widely used as a source of construction materials.

Inselbergs are massive, solitary domes of Precambrian rock rising above the surrounding plains. They are common features of the moister savannas and to a lesser extent of the forests of Africa (see Figure 4.7). It is believed that they

originated as subsurface masses of rock resistant to chemical weathering. These masses remain solid and unaltered, while weathering disintegrates the surrounding area of more open-jointed rocks. The inselberg is exposed later when prolonged cycles of erosion strip away the surrounding weathered rock. The exposed inselberg continues to resist weathering and gradually assumes a dome shape through the successive peeling away of surface layers of rock (see Figure 4.8).

Landforms and Processes in Arid Regions

Arid and semiarid landscapes cover about two-thirds of Africa. The Sahara, the world's largest desert, has an area approximately equal to that of the United States. It forms a broad band some 2,000 km wide across the full width of the continent. Other major desert regions occur in the Horn and in eastern Africa, separated from the Sahara by the major highland region of Ethiopia. These include the Danakil and Ogaden Deserts located primarily in Ethiopia and Somalia and the Chalbi Desert of northern Kenya, as well as large areas that are classified as semiarid. Southern Africa also has major arid regions—notably the Namib Desert that extends for some 2,000 km along the Atlantic

FIGURE 4.7. Inselberg at Abuja, Nigeria. The smooth, rounded form is typical of inselbergs. Photo: author.

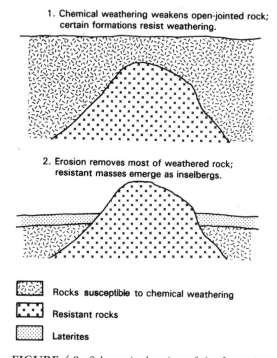

1. Chemical weathering weakens open-jointed rock; certain formations resist weathering.

2. Erosion removes most of weathered rock; resistant masses emerge as inselbergs.

Rocks susceptible to chemical weathering

Resistant rocks

Laterites

FIGURE 4.8. Schematic drawing of the formation of inselbergs. The processes of chemical weathering and erosion that give shape to inselbergs occur gradually over millions of years.

coast, and the semiarid to arid Kalahari Desert centered in Botswana. The Karoo is a dryland region located north of the Cape Fold Mountains in South Africa.

Weathering in arid or semiarid regions is much less active than in hot, moist climates, and most of the weathering is physical weathering. Certain types of rock may be shattered as a result of repeated, large day-to-night fluctuations in temperature. Physical weathering may also occur as a result of intense pressure caused by the progressive growth of salt crystals within porous rocks such as sandstones.

Contrary to popular belief, deserts are not uniform seas of sand dunes. The largest part of the Sahara consists of extensive plains covered with gravel and rocks. Other areas consist of bare rock surfaces. Wind is an important agent in the formation of all these desert types, contributing to the formation of gravel and rock deserts by removing loose sediments from the surface, and creating dunes and other depositional features in sandy desert areas. These

dunes take different forms, depending on local wind direction and sediment supplies. *Barchans* are crescent-shaped dunes, while *seif* dunes are linear ridges extending up to 80 km in length. In mountainous areas such as the Tibesti Massif and Ahaggar Mountains of the central Sahara, the scouring effect of wind-blown sand has created diverse erosional features such as pinnacles and grooves.

Water also plays an important role in the development of arid landscapes. Flat-bottomed watercourses known as *wadis* are normally dry but are filled with water after occasional heavy rains. *Alluvial fans* are formed of sediments carried along mountain watercourses, deposited where stream gradients decline as they enter flatter land. Desert areas also contain large, shallow pans where water accumulates during heavy rainfall. With the evaporation of this water, deposits of mineral salts that had been dissolved in it are left on or near the surface. Residents of the Sahara have exploited these saltpans for centuries as a source of many valued types of mineral salts.

Coastal Geomorphology

Africa's coastline tends to be very straight, with few good natural harbors. Much of the coastline has only a narrow coastal plain that ends abruptly at one of the escarpments running parallel to the coast in several areas. In southern Africa, the Great Escarpment forms an arc 100–200 km from the coastline between Angola and Mozambique, a distance of 5,000 km. The face of the Great Escarpment is an imposing site, exceeding 2,000 m in places. This landform marks the line along which Africa separated from the other constituent parts of southern Gondwanaland, creating new coastlines in what was formerly a midcontinent location.

West Africa between Liberia and Cameroon, and the Indian Ocean coast of South Africa and southern Mozambique, are the regions where processes of coastal deposition have been most active. Where prevailing winds blow parallel to the coast, sand is moved along the coastline by longshore currents. As a result, large sand spits and barrier beaches are deposited, blocking the entrances to most of the few good harbors. Lagoons are often found behind these barrier beaches. At its mouth, the Niger River has deposited a huge *arcuate* (fan-shaped) delta with numerous stream courses along which the Niger flows into the ocean.

Coral reefs fringe most of the coastlines of the Red Sea and Indian Ocean in eastern Africa. Their growth has been possible because of the warm water temperatures, whereas the colder water temperatures along Africa's Atlantic coast are inhospitable for corals. Africa's coral reefs were badly damaged by bleaching during the late 1990s as a result of significantly above-normal water temperatures during an extreme El Niño event. *El Niño* is the name given to periodic shifts in prevailing atmospheric and ocean current circulation patterns that initially develop in the Pacific region and then spread to other parts of the world. It is too early to know the extent of permanent damage to the reef ecosystems of eastern Africa as a result of this El Niño event.

Africa's best harbors are found between Senegal and Liberia, where, as a result of coastal submergence, the mouths of larger rivers become large and often deep inlets. Freetown, the capital of Sierra Leone, is renowned as having one of the very few superior harbors in Africa south of the Sahara. However, coastal submergence has not been the rule. Rather, the creation of the African continent from Gondwanaland resulted in straight coastlines and widespread coastal uplift that are not conducive to the formation of good natural harbors.

Further Reading

The following are general sources on African geomorphology:

Adams, W. M., A. S. Goudie, and A. R. Orme, eds. *The Physical Geography of Africa*. Oxford: Oxford University Press, 1996.

Bridges, E. M. "Africa." In E. M. Bridges, ed. *World Geomorphology*, pp. 30–65. Cambridge, UK: Cambridge University Press, 1990.

Buckle, C. *Landforms in Africa*. London: Longman, 1978.

Faniran, A., and L. K. Jeje. *Humid Tropical Geomorphology*. London: Longman, 1983.

King. L. C. *The Morphology of the Earth*, 2nd ed. Edinburgh: Oliver and Boyd, 1967.

Pritchard, J. M. *Landform and Landscape in Africa*. London: Edward Arnold, 1979.

Thomas, M. F. "Geomorphology and land classification in tropical Africa." In M. F. Thomas and G. H. Whittington, eds. *Environment and Land Use in Africa*, pp. 103–145. London: Methuen, 1969.

The following sources look at aspects of geomorphology in arid and formerly arid regions:

Grove, A. T., and A. Warren. "Quaternary landforms and climate change on the south side of the Sahara." *Geographical Journal*, vol. 134 (1968), pp. 193–208.

Thomas, D. S. G. " 'Relict' desert dune systems: Interpretations and problems." *Journal of Arid Environments*, vol. 20 (1991), pp. 1–14.

Thomas, D. S. G., and Shaw, P. A. *The Kalahari Environment*. Cambridge, UK: Cambridge University Press, 1991.

The following volume is a comprehensive study of Africa's glaciers:

Hastenrath, S. *The Glaciers of Equatorial East Africa*. Dordrecht, The Netherlands: D. Reidel, 1984.

Tectonic processes and the rift valleys are discussed in many sources, including these:

Baker, B. H., Mohr, P. A., and Williams, L. A. *Geology of the Eastern Rift System of Africa*. Boulder, CO: Geological Society of America, 1972.

Nyamweru, C. K. "The African rift system." In W. A. Adams, A. S. Goudie, and A. R. Orme, eds. *The Physical Geography of Africa*, pp. 18–33. Oxford: Oxford University Press, 1996.

Partridge, T. C., and R. R. Maud. "Geomorphic evolution of southern Africa since the Mesozoic." *South African Journal of Geology*, vol. 90 (1987), pp. 179–208.

Internet Sources

The following sites provide links to several sources on African geology and geomorphology:

Geology of Africa. http://geologylinks.freeyellow.com/regafr.html

Zwolinski, Z. *The virtual geomorphology. Regional Geomorphology: Afryka/Africa.* http://main.amu.edu.pl/~sgp/gw.htm

Several sites focus on tectonic processes, especially as they relate to rift valleys:

Geology General, with reference to material and Ethiopian volcanoes page. www.geocities.com/CapeCanaveral/Hall/1760/overview.html

Structural Geology and Tectonics, Duke University. *Turkana Rift, Eastern Branch, East African Rift.* www.eos.duke.edu/Research/Struct/TR.html

Web DoGS. Virtual Plates: Plate Tectonics. www.uky.edu/ArtsSciences/Geology/webdogs/plates/reconstructionc.html

5

Africa's Climate:
Regions, Dynamics, and Change

Virtually all of Africa south of the Sahara lies within 35 degrees of the equator, so tropical climates occur in all but its southern extremity. The climate types found north and south of the equator are, at a very general scale, mirror images of each other, as are the basic patterns of winds and pressure systems. Africa south of the Sahara is affected almost exclusively by tropical air masses. The one exception is its southern extremity, which receives midlatitude westerly winds in winter.

Dynamics of Climate

African climates are best understood in relation to seasonal patterns of air circulation that derive from global circulation systems. Altitude, topography, and ocean currents also affect climate and may be extremely important locally and regionally.

Air Masses

Air masses are large bodies of air that assume particular moisture and temperature character-

istics in their regions of origin. Most of Africa south of the Sahara gets its rainfall from air originating over the Atlantic Ocean and moving inland toward an equatorial zone of low pressure. The main exception is eastern Africa, from Somalia to South Africa, which is affected by moisture-bearing air masses from the Indian Ocean. These are examples of tropical maritime air masses; the names reflect their characteristic high temperatures and very high moisture content resulting from their maritime origins. The Cape region of South Africa receives most of its rainfall in winter (May–August), when it is under the influence of midlatitude westerly winds. This air mass is quite cool and moisture-laden, reflecting its origin over the South Atlantic Ocean.

Tropical continental air masses are dominant during the cooler winter season. In the Northern Hemisphere, tropical continental air masses form over the Sahara Desert—characterized, not surprisingly, by their extreme dryness. As they move southward, they bring hot, dry, and often dusty conditions to the areas they affect. In the Southern Hemisphere, continental air masses develop over the Kalahari and adjacent regions in the south central interior of the continent. Although they are also hot and dry, their temperature and moisture characteristics tend to be less extreme than those of the Saharan air masses. The much larger landmass of the Sahara contributes to the development of more extreme air mass conditions.

Atmospheric Circulation and Seasonality

The dominant pattern of air movement over most of Africa is toward the equator, from the northeast in the Northern Hemisphere and from the southeast in the Southern Hemisphere. To understand the seasonal changes and regional variations in African climates, it is useful to start by considering the patterns of air circulation. The basic sequence is as follows:

- Air movement is from areas of higher atmospheric pressure to lower pressure—in Africa, primarily from subtropical high-pressure areas to the equatorial low-pressure zone.
- Winds converge in this equatorial zone of low pressure, known as the intertropical convergence zone (ITCZ), where air rises, spreads out, and moves poleward in the upper atmosphere.
- This air descends, hot and dry, in the zones of high pressure centered on the Tropics of Cancer and Capricorn. The air then moves away from the tropics, including toward the ITCZ.
- The zones of high and low pressure (in fact, the entire circulation system) shift seasonally in harmony with the apparent movement of the sun's position, north of the equator in the Northern Hemisphere summers and south of the equator in Southern Hemisphere summers.
- The two contrasting air masses described above, tropical continental and tropical maritime, converge at the ITCZ. When tropical maritime air is forced to rise—whether owing to convection, frontal uplift, or relief (the *orographic* effect)—rainfall may well occur. This is because the vertical displacement of air causes it to cool, and as a result reduces its moisture-bearing capacity. As the air cools beyond its dew point (the tempera-

ture at which it becomes saturated), water vapor condenses, and rainfall occurs when atmospheric conditions permit the formation of sufficiently large water droplets.

Figure 5.1a shows the pattern of air pressure, prevailing winds, and rainfall in January when the sun appears overhead in the Southern Hemisphere. Rainfall north of the equator is minimal at this time. South of the equator, onshore southwesterly winds from the Atlantic bring heavy rainfall to the Congo basin, and lesser amounts to the north and south of this basin. In East Africa, south of the ITCZ, large tropical cyclones move onshore from the Indian Ocean. Eastern Madagascar and parts of the coastal mainland between southern Tanzania and South Africa are particularly affected and are likely to receive high winds and torrential rainfall in coastal areas. North of the ITCZ in East Africa, the predominant northeasterly winds bring dry continental air originating over central Asia, so that there is little rainfall.

In July—that is, during summer in the Northern Hemisphere—the ITCZ shifts far to the north and lies over the southern Sahara (Figure 5.1b). Southwesterly winds bring tropical maritime air and, consequently, rainfall to West Africa. These winds from the Atlantic penetrate as far as the Ethiopian Highlands. Almost the entire breadth of the continent receives rainfall at this time.

The occurrence of rainfall is closely related to the cross-sectional structure of the ITCZ (Figure 5.2). The moist tropical maritime air forms a wedge under the dry continental northeasterlies at the ITCZ. Where the wedge is very thin, the weather at the surface is hot and humid, but rainfall is unlikely to occur. Some 400–600 km back from the surface position of the convergence, the wedge is sufficiently thick (perhaps 5,000 m) to permit the development of cumulonimbus clouds and thunderstorms. Further back still, the wedge of moist air continues to become thicker, but there is a more stable atmospheric configuration and little rainfall, despite high humidity and heavy cloud cover.

Rainy seasons are longer near the coast than

ITCZ – Intertropical Convergence Zone

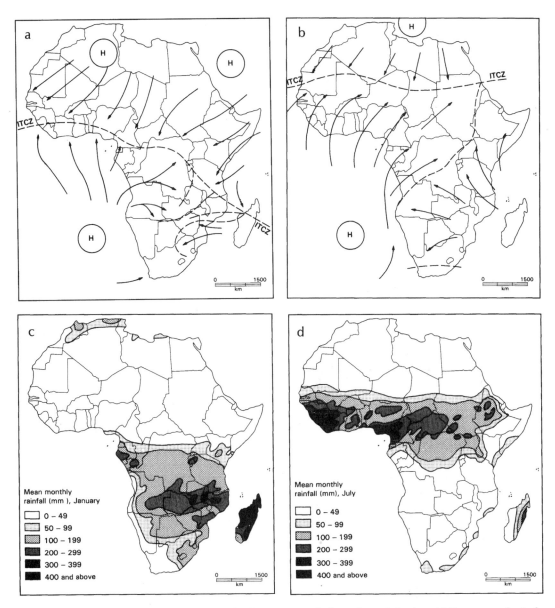

FIGURE 5.1. Seasonal variations in atmospheric pressure, winds, and precipitation. (a) Pressure and winds, January. (b) Pressure and winds, July. (c) Mean monthly precipitation (in millimeters), January. (d) Mean monthly precipitation (in millimeters), July.

farther inland, because coastal regions are under the influence of tropical maritime air for longer periods. Because maximum precipitation occurs in the part of the wedge of medium thickness, places near the coast tend to have a double-maximum pattern of rainfall; that is, there are two periods of higher rainfall each year. The first maximum (see Figure 5.3) occurs in spring

as the ITCZ moves poleward, and the second occurs in early fall as the ITCZ retreats toward the equator. Farther from the equator, where the wedge of moist maritime air does not remain thick and stable long enough to impede rainfall, there is a single midsummer period of heavy precipitation.

Seasonal shifts in the apparent position of

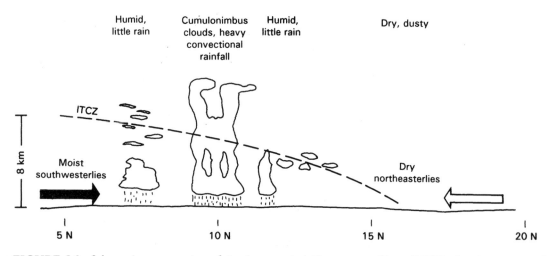

| Humid, little rain | Cumulonimbus clouds, heavy convectional rainfall | Humid, little rain | Dry, dusty |

FIGURE 5.2. Schematic cross-section of the Intertropical Convergence Zone (ITCZ), showing associated weather conditions. After A. T. Grove. *Africa*, 3rd ed. Oxford: Oxford University Press, 1978, p. 14.

the sun and of the ITCZ have only a minor effect on rainfall in equatorial and desert regions. Equatorial regions are under the influence of moist, maritime air almost continuously throughout the year, so have year-round rainfall. In contrast, desert regions such as the central Sahara receive little rainfall at any time, because the ITCZ seldom advances that far inland.

Most rainfall in Africa is *convectional*—that is, the result of local surface-level heating and updrafts that facilitate the development of cumulus and cumulonimbus clouds. Convectional rainfall tends to occur in small systems, and therefore gives rather spotty spatial coverage. Marked variations in the growth of crops may result, particularly in areas of seasonal rainfall.

FIGURE 5.3. Early rainy season storm, northern Nigeria. In savanna environments, the start of a rainy season is a time for celebration. Towering cumulonimbus clouds darken the sky. The calm is broken by strong gusts of wind that raise the dust, then a heavy shower or perhaps a downpour. The air is filled with the smell of earth, as the raindrops moisten the parched earth. Children frolic in the rain. Within days, the dry brown landscape starts to turn green. Photo: author.

For example, one village may have healthy, knee-high grain, while adjacent villages may be awaiting their first substantial rainfall of the year. Clearly, Africans whose livelihoods are closely linked to the land—not only farmers, but herders and the few remaining hunter-gatherers as well—find it essential to pay close attention to rainfall patterns and other climatic variations (see Vignette 5.1 for an example).

Altitude and Relief

Africa has no major mountain ranges comparable to those paralleling the west coast of the Americas from Alaska to southern Chile. Ranges such as the Rockies and Andes have important effects on the movement of air masses and thus on regional climates in North and South America. For example, most areas on the western, windward side of these ranges have much heavier rainfall than nearby areas on the leeward side, where so-called "rain shadow" effects are common.

Similar differences between moister windward slopes and drier leeward slopes may be observed on Africa's highest peaks. For example, parts of the southern slopes of Mount Cameroon have over 10,000 mm of precipitation annually. The windward sides of highland areas such as the Ethiopian Highlands receive somewhat higher rainfall than the leeward sides. Even relatively minor uplands such as Nigeria's Jos Plateau have an orographic effect—increased precipitation that occurs where air is forced to rise over a topographical barrier.

Altitude also has significant effects on temperature; on average, temperatures decline 6.4° C per 1,000 m of elevation. Highland regions in southern and eastern Africa experience comparatively cool, temperate conditions, relative to lower-elevation regions at the same latitude. Indeed, the cooler temperatures prevailing at higher altitudes were a major reason why many white settlers came to Kenya and to Northern and Southern Rhodesia during colonial times. The most dramatic evidence of how altitude affects temperature is to be seen on the slopes of the highest peaks, where vegetation changes from tropical forest or grasslands near the mountain base, to tundra and sometimes glaciers at the summit.

Ocean Currents

Ocean currents have significant effects on climate along the Atlantic coast from South Africa to southern Angola, and from Morocco to Mauritania. These effects are attributable, respectively, to the Benguela and Canaries currents. Both are cold currents, bringing water from cooler regions toward the equator. These cold currents give rise to somewhat cooler temperatures and extremely low precipitation in the nearby coastal areas such as the Namib Desert. The lowest layer of warm, moist air passing over these currents is cooled considerably by the cold waters. This cooling effect often produces dense fogs that blanket the coastline and serve as an important source of moisture for plants and humans in this extremely arid region. However, the cooled air mass seldom produces rainfall as it moves onshore and becomes warmer. Coastal Namibia has only about 25 mm of precipitation per year.

Cloud Cover and Atmospheric Dust

Cloud cover is a key determinant of daily temperature fluctuations. Dense cloud cover reduces daytime temperatures by reflecting considerable incoming solar radiation, and helps to retard the escape of heat from the earth's surface at night. In the absence of dense cloud cover, such as in arid climates or during the dry season in other climates, diurnal temperature fluctuations may be quite large.

The Sahara is the source of an estimated 300 million metric tons of airborne dust each year, 60% of the world's total. Dust is generated in three major regions: (1) the Bodélé Depression to the northeast of Lake Chad; (2) a large area straddling the borders of southern Algeria, northern Mali, and eastern Mauritania; and (3) northeastern Sudan and southern Egypt. Dust from the Mali–Mauritania–Algeria source region affects a wide area along the western flank of West Africa, as well as in North Africa and southern Europe. Dust from the Sudan–Egypt

VIGNETTE 5.1. Ethnoclimatology: The !Kung San Interpret Their Environment

The !Kung San of Botswana are among the very few remaining peoples whose traditional economy has been based on hunting and gathering. This way of life has come under increasing pressure since the 1980s, as a result of the expansion of cattle ranching into !Kung territories. As well, the Botswana government has encouraged the !Kung to settle permanently and farm, and has attempted to limit their hunting rights through its wildlife conservation policies. A declining number of !Kung live as before, moving from place to place and foraging for food.

The !Kung San, like other Africans living close to the land, have well-developed systems of knowledge about the environment, based on generations of careful observation and closely reflecting their society's world view. The reliability of this knowledge about the environment traditionally has been of critical importance for the welfare of the society.

The !Kung recognize five distinct seasons:

- *!Huma* (spring rains), in October and November, characterized by light thunder showers with spectacular lightning.
- *Bara* (main summer rains), from December to March—abundant rainfall, abundant food.
- *Tobe* (autumn), in April and early May, when the end of the rains and still-warm temperatures bring increased evaporation.
- *!Gum* (winter), from May to August, cool and dry with near freezing night temperatures.
- *!Gaa* (spring dry season), from August to October, with very high daytime temperatures but before the onset of rains.

Each of these seasons is characterized traditionally by different patterns of mobility and foraging. !Kung communities traditionally disperse when water is abundant, and congregate around remaining water sources in the dry season. Each season has its characteristic foods. For example, during *!gaa*, the !Kung traditionally make use of a wide variety of plant foods that are ignored at other times when preferred foods are available. Nevertheless, meat is abundant; the intense heat in *!gaa* limits the mobility of animals and makes them easier to kill.

The !Kung recognize different types of winds as harbingers of seasonal change. The onset of the rainy season is signaled by a succession of winds, each with its own name and gender. For example, the strong, dry, "male" *//gebi//gebi!go* winds of September give way to the "female" *//gebi//gebidi*, and with it the first sprinkles of rain in October.

The !Kung inhabit a very marginal environment where the timing, spacing, and amount of rainfall vary greatly from year to year. In evaluating the rains, they place less emphasis on the total amount than on timing and duration—variables that are of critical importance for the maturation of particular species of plants. The onset of rains, and heavy showers during the rainy season (*bara*), are greeted with celebration and prayers.

Studying the ethnoclimatology of societies such as the !Kung San sheds light on their world view, as well as on how they conceptualize annual rhythms and organize their use of the environment as a source of sustenance. Their environmental knowledge is detailed and nuanced, and is solidly based on centuries of observation. Scientists have increasingly come to recognize the value of such culture-based knowledge.

Based on R. Lee. *The !Kung San: Men, Women, and Work in a Foraging Society*. Cambridge, UK: Cambridge University Press, 1979.

source area travels toward the Arabian Peninsula and eastern Mediterranean region.

The name *harmattan* is given to the dust-laden winds that blow from beyond Lake Chad southwesterly across Nigeria and neighboring states numerous times each dry season. Typical episodes last for three to five days, with a dusty haze obliterating the sun, lowering temperatures, and reducing visibility to less than a kilometer. Although the harmattan is more frequent and intense in the drier semidesert and savanna regions near the Sahara, it is also experienced several times annually in coastal cities such as Lagos and Accra.

Climatic Regions

Africa may be divided into several broad climatic regions, characterized by broad uniformity in relation to the characteristics of the climate and the factors that help to account for these characteristics. As Chapter 6 will show, these regions conform broadly to patterns of vegetation, soils, and fauna, reflecting the complex interrelationships of these major components of the large regions known as *biomes*. To give but one example, prevailing climatic conditions in each biome govern the production of biomass, which in turn is the source of organic matter for the soil.

Figure 5.4a shows the major climatic regions; Figure 5.4b provides climate graphs for places that have been selected as "typical" examples of each climate type. These graphs help to illustrate a useful general rule: Equatorial climates have extremely regular patterns of temperature and rainfall, and the degree of variability increases in successive zones away from the equatorial zone. However, through the map in Figure 5.4a and the brief descriptions below focus on common characteristics of the various climate types, there is considerable regional variation and complexity within each of these zones.

Equatorial climates are characterized by heavy rainfall and a dry season that is either very short or absent. Monrovia, Calabar, and

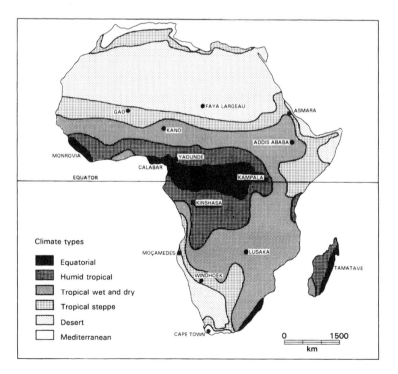

FIGURE 5.4a. Climate types. Climate graphs for representative places in each of the broad climatic regions are shown in Figure 5.4b. After *The Atlas of Africa*. Paris: Atlas Jeune Afrique, 1977, p. 75.

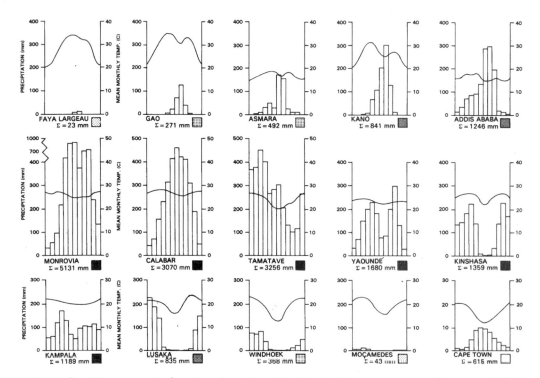

FIGURE 5.4b. Climate graphs for selected stations. The lines indicate temperature; bars are rainfall per month. For the locations of the stations, see Figure 5.4a.

Tamatave are examples of places with an equatorial climate (Figure 5.4b). Rainfall is abundant (e.g., 5,131 mm at Monrovia, 3,070 mm at Calabar, and 3,256 mm at Tamatave) and occurs virtually year-round (10–12 months with 50 mm or more). Temperatures are high, averaging about 25°C. Both annual and diurnal (daily) temperature ranges are very low. Compared to other parts of Africa with an equatorial climate, West African locations tend to have slightly higher total precipitation, as well as seasonal reversals of prevailing winds. These areas are classified as "tropical monsoon" in certain climate classification systems.

Surrounding the equatorial zone is a band of humid tropical climates with less rainfall than the equatorial zone, generally about 1,000–2,000 mm per year. The rainfall tends to peak twice during the year (the double-maximum pattern described previously), with the peaks separated by relatively short but distinct dry seasons. Annual variations in temperature tend to be slightly higher than in equatorial climates. Kinshasa is an example of a place with a

humid tropical climate; its 1,359 mm occurs in a rainy season that lasts for nine months. Yaoundé and Kampala also have humid tropical climates.

Tropical wet-and-dry climates occur on the poleward and eastern flanks of the humid tropical zone. These climates have a lengthy dry season, typically five to eight months in duration. Places such as Kano and Addis Ababa that are located in the Northern Hemisphere receive rainfall between May and September, whereas Lusaka in the Southern Hemisphere has rain from November to March. As distance from the humid tropical zone increases, the duration, amount, and reliability of rainfall decrease. Precipitation generally averages between 500 and 1,000 mm per year; the higher total at Addis Ababa reflects the orographic effect of the Ethiopian Plateau. The annual and diurnal temperature ranges are greater in the tropical wet-and-dry zone than in zones closer to the equator. There are cooler temperatures in Addis Ababa and Lusaka than at Kano because of their greater elevation above sea level.

Between the tropical wet-and-dry and the desert climates, there is a transitional semiarid zone. Places such as Asmara, Gao, and Windhoek that have tropical steppe (semi-desert) climates receive approximately 250–500 mm annually during a rainy season that lasts one to three months. Starting in the 1960s, precipitation in the tropical steppe and adjoining zones has declined significantly, in West African locations by an average of 30%. Not only has the average rainfall decreased, but also there is great year-to-year variability in the timing and amount of rain.

Desert climates have little and unreliable rainfall (e.g., 23 mm at Faya Largeau and 43 mm at Moçamedes). Tropical continental air masses are dominant, with the exception of short periods in summer when brief incursions of moist tropical maritime air bring the possibility of rain. There are extreme diurnal ranges of temperature, and significant annual variations. For example, the differences between mean monthly minimum and maximum temperatures at Faya Largeau range from 15° to 18°C, three times higher than at Calabar in the equatorial zone. The mean monthly maximum temperature exceeds 40°C (104°F) during five months of the year. Moçamedes is somewhat cooler because of its coastal location and the effects of the cold Benguela current.

Mediterranean climates occur in the southern tip of South Africa, as well as north of the Sahara adjacent to the Mediterranean. Rainfall is received in winter (April to September in Cape Town) when the midlatitude westerlies penetrate farthest toward the equator. Precipitation is quite low; Cape Town, for example, receives 615 mm annually.

Climate Change

In the early 1970s, a severe drought in the Sahel received massive media and scientific attention. Initial interpretations of the drought suggested that it was a new phenomenon caused by reduced rainfall and human misuse of the environment, and that the desert was expanding relentlessly as a result. Our understanding of climate change is now much more subtle; it has been enriched not only by scientific studies by climatologists and ecologists, but also by historical studies showing that drought has been a recurring phenomenon in this marginal environment for centuries.

Climate change has been a constant throughout the geological history of the African continent. The presence in southern Africa of *tillites*—rocks formed from till laid down by glaciers hundreds of millions of years ago, in close proximity to Carboniferous-era coal deposits derived from lush tropical vegetation—is evidence of the longevity and magnitude of past changes in the climate.

The evidence of climate change during Quaternary times, especially within the past 20,000 years or so, is striking (Vignette 5.2). Some 20,000 years ago, Africa's deserts were much larger than today, as demonstrated by the presence of several major fields of sand dunes well beyond the current desert margins (Figure 5.5). As European ice caps retreated, so too did desert margins. The mid-Sahara had a savanna environment that supported abundant wildlife and thriving pastoralist societies, known to us by the rock paintings they left behind (Figure 5.6). By about 2,000 years ago, these areas had reverted to desert.

Superimposed on these very large cycles of climate change are smaller ones—namely, cycles that are shorter in duration and smaller in magnitude. Fluctuations in climate over the past 500 years in the West African savannas and certain other areas are reflected in both physical evidence (e.g., fossil pollen, tree rings, and geomorphological features) and historical evidence (oral historical accounts of droughts, floods, and other climate-related phenomena). Climate change in recent centuries has been characterized by decades-long cycles of wetter and drier conditions, and shorter-term cycles of alternating drought and abundance lasting a few years.

Although the evidence of long-term fluctuations demonstrates that change is a fundamental characteristic of African climates, most climatologists are now unwilling to dismiss recent climatic changes as nothing more than normal patterns of fluctuation. In part, this shift in thinking reflects global concerns about how humans may

VIGNETTE 5.2. Quaternary Climate Change on the Saharan Margins

As the last glaciation reached its maximum extent some 18,000 years ago, climate zones in the Northern Hemisphere were displaced toward the equator, and temperature gradients increased. The climate of Africa was considerably colder and drier than at the present. Desert sand dunes formed up to several hundred kilometers beyond the limits of the present-day Sahara and Kalahari Deserts. The major vegetation zones all shifted closer to the equator. Glaciers descended over 1,000 m down the slopes of the high mountains of East Africa.

As the glaciers of Europe and North America began to retreat some 15,000 years ago, Africa's climates became warmer and wetter, and the desert margins retreated. Vegetation covered and stabilized the sand dunes in areas that had previously been desert. What is now the Sahara became a fertile, savanna-like environment with substantial rivers and lakes. The Lake Chad Basin and Kalahari Basin both contained very large lakes; well-preserved old shorelines show that the water level in Lake Chad was at least 40 m above current levels. About 9,000 years ago, the desert margin was about 1,000 km north of its present position (Figure 5.5).

Compelling evidence of these moister environmental conditions exists in the form of rock paintings, fossils, and archeological evidence of human habitation in what is now the center of the Sahara. Many hundreds of colored rock paintings (see Figure 5.6 for an example) record in exquisite detail aspects of the society and ecology of approximately 2,000–10,000 years ago. These paintings show pastoral societies engaged in activities such as herding flocks of animals, hunting, preparing food, and performing rituals. Paintings also show elephant, giraffe, hippopotamus, and numerous other savanna animals. Fossil evidence includes not only animal skeletons, but also pollen indicating the kinds and relative numbers of plant species at different points in time.

About 6,000 years ago, the climate of the Sahara and other comparatively arid environments became progressively drier and less variable. Lake Chad and the other large lakes continued to shrink steadily, and all but the largest rivers disappeared. As the desert margins moved southward, the number and spatial range of savanna animals declined. Human commu-

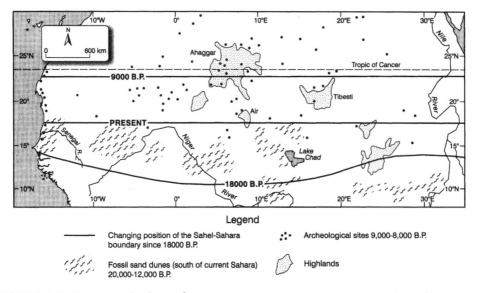

FIGURE 5.5. Quaternary landscape features in the southern Sahara. The map shows fluctuations in the location of the Saharan boundary, as well as some of the landscape features associated with climate changes.

VIGNETTE 5.2. *(cont.)*

FIGURE 5.6. Rock paintings of people and elephants in the Western Cape, South Africa. Rock paintings, often thousands of years old, are found in many parts of Africa. Some of this artwork provides insights into past climatic and ecological changes. Photo: K. Rondi.

nities faced increasingly difficult conditions that forced them to replace their horses with camels, and eventually to abandon most of their territory, which had been rendered unproductive. By some 2,000 years ago, the Sahara had become essentially the desert it is today.

Based primarily on O. Slaymaker and T. Spencer. *Physical Geography and Global Environmental Change*, Chapter 6. New York: Addison Wesley Longman, 1998.

be changing the climate. It also reflects the startling magnitude of changes in the Sahel during the late 20th century. Other arid regions, especially in eastern and southern Africa, have also had reductions in precipitation, greater fluctuations in rainfall, and hotter temperatures. However, the magnitude of these changes is significantly lower in these regions than in the Sahel.

Climatologists have noted a significant difference between 1931–1960 and 1961–1990 rainfall patterns, especially in the Sahel (see Figure 5.7). Rainfall in the Sahel was, on average, 30% less in 1961–1990 than in the previous 30-year period. Not only was there less rainfall, but totals fluctuated greatly from year to year, and the timing of precipitation during the rainy season was increasingly erratic. Asso-

ciated with the less reliable rainfall was an increase in crop failures and hunger. Other climatic changes have been noted, including more extreme temperatures and much more frequent dust storms during the dry season. Figure 5.8 shows the much-increased frequency of dust storms from 1950 to 1983.

Climatologists have debated the causes of the changes that have been observed in recent decades. Three major types of causes have been identified. One hypothesis links the increasing aridity of the Sahel to environmental changes, such as the clearing of woodlands for farming and fuelwood, which may affect climate by increasing albedo and dust content of the atmosphere. A second approach involves correlating cycles of aridity in different parts of Africa to El

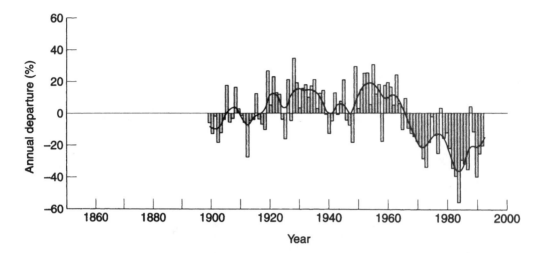

FIGURE 5.7. Rainfall variability in the Sahel, 1900–1993. The data are expressed as percentage deviations above or below the 1951–1980 average. Source: O. Slaymaker and T. Spencer. *Physical Geography and Global Environmental Change.* New York: Addison Wesley Longman, 1998. © 1998 by Addison Wesley Longman. Reprinted by permission.

Niño events—namely, cycles in global oceanic and atmospheric circulation. The third explanation views recent trends as early manifestations of global warming due to pollution of the atmosphere by humans.

Most scientists now agree that human pollution of the atmosphere is causing changes in climate worldwide, and that these human impacts

on climate can be expected to accelerate in the very near future. These changes are linked to increased concentration of carbon dioxide, methane, chlorofluorocarbons, and other so-called "greenhouse" gases in the atmosphere; it is predicted that the concentration of these gases by the year 2050 will be twice what it was before the industrial revolution. The major sources of

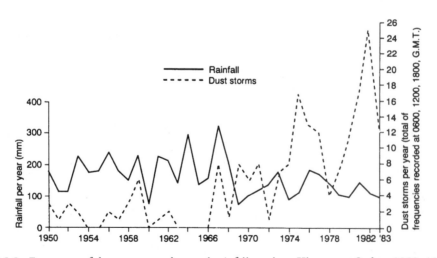

FIGURE 5.8. Frequency of dust storms and annual rainfall totals at Khartoum, Sudan, 1950–1983. Here, as elsewhere along the Saharan boundary, dust storms have become increasingly common. Data source: A. S. Goudie. "Climate: Past and present." In W. M. Adams, A. S. Goudie, and A. R. Orme, eds. *The Physical Geography of Africa.* Oxford: Oxford University Press, 1996. © 1996 by Oxford University Press. Reprinted by permission.

these gases include pollution by automobiles and industry, and the cutting and burning of forests. Unlike earlier periods of climatic change, human beings bear responsibility for creating this present phase.

Africa is not a major source of "greenhouse" gases; none of its countries ranks among the world's 20 largest sources of air pollutants. However, the Intergovernmental Panel on Climate Change (IPCC), in a report published in 2001 (see "Internet Sources"), predicts that Africa is the continent likely to be hit hardest by climate change. Its impacts are predicted to be most serious in desert-margin and savanna regions—which are expected to become hotter and drier, with more frequent and more serious droughts, and much-reduced volumes of water in lakes and rivers. Humid climates will become even warmer and wetter—conditions that will result in large increases in mosquito populations and consequent increases in malaria and other insect-borne diseases. Global warming is expected to result in a rise of up to 1.5 m in sea levels by the mid-21st century. Rising sea levels would threaten to inundate low-lying coastal areas, including much of the Comoros and coastal cities like Lagos and Abidjan.

Perhaps a quarter of the population of Africa south of the Sahara lives in comparatively arid areas where the threat of drought and desertification is already high, and where climatic change would mean diminished agricultural potential and increased hunger. Many millions more who live in coastal areas face the prospect of being displaced by rising sea levels. Because of Africa's poverty, the kinds of public programs that may help people in wealthier countries to cope with rapid changes to their environment are unlikely to be available to Africans. The indifference, or at best the token response, of the international community to Africa's current HIV/AIDS crisis gives little reason to be optimistic about the continent's chances of receiving major international assistance to address the impacts of climate change.

Climate and Human Development

All humans are concerned about the state of the weather and the implications of possible trends in the climate. However, the stakes are particularly high in Africa, where the majority of people earn their livelihood from the land as farmers and herders. The failure of the rains (especially in semiarid environments) may not only spell disaster for rural communities, but may also mean scarcer and more expensive food in urban centers and shortfalls in the production of agricultural raw materials, such as cotton for textile mills or for export. The unfavorable climatic conditions that have been widespread in Africa since the late 1960s have contributed significantly to the persistence of poverty and underdevelopment.

Further Reading

For a detailed, yet very readable introduction to Africa's physical geography, see the following:

Adams, W. M., A. S. Goudie, and A. R. Orme, eds. *The Physical Geography of Africa*. Oxford: Oxford University Press, 1996. (On climate, see especially Chapters 3–5.)

Classical sources on Africa's climates include these:

Griffiths, J. F., ed. *Climates of Africa*. Amsterdam: Elsevier, 1972.
Niewholt, S. *Tropical Climatology*. Chichester, UK: Wiley, 1977.
Thompson, B. W. *The Climate of Africa*. New York: Oxford University Press, 1965.

Climate change in Quaternary and historical times is discussed in the following:

Deacon, J., and N. Lancaster. *Late Quaternary Palaeoenvironments of Southern Africa*. Oxford: Oxford University Press, 1988.
Grove, A. T., and A. Warren. "Quaternary landforms and climate change on the south side of the Sahara." *Geographical Journal*, vol. 134 (1968), pp. 193–208.
Nicholson, S. E. "Climatic variations in the Sahel and other African regions during the past five centuries." *Journal of Arid Environments*, vol. 1 (1978), pp. 3–24.

On contemporary (mid- to late-20th-century) studies of climate change, especially in the Sahel, see these sources:

Fontaine, B., and S. Bigot. "West African rainfall
     deficits and sea-surface temperatures." *Interna-
     tional Journal of Climatology*, vol. 13 (1993), pp.
     271–286.
Hume, M. "Rainfall changes in Africa: 1931–60
     to 1961–90." *International Journal of Climatol-
     ogy*, vol. 12 (1992), pp. 685–699.
McLaren, S. J., and D. R. Knives, eds. *Linking
     Climate Change to Land Surface Change.*
     Dordrecht, The Netherlands: Kluge, 2000. (See
     especially Chapters 1–4.)
Middleton, N. J., and A. S. Goudie. "Saharan
     dust: Sources and trajectories." *Transactions of
     the Institute of British Geographers*, vol. 26
     (2001), pp. 165–181.

Studies of global climate change and its relevance
to Africa include the following:

Hume, M., ed. *Climate Change and Southern Africa:
     An Explanation of Some Potential Impacts and
     Implications in the SADC Region.* Norwich, UK:
     University of East Anglia, 1996.
Rowlands, I. *Climate Change Cooperation in Southern
     Africa.* London: Earthscan, 1998.
Sullivan, M., ed. *The Greenhouse Effect and Its
     Impact on Africa.* London: Institute for African
     Alternatives, 1990.

Internet Sources

The following sites provide access to climate data
for hundreds of African localities:

Africa Data Dissemination Service. http://edcw2ks21.
     cr.usgs.gov/adds/data
World climates. www/climate/index

What's the weather today in South Africa? Consult the South African Weather Service, or CNN:

South African Weather Service. www.weathersa.co.za
CNN.com. www.cnn.com/weather/Africa/sat.html

For current information on climate change, the
IPCC site is especially important:

Global Environmental Change Program, Univer-
     sity of Virginia. *Global Climate Change and
     Africa.* http://africagcc.gecp.virginia.edu/Policy
Intergovernmental Panel on Climate Change
     (IPCC). *Regional Impacts of Climate Change. Chap-
     ter 2: Africa.* www.grida.no/climate/ipcc/regional/
     006.htm

For examples of the ancient rock art showing life
in the Sahara when its climate was moist, see the
following:

Brown, H. University of Alabama at Birming-
     ham. *Archaic Art of North Africa: Saharan Rock
     Art.* www.hp.uab.cdu/imagc_archive/ta/tad.html
The Roland Collection of Films and Videos on
     Art. *Tassali N'Ajjer.* www.roland-collection.com/
     roland-collection/section/26/10.htm

6

Biogeography and Ecology

The Western media have written at length about some aspects of the African environment, in the process repeating certain basic themes. The primary theme has been that there is an environmental crisis in Africa, attributable both to the uncontrollable natural forces (e.g., the assumed steady advance of the desert) and to human mismanagement (e.g., deforestation and declining large-animal populations). This chapter establishes the context for examining these issues by outlining characteristic features of Africa's major terrestrial biomes, and through discussions of deforestation and desertification as ecological concerns. We will return to these themes later in the book, especially in Chapter 24 on the use of flora and fauna as resources.

Ecosystem Basics

An ecosystem incorporates numerous variables that are linked to one another through complex flows of matter and energy. These components are both *biotic* (living organisms such as plants and animals, organized in interdependent communities) and *abiotic* (the nonliving physical and chemical environment). Energy derived from the sun is the basis for ecosystem dynam-

ics. Energy transfer occurs through the operation of food chains, which begin with the growth of plants through photosynthesis. Solar energy converts water and carbon dioxide into *biomass*. This plant life provides sustenance for herbivores, which are in turn consumed by carnivores. Omnivores consume both animals and plants. In addition to the primary food chain, each ecosystem has a decomposer food chain, consisting of microorganisms such as fungi and bacteria that consume dead organic matter. In reality, the structure of food webs is very complex, with each species having a limited range of potential and preferred foods. Food preferences are distinct and complementary; that is, foods preferred by one species are often ignored by others.

The primary productivity of an ecosystem is the amount of biomass that is generated per square meter per year. The productivity of terrestrial ecosystems is greatest where temperatures are highest and rainfall is abundant. As a tropical continent, Africa has relatively few areas where temperature is a major limiting factor. The key determinants of differences in African ecosystem productivity are variations in the amount and timing of precipitation. The primary productivity of tropical rain forests is, on average, about three times as great as that for savannas, and more than 20 times that of semidesert scrub.

African Terrestrial Biomes

A *biome* may be defined as a large, regular region whose climate, vegetation, fauna, and soils are characterized by broad uniformity. A biome is usually named after the predominant vegetation in the region. Emphasis must be placed on the *broad uniformity* of a biome, because within each biome there is considerable variation and complexity, reflecting the diversity of interrelated processes that shape the environment. Moreover, although there is an obvious relation between the elements that constitute a biome, this does not mean that the boundaries between different vegetation types and those separating related climate regions and soil regions correspond exactly with each other.

The associations of vegetation, climate, and soil that characterize Africa's major biomes are summarized in Table 6.1 and are discussed below. Their resulting climate patterns have been shown in Figure 5.4a; a map of vegetation types is provided in Figure 6.1.

Vegetation zones in Africa show a pattern of roughly concentric rings centered on the Congo basin. (Figure 6.2 illustrates four of the eight different types of vegetation regions discussed below.) While rainfall is especially important

for vegetation, factors such as altitude, soil characteristics, and drainage are locally significant. Human use of the environment—selective clearing and planting of species, grazing, and the use of fire—continues to modify vegetation patterns. Therefore, it is important to remember the limitations of "natural" vegetation as a meaningful concept. Large areas of vegetation that initially seem undisturbed have actually been altered significantly by human use.

Tropical Rain Forest

Africa has a vast tropical forest centered in the Congo basin of west central Africa. It extends northward through southern Cameroon and discontinuously along the West African coast as far as Sierra Leone. It is also found in eastern Madagascar. The tropical rain forest occurs in close association with equatorial climates, which are characterized by uniformly high temperatures, heavy rainfall, and a dry season that is either very short or absent. These climatic conditions permit the development of the most biologically diverse of the world's biomes, but they also contribute to the development of soils called Oxisols, which contain few nutrients and little organic matter.

The tropical forest is the biome with the greatest diversity of animal life—a reflection of its very high primary productivity and the countless ecological niches that exist in it. The fauna of Madagascar's forests differ greatly from those of continental African forests; it is esti-

TABLE 6.1. Africa's Major Biomes

Vegetation type	Related climate type	Related soil types
Tropical rain forest	Equatorial	Oxisols
Moist woodland savanna	Humid tropical	Oxisols, Alfisols, Ultisols
Dry parkland savanna	Tropical wet-and-dry	Alfisols (Ustalfs)
Semidesert (Sahel)	Tropical steppe (semiarid)	Alfisols, Aridisols
Desert	Desert	Aridisols
Temperate grassland (veldt)	Subtropical wet-and-dry	Alfisols
Mediterranean	Mediterranean	Alfisols (Xeralfs)
Montane	Highland	Varied, poorly developed

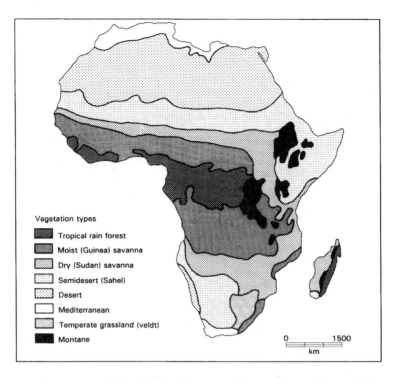

Vegetation types
- ▓ Tropical rain forest
- ▦ Moist (Guinea) savanna
- ▨ Dry (Sudan) savanna
- ▢ Semidesert (Sahel)
- ▦ Desert
- ☐ Mediterranean
- ▥ Temperate grassland (veldt)
- ■ Montane

FIGURE 6.1. Natural vegetation. Note the broad correspondence of vegetation and climate shown on the map, and also summarized in Table 6.1.

mated that 90% of its species are found only there. Madagascar's biodiversity is threatened by the clearance and degradation of more than 90% of its original forests.

Vegetation in the tropical forest is very luxuriant and consists of many different species in close proximity. Vegetation typically occurs in three layers: a shrub layer of relatively low species, a canopy of densely packed trees of medium size, and an emergent layer of isolated tall trees extending far above the canopy. Where the forest has remained undisturbed, the ground level may be quite open but dark. Where the canopy has been disturbed, permitting much more light to penetrate to ground level, a dense and tangled mass of vegetation develops.

Tropical wetland forests occur in very moist environments. Freshwater swamp forests are widespread in the central Congo basin, while saltwater swamp forests occur in several coastal areas. The dominant species are mangroves, which are highly adapted to fluctuating water levels and the brackish water found in estuaries and lagoons.

Human activities, especially farming and lumbering, pose a growing threat to tropical rain forests. The forest ecosystem is very delicate, and indiscriminate clearing can jeopardize the ecosystem's survival. More than 1 million km^2 of Africa's tropical forests have been lost, and the rate of loss is growing. In some areas, however, humans have managed to increase their utilization of forest environments without necessarily destroying them. Indeed, in some cases the expansion of forests may be attributed to human stewardship.

The soils of the tropical rain forest are subject to heavy leaching, a consequence of the region's high temperatures and heavy rainfall. Most mineral nutrients have been leached from the soil, leaving iron-rich laterites behind. Although the forest supplies an abundant supply of biomass, these soils contain little organic matter. The warm, moist environment is ideal for soil bacteria that cause the rapid decomposia

b

c

d

FIGURE 6.2. Views from four vegetation regions. (a) Tropical rain forest, southern Cameroon. (b) Dry parkland savanna, south central Kenya. (c) Semidesert, southern Kenya. (d) Desert, Eritrea. Photos: (a) CIDA (R. Lemoyne); (b) and (c) author; (d) M. Peters.

tion of organic matter. Because these Oxisols are so infertile, subsistence farmers in tropical forest environments practice shifting cultivation, moving their farms every two or three years as the soil's limited store of plant nutrients is exhausted.

Moist Woodland Savanna

Moist woodland savanna forms an almost continuous band adjoining the tropical forest on its northern, eastern, and southern edges. The characteristic vegetation of this biome is a mixture of trees and tall grasses. Where rainfall is relatively high and human influence low, there may be quite a dense canopy of evergreen trees. The effect of agriculture on the savanna ecosystem is especially evident in the "derived savannas" along many parts of the

forest margins, where overly intensive utilization has caused degradation from forest to savanna. Fires, often deliberately set by farmers, herders, and hunters (see Figure 6.3), affect the evolution of savanna vegetation by selectively eliminating species that are not fire-resistant. Moist woodland savanna occurs in conjunction with humid tropical climates. These climates have somewhat less rainfall than equatorial climates; they have a double-maximum pattern of rainfall separated by two short but distinct dry seasons.

Moist woodland savannas north of the equator, often referred to as the *Guinea savanna*, are distinct from those to the south. A type of savanna known as *miombo* woodland covers large expanses of south central Africa. *Miombo* is dominated by three closely related tree species. These species lose their leaves in the dry season

FIGURE 6.3. Vegetation clearance, Côte d'Ivoire. The use of fire to clear brush has some immediate benefits for herders and farmers, but may have major environmental costs, including the elimination of some species and increased soil erosion. Photo: CIDA (M. Faugere).

and are highly susceptible to fire. More than half of *miombo* plant species are found only in this region.

The soils of the woodland savanna are diverse, a reflection of the transition here from forest to relatively dry savanna. Oxisols and Ultisols, both highly leached and relatively infertile, occur where precipitation is above average for this zone. In drier parts, Alfisols are found. The name *Alfisol* reflects the high aluminum (*Al*) and iron (*fi*) content of these soils, which tend to be heavily leached but have somewhat greater organic matter than Oxisols and Ultisols associated with moister environments.

Dry Parkland Savanna

Adjoining the moist woodland savanna on its dry margins is a continuous band of dry savanna vegetation occurring in conjunction with tropical wet and dry climates. These climates have a five- to eight-month dry season and less rainfall than in the woodland savanna zone. As distance from the moist woodland savanna boundary increases, the duration, quantity, and reliability of precipitation decline. The term *Sudan savanna* is commonly used for this type of vegetation, especially

north of the equator. To the south, a drier subtype of *miombo* forest is found.

Many of the characteristics of the dry savanna vegetation reflect the longer dry season and lesser amounts of rainfall, as compared to the moist woodland savanna. Trees are usually more scattered, and the grass is not as tall. The species found here are adapted to the more difficult climate conditions; acacia trees, for example, have very small leaves that are shed seasonally, as well as a tough outer bark and thorns that discourage browsing by animals. The baobab is another of the species usually associated with this biome.

Savanna environments have less diversity of wildlife than tropical forest ecosystems. However, East Africa's dry savanna environments are famous for their huge herds of grazing species and associated predators. Herd densities average more than 40 animals per square kilometer in the Serengeti Plains and several other areas in eastern and southern Africa.

Ustalfs, a type of Alfisol, are the main soil type. These soils are often relatively fertile. Ustalfs usually contain fine dust brought by winds from the desert during every dry season. They are less heavily leached than soils in areas with moister climates and may have a fairly high organic content.

Semidesert (Sahel)

The Sahel is a transitional zone lying between the Sudan savanna and the desert, characterized by a short rainy season yielding an average of 250–500 mm of precipitation. The sparse, unreliable rainfall results in very low ecosystem productivity, and it makes agriculture a very risky activity because crop failures occur frequently. In this harsh environment, the drought-resistant properties of plants are crucial to their survival. Trees tend to have thick bark and small, waxy leaves that do not lose much moisture through transpiration. Some species conserve moisture by shedding their leaves during the dry season. Semidesert vegetation is very vulnerable to damage from fire, overgrazing, and agriculture, and therefore may be prone to desertification. Healthy stands of vegetation are crucial for stabilizing the soil and reducing soil erosion by wind and water.

Desert

Desert climates are characterized by very low and sporadic rainfall, and by substantial daily and seasonal temperature fluctuations. Desert vegetation occurs in the Sahara, Kalahari, and Namib, although different species are typical of the different deserts. In the Namib, plants depend on frequent fogs, generated as winds pass over the cold Benguela Current, for their moisture. Desert plant life must be highly adapted to the sparse and infrequent rainfall and the poorly developed, often saline soils. Vegetation is likely to be denser and more varied along watercourses than in the more open desert. Desert plant life may be remarkably diverse; the South African region of Namaqualand, for example, is renowned for the proliferation of beautiful wildflowers. Many animal species adapt to harsh desert conditions through migration and an ability to go without water for long periods.

The Aridisols of desert environments are poorly developed and have a very low organic content, reflecting the dry climate and lack of vegetation. Salinization and calcification are important soil-forming processes in many dry environments, resulting in the accumulation of layers of salts or calcium carbonate on or below the surface.

Temperate Grassland (Veldt)

Much of the high plateau of eastern South Africa is covered in grasslands, known in the region as the veldt. The veldt is analogous to temperate grasslands found in other continents, such as the prairies of North America. By African standards, the climate is relatively cool, especially in winter when freezing temperatues are common. The precipitation of 450–700 mm per annum is highly concentrated in the summer months. These conditions give rise to vegetation dominated by grasses; trees generally are confined to riversides and other favorable microenvironments. The deep and fertile soils of the veldt, rich in their organic content, form the basis for the prosperous agricultural economy of the region.

Mediterranean

The South African Cape region has a Mediterranean climate, characterized by hot, dry summers and cool, moist winters. The vegetation that develops under these conditions is *xerophytic* (drought-resistant). The vegetation of the Cape region, known as *fynbos*, includes 8,500 species of plants (two-thirds of which are unique to this region) in a very small area. Human impacts threaten the survival of this unique and diverse bioregion. Xeralfs are the soils usually associated with Mediterranean environments. These brownish-colored soils often have quite high natural fertility.

Montane

Africa's highest mountain peaks form a discontinuous zone in which altitude is the primary determinant of climate, vegetation, and soil types. Temperatures decline with increasing altitude, and on mountains' windward sides, rainfall may be very abundant. Vegetation is also zoned vertically—starting at the base of East African mountains, from savanna, to montane forest, to bamboo, heather, alpine tundra, and in some cases ice. Montane forests contain an abundance of species, many of them unusual in appearance and the majority unique to this zone.

Highlands elsewhere in the continent also have flora that differs from that of surrounding lower-elevation areas. Examples include the montane forests of the Ethiopian Plateau, the temperate grasslands of South Africa's highlands, and the isolated Mediterranean species occurring in the Ahaggar Mountains and Tibesti Massif of the central Sahara.

Ecological Concerns

Environmental research, agitation by environmental pressure groups, and global symposia such as the Earth Summits in Rio de Janeiro and Johannesburg have heightened concern about environmental degradation. The destruction of tropical rain forests and desertification have been identified as particularly important threats to the future of the planet. There is evidence that African ecosystems have suffered significant degradation. Nevertheless, newer evidence from contemporary research points to the complexity of patterns and processes of ecological change, and the need to interpret these changes carefully in relation to the incredible diversity of the human geography found in specific regional settings within each of Africa's biomes.

Deforestation

It has been estimated that the global area of tropical forest decreases by an average of 157,000 km^2 annually, an area equal to that of England and Wales. If this rate of deforestation is maintained, it is projected that all tropical forests could disappear by the middle of the 21st century. The clearance of tropical forests has global as well as local implications. Tropical forest ecosystems have incredible biological diversity; when plant species become extinct, humanity is deprived forever of their potential benefits—for example, the discovery of significant new medicines. Moreover, the removal of forest cover facilitates increased soil erosion and

soil compaction, thus reducing agricultural productivity. Regional climatic change, including higher temperatures and lower rainfall, has been observed in deforested environments.

As noted in Chapter 5, deforestation has been linked to predicted changes in the climate of the earth—the so-called "greenhouse effect." With forest clearance, carbon stored in vegetation is released into the atmosphere, while the forest's capacity to create oxygen is reduced. Higher concentrations of greenhouse gases are expected to produce a variety of climatic and ecological effects, including hotter, drier conditions in the savanna. Vegetation will come under increasing stress because of the harsher environmental conditions, and because of the effects of the increasing intensity of human utilization.

Africa's tropical forests are under threat in several areas, including coastal West Africa and Madagascar. Africa as a whole has lost an estimated 5 million ha of tropical forests per year, a loss of 0.8% of the total area each year. Although forest retreat has occurred in parts of West Africa for centuries, the rate of deforestation has accelerated—to 2.1% annually—because of an ever-growing demand for timber and for farmland to support relatively high population densities. Forests in West Africa are also in danger because they occupy fairly small blocks of territory. In contrast, the forests of the Congo basin are less immediately threatened; they occupy a huge area, support low population densities, and are difficult to penetrate because of poor transport development.

Recent research, especially that undertaken by James Fairhead and Melissa Leach, has cast doubt on the prevailing view of forests that are *necessarily* in permanent decline because of human activities (Vignette 6.1). Their studies in southeastern Guinea show, contrary to popular wisdom, that forests in the area are expanding significantly—and did so throughout the 20th century, as a result of human use of the environment. Moreover, their review of previous studies of forest margin environments elsewhere in West Africa suggests that the recent expansion of forests may be quite widespread. Their research does not mean that concern over deforestation is misplaced. Rather, it points to the

VIGNETTE 6.1. Human Impacts on the Forest: Destroyers or Builders?

For a century, European colonial officials, development workers, and academics visiting the forest–savanna boundary zone in Kissidougou District, Guinea concluded that the local people were carelessly destroying their environment. The visitors saw patches of forest surrounded by savanna vegetation, and concluded that these forests were the relics of a former great forest that had been lost due to agriculture and the widespread use of fire. The government of Guinea tried to restrict the use of fire, and development projects were initiated to save the forest. Kissidougou school children were taught that their communities would have to change their destructive ways.

Appearances may be deceiving. British anthropologists James Fairhead and Melissa Leach, undertaking research in the district, repeatedly heard from local informants that now-forested areas had been savannas in living memory. Studies of air photographs confirmed the accuracy of the oral evidence: The forests of Kissidougou were actually expanding into the savanna.

Further research provided explanations for this counterintuitive finding. As a result of human intervention, forest islands grew around each settlement within a generation or so of its creation. Silk cotton and other species of trees grew from seeds in rubbish heaps. Green tree cuttings were used to make fences around household compounds; these cuttings formed roots and became "living fences," and eventually some of them grew into trees. Crops in newly cleared savanna soils were sown in a planned sequence, starting with legumes such as groundnuts, which enriched the soil and facilitated a transition from grassland to forest. Fires were used to convert cleared grasses and brush into nutrients to fertilize crops. By burning vegetation relatively early in the dry season, farmers were able to prevent later fires at a time when the vegetation would be extremely dry, and thus when fires could be very destructive. As the forest aged, it became denser and biologically more diverse, taking on the appearance of a mature, "natural" forest.

Trees within these forest islands were maintained carefully by people living in the settlements. Historically, the dense vegetation of the forest islands had provided protection from enemies' attacks. The forest also provided shade and a variety of valuable products, ranging from palm oil and fruits to firewood, obtained by lopping branches from the trees.

Settlements are relocated occasionally. Once the human communities that created and sustained the forest islands have left, the forests gradually revert to savannas. However, because the population of Kissidougou District has been increasing, there are more settlements and more forest islands. The evidence from air photographs is clear: In Kissidougou, more people mean more forests. The most densely populated parts of the district have the greatest density of forest islands, contrary to the orthodox view that population growth meant environmental degradation.

One of the lessons of Kissidougou is the importance of being open to reconsidering "received wisdom" about the sorry state of the African environment. Although deforestation linked to population growth may be a widespread problem, we should not assume that this is a *necessary* relationship. Another lesson concerns the potential value of indigenous knowledge as the basis for policy directions to address issues of concern, such as environmental degradation.

Based on J. Fairhead and M. Leach. *Misreading the African Landscape: Society and Ecology in a Forest–Savanna Mosaic.* Cambridge, UK: Cambridge University Press, 1996. See also *Second Nature: Building Forests in West Africa's Savannas* [video]. Haywards Heath, UK: Cyrus Productions, 1997. (Available online from IDS Bookstore, UK) dangers of overgeneralization, and to the need to understand the dynamics of human–environmental interaction in specific settings.

Desertification

The Sahelian drought of the 1970s caused greatly increased concern about environmental degradation in general and desertification in particular. Was the Sahelian drought evidence of larger changes in world climate? Was the Sahara literally advancing southward, enveloping communities along the way? To what extent were humans responsible for the apparent growth of desert-like conditions? Could the process of desertification be reversed, or at least be controlled?

Figure 6.4 illustrates the degree of susceptibility to desertification in Africa. Desertification involves a range of changes to ecosystems, including the degradation of vegetation, the loss of soil moisture, and the formation of sand dunes. The cumulative effect of these changes is "the diminution or destruction of the biological potential of the land," ultimately producing desertlike conditions of vegetation and soil in areas beyond the climatic desert. This definition, which came from the United Nations Conference on Desertification in 1977, did not end the debate about the meaning of desertification. Some studies have used the term to mean the *process* of land degradation, whereas others view it as the *end result* of a process of change. All agree that there are both climatic and human dimensions to desertification; climatologists and social scientists often disagree, though, about the relative importance of the two dimensions. Climatologists have also been divided as to whether desertification is linked to climate *variability* (recurring droughts) or to a longer-term process of climate *change* (global warming).

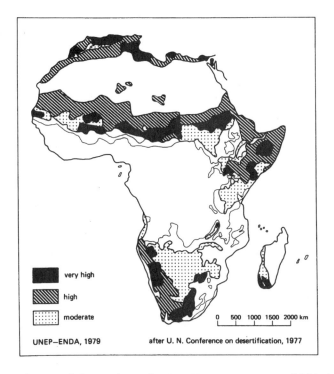

FIGURE 6.4. Degree of susceptibility to desertification. Up to one-quarter of Africa's land area, including large areas not located along the desert fringe, is susceptible to degradation. Source: Environmental Development Action (ENDA). *Environment and Development in Africa*. Oxford: Oxford University Press, 1981, p. 9. © 1981 by United Nations Environment Programme (UNEP) and ENDA. Reprinted by permission.

Most explanations of ecosystem degradation have focused on the pressures exerted by the rapid growth of human and animal populations, and the resultant need to extend and intensify resource utilization (see Vignette 6.2 for one case study). The population of the Sahel and of other dry savanna regions began to increase markedly during the latter part of the colonial era. Farmers and herders were encouraged to occupy more and more marginal environments during three decades of above-average rainfall in the region, by relatively

---

VIGNETTE 6.2. The Causes of Desertification: A Kenyan Case Study

Who is to blame for land degradation in marginal environments? Are the people who inhabit these regions at fault, or does the blame lie elsewhere? The case of the Rendille pastoralists, who live east of Lake Turkana in northern Kenya, is relevant to the questions posed above.

The Rendille occupy a marginal semidesert environment with average rainfall ranging from 200 to 300 mm per year. Much of the area that they occupy shows signs of severe stress. During the droughts of the early 1970s, they lost large numbers of camels, goats, sheep, and cattle upon which they are dependent.

Taken at face value, resource management decisions taken by the Rendille seem to account for the widespread ecological damage. Grazing lands are used unevenly; perhaps one-fifth of the Rendille territory is overgrazed, but considerable grazing land remains unused. This uneven pattern of grazing is related to the following factors:

- The Rendille opt for grazing areas relatively close to springs or easily accessible waterholes, and only as a last resort move to pastures where much effort is needed to get water.
- They avoid areas where cattle raiding occurs most frequently.
- The development of permanent settlements has contributed to uneven use and ecological damage around settlement sites.

Analyzing the behavior of the Rendille from their own perspective suggests other conclusions, however. They seem less the careless destroyers of their own sustenance base than the victims of policies and systems over which they have no control. Far from urban centers and dependent on small, uncertain local markets to sell their animals, the Rendille are forced to maintain large herds and maximize their gains in good years to survive through periodic droughts. They receive poor prices for their livestock—one-third to one-half of the Nairobi price—but must pay very high prices for the products that they purchase. These very unfavorable terms of trade are a strong disincentive for the Rendille to invest extra labor and capital to open up new pastures and watering holes in underutilized areas.

For the Rendille, government policies have been part of the problem, not the solution. When the first development projects were initiated in the late 1960s, the focus was on cattle ranching. Not only were cattle a dubious choice for these vulnerable lands, but government planners virtually ignored the existing Rendille economy with its focus on raising camels, goats, and sheep. Moreover, restrictions on livestock exports and policies designed to ensure low food prices for urban consumers condemned the Rendille to continuing poverty.

In Rendille country, as in other parts of Africa threatened with ecological disaster, policy makers must recognize the need for policy solutions that address both facets of the ecological crisis: Not only do fragile lands need to be protected, but the fragile societies and economies dependent on these resources need careful assistance if they are to survive.

Based on M. O'Leary. "Ecological villains or economic victims: The case of the Rendille of northern Kenya." *Desertification Control*, no. 11 (1984), pp. 17–21.

good prices for groundnuts and other crops, and by the construction of wells and roads in populated areas. After independence, large-scale cattle ranches and irrigation schemes were established in certain marginal, semidesert regions (e.g., in Senegal and Botswana). Feasibility studies often discounted both the wider ecological implications of intensive, large-scale development and the effect of these projects on peasant farmers and pastoralists. Local farmers and pastoralists were displaced from productive lands, and were forced to occupy marginal areas susceptible to ecological damage and crop failure.

In *Working the Sahel* (see "Further Reading"), Mortimore and Adams argue against simplistic cause-and-effect analyses that depict humans as either helpless victims or careless perpetrators of ecological change. They show how the farmers of the Sahel—facing major physical constraints of unreliable rainfall, as well as social and economic constraints of limited labor and capital—have developed sophisticated responses that vary over time and space. These responses may include an *intensification* of local resource use, including new forms of primary production; *diversification* into other income-earning activities; and *migration* to other, less stressed areas. Environmental decision making within Sahelian communities under stress is informed by three general principles: (1) the *diversity* of natural, economic, technical, and social resources that may be mobilized by a household; (2) *flexibility* in the choice of field and grazing locations and cropping strategies; and (3) *adaptability* over the longer term in responding to environmental, social, and economic change. These adaptive strategies that have enabled communities in the Sahel to survive for many generations provide the best hope for the continuing viability of human utilization in this uncertain and challenging environment.

Although widespread degradation of arid environments has occurred, the extent of damage has often been exaggerated by development agencies and by governments. Careful longitudinal research to document the extent and nature of environmental change is in very short supply; published reports of degradation are typically based on inadequate evidence and deeply entrenched assumptions about desertification. In particular, research by Bassett and Bi Zuéli (see "Further Reading") in the savanna of northern Côte d'Ivoire underlines the need for caution in judging human impacts on savanna environments, and in assessing official "calls to arms" to combat desertification.

Sustainable Alternatives

Africa's natural resources are vital to the health and well-being of rural economies and societies—and, in a more general sense, to the well-being of every African nation. What distinguishes Africa from most other global regions threatened by environmental degradation are the large numbers of people living in African regions that are vulnerable to declining productivity, and the scarcity of national resources to address these environmental challenges.

One encouraging development has been the growing understanding among development planners of the importance of vegetation in rural economies. Research on *agroforestry*—the systematic integration of trees into farming systems—has identified promising new techniques to permit development that is ecologically and economically sustainable. The development of *alley cropping*—a sustainable alternative to shifting cultivation for use in humid forest environments—has been a very important innovation. Establishing shelterbelts and helping communities to undertake tree-planting projects help to address the related problems of desertification and fuelwood shortage (Figure 6.5).

What is more important than new technology, however, is a commitment to sustainable approaches to development—approaches that are holistic in orientation; that are sensitive to spatial variations in problems, processes, and possibilities in different settings; that are compatible with the needs, financial resources, and cultural perspectives of ordinary Africans; and that are highly sensitive to possible immediate and longer-term effects on the environment.

FIGURE 6.5. Tending a new windbreak, Burkina Faso. Windbreaks improve crop yields by reducing both soil erosion and evapotranspiration; they also provide a source of fuelwood. Photo: CIDA (D. Barbour).

Further Reading

The following are useful general sources on African biomes:

Adams, W. M., A. S. Goudie, and A. R. Orme, eds. *The Physical Geography of Africa*. Oxford: Oxford University Press, 1996. (An excellent volume— see especially Chapters 10–17 and 19–21.)

Goodall, D. W., ed. *Ecosystems of the World*. Amsterdam: Elsevier, 1982–1993. (This is an excellent series of reference volumes on ecosystems in different parts of the world.)

Lewis, L. A., and L. Berry. *African Environments and Resources*. Boston: Unwin Hyman, 1988.

White, F. *The Vegetation of Africa: A Descriptive Memoir to Accompany the UNESCO/AEFAT/UNSO Vegetation Map of Africa*. Paris: United Nations Educational, Scientific, and Cultural Organization (UNESCO), 1992.

Whitlow, J. R. "The study of vegetation in Africa: A historical review of problems and progress." *Singapore Journal of Tropical Geography*, vol. 5 (1984), pp. 88–101.

Whitmore, T. C. *An Introduction to Tropical Rain Forests*. Oxford: Oxford University Press, 1990.

Several sources examine ecosystems in specific regional settings:

Battistini, R., and G. Richard-Vindard. *Biogeography and Ecology in Madagascar*. Amsterdam: V. Junk, 1972.

Cowling, R. M., ed. *The Ecology of Fynbos: Nutrients, Fire and Diversity*. Cape Town, South Africa: Oxford University Press, 1992.

Lovett, J. C. and S. K. Wasser, eds. *Biogeography and Ecology of the Rain Forests of Eastern Africa*. Cambridge, UK: Cambridge University Press, 1993.

Werger, M. J., ed. *Biogeography and Ecology of Southern Africa*, 2 vols. The Hague, The Netherlands: W. Junk, 1978.

For a variety of perspectives on deforestation, see the following:

Fairhead, J., and M. Leach. *Reframing Deforestation: Global Analysis and Local Realities. Studies in West Africa*. London: Routledge, 1998.

Guppy, N. "Tropical deforestation: A global view. *Foreign Affairs*, vol. 62 (1984), pp. 928–965.

Tropical Forests: A Call for Action. New York: World Resources Institute, 1985.

Whitlow, R. "Man's impact on vegetation: The African Experience." In K. J. Gregory and D. E. Walling, eds. *Human Activity and Environmental Processes*, pp. 353–379. Chichester, UK: Wiley, 1987.

For more information on desertification and human responses to the problem, see these sources:

Bassett, T. J., and K. Bi Zuéli. "Environmental discourses and the Ivoirian savanna." *Annals of* the *Association of American Geographers*, vol. 90 (2000), pp. 67–95.

Desertification Control Bulletin [periodical]. Nairobi, Kenya: United Nations Environment Programme (UNEP).

Swift, J. "Desertification: Narratives, winners, and losers." In M. Leach and R. Mearns, eds. *The Lie of the Land: Challenging Received Wisdom on the African Environment*, pp. 73–90. London: International African Institute, 1996.

Warren, A., and M. Khogali. *Desertification and Drought in the Sudano–Sahelian Region 1985–1991*. New York: United Nations Development Programme/United Nations Sudano–Sahelian Office (UNDP/UNSO), 1992.

Environmental change must be assessed in relation to human production systems and responses to crisis. This theme is especially well explored in writings by Michael Mortimore:

Mortimore, M. *Adapting to Drought: Farmers, Famines, and Desertification in West Africa*. Cambridge, UK: Cambridge University Press, 1989.

Mortimore, M., and W. M. Adams. *Working the Sahel: Environment and Society in Northern Nigeria*. London: Routledge, 1999.

Internet Sources

The *Ecoregions* website provides an outstanding introduction to Africa's ecological zones:

World Wildlife Fund. *Ecoregions*. www.worldwildlife.org/wildworld/profiles/terrestrial_at.html

There are many sites that deal with ecological issues in Africa, among them the following:

Afrol.com. Forests and Deforestation in Africa. www.afrol.com/ms_index.htm

Conserve Africa Foundation. www.conserveafrica.org

The Desert Research Foundation of Namibia. www.drfn.org.na

Direction Nationale de l'Environnement, Guinea. *Environnement et Biodiversité*. www.mirinet.com/gn_env

Forests.org. Forest Conservation Portal. http://forests.org/africa

UNDP. *Drylands Development Centre*. www.undp.org/drylands

United Nations Environment Programme (UNEP). www.unep.org

UNESCO. *Tropical Forests on the World Heritage List*. http://whc.unesco.org/sites/tropical-forests.htm

University of Virginia. *Miombo Network*. www.miombo.gecp.virginia.edu

University of the Western Cape. *UWC's Enviro Facts Index Page*. www.botany.uwc.ac.za/envfacts

Africa in Historical Perspective

All geography is in some sense historical geography. The importance of maintaining a clear sense of historical perspective is especially important in studying the geography of Africa, where, for example, the legacy of colonial rule is so frequently evident in the present-day economic, political, and social circumstances of the continent.

Chapter 7 provides a brief survey of Africa prior to the colonial era. It surveys the origins, organization, and accomplishments of several early empires, refuting the formerly widespread notion that Africa had no history of its own. The chapter also looks at the extent, organization, and effects of the centuries-long slave trade that saw many millions of Africans transported to the New World and elsewhere.

Chapter 8 examines the historical geography of colonialism, starting with the late–19th-century scramble of European powers to carve up the continent. The specific legacies of European rule varied from colony to colony, depending on the choice of economic development and government models. Some colonies were transformed by the alienation of land for white settlement and the development of resource-extracting industries. Others were considered to have little value and were neglected, except as sources of labor. Although the colonial era was, in historical terms, very brief, its continuing legacy has been pervasive.

Chapter 9 explores several themes related to Africa's struggle for independence, and to the subsequent struggles for responsive government, development, and survival. The optimism of the early 1960s has given way to a more sober realization of the complexities of development in a changing world system. Not only is Africa's influence in global affairs extremely limited, but so too is its ability to control even basic aspects of its own destiny. Four decades after colonialism's demise, neocolonialism thrives.

7

The African Past

Available archeological evidence, together with recent genetic and linguistic research, points to Africa as the very cradle of humanity. *Homo sapiens* is thought to have first appeared in Africa some 150,000–200,000 years ago. Comparative studies of DNA, which have enabled scientists to establish in some detail genetic relationships among the world's peoples, support the hypothesis of a common African origin for humankind. About 100,000 years ago, modern humans appear to have swept out of Africa, quickly displacing archaic hominids in other parts of the world.

The evolutionary succession leading to modern humankind has been traced back some 6–7 million years. Much fossil evidence about these hominids, the australopithecines, has been uncovered in South Africa, the rift valleys in eastern Africa, and the Afar Depression in Ethiopia. The australopithecines shared certain physical characteristics—brain size, the shape of teeth and jaws, and skeletal characteristics affecting posture and locomotion—that clearly differentiated them from both the higher apes and the human species *(Homo)* that appeared later. The discovery in 2002 of a hominid skull in Chad—far from the locations of previous discoveries, and (at 6–7 million years) significantly older

than earlier finds—underlines how much we still have to learn about human origins in Africa.

In tracing the evolution of humans, scientists are attaching less and less importance to anatomical change than to the intellectual and cultural development—social organization and the use of tools, for example—that made it possible for humans to utilize increasingly diverse environments. About 2.4 million years ago, a more advanced hominid, *Homo habilis* (known as the "tool maker"), appeared. *Homo habilis* had a larger brain than the australopithecines, used simple stone tools, and lived in encampments. *Homo erectus*, which superseded *Homo habilis* some 1.8 million years ago, had a more erect posture and a still larger brain, and derived a variety of more sophisticated tools. Like *Homo habilis* and the australopithecines, *Homo erectus* apparently lived in savanna environments, particularly near large bodies of water.

Africa has yielded only scattered evidence related to the appearance of early *Homo sapiens* 150,000–200,000 years ago. Over the past 50,000 years, the pace of cultural development of *Homo sapiens* has steadily accelerated. Early advances included the development of more varied and sophisticated stone tools and the first human colonization of tropical rain forests.

The Agricultural and Iron Revolutions

In the totality of human history, the first agricultural revolution is a very recent event, having occurred in Mesopotamia approximately 10,000 B.C. The agricultural revolution in Africa south of the Sahara is even more recent, although the actual date remains a subject of debate. Rock paintings show that pastoralism, initially based on sheep and goat herding and later on cattle rearing, was well established in the central Sahara during the moist climatic phase between 8000 and 4000 B.C. Grinding stones have been found in their settlements, suggesting that they harvested grain. As the climate became progressively drier, communities in the savanna regions south of the Sahara that had depended on fishing turned increasingly to crop cultivation and pastoralism to secure their food supply. Farming and pastoralism became widespread in this savanna belt between 3000 and 1000 B.C. By 1000 B.C. farming was also well established throughout the forest zone. Unlike farmers in the savanna zone, for whom grain crops were predominant, farmers in the forest zone specialized in the cultivation of root crops and bananas.

Agricultural innovation occurred in four "culture hearths" located south of the Sahara: the Ethiopian Plateau, the West African savanna, the West African forest, and the forest–savanna boundary in west central Africa. Within each of these regions, various crops were domesticated and methods of cultivation suited to the local environment were developed. From these foci, crops and agriculture diffused to adjoining areas. Cultigens were also exchanged with other agricultural hearths, especially Egypt and the Middle East. The spread of agriculture frequently accompanied the Bantu migrations into central and southern Africa; these migrations are described later in the chapter.

The list of plants apparently domesticated in Africa is impressive. The following are among the more important:

- Cereals: teff, millet, bulrush millet, sorghum, African rice
- Roots and tubers: yams
- Pulses: Bambara groundnuts (peanuts), cowpeas
- Oil crops: oil palm, castor oil, shea butter
- Starch and sugar plants: enset
- Vegetables: okra, garden eggs (African eggplant)
- Fruits: watermelons, tamarind
- Stimulants: coffee, kola
- Fiber plants: cotton

The next great human revolution, that of making iron, began in Africa about 500 B.C. in Nubia in present-day Sudan. Other early foci of iron making developed at Nok in central Nigeria and in the vicinity of Lake Victoria (Figure 7.1). Unlike other parts of the world, Africa south of the Sahara did not experience a Bronze Age between the Stone and Iron Ages, except in Nubia. The introduction of iron making permitted the construction of improved weaponry and tools, which enabled iron-making peoples to expand territorially at the expense of those using only stone tools and weapons.

Bantu Migrations?

The Bantu peoples, who speak some 450 distinct but closely related languages, occupy the great majority of Africa south and east of the Cameroon–Nigeria border. Other evidence—genetic, cultural, and archeological—also points to the close relationships among the various Bantu speakers. How the Bantu came to dominate so much of Africa has been a point of strenuous and still-unresolved debate among African historians since the 1980s. Three interpretations of the Bantu phenomenon are outlined below.

For many years, most archeologists and historians agreed that the Bantu peoples migrated over several millennia from a common point of origin in southeastern Nigeria, eventually reaching the southern extremity of the continent. Two major streams of Bantu migration were identified. The eastern stream followed the savanna corridors toward the lake district of

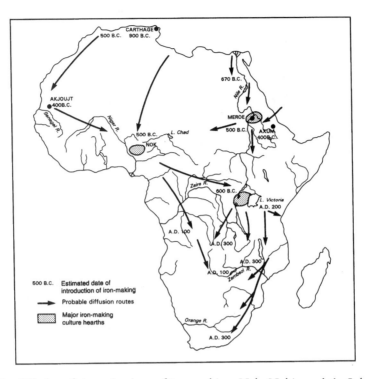

FIGURE 7.1. The diffusion of the technology of iron making. Nok, Nubia, and the Lake Victoria region are the three earliest centers of iron making in Africa south of the Sahara. After K. Shillington. *History of Africa*. London: Macmillan, 1989, p. 38.

East Africa; it then moved southward through East Africa, and beyond into southwestern Africa. The western migratory stream involved the spread of Bantu peoples south through Cameroon into the rain forests of west central Africa. Their migrations brought them into contact with indigenous Stone Age societies who were either forced into retreat or absorbed into the dominant Bantu culture. Distinct new cultures emerged wherever Bantu subgroups settled. The Bantu were farmers who used iron tools and weapons; their migrations were cultural and technological as well as demographic. They brought with them agricultural, herding, and iron-making technologies, as well as new forms of social organization.

Some scholars have challenged this "grand theory" of Bantu migration and conquest, and of the concurrent diffusion of Bantu culture and technology. The Bantu migration theory has been criticized for portraying indigenous peoples as primitive and passive, and thus failing to recognize their potential to achieve cultural

change from within or through an active process of learning from interaction with others. They argue that migration is not the only possible explanation for the wide range of Bantu languages. The fragmentary archeological, linguistic, and genetic evidence that is available does not always support the concept of a single migration sweeping across the continent and bringing multifaceted change. The process of change that shaped Bantu cultures, it is argued, varied significantly over space and time, and took place intermittently over thousands of years.

A third hypothesis allows for the spread of the Bantu peoples and of their culture and technology, but rejects the theory of a grand Bantu migration and conquest. Rather, it sees this process as a gradual expansion, possibly stimulated by population pressure, bringing small bands of Bantu on the move into contact with other groups. The Bantu peoples' knowledge of agricultural and iron-making technologies helped them to colonize new environments.

New cultures—fusions of Bantu and local elements—emerged through intermarriage and diverse types of social and economic interchange.

There is less controversy about several other major migrations in times past. Madagascar was settled more than 1,500 years ago by people from Indonesia. Peoples of Caucasian origin, coming from the Sahara and North Africa as well as from the Arabian Peninsula, occupied several areas. Arab settlement expanded southward along the Nile Valley and beyond to Darfur in present-day western Sudan. In West Africa, the pastoral Fulani extended their grazing territory between the 11th and 16th centuries A.D., occupying much of the savanna from their base in Senegambia eastward to Lake Chad. Indeed, the Fulani still continue to expand their territorial range. These recurrent processes of migration, diffusion, and assimilation have been very important in the evolution of the ethnic and cultural maps of the continent.

African Empires

The story of the empires in Africa south of the Sahara extends almost 3,000 years, beginning with the establishment of the empire of Kush. Other empires later developed in Ethiopia and West Africa. At the time of the colonial conquest, each of the major regions of the continent had several examples of advanced kingdoms or empires (see Figure 7.2). Only a few details pertaining to the most important empires are related in this chapter, but these should serve to disprove the common notion that African development began with the colonial conquest.

Eastern Africa

The Nile River has a long history as a corridor for the movement of peoples, ideas, and trade goods between Egypt and Africa south of the Sahara. About 1000 B.C., the state of Kush in the Nile Valley of present-day Sudan was able to assert its independence from Egypt. Kush

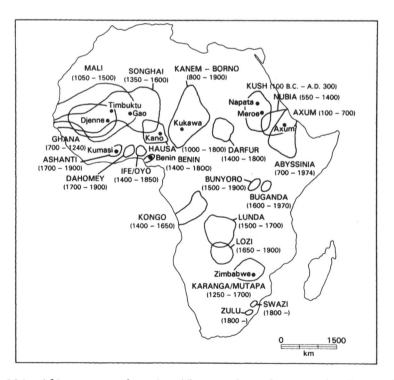

FIGURE 7.2. Major African states and empires. The map shows the names, locations, and approximate dates of existence of the most important precolonial states.

conquered Egypt in the 8th century B.C., and ruled it for several decades. The Kushitic civilization flourished following the rise to ascendancy of Meroë as the capital city, starting in the 6th century B.C. The sophistication of Kush was reflected in its impressive stone architecture, irrigation systems, a large iron industry, its own script, and a well-developed sense of nationhood. The first great empire of Africa south of the Sahara experienced its greatest development during the final three centuries B.C. Its collapse about A.D. 300 seems to have been precipitated by the decline of its agricultural base owing to soil exhaustion, and of its iron industry owing to the overexploitation of forests for charcoal.

Three Christian kingdoms, collectively known as Nubia, were established in the Sudanese Nile Valley in the 6th century A.D. These kingdoms flourished until the 8th century, when the spread of Islam effectively isolated Nubia from the rest of Christendom. The Christian kingdoms of Nubia finally expired in the 15th century.

Long before the emergence of Christian Nubia, the empire of Axum had embraced Christianity. Axum, located in the northern Ethiopian highlands, emerged in the 1st century A.D. through its control of the ivory trade from Africa to Arabia. Axum was predecessor to the Ethiopian or Abyssinian Empire, which for centuries was based at Gondar. In spite of constant pressure from neighboring Islamic states and many centuries of isolation from other Christians, the Ethiopian Empire survived until 1974, when Emperor Haile Selassie was deposed in a military coup. The spectacular 12th-century A.D. churches of Lalibela, which were carved out of solid rock, provide enduring evidence of the vitality of Ethiopian civilization (see Vignette 7.1 and Figure 7.3).

Several city-states were established along the East African coast between Somalia (Mogadishu) and Mozambique (Sofala) between the 8th and 19th centuries A.D. Kilwa, Lamu, Mombasa, and Mogadishu are among the best known of the 40-some major urban centers. Other closely related cities were founded on Madagascar and the Comoro Islands. Historians formerly believed that foreign merchants from Persia and Arabia had developed these cities. However, evidence from newer archeological work, together with a reinterpretation of historical documents, points conclusively to the African origins of these cities. They began as trading and manufacturing centers serving their local hinterlands, then expanded through various trading linkages into the interior; with other cities along the eastern African coast; and later with Persia, Arabia, and India. This major trading network flourished because of the rich diversity of products (especially gold and ivory) available from Africa. The trading system took advantage of seasonally reversing monsoon winds that permitted *dhows* (sailing vessels) to move back and forth with ease along the coast, and to link East Africa and Asia.

The larger city-states of East Africa became increasingly wealthy through trade. Their material wealth was reflected in architecture that featured well-designed mosques, homes, and palaces crafted from coral stone by skilled artisans. Archeologists have found ceramics and other artifacts from as far away as China—evidence of the city-states' wealth and the spatial reach of their trading ties. When the Portuguese first arrived along this coast in the late 15th century, the cities were near their peak. The Portuguese found cities that were not only wealthy, but also literate and cosmopolitan. Merchants from Persia, Arabia, and other parts of eastern Africa had settled alongside the indigenous coastal peoples. Asian settlers brought Islam to the coast. Wherever they settled, migrants—African as well as Asian—introduced new elements to the locally dominant culture. Eventually, a new synthetic language (kiSwahili) and culture (commonly known as the Swahili culture) emerged. Although virtually all visible evidence of the Indian Ocean trading system was confined to the coast, its cultural and economic influence was manifested as far inland as the East African lakes and Zimbabwe.

Western Africa

Between the 9th and 19th centuries, a series of Islamic empires rose to prominence in the savanna of West Africa; the most important of them were Ghana, Mali, Songhai, Kanem-

VIGNETTE 7.1. Landscape and Symbolism: The Rock Churches of Lalibela

The history of Christianity in Ethiopia dates back 17 centuries, to the conversion of King Ezana of Axum. For close to 1,000 years, Axum was a powerful trading state on the route linking Persia to the Middle East. A rich heritage of massive stone monoliths (*stellae*) remains as a powerful testimony to the kingdom's accomplishments.

By the 12th century, the capital of the Ethiopian kingdom had relocated at Roha. It was here some 800 years ago that King Lalibela organized the construction of a dozen churches, which were carved from solid rock. He claimed that he was following instructions from God, received in a vision. After his death, the capital city, Roha, was renamed Lalibela. According to tradition, the churches were built with remarkable speed during King Lalibela's 40-year reign. Archaeological investigations have cast some doubt on this time line; indeed, two of the edifices may have begun as palaces and been converted to churches at a later date.

King Lalibela sought to recreate Jerusalem, and organized the site's landscape accordingly. The churches are clustered in two major groups, representing the earthly Jerusalem and the heavenly Jerusalem. Located between them is a trench representing the River Jordan. Landscape features (including the Mount of Olives and Mount Tabor) and places of worship (including Golgotha and Calvary) are named after holy places in Jerusalem.

a

b

FIGURE 7.3. King Lalibela's legacy. (a) Bet Giorgis, the Church of St. George in Lalibela, Ethiopia. (b) Coptic priest with a historical religious manuscript. Photos: (a) author; (b) E. Peters.

<hr>

VIGNETTE 7.1. *(cont.)*

<hr>

The churches of Lalibela are spectacular architectural achievements. They are of three broad types: Monolithic churches are separated on all four sides from the surrounding bedrock; semimonolithic churches are attached on one side to the bedrock; and cave churches are surrounded on two or three sides and above by unaltered bedrock. The churches vary in size, and each has a range of distinctive design features. Some churches are constructed in an Axumite style, while others show Mediterranean influences. Cross-shaped windows and decorative designs take many forms—Celtic crosses from Ireland, Maltese and Greek crosses from the Mediterranean, and Swastika crosses from India—and attest to medieval Ethiopia's direct and indirect connections to other parts of the Christian world.

Interior design features are also very diverse, and include pillars, arches, and vaulted ceilings in different styles. Some churches are quite plain inside, while others have elaborate geometric designs and portraits of saints, rendered in *bas-relief* and frescoes.

Bet Giorgis, the church of St. George, is the most renowned and arguably the most visually perfect of Lalibela's churches (Figure 7.3). It sits in the middle of an excavation some 20 m square and 12 m deep. The church is carved in the form of a Greek cross, with three nested crosses outlined on its flat roof. The elaborate interior design is illuminated by light passing through small arch-shaped windows.

For centuries, Lalibela has attracted pilgrims from throughout Ethiopia. The churches continue to be used as places for worship and religious contemplation. With the recent construction of an airport and road improvements, increasing numbers of international tourists are coming to marvel at Lalibela's unique religious landscape.

Based on E. Hein and B. Kleidt. *Ethiopia—Christian Africa: Art, Churches and Culture.* Ratingen, Germany: Melina-Verlag, 1999.

<hr>

Borno, the Hausa states, and Sokoto. All had agricultural economies, but their control of one or more of the major trade routes across the Sahara provided the main source of wealth. Slaves, gold, cloth, and ostrich feathers were sent to North Africa in significant quantities; weaponry, coins, and cloth were imported in return. The savanna states also controlled the movement of salt from Saharan mines toward markets in the southern savanna and forest zones.

Ghana, a kingdom located in present-day Senegal and Mali, rose to prominence during the 9th century A.D.. During the 11th century, the rulers and many of the people of Ghana converted to Islam. Accounts of Arab travelers such as Al Bekri, who visited Ghana in 1067, provide insights into the size and splendor of the kingdom. Al Bekri noted, for example, that Ghana could field an army of 200,000 warriors. The kingdom of Ghana experienced a steady decline during the 12th century, however, after the opening of rich new goldfields at Bure, beyond Ghana's borders. As a result, trade routes shifted and Ghana was largely bypassed by traders.

By the mid-13th century, the empire of Mali had emerged in the upper Niger and Senegal Valleys, forged through the skilled leadership of the legendary Sundiata. Mali became extremely rich as a result of its control of the Bure goldfields and valuable salt deposits (see Figure 7.4). Mansa Musa, then emperor of Mali, traveling as a pilgrim through Cairo to Mecca in A.D. 1324, was accompanied by 500 porters each bearing a staff of gold weighing about 2 kg (Vignette 7.2). The Malian state was organized and administered on Islamic principles. Universities were established at Timbuktu and Jenne, well before any existed in northern Europe. Large quantities of books were imported, and scholars from Greece, Egypt, and Arabia were employed.

VIGNETTE 7.2. The Empire of Mali

The 14th-century geographer Al Omari provides us with a graphic description of the empire of Mali (see Figure 7.4), as well as the visit of Emperor Mansa Musa to Cairo in A.D. 1324 while traveling on a pilgrimage to Mecca. Excerpts from Al Omari's writings follow.

The king of this country [Mali] . . . is the most important of the Muslim Negro kings; his land is the largest, his army the most numerous; he is the king who is the most powerful, the richest, the most fortunate, the most feared by his enemies, and the most able to do good to those around him.

[He] presides in his palace . . . where he has a great seat of ebony that is like a throne fit for a large and tall person; on either side it is flanked by elephant tusks turned toward each other. His arms stand near him, being all of gold, sabre, lance, quiver, bow and arrows. Behind him there stand about a score of Turkish or other pages which are bought for him in Cairo; one of them, at his left, holds a silk umbrella surmounted by a dome and a bird of gold. His officers are seated in a circle about him, in two rows, one to the right and one to the left; beyond them sit the chief commanders of his cavalry. . . . Their army numbers one hundred thousand men of whom there are about ten thousand mounted cavalry.

During my first journey to Cairo I heard talk of the arrival of the Sultan Mansa . . . when he came into the Sultan's [of Egypt] presence, we asked him to kiss the ground. But he refused and continued to refuse, saying: "However can this be?" Then a wise man of his suite whispered several words to him that I could not understand. "Very well," he thereupon declared, "I will prostrate myself before Allah who created me and brought me into the world." Having done so, he moved toward the Sultan. The latter rose for a moment to welcome him and asked him to sit beside him: then they had a long conversation.

[He] spread upon Cairo the flood of his generosity; there was no person, officer of the court or holder of any office of the [Cairo] sultanate who did not receive a sum in gold from him. The people of Cairo earned incalculable sums from him, whether by buying and selling or by gifts. So much gold was current in Cairo that it ruined the value of money.

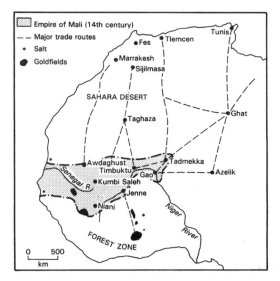

FIGURE 7.4. The empire of Mali in the 14th century. After K. Shillington. *History of Africa*. London: Macmillan, p. 95.

Excerpt from Al Omari. "Mali in the Fourteenth Century." Source for quote: B. Davidson, *The African Past*, pp. 83–87. Harmondsworth, UK: Penguin, 1966.

The decline of Mali during the 15th century coincided with the emergence of Songhai, centered at Gao at the bend in the Niger River. Several independent Hausa states had also developed in northern Nigeria, and the sultanate of Kanem–Borno had established control over the area around Lake Chad.

By the 13th century, advanced states had been established in the forest zone. The Yoruba states of southwestern Nigeria, and the related kingdom of Benin, were the most prominent of them. Their wealth was also based on trade—principally the sale of kola nuts, ivory, and gold to the savanna states. Only after the arrival of Europeans along the coast did slaves become an important component of the forest kingdoms' trade. Although agriculture was the basis of their economies, the forest states were decidedly urban in character. The larger towns were divided into distinct wards and surrounded by walls. European visitors to Benin during the 16th and 17th centuries remarked about the considerable size and orderliness of the city and the grandeur of the palace; they described it as being comparable to the major European cities of the time.

Central and Southern Africa

Relatively little is known of the kingdoms of central and southern Africa, most of which were remote from early European contact. Karanga is best known because of the remarkable ruins of stone towers at its capital city, Great Zimbabwe (Figure 7.5). The kingdom flourished from the 13th to the 15th century A.D.. Its people were skilled metalworkers who mined and crafted gold and copper and smelted iron on a large scale. Archeological evidence shows that Zimbabwe was the center of a flourishing trade in gold six to nine centuries ago. Trade goods from as far as India and China have been found at the ruins of Great Zimbabwe.

Between the 14th and 18th centuries, three major kingdoms—Luba, Lunda, and Kongo—emerged, flourished, and then declined in parts of what is now southern Congo and northern Angola. The prosperity of Kongo depended on a productive agricultural base, metalworking, and a flourishing interregional trade in foodstuffs, metals, and salt. The arrival of the Portuguese, whose main interest was to use the region as a source of slaves, led to the destabilization and finally the collapse of Kongo in the 16th century.

The Slave Trades

For 12 centuries, starting in the 7th century A.D., slaves were a principal export of Africa south of the Sahara (Figure 7.6). Scholars have

FIGURE 7.5. Section of the wall, Great Zimbabwe. These massive stone walls have remained intact for hundreds of years, even though no mortar was used in their construction. Photo: CIDA (B. Paton).

long debated the actual numbers sent, but recent estimates suggest a total on the order of 25–30 million! This brutal trade in human beings not only reduced the populations of many parts of Africa, but also affected local and regional economies, social and political stability, and the environment.

The Trans-Saharan Trade

Trade between North Africa and Africa south of the Sahara increased greatly after the 7th century A.D., following the introduction of the "ship of the desert," the camel. The trans-Saharan route became the conduit for a diverse and mutually beneficial interaction between the savanna states of Africa and the Islamic world. Armaments, books, textiles, and beads moved southward, while gold, ivory, and slaves went northward along a small number of routes. Islamic religious and cultural influences also crossed to the south side of the Sahara. Thus slaves constituted but one element, albeit a very important one, in trans-Saharan commerce.

As many as 9.4 million slaves were exported via the Saharan routes between A.D. 650 and 1900. The journey across the desert on foot was so arduous that it was common for the majority of those sent to perish on the way. Perhaps two-thirds were young women destined to become concubines or house servants in North Africa and Turkey. Male slaves were often employed as soldiers or courtiers; some wielded considerable power and influence in these positions.

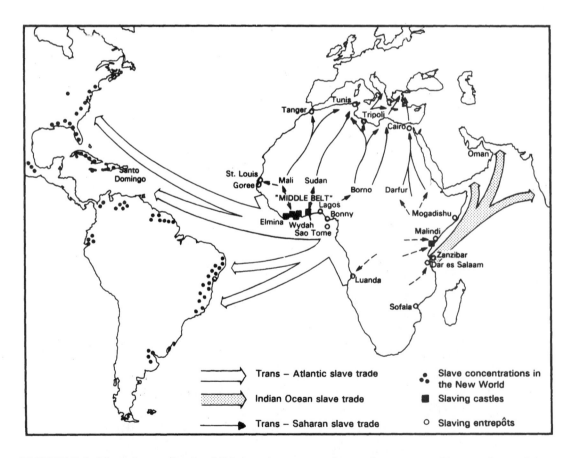

FIGURE 7.6. Three slave trades. In addition to these export slave trade systems, indigenous forms of slavery existed in many African societies.

FIGURE 7.7. Elmina Castle, Ghana. Elmina is the oldest and largest of some 20 fortifications built by various European powers along the Ghanaian coast as the bases for the slave trade. The construction of this mammoth Portuguese castle began in 1482. Photo: author.

East African Slave Trade

An estimated 5 million slaves were exported from eastern Africa as part of the large and diverse maritime trade linking Africa to Arabia, Oman, Persia, India, and even China. Trade along the East Africa coast dates from the early centuries A.D., but its magnitude increased during the 8th century. Initially gold and ivory were the main items of trade, with slaves being of relatively little importance. The sale of slaves later increased in importance, especially during the 18th and 19th centuries, when the slave trade grew dramatically. Slaves from eastern Africa, like those sent across the Sahara, were primarily women and children destined to become concubines and household servants.

Trans-Atlantic Slave Trade

Despite the enormity of the trans-Saharan and East African slave trades, they could not rival the European-controlled slave trade—either in the number of people transported or in the resultant distortion of economic, social, and political structures. Commodities other than slaves ceased to be a factor in African–European trading relationships; virtually the entire com-

mercial economy of the western coast of Africa between the 16th and 19th centuries was organized to facilitate the capture, transportation, and sale of slaves.

Direct European involvement in the acquisition of African slaves began in the 15th century as Portuguese navigators explored the West African coast. The real growth of the trade, however, followed the European conquest of the Americas. African slaves were identified as an ideal source of labor for the mines and plantations of the New World, especially after the decimation of the American indigenous population. By the late 16th century, the English, Danes, Dutch, Swedes, and French had joined the initial slave traders, the Portuguese and Spaniards. The trade continued to grow; approximately 80% of slaves transported across the Atlantic were sent between 1700 and about 1870, when the trade ended. Castles were constructed along parts of the coast, especially in present-day Ghana (see Figure 7.7), as centers for slave trading. Angola was the only other area where substantial permanent bases were maintained. Elsewhere, Europeans relied completely on African intermediaries to assemble slaves, to be traded for manufactured goods during irregular visits by ships to the coast.

The slave trade was one element of the so-called "triangular trade" linking Europe, Africa, and the Americas. European ships carried guns, alcohol, and cheap manufactured goods from Europe to West Africa. These goods would then be exchanged for slaves, and the slaves would be transported to the New World. The money obtained from selling slaves financed the purchase of gold, silver, tobacco, sugar, and rum, which would be transported to Europe. This trading system provided great profits for ship owners and merchants, and also spurred the development of slaving ports such as Liverpool.

The slave trade was a major force in the underdevelopment of West Africa. The most credible estimates of the number of slaves sent across the Atlantic range from 10 to 15 million. Many more died in transit; in slave raids and wars; and from the famine, disease, and economic disruption associated with the trade. Manning, who has written extensively about the demography of the slave trade (see "Further Reading"), estimates that the population of tropical Africa in 1850 was only about half of what it would have been in the absence of slavery and the slave trade.

The demographic effects of the slave trade varied greatly over time and space. It appears that West African kingdoms seldom waged war explicitly to capture slaves prior to the climax of the slave trade in the 18th and early 19th centuries; rather, they sold excess captives obtained in wars fought for other reasons. Thereafter, heightened demand for slaves brought about an increase in the frequency and scale of slave raiding. On some parts of the Atlantic coast of Africa, particularly in Senegal and Angola, the majority of slaves were captured relatively far inland. Elsewhere, such as along the Bight of Benin (coastal Nigeria and Benin), most slaves were captured locally during protracted wars among neighboring kingdoms.

For many smaller and weaker ethnic groups in areas such as the Middle Belt of West Africa, located between the savanna and coastal slaving states, the slave trade brought about significant population decline. However, the populations of some coastal slave-trading states, such as Dahomey, seem to have increased in conjunction with the slave trade. Most of their female and child captives were retained rather than sold; as a result, high rates of fertility prevailed in these societies.

The slave trade disrupted agriculture, manufacturing, and trade, especially in the areas where slave raiding was most severe. People from weaker communities were often forced to abandon their homes and seek refuge in remote or mountainous regions. Millions of young people were removed early in their most productive years; the development of countless communities was retarded by the loss of their energies and skills. The trade goods received in return by the slaving aristocracies were mostly armaments and luxury goods, neither of which provided any impetus for development. In states participating in the slave trade, the growing power and wealth of the aristocrats enabled them to dominate the peasants and to threaten and enslave their weaker neighbors.

In the Americas, slave labor was crucial for the functioning of the enormously profitable mines, as well as on the plantations where sugar, tobacco, indigo, and cotton were produced. Even after the abolition of slavery, the plantation remained the main production unit in many countries; former slaves frequently remained as sharecroppers or indentured laborers. Slaves, however, provided much more than mere labor. Many of them possessed skills (e.g., iron making and weaving) that were greatly needed in the plantation and mining economies. They also had special knowledge about the cultivation of tropical crops and not only grew their own food, but also commonly produced a surplus on small plots allocated to them. Vignette 7.3 provides a closer look at African Muslim slaves in the Americas.

Europe's rulers and merchants benefited materially from the slave-based economic activity. Slavery also fostered the development of racist stereotypes and myths in Europe. Africans were portrayed as shiftless savages; to enslave them was actually rationalized to be a means of rescuing them from their hopeless lives of misery! To justify the enslavement of Africans, Europeans first had to dehumanize them.

VIGNETTE 7.3. African Muslim Slaves in the New World

Who were the people who were enslaved? What were they like? Where did they come from? Popular portrayals of slavery seldom provide real insights into the characteristics and lives of slaves. Hollywood's slaves are typically much alike—simple, emotional folk toiling in the fields.

Diouf's *Servants of Allah*, a compelling book about African Muslims enslaved in the Americas, explodes these myths. Although it is impossible to know how many Muslim slaves were taken to the Americas, it is clear that many were taken to several areas in the United States, the Caribbean, Mexico, and Brazil. The largest number of African Muslim slaves came from Senegambia and nearby regions, such as Sierra Leone and Fouta Djallon in Guinea. Others came from the Bight of Benin (present-day Nigeria). Slaves from specific source areas were sought in particular regions of the Americas. For example, many slaves from Senegambia and Sierra Leone were taken to South Carolina and adjacent states, where their skills in rice and indigo cultivation were much valued. Diouf estimates that 30% of slaves in South Carolina came from Muslim areas.

The Muslim slaves came from diverse backgrounds—not only simple peasants and foot soldiers captured in slave raids and wars, but also artisans, traders, cleric/scholars, and even individuals of noble birth. They brought many skills with them. By virtue of their training in religious schools, a substantial proportion of Muslim slaves were able to read and write in Arabic. Indeed, the literacy rate was probably higher among the Muslim slaves than among the white slave owners. Muslim slaves were sometimes given supervisory and clerical duties because they were literate.

Muslim slaves struggled to preserve their culture and religion. As much as possible, they practiced the tenets of their Muslim faith, such as praying, fasting, and giving alms. Covert Koranic schools were established by some communities to educate their children. Copies of the Koran and other Arabic manuscripts were obtained by various means. These manuscripts were copied and passed on to others. Diouf writes that there was a very substantial trade in copies of the Koran between Africa and the Americas, sometimes conducted by liberated slaves engaged in trans-Atlantic trade. Both these traders and newly arrived slaves facilitated a tenuous flow of news across the Atlantic, and on rare occasions even carried messages from slaves to their families in Africa.

Muslims also put their faith into practice by engaging in many forms of subtle resistance to their enslavement. These ranged from keeping their Muslim names (sometimes covertly) to composing songs and stories of resistance. In short, they refused to forget who they were. In Brazil, Haiti, and elsewhere, Muslim slaves took the lead in organizing armed struggles for freedom. Escaped Brazilian slaves formed free communities organized according to Islamic principles in remote areas. The determined resistance of Muslim slaves created a backlash against them in Brazil and parts of the Caribbean, and resulted in efforts to repatriate some of them to Africa.

By the early 20th century, the active practice of Islam was quickly dying out, due to a variety of pressures that made it difficult to pass the Muslim faith to new generations. Nevertheless, Muslim influences have persisted in a variety of ways—ranging from musical forms and many words in Creole dialects such as Gullah (which is spoken in the Sea Islands of South Carolina and Georgia), to the worship of certain deities and practices in religions such as Candomble and Macumba in Brazil, and Santeria and Voodoo in the Caribbean.

Based on S. A. Diouf. *Servants of Allah: African Muslims Enslaved in the Americas*. New York: New York University Press, 1998.

An intense debate began in Britain in the late 18th century, which culminated in the abolition of slavery. The momentum for abolition developed from several perspectives, including liberal opposition to the treatment of slaves, concern about growing rebelliousness among slaves in certain colonies, and the belief that so-called "legitimate commerce" could be more profitable and successful in supplying the burgeoning demand for tropical raw materials to supply British industry. Gradually, the abolition of slavery was adopted or enforced elsewhere, and by 1870 the Atlantic slave trade came to an end.

Inter-African Slave Trade

In Africa, as in many parts of the world, the institution of slavery has had a very long history. However, the export slave trade was associated with a massive increase in internal slavery, and it altered both the nature of slavery in African societies and the organization of economic activity.

The number of slaves retained within Africa—primarily women and children along the Atlantic coast, and mostly men in the savanna region and in East Africa—increased as the export slave trade grew. Slave raiding made available many different kinds of captives, not just those most in demand for the export slave trade. The collapse of the trans-Atlantic slave trade in the 19th century did not end the wars and raids that had fed the trade. By the end of the 19th century, the population of certain African states was more than half slave, with systems of production primarily based on the use of slave labor. Slaves had become so important in the economy of northern Nigeria, for example, that the British chose not to abolish slavery outright immediately after the colonial conquest for fear of bringing about social and economic chaos.

Not only did slavery within Africa grow, but the nature of the institution changed as well. With wealth and power increasingly determined by the control of slave labor, slaves became increasingly commoditized. Customary rights of slaves within African societies, such as strict limits on the sale of the children of slaves, were progressively eroded.

The Contemporary Significance of African History

It used to be said in some quarters that Africa south of the Sahara had no history of its own. This view, convenient for those intent on the exploitation of Africa, has never withstood close scrutiny. However, in recent decades the careful work of archeologists, historians, and social scientists has shed increased light on the richness of Africa's history and the importance of African contributions to the collective history of humankind. Although the global significance of African history is well recognized, the portrayal of African history and culture, especially in non-African museums, has been hotly disputed. As described in Vignette 2.2 in Chapter 2, Africans question the morality of the continued possession and sometimes the insensitive display in Western museums of African treasures of great historical and cultural significance—more often than not stolen or seized as spoils of war.

Africans have turned to their rich historical traditions as a source of inspiration and identity. The contemporary map of Africa displays several names that are identical to those of ancient African kingdoms. Ghana, Mali, Benin, and Zimbabwe have all been resurrected as country names during the past 30 years. The trend was started by Kwame Nkrumah, who insisted that the colonial name Gold Coast be replaced with Ghana. The fact that ancient Ghana and modern Ghana occupied totally different territories was of no significance. Changing the name to Ghana replaced a name that symbolized British colonial oppression with one symbolizing both new beginnings and ancient roots, thus affirming that the underdevelopment of Africa was neither an original nor a natural state.

For peoples of African descent in the Americas and elsewhere in the African diaspora, African history represents an important reference point. In short, African history is also American history. It is particularly significant that despite the passage of time and past tendencies to ignore or even to attack the African cultural heritage, so many elements of this heritage are still evident throughout the New World. Consider, for example, African influences in the arts, especially music. Spirituals, gospel, blues, jazz, rock and roll, soul, and reggae share common African roots. The themes of the music reflect the life experiences of present-day African Americans, but the structure and rhythm of the music are African. For example, the "call-and-response" (leader and chorus) structure of African work songs and of traditional drumming is commonly found in African American spirituals and jazz. Modern abstract sculpture and painting have also been influenced in important ways by African traditional art. Pablo Picasso, the father of cubism, acknowledged that African traditional sculpture had been his primary source of inspiration.

Further Reading

The following books are among the best general surveys of African history:

Ajayi, J. F. A., and M. Crowder, eds. *Historical Atlas of Africa*. New York: Cambridge University Press, 1985.

Fage, J. D., and R. Oliver, *Cambridge History of Africa*, 7 vols. New York: Cambridge University Press, 1975–1977.

Harris, J. *Africans and their History*, 2nd ed. New York: Penguin USA, 1998.

Shillington, K., *History of Africa*. London: Macmillan, 1989.

Shillington, K. *Encyclopedia of African History*. London: Fitzroy Dearborn, 2003.

United Nations and Educational, Scientific and Cultural Organization (UNESCO). *General History of Africa* (8 vols.). Berkeley: University of California Press, 1986–1999.

Various themes in African prehistory are examined in the following sources:

Berger, L. R., and B. Hilton-Barber. *In the Footsteps of Eve: The Mystery of Human Origins*. New York: Simon and Schuster, 2000.

Larsen, C. S., R. M. Matter, and D. L. Gebo. *Human Origins: The Fossil Record*, 3rd ed. Prospect Heights, IL: Waveland Press, 1998.

Leakey, R., and R. Lewin. *Origins Reconsidered: What Makes Us Human*. New York: Doubleday, 1992.

Newman, J. *The Peopling of Africa: A Geographic Interpretation*. New Haven, CT: Yale University Press, 1995.

The following studies address several key historical themes—empires, Bantu expansion, and Swahili states:

Davidson, B. *African Kingdoms*. New York: Time–Life Books, 1966.

Kusimba, C, and J. C. Vogel. *The Rise and Fall of Swahili States*. Lantham, MD: AltaMira Press, 1999.

Middleton, J. *The World of the Swahili: An African Mercantile Civilization*. New Haven, CT: Yale University Press, 1994.

Vansina, J. *Paths in the Rainforest*. Madison: University of Wisconsin Press, 1990.

Vansina, J. "New linguistic evidence and 'The Bantu expansion.'" *Journal of African History,* vol. 36 (1995), pp. 173–195.

The slave trade (particularly the trans-Atlantic trade) has been the subject of much research that has examined the size and geographical organization of the trade, as well as its effects:

Curtin, P. *The Atlantic Slave Trade: A Census*. Madison: University of Wisconsin Press, 1969.

Inikori, J. E., ed. *Forced Migration: The Impact of the Export Slave Trade on African Society*. New York: Africana, 1982.

Lovejoy, P. *Transformations in Slavery: A History of Slavery in Africa*. Cambridge, UK: Cambridge University Press, 1983.

Manning, P. *Slavery and African Life: Occidental, Oriental and African Slave Trades*. Cambridge, UK: Cambridge University Press, 1990.

The sociocultural history of peoples of African descent in the Americas has been explored by many authors:

Carney, J. *Black Rice: The African Origins of Rice Cultivation in the Americas*. Cambridge, MA: Harvard University Press, 2001.

Fikes, R. "Blacks in Europe, Asia, Canada and Latin America: a bibliographic essay." *A Current Bibliography on African Affairs*, vol. 17 (1984–1985), pp. 113–127.

Thornton, J. *Africa and Africans in the Making of the Atlantic World*, 2nd ed. Cambridge, UK: Cambridge University Press, 1998.

Internet Sources

There are several great comprehensive websites on African history, among them the following:

Boddy-Evans, A. *About African History.* http://africanhistory.about.com

Fung, K., Stanford University. *Africa South of the Sahara. Topics: History.* http://www-sul.stanford.edu/depts/ssrg/africa/history.html

Halsall, P., Fordham University. *Internet African History Sourcebook.* http://www.fordham.edu/halsall/africa/africasbook.html

The following are excellent websites on human origins and development:

Foley, J. *Prominent Humanoid Fossils.* http://www.talkorigins.org/faqs/homs/specimen.html

Institute of Human Origins, Arizona State University. *Becoming Human.* http://www.becominghuman.org

Sealy, E., University of Cape Town. *Archaeology Africa.* http://www.archafrica.uct.ac.za/archhome.htm

8

The Colonial Legacy

European rule in most parts of the African continent had been in existence for only 60–80 years in 1960 as the colonial era moved rapidly toward its end. Though the colonial era was short, it brought profound and lasting changes to the political, economic, and social geographies of Africa. The crises now affecting Africa are seldom comprehensible without reference to specific aspects of the colonial legacy. This is not to imply that the impress of colonialism was identical in all settings. On the contrary, it is important to recognize the variability over time and space of colonial policies and effects. Moreover, colonial rule should not be seen as an omnipotent force; Africans often resisted colonial edicts and found their own ways of adapting to the new reality.

Prelude to Colonization

The 19th century brought about profound changes in the relationship between Africans and Europeans. At the beginning of the century, the slave trade was still in full swing; by the end, the trade had been abolished. At the beginning of the century, Africa away from the coast was virtually unknown to Europeans; by

the end, Europeans had set eyes upon virtually every part of the continent. At the beginning of the century, European political maps of Africa were almost blank; at the end, the maps carried a patchwork of pink, green, yellow, mauve, and orange to identify different colonially controlled areas.

In 1807, Great Britain abolished the African slave trade and moved to impose this policy on other slave-trading countries. (It went on to abolish slavery in the British Empire altogether in 1833.) For a number of African coastal states, the transition from an economy based for centuries on the slave trade to one based on legitimate commerce in primary products such as palm oil and groundnuts was very traumatic. In several states, the coercive powers of traditional rulers were challenged by newly emerging commercial and religious leaders. The resulting unrest made it possible for European trading companies, adventurers, and consular representatives to increase their influence. These trends intensified in West Africa starting in the 1860s, when declining terms of trade in African commodities forced many independent African and European traders to become middlemen working for the large trading companies.

European fascination with Africa began to increase in the late 18th century, particularly after James Bruce returned in 1783 from his quest to find the source of the Blue Nile. Other European explorers followed Bruce, including Park, who twice attempted to follow the Niger River to the sea; Burton and Speke, who followed the White Nile to its source in Lake Victoria; and Livingstone, who explored large parts of central Africa. Much of the funding of this early European exploration came from geographical and scientific societies supported by wealthy individuals and companies. Returning explorers presented their findings to the sponsoring societies and wrote articles for their journals. These journals of discovery generated great interest in a public intent on knowing much more about the world, and among merchants and industrialists eager to discern what trade opportunities might exist in Africa's unknown interior.

During the 19th century, there was also a growing interest in the establishment of Christian missions in Africa, as noted in Chapter 2. The new missions, like the explorers and commercial agents, helped to pave the way for the establishment of formal colonial rule by heightening public interest in Africa and the fate of Africans. Missionaries returned to Europe after visiting Africa and talked passionately about the importance of the "civilizing mission," which involved combating the slave trade, starting schools, and supporting the development of commerce in conjunction with the primary objective of spreading the gospel. The missionary project was fundamental to the European conviction that colonialism was a charitable undertaking by a morally and technologically superior race.

The Scramble for Africa

Prior to 1880, perhaps 90% of Africa south of the Sahara was still ruled by Africans (see Figure 8.1). Two decades later, the only uncolonized states were Liberia and Ethiopia.

Small European enclaves along the coast had existed for centuries, starting with the establishment of slaving castles in the late 15th cen-

tury. Several of these enclaves were consolidated and extended during the 19th century. The French were well established in Senegal and Dahomey; the British in Gambia, Sierra Leone, and South Africa; and the Portuguese in Angola and Mozambique. However, by about 1880, tensions among the major European powers began to be focused increasingly on Africa. The French were angered by Britain's annexation of Egypt to safeguard its interests in the Suez Canal. In South Africa, the British extended their control inland from the Cape of Good Hope, following the discovery of rich diamond deposits at Kimberley. The French sought to expand into the upper Niger region by means of a rail link to Dakar, starting in 1879. King Leopold II of Belgium sent emissaries to annex territory in the Congo basin. The Germans moved on several fronts, proclaiming Togo, Cameroon, Tanganyika (now part of Tanzania), and South-West Africa (now Namibia) as protectorates.

Africa's would-be colonizers held a conference in Berlin in 1884–1885 in an atmosphere of intense distrust to establish ground rules for carving up the continent. Spheres of influence were traded like prizes in some great game of Monopoly. It was decreed at the Berlin Conference that new annexations would not be recognized unless the territory had been effectively occupied. The scramble for Africa (see Figure 8.2) was beginning in earnest; whether "effective control" was established by means of military conquest or through bogus, one-sided "treaties" did not change the end result for Africans.

The French moved inland from several of their existing coastal bases toward Lake Chad—east along the savanna corridor from Senegal, north from Côte d'Ivoire and Dahomey, northeast from Gabon, and south from Algeria. In the process, they hoped to confine and outflank the British and gain control of as much territory as possible. The French also annexed Madagascar and a few smaller islands off the East African coast.

The British claimed far less territory than the French in West Africa, but they managed to secure the most productive and populous areas. In southern Africa, Cecil Rhodes was the driving force behind British expansion, sending a

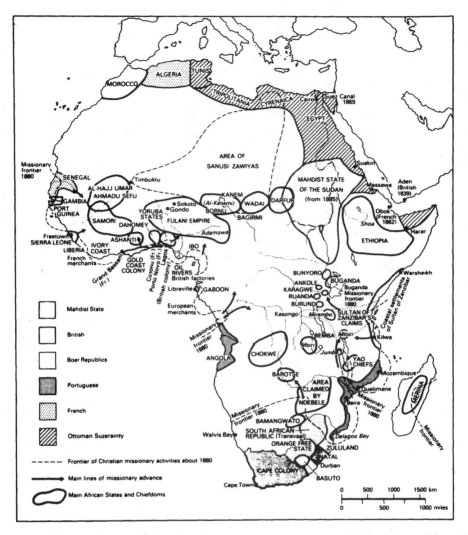

FIGURE 8.1. Africa on the eve of the colonial conquest. European control did not extend beyond a few footholds along the coast. Source: *Africa South of the Sahara, 1991.* London: Europa, 1990, p. 12. © 1990 by Europa Publications Limited. Reprinted by permission.

military unit north to occupy territory and obtain mining concessions in Bechuanaland (now Botswana) and Southern and Northern Rhodesia (now Zimbabwe and Zambia, respectively). Rhodes was responsible for the powerful "Cape to Cairo" metaphor, which envisaged the establishment of continuous British rule and a rail line between Africa's southern and northern extremities. British rule in southern Africa was consolidated as a result of the Boer War of 1899–1902, in which the formerly separate Boer republics of Transvaal and the Orange Free State were defeated and later incorporated

into the Union of South Africa. Britain's long-standing interest in East Africa intensified during the mid-1880s, in response to the newly established German presence in Tanganyika. In East and central Africa, control was initially exercised by commercial interests—the Imperial British East Africa Company and the British South Africa Company, respectively.

The other major colonial powers, the Germans and Portuguese, moved inland from the bases they had established prior to the Berlin Conference. The Italians claimed Eritrea and Somaliland, but failed to conquer Ethiopia as a

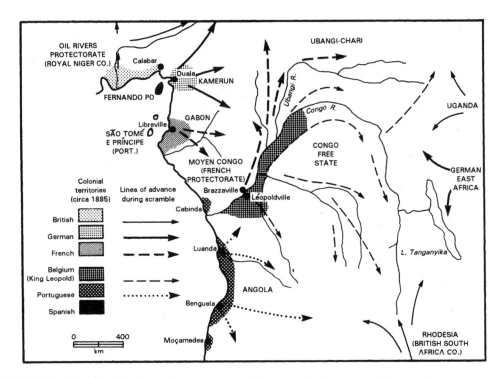

FIGURE 8.2. The scramble for the Congo basin and Angola. Similar territorial scrambles occurred in all parts of the continent in the decade following the Berlin Conference. After K. Shillington. *History of Africa.* London: Macmillan, 1989, p. 312.

result of their resounding defeat at Adowa in 1896 by Ethiopian forces led by Emperor Menelik II. The Spanish also maintained a minor presence on the Guinea coast in present-day Equatorial Guinea. The Congo Free State was also established, with tactics that were exceptionally brutal even for those times (see Vignette 8.1). Only in 1908 did King Leopold relinquish control of "his" Congo Free State to the Belgian government.

The colonial scramble for Africa was not the unproblematic claiming of territory it has sometimes been made out to be. Africans mounted what was often a determined resistance to the advancing colonial regiments. In a few places, such as northern Niger and western Uganda, armed resistance continued into the 1920s. Moreover, rival colonial powers often clashed over the demarcation of boundaries, and political propaganda in the various European nations reflected these clashes (see Figure 8.3). The political decisions eventually made in the capitals of Europe

over the locations of boundaries were frequently arbitrary and ill conceived. The boundaries that the colonial powers created often divided ethnic groups and traditional political units between neighboring countries. To give but one example, the grand sultanate of Kanem–Borno, founded in the 11th century, was divided among the British (Nigeria), Germans (Cameroon), and French (Niger and Chad).

By the early 1900s, the map of colonial spheres of influence was largely fixed (see Figure 8.4). The most important subsequent changes occurred when Germany was forced to relinquish control of its colonies after World War I. Britain, France, Belgium, and South Africa were granted trusteeship of particular German colonies by the League of Nations. With the exception of a few administrative changes, such as the creation of the Federation of Rhodesia and Nyasaland in 1953, the political map of Africa did not change significantly for the rest of the colonial era.

VIGNETTE 8.1. Leopold's Congolese Legacy

Driven by reckless ambition, greed, and vanity, King Leopold II of Belgium perused the world map looking for likely colonial opportunities. For tiny Belgium to be taken seriously, he reasoned, it must have colonies that would generate wealth for him and for his country. He settled on the Congo basin, and set out to recruit the U.S. explorer Henry Morton Stanley, who had just returned from Africa, as his emissary. In 1879, Stanley sailed to Africa, under contract to Leopold to claim as much of the Congo as quickly as possible. In Europe, Leopold sought to legitimize his conquest as a means of controlling slave raiding and bringing Christian civilization to the "wretched" people of the region.

Having succeeded in gaining initial support for the Congo Free State, Leopold's attention turned to the quest for profit, initially with ivory. Not only was ivory available in huge quantities, but it also was a low-weight, high-value commodity that could be exported profitably, despite the rudimentary nature of the transportation system. Leopold's agents scoured the country for ivory, for which Africans received little and often nothing. Countless thousands of people were forcibly recruited as porters to carry the ivory many hundreds of kilometers from the interior to river stations for export. Large numbers of them died in the process.

The horror of the ivory trade pales in comparison to the next phase of exploitation—namely, the rubber trade that started in the 1890s, when wild trees were the only significant source of rubber and the Congo had the largest concentration of these trees. Rubber trees were widely scattered in the forest, and the work of collecting rubber was very difficult. Villagers were recruited by force to collect rubber, and terror tactics were routinely used to ensure compliance. One method was to hold family members hostage, in chains, until the arbitrarily set rubber quota was met. Lashes with a rawhide whip, cutting off hands and feet, and instant executions were used routinely to punish and to terrorize others into submission. The wild rubber economy abated only after the supply of cheaper, higher-quality cultivated rubber from other countries began to reach world markets.

The opportunity to make a fortune with "no holds barred" attracted many brutal and unscrupulous individuals to the Congo. The terrorizing activities of commercial and government agents received support from militias, composed mainly of African recruits. Many missionaries turned a blind eye to what was happening, convincing themselves that the end (the "civilizing mission" and the quest for souls) justified the means. A few did speak out, putting themselves and their families at considerable risk, but as a result helping to inform the European public of the dark secrets of Leopold's Congo. Although Leopold was forced ultimately to cede title to the Belgian government, Africans continued to suffer under the new administration.

The brutal lawlessness of the Congo Free State of the 1890s was chronicled in the famous novel *Heart of Darkness*, written by Joseph Conrad after he made a six-month trip to the country. How dark was the Belgian subjugation of the Congo? Historical demographers now estimate that between 1880 and 1920, 10 million Congolese—half of the population—perished from the ivory and rubber trades, and from other direct and indirect effects of colonial subjugation.

In his concluding chapters, Hochschild reminds us that what happened in the Congo—the forced labor, pillaging of resources, and huge losses of population, all done in the name of civilization and development—was proportionally just as severe in several other colonies ruled by other powers, and continued well into the 1930s. Moreover, he asks why the world has almost totally forgotten about what was certainly among the most horrible holocausts of recent times.

Based on A. Hochschild. *King Leopold's Ghost: A Story of Greed, Terror, and Heroism in Colonial Africa.* Boston: Houghton Mifflin, 1998.

a   b

FIGURE 8.3. "Our colonialism, their colonialism." (a) "John Bull" reluctantly accepts responsibility for the orphan Uganda abandoned on his doorstep. (b) The German eagle swoops in on defenceless villagers. Source: *Punch*—(a) April 21, 1894; (b) April 26, 1890.

The Colonial State

The typical colonial state was run by a small cadre of administrative and military officers and as an extension of the European metropolitan state. The broad outlines of colonial policy were developed in Europe and generally reflected the political climate of Europe rather than the needs of Africa. The role of colonial officials was to interpret and implement policy directives from the metropole in relation to the particular situations found in specific regional settings. The colonial state focused on maintaining law and order and promoting kinds of development deemed to be in the interest of the metropole. The goals did not include fostering the development of modern, self-reliant nation-states—a notion inconceivable in light of the racist assumptions that underpinned colonialism (Vignette 8.2). The structure of colonial states varied greatly. In colonies with a large white-settler presence, most notably Kenya and Rhodesia, the settlers used their considerable power to consolidate and legitimate their special privileges. Restrictions were placed on the economic and social choices available to Africans, as well as on those of Asian and Arab populations. For example, the production of certain crops was reserved exclusively for settlers. On the other hand, colonial governors in the settler-dominated colonies sometimes acted as a brake on settler self-interest. The overexploitation of African labor, for example, could be counterproductive for the colony and even for European employers if it meant that Africans did not have sufficient time and land to pro-

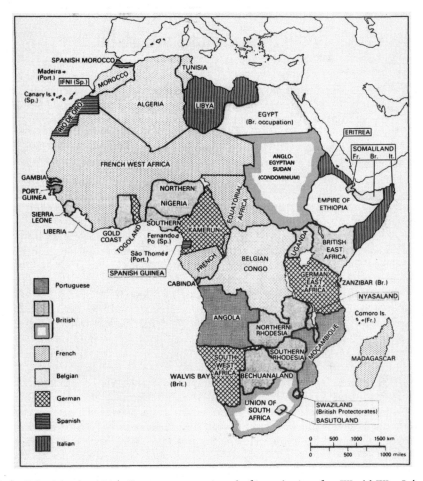

FIGURE 8.4. Colonial rule, 1914. Germany was stripped of its colonies after World War I, but otherwise the colonial map remained essentially the same. Source: *Africa South of the Sahara, 1991.* London: Europa, 1990, p. 14. © 1990 by Europa Publications Limited. Reprinted by permission.

duce enough food for themselves and their families.

At the other extreme were regions where the British implemented a system of indirect rule, which involved keeping and modifying traditional political structures to suit European needs. Indirect rule was a pragmatic approach implemented in settings where few European officials were available and where there were well-established indigenous political systems. Local authorities could be given responsibility for unpopular measures such as tax collection and labor recruitment for the colonial state's projects. Indirect rule was used as a reason for paying little attention to the development of social and economic infrastructures, on the pre-

text that local authorities were responsible for financing and implementing projects and services—for example, primary education. However, the apparent powers retained by the indigenous rulers were an illusion. These rulers were handpicked by their colonial overseers, were told what to say and do, and were replaced if they were incompetent or too independent minded. Indirectly ruled areas such as northern Nigeria emerged from the colonial era experiencing very little development. This neglect, combined with the use of different administrative models in other parts of these same countries, has contributed in no small way to a continuing history of interregional misunderstanding and occasional violence.

VIGNETTE 8.2. "Africa Decivilized": Accounting for the Ruins of Zimbabwe

Colonial rule was justified, in large part, on the myth that the "dark continent" had no history worthy of the name. If Africans were believed to be incapable of indigenous development, colonialism could be considered a charitable undertaking, rather than an exercise in conquest and exploitation.

Where there was tangible evidence of past achievements, it tended to be discounted or ignored, or even falsified. The misinterpretation—indeed, the deliberate distortion—of the history of the ruins of Great Zimbabwe provides an interesting example of the "decivilization" of Africa under colonial rule.

Cecil Rhodes and other early European visitors to Zimbabwe were fascinated by the ancient gold workings and awe-inspiring stone structures they found. Local African legends spoke of the ruins as having been constructed by African ancestors countless generations before. But Rhodes ignored this explanation, and instead alluded to legends about long-distance voyages by the ancient Egyptians and Phoenicians. He paid archeologists to study the ruins and to conclude that the seat of power of Solomon and the Queen of Sheba had been found.

Radiocarbon dating and other archeological evidence revealed that Great Zimbabwe had flourished after 1000 A.D., far too late for the Queen of Sheba myth to have any basis. For decades after it had been demonstrated conclusively that Africans had constructed this edifice, Rhodesian history books, museum literature, and other government publications continued to claim otherwise. Archeology was deliberately censored; to acknowledge African history and African achievement was too painful for those who wished to maintain white domination.

The history of Great Zimbabwe became an important symbol for both sides in the struggle for black majority rule after white Rhodesia unilaterally declared its independence in 1965. For Africans, Zimbabwe was a potent symbol of African achievement, as well as the chosen name for the new country for which many were prepared to sacrifice their lives. For the white Rhodesian state, Zimbabwe symbolized the unthinkable—namely, that Africans could (and would) prevail.

The following excerpt from "Zimbabwe: Bantu theory v. truth: Link with Arabia," which appeared in a 1972 issue of *Property and Finance* magazine, demonstrates the continuing denial of truth and acknowledges why it was considered so important to do so.

In these days of deliberate subversion of civilized authority, the international New Liberalism attempts to mould all aspects of life including the sciences, to its glorification of the Negro. Archeology, the study of antiquities, is a natural victim, for it can be used as a means of creating an artificial cultural respectability for Black nationalism and, accordingly, a justification of Black rule. Against all objective evidence, Zimbabwe is again being promoted as a Bantu achievement.

Based on J. Frederickse. *None but Ourselves: Masses vs. the Media in the Making of Zimbabwe.* London: Heinemann, 1982.

Where there was no history of chieftaincy, as among the Igbo in Nigeria, the colonial powers often created chiefs and used them to perform unpopular tasks. These new chiefs were regarded with suspicion, especially in the French and Belgian colonies, where their role as colonial puppets was particularly obvious. Moreover, these "chiefs" were often outsiders or low-status individuals chosen only because they could be counted on for total obedience.

To maintain law and order, the colonial administrations relied heavily on African recruits commanded by European officers; to run the lower levels of the bureaucracy, the colonial administrations relied on educated Africans. These supporting cadres were often recruited from subordinate classes, such as former slave families, or from other regions. As such, these cadres owed their status to the colonial state and could be counted upon to implement its repressive policies, often enthusiastically. The French, Portuguese, and Belgians attempted to consolidate African support for the colonial state by granting special "assimilated" status to Africans who were Western-educated. These educated Africans received certain social and political privileges denied to the African masses, but remained clearly subordinate to Europeans in the colonial social hierarchy.

Colonial Economic Policy

Colonial economic policy is often described in relation to the introduction of new crops and new modes of production and to the construction of infrastructure. Without doubt, developments of this sort represent the most tangible legacy of colonial economic policy. Railroads and roads were constructed, often across difficult terrain. Mines were opened and plantations established. Africa became a major source of cocoa, rubber, groundnuts, palm oil, coffee, and other agricultural products; tropical woods; and minerals, especially copper, diamonds, and gold. New cities were established to serve as administrative and commercial centers. Western education and health care were introduced.

The economic contribution of colonialism was long considered to be the incorporation of Africans into the modern world economy as productive workers and farmers. The growth of production in colonial economies was interpreted as evidence of progress in the broad "civilizing mission" used to justify the colonial project (as illustrated in Figure 8.5). Conversely, the people in areas that did not become important sources of production were portrayed as being backward and hence as not meriting much attention.

The spatial pattern of colonial development that evolved was highly uneven and reflected expectations of profit rather than altruism. It is useful to generalize about this uneven develop-

ment in relation to three approaches—namely, direct European investment, African smallholder production, and labor supply—that created distinctive socioeconomic structures and landscapes in the areas affected by each.

Direct European investment occurred in a relatively few areas of the continent. These enclaves of European capital included such mining areas as the copper belts of Northern Rhodesia and the Belgian Congo, such commercial plantations as the sisal farms of Tanganyika, and such areas of European settlement as Kenya's so-called "White Highlands." The success of European ventures in these areas depended on three critical elements: control of the most desirable land, access to overseas markets, and a secured supply of cheap African labor. These elements were integrally linked.

As white farmers gained legal title to large tracts of desirable land, Africans who had occupied these lands for countless generations were forced to move to newly created reserve lands that were often located in areas with less fertile soils and that were, as a rule, overcrowded. In Kenya, the imposition of soil conservation programs on African reserve lands deprived farmers of even more of their scarce land and time, causing the women who had been most adversely affected to mount a campaign of resistance (Vignette 8.3). To make ends meet, most families in the reserve lands had to rely on selling their labor to European farmers. The levying of taxes on all Africans, as well as legislated restrictions on Africans' access to commercial markets for livestock and cash crops, further ensured a ready supply of labor. Even then, the colonial state often had to resort to forced-labor recruitment to guarantee enough workers for commercial European ventures (see Figure 8.6).

Perhaps the greatest contribution of the colonial state to the development of these enclaves of European capitalism was the construction of railroads and ports to permit the mineral and agricultural products to reach overseas markets (see Figure 8.7). The development of this infrastructure to benefit European capital was often very costly, not only in money but also in suffering and loss of life.

In countries such as Ghana, Senegal, and Nigeria, which were unattractive to European set-

a

b

FIGURE 8.5. Portraits of colonialism. (a) "The black man's burden," German East Africa (Tanzania), circa 1910. (b) "The shepherd and the sheep." This early-20th-century postcard illustrates the often patronizing worldview of early Christian missionaries. Photos: (a) O. Haeckel; (b) Mission des Pères Blancs.

tlers, quite densely populated, and characterized by well-developed indigenous farming systems, a different colonial development strategy emerged. African farmers were encouraged—if necessary, by coercion—to grow crops for export to Europe. For African farmers, the incentive to grow cash crops was both negative (the need for cash to pay taxes) and positive (the opportunity to accumulate wealth and purchase manufactured goods). Non-African entrepreneurs, including both large trading companies and Asian and Arab family businesses, profited from the expansion of African smallholder production. They purchased and exported the cash crops; imported and sold consumer goods; and often financed export crop production with advances of seed or cash loans, typically charging very high rates of interest.

Capitalists and the colonial state often reaped great profits, even when the course of

development did not conform to what had been planned. Cotton had long been cultivated in the Kano region in northern Nigeria, where it formed the basis for a thriving textile industry. The British hoped that the region would become a major source of raw cotton for British textile mills. However, Hausa farmers did not respond as the British hoped, largely because of the difficulty of integrating large-scale cotton production into the farmers' existing agricultural systems. However, when local traders began to promote groundnuts as a cash crop and sent representatives into the countryside to advance credit for farmers to grow the crop, the response was overwhelming. Groundnut production in Kano Province increased tenfold between 1912 and 1913. The government was so unprepared for the deluge of groundnuts that it was unable to transport the entire crop to the port of Lagos before the next harvest.

VIGNETTE 8.3. Resistance under Colonialism: A Kenya Example

Africans responded to colonial rule with many forms of active and passive resistance. The revolt of Kenyan women against imposed soil conservation schemes provides an example of such resistance. This protest arose as a result of the combined effects of environmental deterioration, changing class and gender roles, and contradictions in colonial policy.

Starting in the 1930s, the alleged degradation of land farmed by Africans became an issue of major concern. It was claimed that poor farming practices were causing increased soil erosion, and that enforced conservation programs were needed to protect the environment. Behind the overt concerns about conservation lay the real issue—namely, the struggle by white settlers to consolidate and legitimate their control of the White Highlands.

Following the imposition of colonial rule, African lands were indeed farmed much more intensively and without many traditional conservation practices. However, these changes reflected the pressures of accommodating a growing population on a much reduced land base. Traditional strategies such as crop rotation, intercropping, and extended fallowing had to be sacrificed in order to address immediate needs for sustenance. The growing substitution of Irish potatoes for sweet potatoes and of maize for millet increased the susceptibility of land to erosion. Moreover, as the interests and responsibilities of men and women changed in the now monetarized economy, women's traditional rights to land increasingly became the focus of competition.

The response of the colonial state was to institute soil conservation programs such as planting lines of Napier grass on erosion-prone slopes and making compost pits. What started as a voluntary program became increasingly compulsory, and chiefs were made responsible for meeting established targets. After World War II, the program's focus shifted to the construction of bench terraces along the contours of slopes. The new approach was strongly opposed by women because it required so much of their land and their time. For every 1-m drop between contour terraces, up to 2 m of land was lost. The colonial state insisted that several days of unpaid forced labor be devoted each month to the onerous work of building terraces.

Women mounted widespread acts of resistance, both visible and invisible, against compulsory soil conservation work. In April 1948, this resistance escalated into a full-scale "Women's Revolt" in Murang'a District. The women arrived, 2,500 strong, at the district headquarters to announce their refusal to do more soil conservation work, stating that they had "quite enough to do at home." Attempts to persuade them to return to work were rebuffed. A few weeks later, after their protests continued to escalate, the women involved were arrested and fined.

Although this protest was remarkable for its size and intensity, it was not an isolated incident. Throughout Africa, ordinary people mounted protracted and often subtle resistance to colonial rule; in doing so, they certainly hastened the end of colonialism.

Based on F. Mackenzie. "The political economy of the environment, gender and resistance under colonialism in Murang'a District, Kenya." *Canadian Journal of African Studies*, vol. 25 (1991), pp. 226–256. For an understanding of the broader historical context of this revolt, see F. Mackenzie. *Land, Ecology, and Resistance in Kenya*. Portsmouth, NH: Heinemann, 1998.

The effect of introducing cash crops into smallholder farming systems was not always positive, or even benign. These crops required heavy investments of labor and often had to be grown on the best land. Certain crops, notably cotton, make heavy demands on soil nutrients and facilitate increased soil erosion. Although the expansion of cash crops often brought higher incomes, it also increased malnutrition because of the reallocation of land and labor from food to cash crop production.

Many parts of Africa were perceived by colo-

FIGURE 8.6. Preparing a new plantation, Nyasaland (Malawi), 1920s. The production of cash crops—in this case, probably tobacco—involved the use of forced labor in many areas. Photo: Nyasaland Pharmacies.

nial authorities to have little potential for modern development. They were far from ports and railroads, and typically had rather infertile soils or semiarid climates. The people in these regions were often portrayed as "backward." Still, these areas were expected to contribute their share to the upkeep of the colonial state. Their contribution was in the form of migrant labor for areas of mining and export-oriented agriculture, often located many hundreds of kilometers away. Tax-

ation and forced-labor recruitment were used to ensure that labor was made available, even when wages were extremely low and working conditions brutal. Labor reserve areas such as Mali, Niger, Chad, and Basutoland (now Lesotho) received very little indeed by way of infrastructural development. Moreover, local economies often suffered because of the loss of labor normally used for food production.

A notable example of the use of forced labor

FIGURE 8.7. Steam-powered boats, Kinshasa, Belgian Congo (now the Democratic Republic of the Congo), circa 1930. The one boat is loaded with sacks of produce, quite likely brought from up-country for export. Photo: E. Nogueira.

occurred in the construction of the 450-km-long Congo–Brazzaville railway in French Equatorial Africa between 1921 and 1934. The project was extremely difficult because of the forested, mountainous terrain crossed by innumerable large rivers, and because of horrid working conditions—sweltering heat, torrential rains, and frequent epidemics of malaria and other diseases. Of the more than 120,000 African workers (virtually all forcibly recruited) who built the railway, it is estimated that half perished. One-fifth of the unwilling recruits came from the Sara people of Southern Chad. All of this suffering and the loss of 10,000 young men contributed nothing to their region's development; the railway ended 1,200 km from the Sara homeland.

Evaluating the Colonial Legacy

Africa's colonizers saw themselves as benign interlopers, carrying the torch of civilization to a primitive, "dark" continent. They emphasized the imposition of peace among warring rivals; the introduction of Western medicine and education; the transformation of African economies through the development of mining and of commercial agriculture; and the creation of legal and administrative systems. The development process was seen as difficult and often frustrating, but ultimately rewarding because of the many benefits that the "civilizing mission" brought to Africans.

Writers within the modernization perspective, which prevailed in development studies during the 1960s, continued to see the colonial legacy, as well as the path of development in the postcolonial era, in much the same way. This is illustrated in the following excerpt, which portrays colonial officials and entrepreneurs as "eager beavers," diligently creating opportunity and transforming African society:

Roads and railways link the administrative nodes and provide, in turn, channels through which modernizing innovations seep. . . . Modernization in all its innovative aspects is distributed from the major sources through the tarred arteries and laterite capillaries of the land and society.

Some innovations have economic implications—coffee revolutionizes Kilimanjaro, sisal spreads around Tanga and Morogoro, cotton seeps through Sukumaland—and commodity flows swirl through the road network, feeding back information to the administrators who upgrade, realign, and tar the dirt tracks of the previous year. (P. Gould. "Tanzania 1920–63: The spatial impress of the modernization process." *World Politics*, vol. 22 [1970], pp. 149–170.)

Writers from the dependency school challenged vigorously the view that the essence of the colonial legacy was the dynamic transformation of the "blank map" of Africa. They pointed to the undermining of indigenous economies under colonialism (see Figure 8.8) and societies, and argued that whatever changes had taken place were implemented primarily for the benefit of Europeans, not Africans. Colonialism developed Europe and underdeveloped Africa. Walter Rodney's 1974 book *How Europe Underdeveloped Africa* (see "Further Reading") was particularly influential in the radical reinterpretation of the colonial era. Rodney argued:

Colonial Africa fell within that part of the international capitalist economy from which surplus was drawn to feed the metropolitan sector. As seen earlier, exploitation of land and labor is essential for human advance, but only on the assumption that the product is made available within the area where the exploitation takes place. Colonialism was not merely a system of exploitation, but one whose essential purpose was to repatriate the profits to the so-called "mother country." From an African viewpoint, that amounted to consistent expatriation of surplus produced by African labour out of African resources. It meant the development of Europe as a part of the same dialectical process in which Africa was underdeveloped. (p. 162)

Although more recent evaluations of the colonial legacy tend to be subtler than Rodney's, the interpretation is usually the same. Europe may have transformed Africa, but the process was far from unproblematic. Colonialism built, but it also destroyed, inflicting profound damage on indigenous societies and the environment. Whatever benefits colonial Africa realized from, for example, the introduc-

FIGURE 8.8. Indigo dyeing, Kano, Nigeria. Kano's textile industry, like most precolonial industries, was pushed toward extinction by the importation of European manufactured cloth. Photo: author.

tion of modern medicine were trivial compared to the benefits reaped by Europe from the exploitation of African labor and resources. Moreover, far from being an altruistic undertaking, the colonial project was underpinned by a racist ideology and enforced by using Europe's superior military technology to maintain dominance over the African people.

It is perhaps ironic that it took European colonialism to inform Africans that they were African. European powers created not only the map of Africa as we know it, but also a sense of common identity among Africans, in reaction to the domination and humiliation of colonialism. This sense of identity was manifested in a continent-wide movement for independence and calls in some quarters for an African political union. Ironically, these calls for union were frustrated by other African leaders, who insisted that the political map of independent Africa should conform closely to that established under colonialism.

Further Reading

General studies of the history of colonialism and its effects on African societies include the following:

Amin, S. "Underdevelopment and dependence in black Africa: its origins and contemporary forms." *Journal of Modern African Studies*, vol. 10 (1972), pp. 503–524.

Fanon, F. *A Dying Colonialism*. New York: Grove Press, 1967.

Freund, B. *The Making of Contemporary Africa: The Development of African Society since 1800*, 2nd ed. Boulder, CO: Lynne Rienner, 1998.

Gann, L. H., and P. Duignan, eds. *Colonialism in Africa, 1870-1960*, 5 vols. London: Cambridge University Press, 1969–1975.

Maddox, G. *The Colonial Epoch in Africa*. New York: Garland Press, 1993.

Rodney, W. *How Europe Underdeveloped Africa*. Washington, DC: Howard University Press, 1974.

The implementation of colonial rule varied considerably among the colonial powers and reflected the circumstances encountered in particular regions. The following are selected studies of various forms of colonial rule:

Berman, B. *Control and Crisis in Colonial Kenya: The Dialectic of Domination*. London: James Currey, 1990.

Clarence-Smith, G. *The Third Portuguese Empire*. Manchester, UK: Manchester University Press, 1985.

Hochschild, A. *King Leopold's Ghost: A Story of Greed, Terror, and Heroism in Colonial Africa*. Boston: Houghton Mifflin, 1998.

Miles, W. F. S. *Hausaland Divided: Colonialism and Independence in Nigeria and Niger.* Ithaca, NY: Cornell University Press, 1994.

Suret-Canale, J. *French Colonialism in Tropical Africa, 1900–1945.* New York: Universe, 1971.

There is a large literature on the transformation of African economies and societies under colonialism in particular settings. See, for example, the following sources:

Arrighi, G. "Labour supplies in historical perspective: A study of the proletarianization of the peasantry in Rhodesia." *Journal of Development Studies*, vol. 6 (1970), pp. 197–234.

Azevedo, M. "The human price of development: The Brazzaville railroad and the Sara of Chad." *African Studies Review*, vol. 24 (1981), pp. 1–19.

Chimhundu, H. "Early missionaries and the ethnolinguistic factor during the invention of tribalism in Zimbabwe." *Journal of African History*, vol. 33 (1992), pp. 87–110.

Clark, L. *Through African Eyes: Cultures in Change.* New York: Praeger, 1970.

Shenton, R. W. *The Development of Capitalism in Northern Nigeria.* Toronto: University of Toronto Press, 1986.

The "Social History of Africa" series, published by Heinemann, has a wealth of studies that provide insights into the impress of colonialism in specific settings. Books in the series include these:

Echenberg, M. *Black Death, White Medicine: Bubonic Plague and the Politics of Public Health in Senegal, 1914–1945.* Portsmouth, NH: Heinemann, 2001.

Isaacman, A. *Cotton Is the Mother of Poverty: Peasants, Work, and Rural Struggle in Colonial Mozambique, 1938–1961.* Portsmouth, NH: Heinemann, 1995.

Sikainga, A. *"City of Steel and Fire": A Social History of Atbara, Sudan's Railway Town, 1906–1984.* Portsmouth, NH: Heinemann, 2002.

Internet Sources

In addition to the general Internet sources listed at the end of Chapter 7, see the following:

Central Oregon Community College. *African Timelines* [see especially Parts III and IV]. www.cocc.edu/cagatucci/classes/hum211/timelines/htimelinetoc.htm

Smithsonian National Museum of Natural History. *African Voices.* www.mnh.si.edu/africanvoices

9

Independent Africa: The Struggle Continues

The year 1960 was an important watershed in the history of Africa south of the Sahara: The number of independent countries increased in that year from 5 to 22. Nevertheless, it is important to put 1960 into perspective. The events of that year would have been impossible but for decades of determined resistance to colonial rule. Moreover, 1960 was only a beginning. It took many more years before countries like Angola, Zimbabwe, and Namibia were able to achieve independence; in South Africa, the struggle for majority rule was not completed until 1994. (Figure 9.1 illustrates the time course of the transition to independence in Africa.) Finally, gaining independence has proved to be only one step along an immensely difficult path toward stability and development.

The Struggle for Independence

The struggle for independence proceeded on two fronts. On the one hand, there was an intellectual battle against colonial rule, led initially by a number of Africans studying in Europe and the United States who became involved in the Pan-African movement of W. E. B. Du Bois and Marcus Garvey. Pan-Africanism and its credo, "Africa for the Africans," were advanced first through a series of six international meetings held between 1900 and 1946, and later through the development of the Organization of African Unity (now the African Union; see Vignette 9.1). The second front in the struggle against colonialism consisted of an uncoordinated but sustained pattern of armed and passive resistance mounted by ordinary Africans. Later, and especially after World War II, these two forms of resistance merged. Intellectual leaders such as Kwame Nkrumah in the Gold Coast (Ghana) and Nnamdi Azikiwe in Nigeria returned to Africa and established political parties, trade unions, and independent newspapers that worked for closer cooperation among those involved in the fight for justice and self-determination.

Although the achievement of independence in African nations was sometimes as dramatic as Togo's independence monument (Figure 9.2) suggests, the actual paths of the independence struggle varied greatly—depending, among

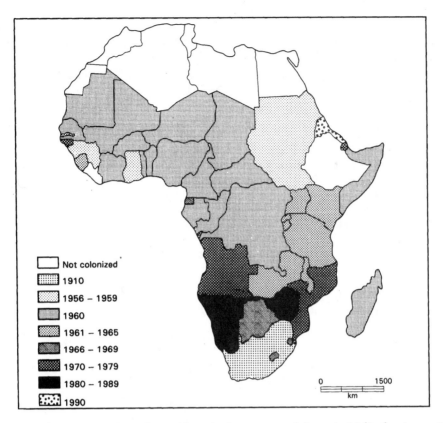

FIGURE 9.1. Transition to independence. Note the importance of the early 1960s for the attainment of political independence.

other things, on the nature of the local colonial state and the extent to which independent thinking and dissent were tolerated. In a few countries, the quest for independence involved protracted armed struggle against the colonizers. In approximately one-quarter of the colonies, large organized campaigns of political agitation, protest, and civil disobedience preceded independence. Elsewhere, political organization was less advanced; independence tended to come as a result not of intense local pressure, but of changing colonial policies in response to events elsewhere. Several of these themes are evident in the following sketches outlining the path to independence in five colonies.

Gold Coast

In the Gold Coast, the struggle for independence began in earnest in 1947 following the return of Nkrumah from the United States, where he had studied and was active in the Pan-African movement. After joining the United Gold Coast Convention (UGCC), he organized a campaign of passive resistance, with self-government as the ultimate goal. In 1948, the violent suppression by the police of a peaceful demonstration of ex-servicemen, followed by the jailing of Nkrumah and other leading citizens, served only to radicalize the movement for self-government. After his release from prison, Nkrumah left the UGCC to form a new political organization, the Convention People's Party (CPP), and to pursue a more militant strategy to achieve "self-government now." Nkrumah was again imprisoned, but the CPP still won a resounding 90% of the vote in municipal elections in 1951. Recognizing the implications of this mandate, the British granted internal self-government, followed in 1957 by full independence.

VIGNETTE 9.1. From Organization of African Unity to African Union

The Pan-African movement came home to Africa with the launching of the First Conference of Independent African States, held in Accra in April 1958. Representatives from Ghana, Liberia, Ethiopia, and Sudan, as well as four North African states, were present. The movement was guided by nine principles:

1. Africa for the Africans and total rejection of colonialism
2. Political unification (United States of Africa) as a goal
3. A renaissance of African morale and culture
4. African nationalism supplanting the tribalism of the past
5. Economic development according to socialist principles
6. Democracy as the most desirable method of government
7. Rejection of violence, except to resist military repression
8. Solidarity of black peoples everywhere
9. Positive neutrality, except in regard to African interests

For Ghana's Kwame Nkrumah, national independence was only a necessary first step toward the ultimate objective of African unity. In an attempt to put theory into practice, Ghana, Guinea, and later Mali agreed to unite in 1959. However, this union agreement was never implemented, and other African states increasingly expressed concern about the need to protect their sovereignty.

With many states achieving independence in 1960, the question of unity became much more complex. Two major groups emerged in 1961: the Casablanca powers and the Monrovia powers. The Casablanca group brought together seven of the more radical states, led by Ghana. Four months later, representatives from 20 states met in Monrovia and put forward an alternate vision of Africa's future. This group included delegates from a dozen former French colonies and eight other states, the most prominent being Nigeria and Ethiopia. They argued strongly against formal political union, focusing instead on the need for a mutual recognition of equality and sovereignty; noninterference in each other's internal political affairs; and cooperation in economic, cultural, and scientific fields.

Nigeria, Guinea, and Ethiopia took the lead in bringing the two sides together at the first successful summit of African leaders at Addis Ababa in May 1963. Nkrumah renewed his call for a revolutionary union of all African states, modeled on the United States or the Soviet Union, but was rebuffed. Instead, delegates were inspired by the appeal of Algeria's Ben Bella to set aside bickering and unite in support for Africans still struggling to achieve independence. The outcome of the meeting was the signing of the charter for the Organization of African Unity (OAU).

In its four-decade history, the OAU had its share of successes and failures. It played an active role in the struggles against Portuguese colonialism and apartheid, and was instrumental in mediating several disputes between neighboring states. However, it proved itself incapable of providing any real leadership on such issues as Africa's slide into debt bondage, or reconciliation in recent civil wars in countries such as Somalia, Sudan, Rwanda, and Sierra Leone. Its influence was limited by the strong tradition of noninterference in internal affairs, and its leadership was often weak.

In July 2002, African leaders met in South Africa to inaugurate the African Union as a "reborn" OAU. They spoke optimistically of a more proactive organization designed to respond to the challenges facing Africa in the 21st century. One of the cornerstones of the new Union will be the New Economic Partnership for Africa's Development (NEPAD), with the objective of strengthening governance and achieving greater economic development. The African Union faces several major challenges, among them leadership (the same cast of national leaders, some of them seriously flawed, remains in power) and a lack of serious commitment by the world's richest countries to bring about fundamental changes in trade and aid relationships to facilitate African development.

Côte d'Ivoire

The path to independence was rather less dramatic in Côte d'Ivoire. Félix Houphouet-Boigny, a wealthy farmer, became the leading Ivoirian politician and served in several French cabinets during the 1950s. He supported a gradual transition to self-government and strongly opposed the aspirations of several other leaders from French West Africa for a rapid transition to independence and a federation of West African states. Fearing that its wealth would be used to subsidize the poorer Sahelian regions of French West Africa, Côte d'Ivoire opted for independence in 1960 to forestall any move toward a regional federation. Houphouet-Boigny remained president for over 30 years, until his death at age 88 in 1993. Throughout his years in power, he remained perhaps Africa's staunchest advocate of conservative positions on economic, political, and social matters.

Kenya

Kenya had a large number of white settlers who farmed much of the best land and used their control of the colonial state to advance

FIGURE 9.2. Independence monument, Lomé, Togo. This impressive monument portrays independence as breaking the chains of bondage. Photo: author.

their own interests. British military expeditions succeeded in quelling armed resistance between 1896 and 1905, but Kenyans continued to engage in passive resistance and staged many organized protests against land, labor, and taxation policies of the colonial state. Political organizations, principally the Young Kikuyu Association and after 1944 the Kenya African Union, spearheaded the protests. From 1952 to 1958, the Mau Mau secret society launched guerrilla attacks on white settlers and black allies of the British. The British responded with a series of repressive measures to end the rebellion and jailed thousands of Africans. Although the Mau Mau revolt did not result in immediate independence, it convinced the British that change was inevitable. In 1963, Kenya became independent under the leadership of Jomo Kenyatta, who had been imprisoned for his alleged involvement in the Mau Mau.

Namibia

Namibia achieved independence only in 1990 after a quarter-century of armed struggle by the South West African People's Organization (SWAPO). The German colony of South-West Africa was made a South African trust territory after World War I. South Africa virtually annexed the territory in 1949, and it proceeded to implement many elements of apartheid.

Starting in the 1940s, several major strikes were launched by African workers to protest social and labor conditions. SWAPO began its war of independence in 1966, the same year that the United Nations declared South African rule illegal. Guerrilla activities were concentrated in the northern part of the country near the border with Angola, where SWAPO had established several bases. South Africa waged a protracted and bloody counterinsurgency campaign in both Namibia and Angola, but could not defeat SWAPO. Meanwhile, there was growing international pressure on South Africa to get out of Namibia. South Africa finally agreed in 1989 to hold elections leading to independence. Their strategy of last resort was to work to elect a compliant regime that would not seriously challenge South African interests. SWAPO won a convincing victory, however, and shortly thereafter Namibia gained its hard-won independence.

Eritrea

Africa has had numerous secessionist movements, among the most important being the attempted secessions of Katanga from the Congo-Kinshasa (later Zaire and now the Democratic Republic of the Congo) in 1960 and 1961, Biafra from Nigeria in the late 1960s, and Somaliland from Somalia commencing in the early 1990s. However, Eritrea represents the only country to have succeeded in winning internationally recognized independence in this way. Formerly a colony of Italy, Eritrea was awarded to Ethiopia by the United Nations in 1952. The armed struggle for independence began in 1961 and lasted three decades. The Eritrean People's Liberation Front continued to control all but the largest towns, despite massive, protracted campaigns by the Soviet-backed Ethiopian military to crush the rebellion. The collapse of the Ethiopian government in 1991 resulted in large part from the debilitating effects of its unsuccessful Eritrean campaigns. The new regime adopted a more conciliatory stance and agreed to a referendum on independence. The referendum of April 1993 produced a 99% vote in favor of independence, setting the stage for Eritrean independence a month later. After five years of mostly peaceful coexistence, Eritrea and Ethiopia went to war in 1998 over a disputed border area. The two-year war cost a total of 100,000 lives, displaced 750,000 Eritreans from their homes, caused much destruction and economic disruption, and diverted huge sums from development into war-related expenditures.

The Struggle for Responsible Government

Throughout the world, relatively few governments can be said to be truly representative and responsive to the needs and aspirations of their citizens. This dearth of responsible governments is reflected in the small number of democratically elected regimes, the many military

dictatorships, and the brutal terrorism that certain regimes have inflicted on their own citizens. Much of the weakness in African political institutions can be traced to the colonial period, particularly the transition to independence. Nowhere had Africans had a truly effective voice either in government or in administration. Moreover, there were concerted efforts by colonial powers to hand-pick new leaders who would not seriously challenge the interests of the metropole. Independence came quickly, with little advance preparation and with unworkable European-inspired constitutions.

In the Belgian Congo (again, now the Democratic Republic of the Congo), political chaos erupted almost immediately after the Belgians precipitously declared it independent. The United Nations intervened to end the attempted secession in Katanga Province, but several more years of violence ensued. In many Western circles, the Congo instantly became the symbol for Africa's unreadiness for independence—the evidence that chaos and corruption would be the hallmark of independent states. The political history of what then became Zaire under the military dictatorship of Mobutu Sese-Seko, who ruled from 1965 to 1997, is encapsulated in the following quotation: "The Mobutu regime drifts in a sea of corrupt incompetence unable to defend itself, propped up by the United States and France who are intent on safeguarding their own political and economic interests" (I. L. L. Griffiths. *An Atlas of African Affairs*. London: Methuen, 1984, p. 75). This reference to foreign support points to a telling characteristic of many authoritarian regimes—namely, that their hold on power is as dependent on external backing as it is on local support. In Zaire, foreign troops intervened three times to quell anti-Mobutu uprisings.

Following independence, African regimes faced a rising tide of expectations with few resources and little time to respond. In these circumstances, many opted for single-party states, believing that interparty maneuvering was a luxury they could not afford at such a crucial time in their history. Institutions such as the military, trade unions, and civil service were typically brought under closer control, and decision making became increasingly concentrated in a few hands. In several countries, the course of development was closely controlled by a charismatic leader such as Nkrumah or Kenyatta, whose legitimacy was based on leadership of the anticolonial struggle.

The absence of a political opposition has provided a ready excuse for sections of the military to intervene and overthrow allegedly ineffective or corrupt regimes. Between 1960 and 2002, there were over 85 successful coups d'état in 35 countries and many more attempts that did not succeed (see Figure 9.3). The record of Africa's military governments has been as mixed as that of its civilian regimes. Some military regimes have provided relatively stable governments and have managed to balance regional, ethnic, and religious interests more effectively than the civilian regimes that they replaced. Other military regimes—exemplified by Idi Amin's regime in Uganda—have been thoroughly inept and have relied on terror to retain power.

Despite the preeminent role of the military in bringing about change, not all transfers of power have been accomplished by force. Ghana and Nigeria, among other countries, have had military regimes that relinquished power to elected civilian governments; in these countries, however, there has been a tendency for democracy to be subverted by renewed military intervention. Some countries, including Senegal and Tanzania, have had orderly transitions of power from one regime to another and no military coups.

Many African countries have had, and continue to have, one-party rule. Generally, the dominance of the ruling party is enshrined in the national constitution. The legitimacy of these regimes has often been a source of disagreement between African and Western observers. Leaders such as Julius Nyerere argued that one-party rule encouraged all citizens to work together toward common development goals, whereas they contended that multiparty systems encouraged ethnic and regional factionalism. The ruling parties of several one-party

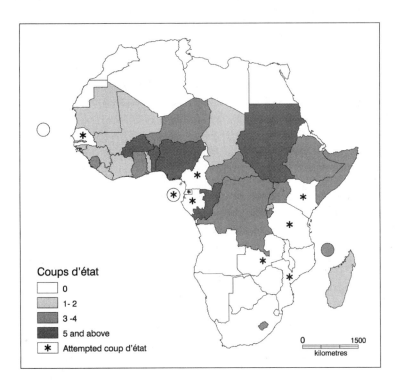

FIGURE 9.3. Successful coups d'état. Coups have been much less common in East and southern Africa than elsewhere in the continent. Primary data source: C. Cook and D. Killingray. *African Political Facts since 1945*, 2nd ed. London: Macmillan, 1991. Updated to 2002.

states, including Tanzania under Nyerere and Uganda under Yoweri Museveni, were linked to the grassroots through a well-developed network of community councils. Parliamentarians and party officials were elected from slates of party members. Other one-party states were dictatorships, with few trappings of popular mobilization or participation in the political process. Examples include Hastings Kamuzu Banda's government in Malawi (1964–1994) and Houphouet-Boigny's regime in Côte d'Ivoire (1960–1993).

The 1990s brought significant political change to the political landscape. Responding to internal crises and to international events such as the replacement of communist regimes in Eastern Europe, more and more Africans began to agitate for responsible, democratically elected governments. In the majority of countries, those in power have agreed, often reluctantly, to permit opposition groups to organize and engage in political activities. Incumbents have been voted out of office in several countries, including Zambia, Cape Verde, Benin, and Madagascar. The number of democratically elected governments continued to grow throughout the 1990s.

The progress toward democratization is exemplified by changes that have occurred since the early 1990s in southern Africa (see Vignette 9.2 and Figure 9.4). Prior to the demise of apartheid, South African efforts to destabilize the governments of neighboring countries retarded the development of democratic institutions in these countries. Military intervention by South Africa and guerrilla warfare supported by South Africa caused a great many deaths, forced millions to flee their homes, destroyed economic and social infrastructures, and paralyzed activities throughout the countryside. The case of southern Africa reminds us that democratization may reflect not only factors internal to a particular country but also those that are external—namely, relations with neighboring countries, former colonial masters, world superpowers, and international financial institutions.

The struggle for democracy has been less successful in other countries. In Zaire, Mobutu was toppled as a result of a rebellion led by Laurent Kabila in 1997. However, the change of regime brought neither peace nor competent governance to the country, which Kabila renamed the Democratic Republic of the Congo. Most of the eastern and northern Congo was captured by rebels sponsored by Rwanda and Uganda, and Kabila's tenuous hold on power was only assured by the presence of foreign troops from Zimbabwe, Angola, and three other countries. After Laurent Kabila was assassinated in 2001, his son assumed power. The Democratic Republic of the Congo is one of several examples of countries where the state has ceased to have the capacity to govern.

Figure 9.5 provides an overview of the state of freedom in African nations as of 2002, according to the human rights group Freedom House. Freedom House assesses freedom according to a range of indicators of political freedom and civil liberties, and aims to reflect actual conditions rather than hypothetical rights as stated in constitutions and official pronouncements. Countries are listed as "free," "partly free," or "not free." Clearly, any such index is subject to disagreement about its relevance and accuracy. The record of change in recent years has been mixed: Some 8 countries regressed politically during the 1990s, while 22 others became more democratic. Nevertheless, if one compares the state of political freedom in 2002 to the situation 30 years before, there has been great progress. Whereas only 5% of countries were categorized as free in 1972, this figure had risen to 23% in 2002. The proportion of countries listed as not free had declined from 72% in 1972 to 31% in 2002. Even in many of the countries shown as only partly free, there has been progress toward more open, democratic societies.

When assessed in the context of other factors at the same time—weak and stagnant economies, declining development assistance, and hardship caused by structural adjustment policies—Africa's progress toward democratization during the 1990s has been noteworthy. Unfortunately, the struggle for democracy may ultimately yield little in the way of concrete results. The record of failure over four decades of rule by countless African governments provides little encouragement that future democratically

VIGNETTE 9.2. A Democracy Dividend?

The apartheid regime in South Africa worked tirelessly to weaken governments in the region. South African destabilization attacked democracy in various ways: Political leaders were assassinated, and South African troops launched invasions. Attacks by South-African-backed guerrillas not only destroyed economic and social infrastructure, but also thwarted the development of civil society institutions. Resources were diverted from development into defense spending, and governments were given a pretext for the introduction of repressive measures.

What has happened to the state of democracy in southern Africa since the fall of apartheid? Has South Africa's political transition created conditions that are conducive to democratization at home and in other countries in the region? What follows is a brief synopsis of trends in democratization:

- South Africa itself has been widely praised for the way it has overcome the legacy of apartheid to create a very liberal constitution, and to build responsive democratic institutions that seem to have become more robust with time.
- Lesotho has been a democratization success story; over the past decade it has gone from being classified by Freedom House as not free, to partly free, and in 2002 to free. Under apartheid, the Lesotho government was repeatedly harassed by South African threats and incursions, which succeeded in undercutting democracy.
- Botswana has maintained strong democratic institutions since independence. Occasional South African interventions did not succeed in undermining Botswana's democracy. Namibia has also been classified as free since it achieved independence in 1990.
- Swaziland is considered by Freedom House to be not free. The Swazi monarchy has resisted calls to permit full electoral democracy in the country.
- Mozambique has benefited greatly from the end to South Africa's protracted support of the Renamo guerrilla movement, which for more than a decade devastated every aspect of its society and economy. Following the signing of the peace accord in 1992, free elections were held (Figure 9.4), and there has since been sustained progress in the strengthening of democratic institutions.
- In Angola, the civil war that began at the time of independence in 1975 continued until 2002, a decade after South African destabilization came to an end. The country has only

(cont.)

FIGURE 9.4. Women waiting to vote in Mozambique's first postconflict election, in 1994. Photo: CIDA (B. Paton).

VIGNETTE 9.1. (cont.)

begun to rebuild its society and economy, but there is greater reason for optimism now than at any point since independence.

- Zimbabwe has attracted much attention in recent years for the erosion of its democratic institutions. While elections are contested, President Mugabe's government has abused press freedom, undermined judicial independence, and persecuted its political opponents.
- During the 1990s, both Zambia and Malawi had elections that resulted in peaceful transfers of power to opposition groups. There has been a strengthening of democratic institutions over the past decade, albeit with some setbacks.

Although the record of democratic change since the early 1990s has been mixed, there has been significant progress toward the development of more democratic institutions in the region. Of the 10 countries listed above, 5 made notable progress toward democratization soon after the political changes took place in South Africa. It is only in Zimbabwe that democratic institutions have been eroded over the past decade; these changes were independent of events in South Africa.

Data source for freedom ratings: Freedom House. *Freedom in the World 2002*. New York: Freedom House, 2002.

elected regimes will achieve significant progress. All governments—civilian or military, one-party or multiparty—face the daunting task of addressing the urgent expectations and needs of a growing population with limited and generally declining resources.

The Struggle for Development

The time of independence was characterized by an optimism that reflected economic trends and ideas about progress not only in Africa, but also in other parts of the world. During the 1950s, production and consumption increased greatly in North America and Europe, spurred by postwar reconstruction and a mood of optimism. Demand for African raw materials rose; so too did the prices of these commodities. A wave of new investment led to the opening of mines and plantations, and in some colonies led to the first industries' producing goods for local markets. The Alliance for Progress was launched in Latin America, with the promise of an economic revolution like the one achieved by the Marshall Plan in Europe. It was confidently

predicted that Africa would now be able to follow this same path toward development.

Needless to say, results were often very discouraging. Commodity prices leveled off or fell, and demand for a number of products failed to keep pace with rising production. Moreover, the attempts of many governments to stimulate growth and to diversify their economies—for example, by establishing industries—were ill conceived and ineptly implemented. More and more coffee (or other commodities) from more and more countries simply brought about lower prices and created large unsold surpluses. Development initiatives were also often severely handicapped by the lack of basic infrastructure, such as a good road network, a reliable and cheap supply of electrical energy, and an educational sector capable of training skilled people of all kinds.

African countries have experimented since independence with several models of political, economic, and social development. These approaches may be grouped into three broad categories: (1) capitalism, (2) populist socialism, and (3) Afro-Marxism. These strategies are outlined briefly at this point and are then de-

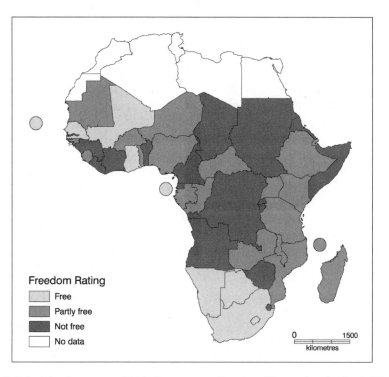

FIGURE 9.5. Political freedom score, 2002. Data source: Freedom House. *Freedom in the World 2002.* New York: Freedom House, 2002.

scribed in more detail in Chapter 25. For a variety of reasons, especially the intervention of global financial institutions and the growing assertiveness of major aid donors, the development policies of African states have become increasingly alike since 1990. African socialism and Afro-Marxism represent ideologies that were followed in several countries prior to 1990, and although many legacies of these past approaches are still readily apparent today, these ideologies are no longer in current practice.

Prior to independence, modern economies were organized on capitalist principles. Major investments in trade, commercial agriculture, resource extraction, and manufacturing had been undertaken during the colonial period by large European trading companies. Independent entrepreneurs from Europe, the Middle East, and the Indian subcontinent were very active in the agricultural and trading sectors—often with great success, as the growth and diversification of Asian enterprises in East Africa show. In many colonies, African capitalists expanded the range and size of their enterprises; indigenous entrepreneurs achieved particular success in Nigeria, Ghana, and Senegal.

Countries such as Côte d'Ivoire, Kenya, and Malawi followed an explicitly capitalist development strategy. They emphasized economic growth rather than equity as a primary development objective. Compared to other African countries, they developed quite open economies. Private investment, both domestic and foreign, was generally welcomed. Still, there have been limits to this openness. Capitalist governments often became partners with capitalist investors in joint ventures; and in Kenya and several other countries, legislation was passed to closely regulate the economic role of entrepreneurs of non-African origin.

Several countries searched for an alternative to capitalism, which was criticized as being exploitative and incompatible with African traditions of cooperative production and communal ownership of land. African socialism was proposed as a strategy that would bring about more equitable development. The state assumed a dominant role in the economies of Ghana, Tanzania, and other African socialist countries. The role of capitalist countries was severely limited, and parastatal companies were formed to undertake high-priority economic projects. Priority was given to rural development, with the objective of reducing social and economic disparities between city and countryside.

Mozambique, Angola, and Ethiopia, among others, adopted Afro-Marxism as a state ideology. In several of these countries, this commitment to Marxism was forged in the context of armed struggle for independence. Economic strategies aimed at gaining control of the "commanding heights" of the economy through state direction. State farms and state factories were organized on the lines of enterprises in the Soviet Union and other communist states. For the most part shunned by the West, Afro-Marxist states became very dependent on trade and aid linkages with communist states. However, a number of factors have doomed Afro-Marxism as an official ideology, including its failure to bring about real economic progress, the demise of communism in Europe and the Soviet Union, and intense pressure for economic liberalization as a precondition for loans and development aid.

Africa's failure to "develop," let alone to keep pace with growth in other parts of the world, is widely recognized (and poignantly illustrated in Figure 9.6). Yet there does not seem to be the political will on the part of either the world's major economic powers or the majority of African states to break the current cycle of underdevelopment. This state of indecision was well illustrated in 2002 when African leaders, at the behest of the West as a condition for increased development assistance, created a proposal for a New Economic Partnership for Africa's Development (NEPAD). The NEPAD strategy was filled with optimistic generalities, but it sidestepped some key issues such as HIV/AIDS and contained little that was bold and new. Similar criticisms apply to Western leaders, who failed to commit themselves to the much-increased support for African development that they had been saying was necessary. In spite of all their rhetoric of concern, Africa had again been deemed too peripheral to the major interests of the West to warrant meaningful attention.

The Struggle Ahead

African countries can point to some notable accomplishments in the years since independ-

FIGURE 9.6. Photographer's signboard, Hadejia, Nigeria. The name "Struggling Man Joe" epitomizes the daily struggle of hundreds of millions of ordinary Africans to survive and to progress. Photo: author.

ence. All countries have greatly expanded their health and education systems, with greatly improved accessibility and a far greater range of services than those available in 1960. For Mozambique and Angola, the political changes in South Africa in the 1990s created new opportunities for internal peace, with the end of the long-standing fight against destabilization by South Africa (see Figure 9.7b). In South Africa itself, the painful process of healing the deep wounds left by racism and apartheid was pursued vigorously under the charismatic leadership of Nelson Mandela.

Nevertheless, Africa has increasingly come to be seen as a continent in crisis. The typical African state has been severely weakened by its declining ability to address the basic needs, much less the aspirations, of its citizens. With the stagnation or decline of both export and aid revenues and the growing cost of debt servicing, few governments have the means to properly maintain existing services and infrastructures, much less to commence significant new development initiatives. Many nations, such as Ethiopia, remain debilitated by past and present regional conflicts (see Figure 9.7a). Africa's struggle with the HIV/AIDS epidemic, which had infected 30 million Africans as of 2002, has further contributed to the sense of crisis. The impact of the disease extends far beyond morbidity and mortality. AIDS is having a devastating economic and social impact on households, communities, regions, and nations.

Although the intervention of the World Bank and the International Monetary Fund (IMF) has provided some immediate support for African states in trouble, their assistance has come with a very high price. The structural adjustment measures that they have imposed—including currency devaluation, severe cutbacks in social spending, the introduction of user fees for education and health, and the removal of food subsidies—have been immensely unpopular with the public and in some cases have led to strikes and protests by frustrated citizens. In several countries, governments were forced to adopt policies that were fundamentally opposed to their own long-established strategies and philosophies of development. Even where structural adjustment has been seen to be successful—in Ghana and Uganda, for example—the gains have been modest. The relation between African governments and international financial institutions has been seen (justifiably) as a new form of neocolonialism, more concerned with debt repayment than with African development.

In a controversial 1994 article entitled "The Coming Anarchy" (see "Further Reading") Robert Kaplan painted a picture of a chaotic Africa that was increasingly unable to function. Kaplan's vision bears little resemblance to the reality of most of the continent. Nevertheless, in several nations, the state has essentially ceased to exist. Somalia has not had an effective national government for more than a decade; power has been exercised by numerous regional warlords, and in the north by unchallenged secessionist regimes in Somaliland and Puntland. Liberia and Sierra Leone went through prolonged periods during the 1990s without an effective central government, and while Sierra Leone is relatively peaceful as this book goes to press (though the structures of state remain extremely fragile) Liberia's condition remains chaotic. The Democratic Republic of the Congo has endured close to a decade of civil war, with heavy involvement by several outside powers. Where there has been a collapse of the state, organized economic activity has come to a virtual standstill, countless civilians have died, and millions have endured great hardship. These examples of the collapse of the state are cause for concern, because they show the potential effects of the West's long-standing failure to provide adequate support for African development and democratization. These cases also demonstrate that national disputes may easily lead to regional destabilization. For example, consider the disruptive effects of Liberia's civil war on its neighbors, Sierra Leone, Côte d'Ivoire, and Guinea.

In the prevailing environment of pessimism about Africa's current situation and future prospects, the continent's success stories are too often forgotten. Botswana is a prime example. When it achieved independence in 1966, it was one of Africa's poorest countries, with virtually

FIGURE 9.7. African countries have too often served as political pawns in regional conflicts and superpower struggles. (a) Tank graveyard near Massawa, Ethiopia (now Eritrea). For Ethiopia, wars have meant untold misery and a tragic diversion of scarce resources from development needs. (b) Government troops on patrol in Mozambique, late 1980s. South-African-sponsored destabilization wrought havoc on Mozambique's society and economy for almost two decades, although this has come to an end with the abolition of apartheid. Photos: (a) M. Peters; (b) CIDA (B. Paton).

no export economy and a small population dependent on earnings as labor migrants in South Africa. Several major mineral discoveries changed its economic prospects; Botswana's per capita income of $3,300 per year is the highest in Africa. Botswana's successes are not limited to the economic sphere; since independence, it has remained a healthy democracy with strong public administration.

The primary challenge for Africa in the first years of the new millennium is to learn from successes such as those of Botswana and other countries, including Uganda. Uganda has demonstrated that effective health promotion strategies, backed by firm government commitment to combat HIV/AIDS, can bring about significant reductions in infection rates within a few years. Countries such as Nigeria and Ghana, where the HIV/AIDS epidemic is at a relatively

early stage in its development, have the opportunity to apply the lessons of Uganda to slow its development and (it is hoped) to avert the massive infection rates that have occurred in southern Africa.

The attainment of independence was a major accomplishment for Africa, particularly in countries where armed struggle was needed to bring about decolonization. In the subsequent struggle for stability and development, there have been some successes and many disappointments. Recent trends, particularly the involvement of the IMF and World Bank, have eroded African sovereignty. During the early 1970s, the motto of the independence movements in the colonies then ruled by Portugal was "*a luta continua*"—"the struggle continues." This motto epitomizes the situation that still prevails in Africa south of the Sahara.

Further Reading

On the struggle for independence, see the following sources:

Kenyatta, J. *Facing Mount Kenya*. London: Secker and Warburg, 1953.

Legum, C. *Pan-Africanism: A Short Political Guide*. New York: Praeger, 1965.

Nkrumah, K. *I Speak of Freedom*. New York: Praeger, 1961.

Wallerstein, I. *Africa: The Politics of Independence*. New York: Vintage, 1961.

The following sources provide useful overviews of the African situation at various points in time:

"Africa: A Generation after Independence." Special theme issue of *Daedalus*, vol. 111, no. 2 (1982).

Crowder, M. "Whose dream was it anyway?: Twenty-five years of African independence." *African Affairs*, vol. 86 (1987), pp. 7–24.

Davidson, B. *The Black Man's Burden: Africa and the Curse of the Nation State*. New York: Times Books, 1992.

Legum, C., I. W. Zartman, S. Langdon, and L. Mytelka. *Africa in the 1980s: A Continent in Crisis*. New York: McGraw-Hill, 1979.

Mamdani, M. *Citizen and Subject: Contemporary Africa and the Legacy of Late Colonialism*. Princeton, NJ: Princeton University Press, 1996.

Ominde, B. *A Political Economy of the African Crisis*. London: Zed Books, 1988.

Much has been written about African governance. See, these sources for example:

Harbeson, J., D. Rothchild, and N. Chazan, eds. *Civil Society and the State in Africa*. Boulder, CO: Lynne Rienner, 1994.

Hyden, G., and M. Bratton, eds. *Governance and Politics in Africa*. Boulder, CO: Lynne Rienner, 1992.

Joseph, R., ed. *State, Conflict, and Democracy in Africa*. Boulder, CO: Lynne Rienner, 1999.

Sandbrook, R. *Closing the Circle: Democratization and Development in Africa*. London: Zed Books, 2000.

United Nations Development Programme (UNDP). "Democratic Governance for Human Development." In *Human Development Report 2002*, pp. 51–61. New York: UNDP, 2002. (Available online at http://hdr.undp.org/reports/global/2002/en)

The contemporary state of African development has been the subject of many studies, especially since 1990. Examples include the following:

Callaghy, T. "Africa: Back to the future." In L. Diamond and M. F. Plattner, eds. *Economic Reform and Democracy*, pp. 140–152. Baltimore: Johns Hopkins University Press, 1995.

Hope, K. R., ed. *Structural Adjustment Policies, Reconstruction, and Development in Africa*. Wayne, NJ: Avery Group, 1997.

Kaplan, R. "The coming anarchy." *The Atlantic Monthly*, vol. 273, no. 2 (1994), pp. 44–76.

Leonard, D. K., S. Strauss, and S. G. Mezey. *Africa's Stalled Development: International Causes and Cures*. Boulder, CO: Lynne Rienner, 2003.

Samatar, A. I. *An African Miracle: State and Class Leadership and Colonial Legacy in Botswana Development*. Portsmouth, NH: Heinemann, 1999.

Van de Walle, N. *African Economies and the Politics of Permanent Crisis, 1979–1999*. Cambridge, UK: Cambridge University Press, 1999.

Wallace, L. *Africa: Adjusting to the Challenges of Globalization*. Washington, DC: International Monetary Fund, 2000.

Internet Sources

The following sites may be used as starting points for research on development-related topics:

Stanford University. *Africa South of the Sahara. Topics: Development*. www-sul.stanford.edu/depts/ssrg/africa/devel.html

University of Sussex. *Eldis: Development Gateway*. www.eldis.org/search

The following websites for major development organizations, mostly international, provide access to a wealth of information on Africa:

African Development Bank. www.afdp.org

Economic Community of West African States (ECOWAS). www.ecowas.int

The New Economic Partnership for Africa's Development (NEPAD). www.touchtech.biz/nepad

Southern African Development Community (SADC): Towards a Common Future. www.sadc.int

United Nations Development Programme (UNDP). www.undp.org (see in particular Millenium Development Goals, annual *Human Development Reports*, and other publications)

United Nations Economic Commission for Africa (UNECA). www.uneca.org (also provides access to UNECA Subregional Development Centres and to specialized groups such as African Development Forum and African Knowledge Networks Forum)

U.S. Agency for International Development (USAID). *USAID in Africa.* www.usaid.gov/locations/sub-saharan_africa

World Bank. *The World Bank Group.* www.worldbank.org (see especially annual *World Development Report, World Development Indicators*, and specific reports)

For information on the current state of democratization, see the following source:

Freedom House. *Welcome to Freedom House.* www.freedomhouse.org

Dynamics of Population

Geographers have long recognized the importance of examining carefully the dynamics of population, which encompass distribution, growth, and mobility. Where people live and where populations are growing, whether through natural increase or migration, often reflects the relative health—ecological, economic, social, and political—of different places. At the same time, rapid population change may both reflect and contribute to the destabilization of what previously appeared to be relatively stable situations.

Many of the characteristics of Africa's present population distribution and patterns of change reflect processes that have operated for decades, centuries, or even millennia. For example, population distribution reflects in certain ways not only postindependence migrations, like those from the countryside to the city, but also the colonial reshaping of Africa's political and economic map; it even reflects major historical migrations.

Chapter 10 looks at the uneven distribution of population. Some of the explanations that have been used to account for population distributions, including both environmental and historical factors, are noted. The uneven distribution of population may have a variety of effects, ranging from environmental degradation in vulnerable environments to the limited prospects for diversified development that exist for many countries with small national populations.

In Chapter 11, the focus turns to the rapid growth of African populations, and the implications of this growth for the continent's future. With a growth rate of 3% per year, population is expanding faster in Africa than anywhere else in the world. Linkages are often made between population growth and various problems, whether damage to ecosystems or food shortages. However, although few would deny that *rapid population growth has serious implications, the assertion that such growth is the primary cause of these problems remains open to debate. Over the past decade, HIV/AIDS has had an increasingly significant impact on population growth, although the nature and extent of this impact vary regionally within the continent.*

Chapter 12 focuses on population mobility. Several distinct explanations have been advanced to account for Africa's high rates of mobility and the resultant differential effects on various regions, depending on whether they are source areas or destinations for migrants. Although the flow of labor migrants to islands of economic development, first established in colonial times, continues to be very important, the flight of refugees to safety from areas of political or ecological distress has become an increasingly prominent form of migration.

10

Population Distribution

The population of Africa south of the Sahara is very unevenly distributed—a situation that reflects a variety of ecological and historical–cultural factors. Mean national population densities (Figure 10.1) range from 2 persons per square kilometer in Namibia to 337 per square kilometer in Rwanda (Figure 10.2). The least densely populated countries, those with under 10 persons per square kilometer, account for just 6.4% of the total population but occupy 30.8% of the total area (Table 10.1). These 9 countries that have very low densities are diverse, as the following partial listing will show: Mali, Gabon, Congo, Central African Republic, and Namibia. They occur throughout the continent and in all of the major biomes—tropical forest, savanna, semidesert, and desert. Conversely, countries with the highest densities (over 80 per square kilometer) account for about one-third of the population but only 6.7% of the area. This group of 11 countries is dominated by Nigeria, which accounts for 60% of the high-density group's total population and 56% of its area.

Areas of high or low population density seldom, if ever, correspond to national units. It is not uncommon for regions of exceptional density to cut across national boundaries. Moreover, substantial regional variations in population density are found *within* virtually all nations. Vignette 10.1 (with Figure 10.3) looks at the uneven distribution of population in one country, Zambia, and touches upon some of the factors that help to explain these spatial variations in population density.

Localized pockets of high population density are to be found in most countries, especially in the vicinity of major cities. Of greater interest are several larger clusters of high density, the most important of which are the following:

- Several areas in Nigeria, particularly the Igbo homeland in southeastern Nigeria, the Yoruba heartland in southwestern Nigeria, and the closely settled zones surrounding the Hausa cities of Kano, Zaria, and Sokoto in northern Nigeria
- A zone extending from Burundi and Rwanda along the western and northern shorelines of Lake Victoria through southern Uganda and western Kenya
- Other localized pockets in eastern and central Africa found in southern Malawi, northeastern Tanzania, the hinterland of Nairobi in Kenya, and central Ethiopia
- Several pockets in South Africa, particularly in the areas that were established by the

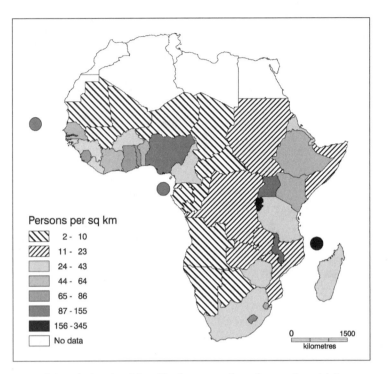

FIGURE 10.1. National population densities. The large number of countries with low to very low densities is especially striking. Data source: World Bank. *World Development Report 2002.* New York: Oxford University Press, 2002.

apartheid regime as "homelands" for the black population

- The island nations of Cape Verde, Comoros, and São Tomé e Príncipe

The following areas are characterized by particularly low densities and only localized concentrations of settlement:

- The Sahara Desert, together with the Sahelian zone along its southern fringe
- The desert and semidesert regions of Botswana, Namibia, Angola, and western South Africa
- Large parts of west central Africa between Chad and Angola, including most of the tropical forest zone in Gabon, Congo, and the Democratic Republic of the Congo, and varied savanna environments in Central African Republic, southern Sudan, and Angola

It is often hard to ascertain the specific causes and effects of exceptional densities in particular regions. The following section identifies some of the factors that help to account for variations in population density.

The Ecology of Population Density

Population distribution is influenced by many factors; some of these pertain to the physical environment and its capacity to support population, and others reflect social and political factors. Often present-day distributions reflect events or situations that occurred decades or even centuries in the past. For example, the slave trade caused a loss of population in West Africa's middle belt, from which many slaves were taken, while certain highland areas, such as Nigeria's Jos Plateau, served as defensive refuges from the slave trade and became more densely populated. The *redistribution* of population may well occur as circumstances change, but redistribution tends to be a very prolonged process.

FIGURE 10.2. Densely settled rural landscape, Rwanda. Despite the mountainous terrain, this is one of the most heavily populated regions of Africa, with average densities of over 300 per square kilometer Photo: CIDA.

Precipitation—specifically, the amount, seasonal distribution, and reliability of rainfall in particular regions—is the most important of the climatic factors related to the distribution of population. Where rainfall is low and the rainy season short and uncertain, as in drier savanna and semidesert regions, low densities are the rule. The reverse is not necessarily true; the majority of African tropical forest and moist savanna environments, where adequate precipitation is not a limiting factor, also have low population densities.

Human tropical environments with low population densities often have infertile laterite soils that are unsuitable for intensive cultivation. The primary exceptions are scattered

pockets of fertile volcanic soils that can be farmed intensively on a permanent basis. These often support very dense populations, as in Rwanda and Burundi. Lowlands with relatively young alluvial soils may also support densities that are above average for their region. Such is the case in the valleys of the Gambia, Senegal, and upper Niger Rivers of West Africa.

The relation between soil fertility and population is not one-way; human activities may enhance or reduce the "natural" potential of the environment. Overcultivation and overgrazing, especially on marginal and hilly land, may increase rates of wind and water erosion. In tropical forest regions suitable for shifting cultivation, overly frequent clearance and cropping

TABLE 10.1. National Population Densities in Relation to Area, 2000

Density/km²	Percentage of population	Cumulative percentage	Percentage of area	Cumulative percentage
0–4	1.0	1.0	11.2	11.2
5–9	5.4	6.4	19.6	30.8
10–19	9.3	15.7	21.2	52.0
20–39	29.5	45.2	29.4	81.4
40–79	22.4	67.6	11.9	93.3
80–159	29.1	97.6	6.5	98.8
160 and over	2.4	100.0	0.2	100.0

Data source: World Bank. *World Development Report 2002.* New York: Oxford University Press, 2002.

VIGNETTE 10.1. Case Study: Population Distribution in Zambia

Patterns of population density within each country reflect the influence of unique combinations of physical and historical–cultural influences. Zambia's 9.9 million people are spread over 753,000 km², for an average density of only 13 people per square kilometer. The population, however, is very unevenly distributed.

The largest area of relatively high density follows the main railway line from Livingstone (near Victoria Falls) northeast to Lusaka and beyond to the Copper Belt near the border of the Democratic Republic of the Congo (Figure 10.3). About half of all Zambians live within about 40 km of this railway; Zambians refer to this linear region as the "line of rail." Compared to Nigeria, Rwanda, or central Ethiopia, the densities of 15–50 per square kilometer typical of the line of rail are relatively low. However, within Zambia there is a marked contrast between the line of rail and most of the rest of the country, including virtually uninhabited valleys of the Kafue and Luangwa Rivers, which flank it on either side.

Virtually all of Zambia's modern economy is found along the line of rail—the larger cities, the northern copper-mining areas, and the commercial farming region near Lusaka. This con-

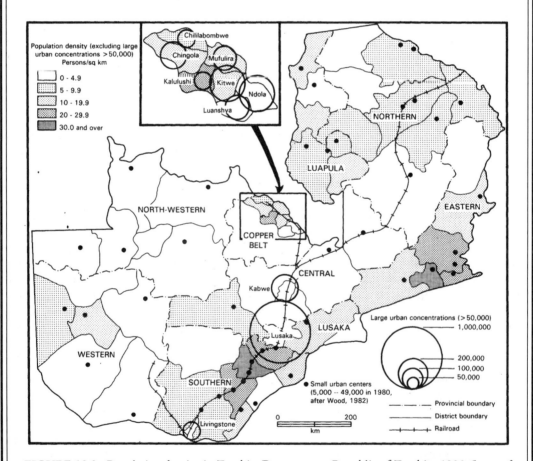

FIGURE 10.3. Population density in Zambia. Data sources: Republic of Zambia. *1990 Census of Population, Housing and Agriculture: Preliminary Report.* Lusaka: Central Statistical Office, 1991. A. P. Wood. "Population trends in Zambia: A review of the 1980 census." In A. Findlay, ed. *Recent National Population Change*, pp. 102–125. Durham, UK: Institute of British Geographers, 1982.

VIGNETTE 10.1. (*cont.*)

---

centration of modern economic activity, which dates from the early colonial era, has been a powerful magnet for young people (especially young men) from Zambia's periphery.

The line of rail supports higher densities, in part, because it has relatively fertile and well-drained soils. In the low-density river valleys, soils tend to be either very shallow and infertile or poorly drained. Moreover, the valleys are infested with tsetse, while the area along the line of rail is free of tsetse. Poor soils and a legacy of tsetse-transmitted sleeping sickness are of widespread significance as explanations of the very low densities (under 5 per square kilometer) found in most of peripheral Zambia.

High densities are found in the southeast region close to Malawi and Mozambique, where comparatively fertile soils are intensively cultivated. The high densities also reflect the legacy of Portuguese colonialism and the flight of Mozambicans from a regimen of forced labor; population densities are much lower across the border in Mozambique, despite a physical environment and an ethnic profile closely resembling those of southeastern Zambia. Small areas of quite dense settlement also occur in the northwest, near Lake Mweru on the border of the Democratic Republic of the Congo.

The regional imbalances in population and the very large area of extremely sparse population are major constraints on Zambia's development. The overwhelming concentration of modern economic activity has fostered a persistent migration of Zambians from the resource-poor and opportunity-poor periphery to the heartland, which has helped to perpetuate regional disparities in development. Because of the prolonged decline of Zambia's economy, funds to improve transportation and services in the low-density periphery are very scarce indeed.

---

results in the degradation of vegetation and soils. On the other hand, well-chosen methods that include appropriate use of manure and ground cover, intercropping, fallowing, and other fertility-enhancing techniques increase yields and ensure the long-term usefulness of the land. In the Kano region in Nigeria, for example, intensive application of manures and the use of methods such as intercropping explain the ability of this fairly dry area to support rural population densities as high as 400 persons per square kilometer. In Machakos District, Kenya, hillsides that seemed hopelessly eroded and infertile in the 1930s are now lush, productive farmland that supports a far denser population, as a result of very intensive rehabilitation and appropriate cultivation techniques.

Patterns of risk from debilitating and often deadly diseases such as sleeping sickness and river blindness are often reflected in the distribution of population. Vignette 10.2 (with Figure 10.4) illustrates on a relatively local scale how otherwise attractive environments may be depopulated because of the effect of disease. Sleeping sickness has been identified as a primary cause of low population densities in parts of the savanna and tropical forest with heavy infestations of tsetse. Indeed, maintaining high densities and clearing the bush were strategies used by precolonial African societies for tsetse control.

Ecological changes implemented under colonialism have affected population distributions. Diseases as well as people migrated along the transportation corridors, and newly created forest reserves provided environments where disease vectors could proliferate. In certain areas, long-established communities were abandoned as a result of new disease hazards. However, the opposite trend occurred in some areas; that is, previously avoided areas became accessible to new settlers, owing to locally successful vector and disease control programs. The political–economic map was reshaped under colonial

VIGNETTE 10.2. River Blindness and Population Distribution in Ghana

John M. Hunter's studies of river blindness in northern Ghana have provided valuable insights into the effect of ecological hazards on the health of local societies. Since this research was first undertaken in the late 1960s, new hope has emerged that the disease and its effects can be substantially controlled.

River blindness—more formally known as onchocerciasis—is caused by a parasitic worm called *Onchocerca volvulus*, which is transmitted from person to person by the bite of the fly called *Simulium damnosum*. Adult worms breed within the human subject; in a heavily infected person, their numbers may be in the billions. Eventually, the parasites may invade the eye socket and, if sufficiently numerous, cause blindness. Because the flies that carry the worms breed in fast-flowing water and usually have a restricted flight range, the probability of being bitten increases as one approaches a riverine breeding site.

Hunter's research in the 1960s showed that between 1 and 27% of Nangodi's population was heavily infected, and that the highest rates were in chiefdoms closest to the Red Volta River. The study showed how high rates of infection threatened the viability of communities. Heavily infected individuals cannot participate fully in the work of the household, so that food production and family incomes decline, and the workload (including the care of the blind family members) has to be done by fewer people. The people of Nangodi responded by abandoning severely affected villages for locations that were farther from the river. In some of these ar-

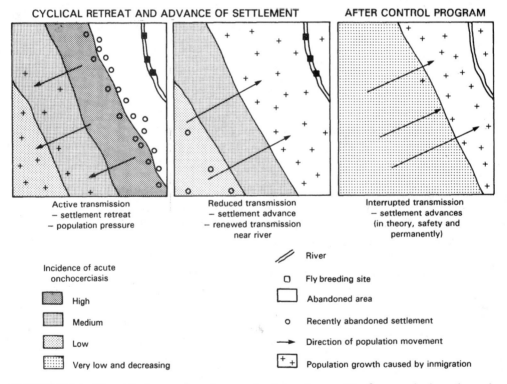

FIGURE 10.4. River blindness and settlement: A schematic map. The first panels show the cycle of settlement advances and retreat described by Hunter. The last panel suggests that disease control programs may eliminate the rationale for this cycle—that is, permit permanent settlement of the valleys.

VIGNETTE 10.2. (cont.)

eas of refuge, population densities increased to over 500 per square kilometer. Intensified land use contributed to soil exhaustion, erosion, and hunger. Emigration, particularly of young males, became necessary for household survival.

Hunter identified a long-term cycle of settlement advance and retreat in Nangodi (Figure 10.4). As the disease becomes more prevalent, the most seriously threatened villages cannot sustain themselves and are abandoned. The displaced people add to the pressure on already overutilized resources elsewhere in the district. Eventually, population pressure, the attraction of fertile river bottomlands, and the apparent reduction of the disease threat lead to a decision to resettle formerly abandoned lands. With more people living near the fly's riverine breeding sites, a renewed cycle of infection and socioeconomic crisis becomes likely.

Starting in the mid-1970s, an international campaign was launched to control river blindness by using chemical sprays to destroy the fly vector at its larval stage. Chemotherapy was also used to kill parasite worms living in the bodies of human subjects. The Onchocerciasis Control Programme (OCP) was conducted initially in 7 countries, and was expanded to 11 in 1988. Transmission has been effectively interrupted in virtually all of the project area in Ghana and other OCP countries. As a result, some 25 million ha of riverine lands have been rendered safe for resettlement. People are now returning to the fertile river valleys to establish farms and set up settlements near the treated rivers. Using the OCP as a model, The African Programme for Onchocerciasis Control (APOC) has been extended to 19 other African countries where there are pockets of the disease.

The immediate impact of the OCP has been exceptional; indeed, it has been widely cited as one of the most effective development initiatives in Africa's history. Nevertheless, some doubt remains about the longer-term sustainability of the success. Will the flies eventually become resistant to the chemical sprays, or the worms to the drugs used to treat infected individuals? Could this result in a renewed cycle of infection and the reabandonment of the valleys?

Based primarily on J. M. Hunter, "Population pressure in a part of the West African savanna: A study of Nangodi, northeastern Ghana." *Annals of the Association of American Geographers*, vol. 57 (1967), pp. 101–114 and Onchocerciasis Control Progamme in West Africa. www.who.int/ocp

rule. Many new cities were developed as administrative and trade centers, and the economic landscape was transformed through the development of new mines and cash crop areas. Migration increased in response to new opportunities and new pressures, and was facilitated by greater political stability and improved transportation.

Among the more recent developments that continue to change the distribution of population, two of the most important have been disparities in development and political/military conflicts. Development disparities—the concentration of modern economic development and services in the largest cities and a few favored areas—have caused increased rural–urban, interregional, and international migration.

Although resettlement has sometimes been undertaken deliberately, as in the construction of large dams and reservoirs, more often migration has been the inevitable but unplanned consequence of unbalanced development strategies. Instability and conflicts, whether political or ethnic in nature, have resulted in a massive dislocation of people; millions have fled from their homes as refugees to adjacent countries or to safer places within their own countries.

In South Africa, official policies of racial segregation and compulsory relocation created pockets of very high density in the homelands (Figure 10.5). Under apartheid, a mere 13% of the land was allocated to the three-quarters of the population that was black. An estimated 3.5 million people were forcibly removed from

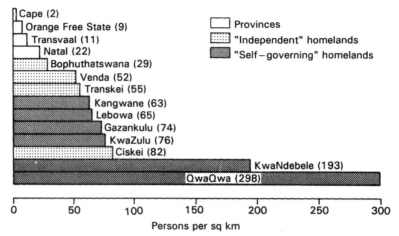

FIGURE 10.5. Rural population densities in South Africa's apartheid homelands. Densities were especially high in comparison to the rest of South Africa. After F. Wilson and M. Ramphale. *Uprooting Poverty: The South African Challenge.* Cape Town, South Africa: David Phillip, 1989, p. 36.

their homes between 1960 and 1983, most into the already overcrowded homelands. The population of the 10 homelands increased from 7 million in 1970 to 16.9 million in 1991. Overall, 20–40% of the rural population in the former homelands has no access to land or livestock. Most plots are too small and too infertile to provide adequate subsistence, and intensive cultivation of marginal land has accelerated soil erosion.

The demise of apartheid ended forced removals and restrictions on where the nonwhite population could live and work. Although out-migration rates from poorer rural areas such as the former homelands have increased, the legacy of four decades of forced removals will not end soon, if only because the extreme poverty severely limits the choices available to the rural poor. According to the 1996 census, poverty rates in the former homelands ranged from 48% in KwaNdebele to 92% in Transkei. Up to 40% of infants and small children are malnourished. Well-intentioned programs to improve housing, redistribute land, and strengthen social programs have met only a fraction of the need.

In Zimbabwe, as in South Africa, official land apportionment policies became a primary determinant of population distribution. By 1930, over 50% of all land in what was then

Southern Rhodesia had been set aside for Europeans, despite the fact that Europeans never accounted for more than 5% of the population. The most fertile areas in the central part of the country were reserved for white settlement. Africans living in areas designated for European use were faced with the choice of moving to the reserves or remaining as laborers or rent-paying tenants on European farms. Population pressure and land degradation increased steadily on the reserves, which included large areas that were marginal and unable to support many people. The injustice of this situation culminated in civil war during the 1970s, ending in 1980 with the creation of an independent black state.

Population pressure and land apportionment were issues that, for political reasons, could not be ignored by the new government. Vacant and underutilized European farms were purchased for the resettlement of landless African families. The former large farms were subdivided into village units in which each settler family was allocated a plot of land in the village, 5 ha of arable land, and communal grazing rights. Large, unorganized movements of African squatters into formerly European lands occurred in some areas.

Some 20 years after independence, issues related to the inequitable distribution of land and wealth surfaced again. The state vowed to seize large areas of land farmed by whites for redistribution to landless blacks, and expressed support for a series of land invasions. The governing party manipulated the issue of land ownership for its own political purposes. Nevertheless, this has only taken place because the initial land reform was inadequate to reduce regional disparities and to address problems of land degradation and poverty in former reserve areas.

Urban and Rural Settlement

The African population is still predominantly rural; as of 2000, only 38% of the population resided in urban centers. East Africa, with 23% of its population in cities, is considerably less urbanized than West Africa (with 40%), central Africa (with 38%), or southern Africa (with 43%). Individual countries (see Figure 10.6) fall along a continuum ranging from Burundi and Rwanda (under 10% urban) to Botswana and Djibouti (over 70%).

Average annual growth rates for African urban populations have been about 4%, well above rates in Asia and Latin America. The urban population of West Africa rose from 10.6 million in 1960 to 56.3 million in 1990 and 97.0 million in 2000. Urbanization has increased at an equally rapid pace in East Africa, from 5.7 million in 1960 to 32.7 million in 1990 and 45.0 million in 2000. In central Africa, urban populations increased from 6.3 to 34.6 million between 1960 and 2000. Rates of urban growth seem to have declined slightly during the 1990s, but remain very high. The slowing of urban growth appears to be related to the effects of structural adjustment on urban opportunities and the cost of living in cities. In southern and central Africa, the HIV/AIDS epidemic has contributed to a reduction in urban growth.

Very few countries have well-developed urban systems with a substantial number of cities of varying sizes and functions. The countries with the most extensive urban systems are Ni-

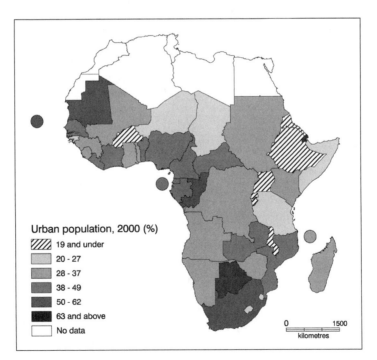

Urban population, 2000 (%)

- 19 and under
- 20 - 27
- 28 - 37
- 38 - 49
- 50 - 62
- 63 and above
- No data

FIGURE 10.6. Urbanization as a percentage of the total population, 2000. Africa is the least urbanized of the continents, but rates of urbanization vary greatly from country to country. Data source: United Nations Development Programme (UNDP). *Human Development Report, 2002.* New York: Oxford University Press, 2002.

geria and South Africa. Uganda is typical of the majority of countries, in that it has a *primate city* system. The population of Kampala, Uganda's largest city, is almost ten times that of the second-ranked city, Jinja. In a primate system, the primate city has a much larger population than other cities, as well as a disproportionate share of modern economic and political functions. Primacy is partly attributable to the recency of urbanization in most of Africa. It has been argued that such concentrations of population, wealth, and power in a single city perpetuate

and deepen regional disparities and promote accelerated rural–urban migration.

Rural settlement patterns are a reflection of the ways in which particular cultural groups approach the social organization of space (Figure 10.7). Some rural dwellers live in nucleated settlements varying in size from a few families to several thousand inhabitants. Elsewhere, rural settlement is dispersed, with each individual family living on its own plot of land (for an example, see Figure 10.8). The dispersed farm compounds may each have an extended family

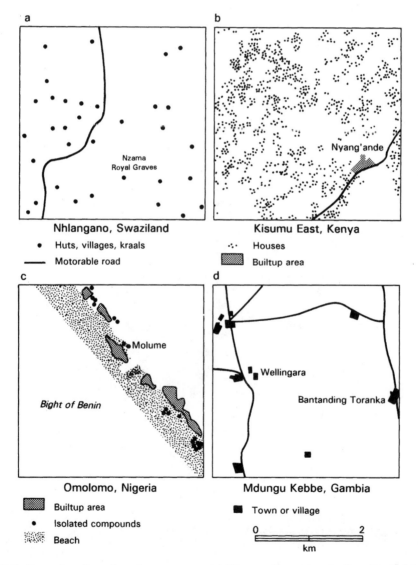

FIGURE 10.7. Examples of rural settlement types. (a) Dispersed compounds, Swaziland. (b) Very dense dispersed settlement, Western Kenya. (c) Linear pattern of nucleated fishing villages, Nigeria. (d) Nucleated agricultural settlements, Gambia.

FIGURE 10.8. Dispersed rural settlement in KwaZulu–Natal, South Africa. Photo: K. Child.

group consisting of several nuclear families belonging to a single lineage.

Patterns of rural settlement are important in the provision of education and health care, and in the general integration of rural people into national life. It is much more difficult to provide services to dispersed compounds than to village-based populations. The apparent advantages of nucleated settlements have led several governments to group dispersed rural populations in villages. Developing villages, however, has tended to be infinitely more complicated in practice than in theory. Rural dwellers have often resisted being resettled in villages because they prefer their ancestral lands, or prefer to live on their farms, or distrust the government. Official plans for resettlement have often been too ambitious, given the scarcity of needed resources.

The problems of service delivery and integration are most acute among pastoral nomads. Many countries have programs to encourage pastoralists to settle permanently and to send their children to school, but many pastoral peoples have jealously guarded their independence and resisted resettlement initiatives. On the other hand, growing crises of civil unrest, drought, disease, and environmental degradation, as well as the loss of access to traditional grazing lands, threaten the viability of pastoral ways of life in several countries and have forced many pastoralists to adopt a sedentary lifestyle.

National Populations

It is ironic, given the widespread concern about rapid population growth and high population densities, that the small size of national populations is a developmental constraint in many countries. Only 18 out of 46 countries in Africa south of the Sahara have national populations greater than 10 million, while 5 have fewer than 1 million citizens (Table 10.2). Nigeria alone has a population that is over 20 million larger than the total of 102.3 million living in the 29 countries with fewer than 10 million each.

Figure 10.9 assigns to each country an area proportional to its national population. The overall shape of the cartogram, as compared to the normal base map of Africa south of the Sahara, is distorted significantly by the position of the large and high-density population of Nigeria adjacent to the sparsely populated states of west central Africa. The cartogram is a vivid illustration of the dominance of Nigeria's population. Elsewhere, the block of relatively populous countries in East Africa, together with the Democratic Republic of the Congo, is in marked contrast with the plethora of small-population states in southern, west central, and West Africa.

Countries that are both small in population and poverty-stricken may be unable to undertake many types of large-scale development. Their domestic markets cannot support most

TABLE 10.2. National Populations, 2000

Population (in thousands)	No. of countries	Examples of countries
0–999	5	Cape Verde, Comoros, Equatorial Guinea
1,000–4,999	14	Congo, Liberia, Togo
5,000–9,999	9	Rwanda, Senegal, Somalia
10,000–19,999	10	Ghana, Mozambique, Zimbabwe
20,000–49,999	5	South Africa, Sudan, Tanzania
50,000–99,999	2	Dem. Rep. Congo, Ethiopia
100,000 and over	1	Nigeria

large-scale or specialized industries. They often remain dependent on foreign aid to balance their budgets, and on other countries for such professional services as higher education and specialist health care. In general, Africa's small-population states do not possess the special advantages of wealth, location, local capital, and skilled labor that have permitted some small countries elsewhere in the Third World—Singapore and the Bahamas, for example—to achieve notable economic success.

Governments are increasingly aware of the importance of population size and population distribution for national, regional, and local de-

velopment. They are concerned about the rapid growth of cities and the concurrent decline of many rural economies, as well as the apparent relation between population pressure and environmental degradation in many parts of rural Africa. They express interest in the development of sparsely populated regions, some of which have considerable potential. Unfortunately, the political will to address these issues seriously has only rarely been present, and the resources to bring about effective and comprehensive change have been even less available.

Problems of population distribution, however, represent only one component of the larger

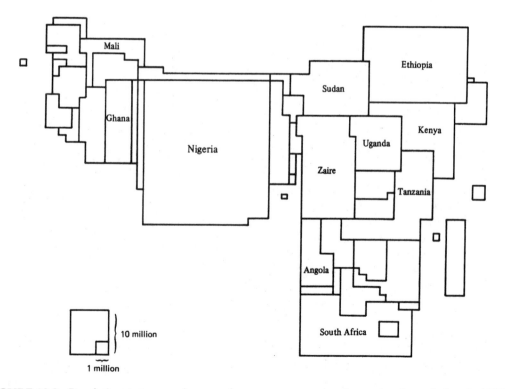

FIGURE 10.9. Population cartogram. Areas on the map are proportional to national populations in 1990.

population picture. Comprehensive population policies must also address problems of rapid growth—on average, 2.6% annually—in African populations, as well as the dramatic migrations that are redistributing the African population. The following chapter examines patterns of population growth, reasons for rapid population growth, and factors that are bringing about somewhat slower rates of population increase.

Further Reading

Official data on population distribution in Africa are very uneven in their coverage, quality, and accessibility. Academic studies of population distribution are similarly uneven. For some countries, you may be able to obtain census data from national census reports or annual statistical handbooks. Published population distribution maps are available for some countries.

For insights into the historical background of current population distributions, see this source:

Newman, J. *The Peopling of Africa: A Geographical Interpretation.* New Haven: Yale University Press, 1995.

Various aspects of population distribution are considered in the following sources:

Barbour, K. M., J. S. Oguntoyinbo, J. O. C. Onyemelukwe, and J. C. Nwafor. *Nigeria in Maps.* New York: Africana, 1982. (Look for other volumes in the "In Maps" series of national atlases on Liberia, Malawi, Tanzania, Zambia, and Sierra Leone.)

Davies, H. "Population growth, distribution, and density change, and urbanization in Zimbabwe: A preliminary assessment following the 1982 census." *African Urban Notes*, vol. 2 (1987), pp. 13–23.

Kloos, H., and A. Adugna. "The Ethiopian population: growth and distribution." *Geographical Journal*, vol. 155 (1989), pp. 35–51.

Moss, R. P., and R. J. Rathbone. *The Population Factor in African Studies.* London: University of London Press, 1975.

Ominde, S. H. *Population and Development in Kenya.* London: Heinemann, 1984.

Simpkins, C. *Four Essays on the Past, Present, and Future Distribution of the Black Population of South Africa.* Cape Town: South Africa Labour and Development Research Unit, 1983.

United Nations Economic Commission for Africa (UNECA). *Population Distribution and Urbanization in ECA Member States.* New York: United Nations Economic Commission for Africa (UNECA), 1983.

Zinyama, L., and R. Whitlow. "Changing patterns of population distribution in Zimbabwe." *Geojournal*, vol. 13 (1986), pp. 365–384.

Issues related to population pressure are discussed in the following sources:

Higgins, G. M. *Potential Population Supporting Capacities of Lands in the Developing World.* Rome: Food and Agriculture Organization of the United Nations, 1982. (This book contains a technical report and maps portfolio.)

Prothero, R. M. *People and Land in Africa South of the Sahara.* London: Oxford University Press, 1972.

Tiffen, M., M. Mortimore, and F. Gichuki. *More People, Less Erosion.* Chichester, UK: Wiley, 1994.

Internet Sources

For data on African population and factors affecting its distribution, see the following websites:

Deichmann, U. *African Population Database Documentation.* http://grid2.cr.usgs.gov/globalpop/africa

UNEC. *Population Information Africa (POPIA).* www.uneca.org/popia

United Nations Environment Programme (UNEP). Consultative Group for International Agricultural Research (CGIAR) Meta-Datasets for Africa. www.grida.no/cgiar/htmls/africa.htm

For further information on onchocerciasis control, see the World Health Organization (WHO) site:

WHO. *Empowering Communities to Eliminate Onchocerciasis as a Public Health Problem.* www.who.int/pbd/Oncho-brochure.pdf

11

Population Growth

Population growth is occurring much faster in Africa than in any other continent. The average annual growth in the 1980–2000 period was 2.7% in Africa south of the Sahara, compared to 2.0% in south Asia, 1.8% in Latin America and the Caribbean, 1.4% in east Asia, and 0.3% in Europe. São Tomé e Principe, Cape Verde, and Somalia were the only countries in Africa south of the Sahara that grew at less than 2.0% annually. At current rates of natural increase, Africa's population would double every 26 years.

Such rates of growth have serious implications for development planning. Among the most obvious are the employment needs of young people reaching adulthood and the continuing growth in demand for services such as health care and education. In the many areas where rural economies are under pressure, population growth increases the incentive of young people to move to the city. Without steady economic growth, the increasing demand for essential services and jobs cannot be met.

At the beginning of the new millennium, new questions were being asked about African population growth, particularly in relation to the impact of the HIV/AIDS epidemic. Is it possible that AIDS could bring about demographic collapse? If so, would this collapse occur uniformly, or would it happen primarily in certain settings (e.g., large cities) and regions? In any case, how would it alter the structure of the African population?

Patterns of Population Growth

Populations may grow in two ways: (1) by natural increase—that is, births minus deaths; or (2) positive net migration—that is, *inmigration* (arriving at a place) minus *outmigration* (leaving a place). National population changes in many African countries have been almost entirely attributable to natural increase (see Figure 11.1), with net migration across international frontiers being relatively unimportant. Some African countries, however, have experienced considerable gains or losses because of refugee movements or international migration of labor to areas of economic opportunity. The estimated 370,000 refugees from Sierra Leone living in Guinea in 2000 represented about 8% of Sierra Leone's population and about 5% of Guinea's population. The 1.2 million labor migrants from Burkina Faso in Côte d'Ivoire account for almost one-sixth of Burkina Faso's total population and perhaps one-eighth of the Ivoirian population.

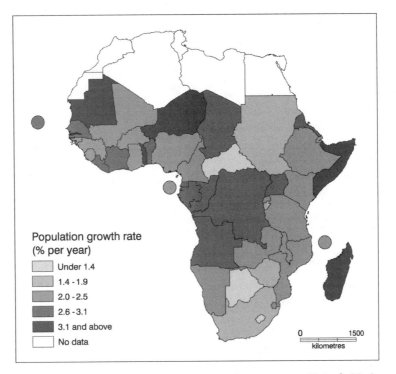

FIGURE 11.1. Crude rate of natural increase, 2000. Data source: United Nations Development Programme (UNDP). *Human Development Report 2002*. New York: Oxford University Press, 2002.

Prior to the 1990s, crude birth rates were very stable. Between 1970 and 1990, birth rates declined by more than 10% in only 8 African countries. Birth rates actually increased in 10 countries (e.g., Gabon and Ethiopia) during this period. During the 1990s, reductions of at least 10% in the crude birth rate occurred in 30 of the 46 countries. Indeed, birth rates declined by 20% or more in 7 countries, including Ghana, Kenya, and Sudan.

With few exceptions, crude death rates declined by 15–35% between 1970 and 1990. This downward trend has since been reversed in several countries, due to the impact of the HIV/AIDS epidemic. During the 1990s, death rates rose in about half of Africa's countries. However, even when the impact of AIDS is discounted, the reduction in mortality in recent decades has been less spectacular in Africa than in other parts of the Third World.

Mortality rates for African children under the age of 5—among the most sensitive measures of underdevelopment—remain very high. Between 1970 and 1991, national rates of in-

fant mortality declined by an average of about 30%. Childhood vaccination programs made possible much of this reduction, especially since the 1980s, when the United Nations Children's Fund–World Health Organization (UNICEF–WHO) Expanded Immunization Program was implemented; measles, whooping cough, tetanus, and polio were the primary target diseases. Infant mortality rates began to increase again during the 1990s in several countries seriously affected by the HIV/AIDS epidemic (e.g., South Africa) and civil wars (e.g., Sierra Leone).

Because of the high levels of infant mortality and deaths from AIDS, life expectancies are significantly lower in Africa than in other parts of the world. Life expectancy at birth is below 50 years in 30 countries and below 60 years in all countries except Comoros, São Tomé e Principe, and Cape Verde. Life expectancy has fallen below 40 years in Sierra Leone, Botswana, Zimbabwe, Zambia, and Malawi. By way of comparison, Haiti is the only country in the Western Hemisphere where life expectancy is less than 60 years.

Africa and the Demographic Transition

Demographers often relate changes in a population's birth and death rates to a model known as the *demographic transition*. It states that populations pass through a series of predictable stages—from an initial stage when both birth and death rates are high, to transitional stages when first the death rate and later the birth rate declines, to a final stage in which both birth and death rates stabilize at a new low level. Natural increase is relatively low during both the initial and final stages, but relatively high in the middle, transitional stages, when there is a significant excess of births relative to deaths. The demographic transition model was based on the pattern of demographic change in Europe a century ago; the pattern has since been replicated in many other parts of the world.

Prior to 1990, almost all African countries could be considered to be in early transitional stages; their death rates had fallen considerably, but their birth rates remained almost uniformly high. As noted above, patterns of change, especially in crude mortality rates, have been extremely varied since 1990: Some countries have achieved significant declines, and others have shown dramatic increases. Such diversity fits uneasily into the simple demographic transition model, with its prediction of unilinear change toward a uniform outcome.

If we focus on the decade of the 1990s, several distinct patterns of demographic transition may be identified. Table 11.1 identifies five such patterns, and provides examples of countries within each grouping. Each of these contemporary patterns is outlined below.

The demographic characteristics of countries in *early transition*, such as Chad and Niger, changed little during the 1990s. Their crude birth rates were among the highest in Africa and did not decline during the decade. Their crude death rates continued to decline, but in each case remained well above average for Africa south of the Sahara. The demographic iner-

TABLE 11.1. Natural Increase in Selected African Countries

Country	CBR[a] 1990	CBR[a] 2000	% change in CBR 1970–1990	% change in CBR 1990–2000	CDR[b] 1990	CDR[b] 2000	% change in CDR 1970–1990	% change in CDR 1990–2000	Pop. growth 1980–2000 (%/year)
Early transition									
Chad	44	45	−2	+2	18	16	−31	−11	2.7
Niger	52	51	+4	−2	19	19	−32	0	3.3
Accelerating transition									
Ghana	45	30	−2	−33	13	11	−19	−18	2.9
Sudan	44	34	−6	−23	15	11	−32	−27	2.4
Interrupted transition									
Cameroon	42	37	−2	−12	12	14	−33	+17	2.7
Côte d'Ivoire	46	37	−10	−20	14	17	−30	+18	3.3
Reversed transition									
South Africa	31	26	−20	−16	9	16	−36	+77	2.2
Zimbabwe	36	30	−32	−17	8	18	−50	+125	2.9
Restabilized transition									
Tanzania	46	39	−6	−15	18	17	−18	−6	3.0
Uganda	52	45	+4	−15	18	19	+12	+5	2.8

[a]CBR, crude birth rate (births per 1,000 people per year).
[b]CDR, crude death rate (deaths per 1,000 people per year).
Data sources: World Bank, *World Development Indicators 2002. Washington, DC: World Bank, 2002. World Bank. World Development Report 1993. New York: Oxford University Press, 1993.*

tia in several of the countries in this grouping was in part attributable to the effects of civil war and the collapse of the state. Included are several of Africa's poorest nations.

Ghana and Sudan are among several nations (the majority of which are located in West Africa) that experienced an *accelerated transition* during the 1990s. In Ghana, crude birth rates fell dramatically from 45 per 1,000 in 1990 to 30 a decade later. This decline was in sharp contrast to the 1970–1990 period, when only minor changes occurred in the birth rate. Ghana's crude death rate also declined significantly, by 19%, during the 1990s. In comparison to countries in the early transition group, the nations in accelerated transition tend to be more urbanized, more politically stable, and more developed. Although HIV/AIDS is present in all of them, it had only minor demographic impacts during the 1990s.

The term *interrupted transition* is used to describe the demography of several countries in western Africa where the shift to lower birth and death rates stalled during the 1990s. Though crude birth rates declined moderately (usually by 10–20% during the decade), the prolonged decline in crude death rates was reversed. The slight increases in the death rate signal the beginning of HIV/AIDS as a growing demographic concern. Several of these countries had been among the first to achieve reductions in their birth rates (during the 1980s), and most have death rates below the African average.

A number of southern African countries entered a period of major demographic crisis during the 1990s. Zimbabwe, for example, had an increase in the crude death rate from 8 per 1,000 in 1990 to 18 only 10 years later. Within a decade, Zimbabwe went from having the second lowest death rate to one of the highest. Life expectancy at birth declined from 60 to 40 years. This pattern of *reversed transition* reflects the devastating impact of HIV/AIDS in these countries.

In 1990, the prevalence of HIV/AIDS was higher in eastern and central Africa than in southern Africa. In several of these countries, the number infected by HIV remains very high, but the percentage of the population that is infected is no longer growing. In countries such as Uganda and Tanzania, the demographic transition appears to have *restabilized*. The death rates, which increased rapidly during the late 1980s, did not grow appreciably during the 1990s. Crude birth rates tend to be quite high, but are starting to decline.

Despite the divergent patterns of change in the past decade, the populations of all African countries are continuing to grow quite rapidly. Annual growth rates for the 10 countries in Table 11.1 were between 2.2% and 3.3% for the period 1980–2000. Rates of growth are now declining in the countries where HIV/AIDS has had the greatest impact, but they remain above 2% in all but Botswana, Zimbabwe, and South Africa.

The Demographic Impacts of HIV/AIDS

It is clear that HIV/AIDS is having significant impacts on Africa's population structure and growth, and that there are significant regional differences in these impacts. Nevertheless, both the regional patterns of infection and the patterns of incidence in the population—by age and gender, as well as by occupation and place of residence—continue to change with astonishing rapidity.

Between 1997 and 2002, the estimated number of Africans infected with HIV rose from 10.5 million to 29.4 million. According to the Joint United Nations Programme on HIV/AIDS (UNAIDS), 70% of the global population currently living with HIV/AIDS, and 77% of all AIDS deaths, are in Africa south of the Sahara. In 2002, an estimated 2.4 million Africans died from AIDS.

The dominant patterns of spread of HIV are quite different in Africa from those in Europe and North America. The most common method of transmission is heterosexual sexual activity. Women are more likely to be infected than men—58% of Africans living with HIV/AIDS are female—and women are infected at an earlier age on average than men. The second major means of transmission is from HIV-positive mother to child, either prior to birth or by way of breastfeeding. Physical stress related to pregnancy and childbirth is the major reason why women have a shorter life expectancy than men from the initial time of infection.

Young Africans are at serious risk of becoming infected with HIV. Girls are considerably greater risk than boys, largely because they are vulnerable to sexual exploitation by older men (see Vignette 11.1). Many young people have little knowledge of AIDS or of how to prevent it. In some countries, this ignorance about HIV/AIDS is the result of the government's unwillingness or inability to develop a strong public health response to the epidemic. Before success can be achieved, the pervasive atmosphere of shame, stigma, and silence needs to be broken (see Figure 11.2).

In the early 1990s, HIV/AIDS was concentrated primarily in urban centers of what was then called the "AIDS belt," which stretched across central Africa. High-risk groups, such as sex trade workers and persons with other sexually transmitted diseases, had particularly high rates of infection. By the late 1990s, the locus of the epidemic had shifted southward into Zimbabwe, South Africa, Botswana, and other countries in the region, and had become widely dispersed through the entire adult population. In contrast, rates of infection had leveled off in several of the central African countries, such as Uganda, where the largest HIV/AIDS epidemics had first occurred. As of 2002, rates of HIV infection in most West African countries were estimated to be 3–10%—a sufficiently large nucleus to create the potential for explosive growth over the next decade in the region (Figure 11.3).

Zimbabwe and Botswana are among four countries where HIV/AIDS has struck the hardest, with over 30% of all adults being affected by HIV. Based on current patterns of transmission, over 60% of 15-year-old boys in these countries will be infected with HIV in their lifetimes. Life expectancy in Zimbabwe has fallen by one-third, from 60 to 40, in just one decade. Rates of fertility have also declined because of the death of so many young women during their childbearing years.

Many have assumed that HIV/AIDS would cause population decline in Africa south of the Sahara, but even in the most seriously affected countries such as Zimbabwe, births continue to exceed deaths. Such a pattern of continuing population growth in the face of rising levels of HIV/AIDS had been predicted by some demographers as early as 1990. Ironically, the very high rates of fertility that have so often been criticized for creating an African population explosion now appear to be responsible for staving off demographic collapse due to the HIV/AIDS epidemic. Nevertheless, the demographic, economic, and social impacts have been massive. To give but one example, an estimated 11 million African children are "AIDS orphans" who have lost one or both parents to the disease.

The case of Uganda, however, gives some cause for hope. Between 1990 and 2000, HIV rates among adolescent girls in Kampala fell from 22% to 7%, primarily as a result of behavioral changes fostered by intensive HIV/AIDS education programs targeted particularly at young people. The strong support of the Ugandan government for these measures was instrumental in their success. In countries where governments have tackled these issues with less determination, rates of HIV infection have remained high.

Age Structure

High rates of population growth are clearly reflected in the age structure of the population (Figure 11.4). The African population is typically very young (Figure 11.5); between 40 and 50% of the population consists of children aged 0–15 years, while less than 5% of the population is over 65 years of age. By way of comparison, children make up 20% and adults over 65 years of age 12% of the U.S. population.

A gradual change in the shape of the population pyramid in countries such as Botswana and South Africa that have been hit hard by HIV/AIDS is but one impact of the epidemic (see Figure 11.4). A "chimney-shaped" population pyramid is gradually emerging. Compared to the typical African pyramid, the HIV/AIDS-affected pyramid has a narrower base (fewer births, and more deaths of young children from AIDS). The shape is also affected by the much higher mortality among young adults because

VIGNETTE 11.1. The Challenge of Educating Young Africans about HIV/AIDS

In the quest to slow, and ultimately to control, the spread of HIV/AIDS in Africa, what happens to young people is of critical importance. Without adequate health education and support systems, HIV may spread rapidly through the adolescent and young adult population. The premature death of young adults robs the society of their reproductive potential, as well as of the full range of their contributions as human beings.

In most African countries, a majority of young people lack sufficient information about HIV/AIDS and its spread. Among the misconceptions are that AIDS-infected individuals may be identified by appearance alone, that AIDS is caused by witchcraft, and that the risk of becoming infected is extremely small. Even youth who are aware that condoms offer protection may not know where to find them or how to use them properly. A survey in Somalia found that three-quarters of adolescent girls had not heard of AIDS, and 99% did not know how to avoid infection.

Young women are especially vulnerable. In major urban centers in East Africa, about one-fifth of adolescent females (aged 15–19) are already HIV-positive, compared to only 3–7% of 15- to 19-year-old boys in these cities. These data reflect a particular sexual mixing pattern, in which older men target younger females because they are thought to be "safer" as partners. Young women who are orphans or are from poorer families are vulnerable to the advances of well-to-do "sugar daddies." Others with few alternative sources of income work in the sex trade at great risk to their health; 70% of adolescent sex workers in Abidjan, Côte d'Ivoire are HIV-positive. Marriage is not necessarily a safe alternative, especially for girls marrying older men; a Kenyan survey found that half of the men marrying women 10 years or more younger than themselves were HIV-positive.

Many young Africans are at particular risk because of circumstances in which they become sexually active. The younger the age of engaging in sexual activity, the less likely it is that a condom will be used. This may be because of lack of information about, or easy access to, condoms, or it may be due to a youthful sense of invulnerability. Often it is a question of power related to age and gender. A significant proportion of girls' early sexual encounters are forced. Younger girls are more susceptible to vaginal injury as a result of sex; these injuries greatly increase the risk of becoming infected.

(cont.)

a   b

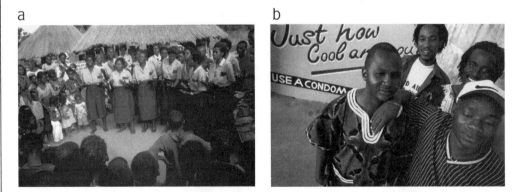

FIGURE 11.2. Educating youth about HIV/AIDS in Zambia. (a) Peer educators engage in a street theatre performance outside a bar. (b) Workers in an AIDS project pose outside a Community Youth Center; the message on the wall asks, "Just how cool are you? Use a condom." Photos: CIDA (D. Trattles).

VIGNETTE 11.1. *(cont.)*

The success of countries such as Uganda in reversing HIV infection rates among young adults, and Senegal in slowing the onset of the epidemic, shows the way forward. The governments of these countries have tackled the challenge of HIV/AIDS with determination. They have worked with religious and community leaders to end the silence about AIDS and the stigmatization of its victims. Their programs have especially targeted youth and older children with information about HIV and strong messages about the avoidance of early sex, unprotected sex, and sex with multiple partners.

The most effective education strategies are those that "speak to youth in their own language." These programs often involve young people as peer counselors, and utilize innovative strategies such as street theatre (see Figure 11.2). Educational programs are unlikely to succeed where infrastructural support is weak. Knowledge about condoms means little if no condoms are available. Youth-friendly programs initiated throughout the health system are important to create a welcoming environment for young people seeking information or an HIV test, or requiring treatment for their illness.

Based on *Young People and HIV/AIDS: Opportunity in Crisis*. New York and Geneva, Switzerland: UNICEF, UNAIDS, and WHO, 2002.

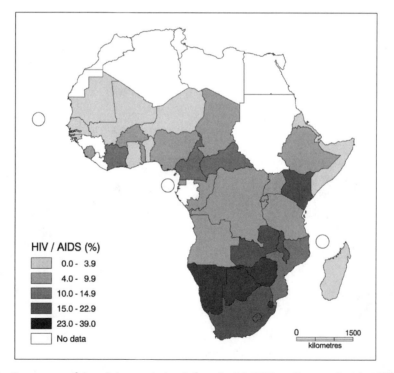

FIGURE 11.3. Percentage of the adult population infected with HIV or diagnosed with AIDS, 2001. Rates of infection are highest in Southern Africa. Data source: Population Resource Center. *AIDS in Africa.* www.prcdc.org/summaries/aidsinafrica.html

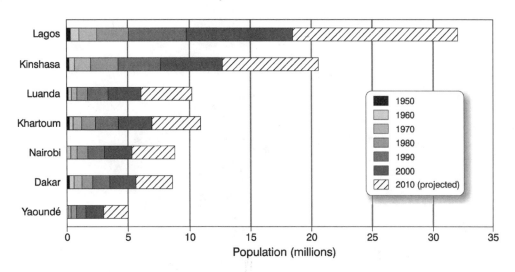

FIGURE 11.4. Population pyramids in the context of demographic change. (a) The broad- based pyramid for Benin is typical for African countries with very high fertility. (b) South Africa's current population pyramid reflects its lower fertility and slower population growth. The dark-shaded pyramid shows the impact that AIDS will have on South Africa's population pyramid by 2009; large reductions in young children and younger adults are projected. Sources: (a) U.S. Bureau of the Census. *IDB Population Pyramids*. www. census.gov/ipc/www/idbpyr.html (b) D. Bourne. "Demographic implications for development in South Africa as a result of the AIDS epidemic: A graphical review." *Urban Health and Development Bulletin*, vol. 3 (June 2000).

of AIDS. Some increase is anticipated in the proportion of the population aged 50 and above; HIV/AIDS has little impact on the older age cohorts.

Cultural, Ecological, and Socioeonomic Determinants of Fertility and Mortality

How do we explain the dominant patterns of population change in Africa south of the Sahara—namely, high levels of fertility that have begun to decline in many countries, and comparatively high rates of mortality? This section explores some of the ways in which aspects of culture, physical environment, development, and underdevelopment have affected fertility and mortality.

Determinants of Fertility

Based on recent age-specific fertility rates, women in Africa south of the Sahara give birth an average of 5.4 times (Figure 11.6). Total fertility rates remain high, but they have started to decline, especially among the urban, educated population. In 1980, the total fertility rate for Africa south of the Sahara was 20% higher, at 6.7 births per woman.

High fertility rates are in accord with the widespread preference in Africa for large families. According to the World Fertility Survey, the average desired family size among women

FIGURE 11.5. Children, Ogun State, Nigeria. With a median age below 20 years in most countries, Africa has the world's most youthful population. Photo: author.

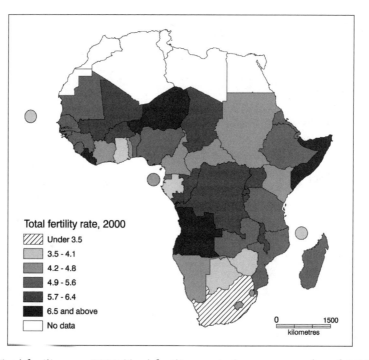

FIGURE 11.6. Total fertility rate, 2000. Total fertility rate is the average number of children per woman, based on current age-specific fertility rates. Data source: UNDP. *Human Development Report 2002*. New York: Oxford University Press, 2002.

in Africa south of the Sahara is 6.5 children. The World Fertility Survey found that young women (15–19 years) were as likely as older women (45–49 years) to prefer large families. Sociocultural factors, such as the prestige associated with large families and the deeply ingrained belief that fertility is the key to the survival and prosperity of any society, underlie the desire for large families. Many young couples face sustained family pressure, especially from the husbands' mothers, to have several children.

The desire for large families helps to account for a high rate of marriage among young adult women. Traditional African societies have strongly favored the early marriage of women; puberty is considered to be the ideal age for marriage in many societies. Consequently, most women spend the great majority of their fertile years (approximately 15–45 years of age) married. This maximizes the potential fertility of individual women and collectively of the entire society.

Certain cultural beliefs and practices do help to limit fertility. For example, it is customary in African societies for women to breastfeed

their infant children for 18–30 months, and to abstain from sexual intercourse during this time, on the grounds that becoming pregnant again too soon would jeopardize their infants' chances of survival. Unfortunately, the practice of this method of birth control has declined, especially in urban areas, because of the widespread use of infant formula and of modern methods of contraception. Both help to undermine the rationale for sexual abstinence.

High fertility often makes economic sense, especially in rural areas. The economic benefits of high fertility are in the form of help that children can give with work on the farm and in the household, and the security that children provide for parents in their old age. In societies where children are both wanted and economically important, childlessness is considered to be a major tragedy.

High rates of fertility in poor communities are in part a response to high rates of infant and childhood mortality. It is not uncommon, especially in poorer rural communities, to encounter women who have watched half of their offspring die. In regions where the survival of children to adulthood remains uncertain, Africans will continue to have many children. Under-5 mortality rates such as 150 per 1,000 live births in Côte d'Ivoire and 148 in Gabon (see Figure 11.7) show that nations that are better off also continue to have unacceptably high infant and child mortality. The key is not wealth, but the uses to which that wealth is put—for example, the extent to which it is used to improve living conditions, basic services, and opportunities for the poor

Although high fertility is the norm in African societies, smaller families are becoming increasingly common among urbanites with above-average incomes and education levels. Urbanites have comparatively easy access to information and technology for all aspects of child rearing and birth control, and they consider smaller families to be the key to obtaining better educational chances for their children and better opportunities for material progress themselves. Increasing the educational opportunities available to girls has been recognized as one of the most effective strategies to slow population growth.

Modern birth control makes it easier for up-wardly mobile women to combine child rearing and a career. Better-educated women are more likely to be employed outside the home; employment provides a strong incentive for women to have fewer children and to make use of family planning to delay having children or achieve better control over the timing of pregnancies (see Vignette 11.2 and Figure 11.8).

Determinants of Mortality

Conventional explanations of Africa's high mortality have focused on extensive poverty and an unhealthy environment. During colonial times, the pervasiveness of environmental health risks was emphasized. Disease, malnutrition, and poverty were seen as parts of a vicious circle preventing development. This pessimistic assessment of the African condition helped to legitimate colonial rule—first, by singling out disease vectors and parasites as the main causes of underdevelopment; and second, by casting Europeans in the role of saviors who would eventually liberate Africa from its disease burden.

Now as then, even the most "environmental"

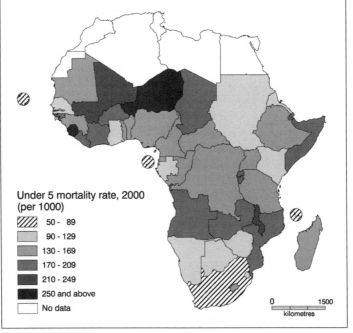

FIGURE 11.7. Under-5 mortality rate, 2000. Mortality rates for the most disadvantaged groups within each country are often well above the already high national values. Data source: UNDP. *Human Development Report 2002*. New York: Oxford University Press, 2002.

VIGNETTE 11.2. Modern Contraception in Africa: A Status Report

Thus far, the majority of Africans have shown little interest in modern birth control technology. In most countries, fewer than 5% of women use modern contraceptives (Figure 11.8a). These low rates contrast strikingly with those from other developing nations—80% in China, 57% in Brazil, and 46% in Egypt, for example. However, utilization rates are beginning to rise in a number of countries, the majority of them in southern Africa. Whereas in 1990 the utilization of modern birth control exceeded 15% of women of reproductive age in only three countries, a decade later this threshold had been reached in 10 countries.

Starting in the 1970s, family-planning programs were introduced into several African countries as part of larger foreign aid packages. Many governments were ambivalent about family planning and did not commit the resources needed for success. Moreover, these early programs tended to be poorly designed—for example, paying little attention to the cultural context of high fertility. As such, they were susceptible to rumors that they were neocolonial and un-African or that the use of contraceptives had caused health problems among women. Such rumors increased popular resistance to birth control.

The common assumption that modern birth control is introduced into a vacuum ignores important mechanisms whereby individuals and societies have traditionally limited fertility. Whereas modern contraception is used by relatively few, most women of childbearing age practice some form of traditional birth control. Extended sexual abstinence after birth, and the fertility-limiting effects of prolonged breastfeeding, provide very effective traditional means of spacing births. However, traditional birth-limiting strategies are practiced less carefully, especially among modernizing urban dwellers.

The experience of Zimbabwe is noteworthy because of the country's success in promoting

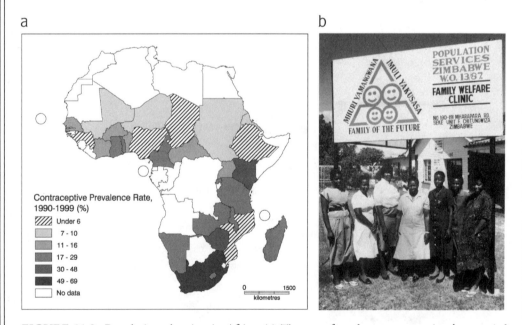

FIGURE 11.8. Population planning in Africa. (a) The use of modern contraception by married women of reproductive age has increased, but remains low compared to other parts of the world. Data source: UNDP. *Human Development Report 2002.* New York: Oxford University Press, 2002. (b) Family-planning clinic, Zimbabwe. Photo: CIDA (D. Barbour).

VIGNETTE 11.2. *(cont.)*

---

family planning during the 1980s. Zimbabwe's crude birth rate fell from 53 births per 1,000 people in 1970 to 36 in 1991 and to 30 in 2000. This transition to lower fertility has been attributed to the country's strong commitment to family planning (see Figure 11.8b), in conjunction with broad-based economic development, enhanced opportunities for women, and improved health care. Since the late 1980s, HIV/AIDS has become an increasingly important factor as well.

Because birth control was promoted as a means of birth *spacing* rather than birth *limitation,* much of the cultural resistance to the technology was overcome. Increasing numbers of Zimbabwean women took advantage of the opportunity to exert more control over the timing of births; as they did so, the number of births fell. The threat of HIV/AIDS has provided an additional incentive for using condoms as protection against both HIV and unplanned pregnancy. As of 2000, 42% of Zimbabwean women of reproductive age had adopted some form of modern contraception.

The Zimbabwean experience shows that an African fertility transition may occur when effective population planning occurs as part of a sound strategy of socioeconomic development, and when it is perceived to address people's needs. An increasing number of other African countries are following in Zimbabwe's footsteps, among them Ghana and Kenya. Limiting the size of families is increasingly seen, especially in middle-class urban families, to be the key to ensuring the best possible opportunities for children. In poorer families, the challenge of trying to make ends meet in the face of declining real wages under structural adjustment has been an incentive to reduce fertility.

---

of diseases reflect the political and economic milieu in which they occur. Malaria serves as an example of a disease that can only be understood in relation to political–economic conditions. Malaria kills 500,000 African children each year and contributes to the deaths of many more who succumb to the opportunistic infections that often follow malaria. It is less often fatal to adults, but recurrent bouts of fever, which occur as new batches of malarial parasites enter the blood stream, cause weakness, susceptibility to other infections, and decreased productivity.

Malaria is transmitted from person to person by the female anopheles mosquito. Sources of stagnant water where mosquito larvae may develop—small ponds and rain barrels, for example—are ubiquitous in poor communities that are crowded, poorly drained, and deprived of modern amenities. Such communities are likely to have not only the highest rates of infection, but also inadequate or nonexistent health care facilities where cases of malaria may be treated.

During the 1960s, WHO launched a global campaign to eradicate malaria. Breeding sites were drained or sprayed with a thin film of oil to suffocate the mosquito larvae, and homes were sprayed with insecticides. Antimalarial drugs were used to treat malarial patients and were given as prophylactic measures to susceptible populations. This chemical onslaught failed, primarily because the mosquitoes developed resistance to pesticides and the malarial parasites became resistant to the widely used antimalarial drugs. Moreover, budgetary cutbacks resulted in malaria control programs that were intermittent at best. The strains of malaria now prevalent are much deadlier and harder to control than those of the past.

Until quite recently, childhood vaccination programs against easily prevented childhood diseases such as measles were poorly and sporadically organized. The vaccine that prevents these deaths costs pennies per child. Again, it was the rural poor who suffered most from this neglect. Rates of vaccination coverage have increased markedly, thanks to the Expanded Immunization Program sponsored by UNICEF and WHO.

Although the long-established infectious and "environmental" diseases continue as major causes of death, excess mortality is increasingly related to the sale of potentially lethal products in the African marketplace and to the deepening crises of underdevelopment. These newer causes of disease and death are included in Table 11.2 (see numbers 8 to 10), which identifies several types of health problems linked to underdevelopment. Increasing proportions of illnesses and deaths in African societies are linked to modernization and global capitalism. Though urban areas are most heavily affected, it would be difficult to find rural areas that remain untouched. Third World markets have become increasingly important for transnational companies, as demand for products such

as tobacco and infant formula are shrinking in the industrialized countries because of slow-growing populations, changing lifestyles, and health education.

The use of infant formula has continued to grow in response to aggressive marketing campaigns, despite the poor economic situation that makes formula prohibitively expensive for most Africans. Advertisements for infant formula promise healthier babies, but the reality is exactly the opposite for poor mothers unable to prepare sufficient, safe, sterile formula for their babies. Ironically, although the inappropriate use of formula is recognized as being among the most important cause of infant deaths, WHO endorsed bottle feeding in 2001 for mothers infected by HIV/AIDS. The risks associated with bottle feeding were considered the lesser of two evils, compared to the risks of infecting babies with HIV through breast milk.

TABLE 11.2. Diseases of Underdevelopment

Development processes related to increased health risks	Examples of the resultant health problems
1. Changes in society–environment relationships (e.g., resettlement in a different ecological milieu)	Epidemics of infectious, parasitic, and animal diseases (e.g., sleeping sickness, malaria, river blindness, HIV/AIDS, and Ebola)
2. Increased population movements owing to improved transportation and quest for opportunity	Previously uninfected people exposed to tuberculosis, malaria, HIV/AIDS, and other infections
3. Changes in water flow and use (e.g., with dams and irrigation schemes)	Increased risk of water-borne diseases (e.g., schistosomiasis and guinea worm)
4. Changes in vegetation cover affecting habitat of disease vectors	Increased risk of vectored diseases (e.g., sleeping sickness, river blindness, and yellow fever)
5. Microenvironmental changes (e.g., in house construction and neighborhood density)	Cerebrospinal meningitis, tuberculosis, cholera, and other infectious diseases
6. Changes in value systems and erosion of traditional values	Stress, mental illness, suicide, malnutrition
7. More environmental pollution and introduction of industrial and agricultural chemicals	Cancers, respiratory diseases, poisoning of food chains (e.g., fish), fetal abnormalities
8. Marketing of potentially harmful consumer goods (e.g., tobacco, infant formulas, alcohol, drugs)	Malnutrition, infant diarrhea, addiction, cancers, respiratory diseases, allergic reactions
9. Increased armed conflict within and between nations, directly and indirectly affecting health	Hunger and malnutrition from economic disruption, psychosocial and physical trauma (e.g., landmines, torture)
10. Debt crises forcing states to cut social services, raise taxes, reduce public sector employment	Hunger and malnutrition, psychosocial stress, shortages and high cost of essential drugs, user fees limit access to services

Note. Based on C. C. Hughes and J. M. Hunter. "Disease and development in tropical Africa." *Social Science and Medicine,* vol. 3 (1970), pp. 443–493. R. Stock. "Disease and development or the underdevelopment of health." *Social Science and Medicine,* vol. 23 (1986), pp. 689–700.

Tobacco sales in Africa have increased five-fold in 20 years, in response to heavy advertising; the inevitable proliferation of lung cancer and other smoking-related diseases has begun to occur. The cigarettes sold in Africa tend to have very high levels of tar and nicotine. The risk of respiratory illness is especially great for smokers who live and work where they are exposed to serious air pollution.

Conflict and war clearly have the potential to have a major impact on population. The toll of the 1994 genocide in Rwanda amounted to almost 10% of the national population (not counting those who perished later, after fleeing to the Democratic Republic of the Congo). Between 1995 and 2002, an estimated 3 million perished in a complex series of conflicts in the Democratic Republic of the Congo. The health costs of insurrections and wars are staggering—communities abandoned for the relative safety of refugee camps, crops unsown or unharvested because of fighting, health centers destroyed, medicines diverted to treat military casualties, and so on. To give one example, Angola leads the world in amputations per capita, the legacy of over 30 years of anticolonial struggle and externally funded insurrection.

Population Growth: Implications for Development

The plight of nations with rapidly growing populations has been the subject of strenuous debate. Though all parties agree that rapid population growth is a significant phenomenon, there is no agreement about the nature of the problem or appropriate solutions. Two of the most influential perspectives on African population growth have come from the neo-Malthusian and Boserupian schools. Whereas neo-Malthusians see population growth as the root of environmental degradation and underdevelopment, Boserupians argue that population growth serves to stimulate development.

Thomas Malthus, a late-18th-century economist, developed a theory of population growth and resource scarcity that is still very influential, especially in conservative circles. Malthus predicted inevitable misery as the geometric growth of population outstripped food supplies

and caused irreparable environmental damage. He stated that without preventative checks to reduce fertility, population growth would be halted by rising mortality. For modern neo-Malthusian writers such as Paul Ehrlich and Garrett Hardin, a classic example of the catastrophe predicted by Malthus is now emerging in Africa. The solutions proposed by neo-Malthusians are often drastic; Hardin, for example, has advocated what he calls "lifeboat ethics," a survival-of-the-fittest doctrine that argues for the abandonment of most poverty-stricken states and the use of drastic salvage measures in other Third World countries that have strategic importance.

In contrast, a 20th-century Danish agricultural economist, Ester Boserup, argued that population growth actually stimulates societies to intensify their use of the land. The transition from less intensive methods of production, such as shifting cultivation, to progressively more intensive production systems was attributed to pressures exerted by rising populations to use their resource base more effectively. More recent research in Guinea and Kenya, undertaken by members of the Boserupian school, found that humans not only used the land more intensively as populations grew, but also increasingly applied conservation methods that brought environmental benefits such as increased forest density and reduced soil erosion.

Clearly, with populations doubling every 25 years, African nations face an enormous challenge. This is especially true in poor, ecologically vulnerable, and already densely populated countries such as Burundi and Burkina Faso. However, rapid population growth was not, and is not, the preeminent problem; the present crises would in all probability have occurred even if the population had remained stable. Focusing on population control as *the* key to solving Africa's problems has been inappropriate and futile.

Further Reading

The nature and significance of population growth are examined in the following sources:

Goliber, T. J. "Africa's expanding population: Old

problems, new policies." *Population Bulletin,* vol. 44, no. 3 (1989).

Kelley, A. C., and C. E. Noble. *Kenya at the Demographic Turning Point?* Washington, DC: World Bank, 1990.

Merrick, T. W. "Population and poverty: New views on an old controversy." *International Family Planning Perspectives,* vol. 28 (2002), pp. 41–46.

United Nations. *Population Growth and Policies in Africa South of the Sahara.* New York: United Nations, 1986.

For additional sources focusing on African fertility, see the following sources:

Brand, S. *Mediating Means and Fate: A Socio-Political Analysis of Fertility and Demographic Change in Bamako, Mali.* Leiden, The Netherlands: Brill Academic, 2001.

Caldwell, J. C. "The economic rationality of high fertility: An investigation illustrated with Nigerian survey data." *Population Studies,* vol. 31 (1977), pp. 5–28.

Cochrane, S., and S. Farid. *Fertility in Sub-Saharan Africa: Analysis and Explanation.* Washington, DC: World Bank, 1990.

Doenges, C. E., and J. L. Newman. "Impaired fertility in tropical Africa." *Geographical Review,* vol. 79 (1989), pp. 101–111.

Riedmann, A. *Science That Colonizes: A Critique of Fertility Studies in Africa.* Philadelphia: Temple University Press, 1993.

Scribner, S. *Policies Affecting Fertility and Contraceptive Use: An Assessment of Twelve Sub-Saharan Countries.* Washington, DC: World Bank, 1995.

The following studies look at patterns of disease and death and their explanation:

Booker, S., and W. Minter. "Global apartheid." *The Nation,* July 9, 2001, pp. 11–17.

Ewbank, D. C., and J. N. Gribble, eds. *Effects of Health Programs on Child Mortality in Sub-Saharan Africa.* Washington, DC: National Academy Press, 1993.

Feachim, R. and D. Jamison, eds. *Disease and Mortality in Sub-Saharan Africa.* Oxford: Oxford University Press, 1993.

Packard, R. "Industrial production, health and disease in sub-Saharan Africa." *Social Science and Medicine,* vol. 28 (1989), pp. 475–496.

Turshen, M. *The Political Ecology of Disease in Tanzania.* New Brunswick, NJ: Rutgers University Press, 1984.

For insights into the demographic impacts of HIV/AIDS, see these sources:

Setel, P. W. *A Plague of Paradoxes: AIDS, Culture, and Demography in Northern Tanzania.* Chicago: University of Chicago Press, 1999.

Torrey, B., and P. Way. "Seroprevalence of HIV in Africa, Winter, 1990." *CIR Staff Paper 55.* Washington, DC: U.S. Bureau of the Census, 1990.

Way, P., and K. Stanecki. "The demographic impact of an AIDS epidemic on an African country." *CIR Staff Paper 58.* Washington, DC: U.S. Bureau of the Census, 1991.

Internet Sources

The following websites may be consulted to obtain current demographic data:

Johns Hopkins Center for Communication Programs. *Popline Digital Services.* www.jhuccp.org/popinform/basic.html

United Nations Department of Economic and Social Affairs. *Demographic, Social, and Housing Statistics.* http://unstats.un.org/unsd

United Nations Population Fund (UNFPA). www.unfpa.org

U.S. Bureau of the Census. *IDB Population Pyramids.* www.census.gov/ipc/www/idbpyr.html

The major nongovernmental organizations concerned with population all maintain major websites with descriptions of research and current programs in Africa and elsewhere:

International Planned Parenthood Federation. http://ippf.org

Population Council. www.popcouncil.org

Population Reference Bureau. *PopNet.* www.popnet.org

Reproductive Health Gateway. www.rhgateway.org

There are many sites that focus on health issues, especially with HIV/AIDS:

Joint United Nations Programme on HIV/AIDS (UNAIDS). www.unaids.org

World Health Organization. www.who.int/en

12

Population Mobility

Africans are highly mobile, and their oral histories indicate that this has been so since time immemorial. The effects of migration extend far beyond gains or losses in population numbers: Migrants bring with them their fertility, their wealth (or lack of it), their skills, their culture, and a host of personal characteristics, modifying in the process both the communities of destination and of origin.

Explanations of African Population Mobility

The reasons why people migrate are extremely varied—sometimes the quest for new opportunities, sometimes the flight to safety from turmoil or ecological disaster, and sometimes the observance of social or religious custom. Three types of explanations are often advanced to account for Africa's high mobility rates:

- Some writers have examined population mobility as an important dimension of African culture, viewing modern migrations as a continuation of the continent's long tradition of population mobility.
- Others have focused on the importance of

perceived economic opportunities as a stimulus for migration.
- Still others have argued that migration in Africa occurs primarily as a response to forces that essentially compel people to move.

Migration and Culture

Africans' long history of mobility is reflected not only in countless local legends of the migration of ancestors, but also in major migrations that occurred over several centuries and that have been verified. Studies of the interrelation among languages, patterns of diffusion of particular cultural practices and technologies, and genetic analysis not only prove that past migrations occurred, but also help to unravel the complex relation between patterns of mobility and the sociocultural evolution of the continent. Several more recent large-scale migrations, such as the exodus of the Ndebele from South Africa to Zimbabwe in the mid-19th century, are part of the modern historical record.

More localized forms of circulation associated with seasonal rhythms in traditional economies are also rooted in antiquity. For example, many pastoralists move between seasonal pastures in order to have access to adequate water and fodder for their animals and to minimize risks from ecological dangers. Another example is provided by the seasonal movement of many agriculturalists from their home farms to riverine sites where irrigation is possible. Other types of mobility associated with trade and craft production also existed long before European rule was imposed.

For many Africans, religious and social obligations have been an important stimulus for migration. Since the 14th-century reign of Emperor Mansa Musa of Mali, West African Muslims have been traveling to Mecca and other holy places of Islam. Many West Africans have settled permanently along the entire savanna corridor through Sudan once followed by most pilgrims (see Vignette 12.1 and Figure 12.1). The pilgrimage is an example of a long-established migration stream that grew significantly in the more stable political environment of the colonial era.

The scale and diversity of past migrations have been cited as evidence that Africans have always been mobile and that they have a propensity to migrate. The interpretation of migration as traditional has occasionally been extended to the analysis of contemporary labor migrations, which have been characterized at times as a modern rite of passage for young African males—essentially a part of their initiation into manhood.

Migration as a Response to Economic Opportunities

The economic context within which population mobility has taken place, especially during the colonial and postcolonial eras, has received much attention. The economic explanations of migration focus on disparities of development between stagnant migrant source areas and the "islands of economic development" to which migrants are attracted. Economic models conceptualize mobility as a voluntary response by individuals who are motivated to take advantage of opportunities, particularly the availability of jobs. It is generally assumed that migrants have sufficient knowledge of opportunities at potential destinations to be able to make rational choices.

One of the leading economic theorists of migration, Michael Todaro (see "Further Reading"), claimed that artificially high urban wages maintained by government policy attract migrants to the city from the rural periphery. Although this migration may be perfectly rational for the individual, it can be counterproductive for the society and economy because too many migrants respond to the high urban wages. Todaro advocated that wage differentials be eliminated to reduce rural-to-urban migration.

The Todaro model formalizes earlier descriptive models of migration behavior, notably the push–pull and "bright lights" models. The first of these draws an analogy between the forces underlying migration and the physical forces of push and pull. Push factors are those that encourage the potential migrant to leave an unsatisfactory home environment, while pull factors are attractive aspects of a place of opportunity. The attraction of the "bright lights" of the city is another simple analogy used to explain migration behavior. The "bright lights" are seen as being especially influential for young people who have attended school and developed aspirations for a modern lifestyle.

Structural adjustment policies implemented during the 1980s and 1990s had a major impact on urban economies. The "bright lights" began to dim: Structural adjustment caused large job losses in the public sector and industry, and the subsequent influx of workers into the informal sector made it more difficult to succeed there. At the same time, market deregulation caused the prices of food and housing in urban areas to increase dramatically. Many urban-based migrants returned to their rural homes because they could not make ends meet.

The reciprocal nature of rural–urban linkages is increasingly recognized. Many extended families are based partly in the city and partly in the countryside, often deliberately so. These families benefit collectively from access to better and more diverse services in the city, as well as from access to cheaper food, housing, and land in the rural areas. The perspective that the rural–urban flow of people and resources is reciprocal differs from earlier approaches, such as that of Todaro, in which rural households were seen as depend-

VIGNETTE 12.1. The Overland Pilgrimage to Mecca

Each year, hundreds of thousands of West Africans travel by air to Saudi Arabia as pilgrims to Islam's holy places. A few thousand still follow the overland route from Nigeria through Chad and Sudan to the Red Sea, a route used by pilgrims for three centuries (Figure 12.1). The overland pilgrimage reached its maximum extent between the 1920s and 1950s: The consolidation of colonial rule made the journey safer; the introduction of trucks made it faster and easier; and the opportunities for work along the way made it financially attainable.

The significance of the pilgrimage extends beyond its religious meanings. The pilgrimage route has served as a cultural conduit along which innovations such as new crops, farming techniques, and styles of architecture have spread, both eastward and westward. The overland journey is seldom completed in less than three years and may be spread over two to three decades. Pilgrims traditionally have taken jobs along the way to finance the journey. They work as farm laborers, porters, petty traders, or barbers, or in other mostly menial jobs. Their economic role has been especially significant in the Gezira irrigation scheme.

West Africans, or "Fellata" as they are called in Sudan, have established many villages along the pilgrimage route and also have their own neighborhoods (zongos) in the larger towns. Pilgrims en route stay in these communities and obtain information about the journey and employment opportunities. Hausa is the lingua franca of the Fellata, reflecting the predominance of northern Nigerians among the pilgrims. There are more than 1 million Fellata in Sudan, representing some 5% of the national population. Although many have been there for several generations, the Fellata are regarded as temporary residents by the Sudanese government. As outsiders who often speak Arabic poorly and who willingly accept unattractive jobs, they have low social status. Both the Nigerian and Sudanese governments have occasionally found it politically expedient to demand the repatriation of the Fellata. However, only token repatriations have occurred.

The diverse effects of the overland pilgrimage illustrate an important facet of African population mobility—namely, that the characteristics and meanings of particular migrations are seldom as simple as they first appear.

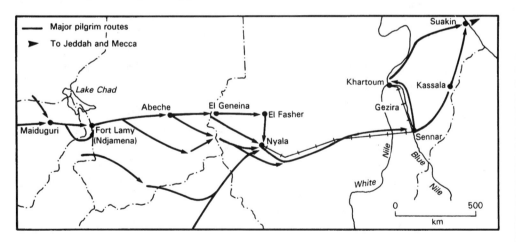

FIGURE 12.1. The overland pilgrimage from West Africa to Mecca. After J. S. Birks. "Overland pilgrimage in the savanna lands of Africa." In L. A. Kosinski and R. M. Prothero, eds. *People on the Move*. London: Methuen, 1975, p. 303.

ent on transfers from urban centers. The reciprocal flow of people and resources is probably not new; rather, its importance has grown and has finally been recognized.

Involuntary Migration

Individual responses to perceived opportunities certainly are reflected in migrant behavior. Political economists, however, question whether the primary determinant of mobility is the perception of opportunity or the structural factors impinging on the lives of potential migrants. They point to the historical effects of deliberate colonial policies, such as the imposition of a head tax, which forced many to migrate. In regions located far from the coast and from transportation routes, growing cash crops was not a viable option. The function in the colonial economy of these labor reserves was to provide a source of cheap, captive labor for more commercially developed regions. The expropriation of lands for European farms and forest reserves undermined the viability of indigenous economies; many African producers were forced to relocate onto smaller farms in less fertile areas or to become labor migrants.

South Africa's notorious migrant labor system emerged, following the discovery of gold in the Witwatersrand, to satisfy the insatiable demand of the mining industry for cheap, unskilled labor. By 1915, some 200,000 migrant workers were employed in 60 mines. Workers from Swaziland were gradually incorporated into this system between 1890 and 1920. Initially, mine recruiters had little success in recruiting Swazi laborers because that country's rural economy was quite strong. The first surge of Swazi migrants in 1898–1899 was a response to a series of ecological crises—four years of drought, locust invasions, and a rinderpest epidemic that killed 80–90% of the cattle. As well, the colonial began to pursue tax evaders aggressively, and Swazi chiefs began to encourage labor migration as a means of raising money. By 1920, the growth of labor migration had weakened traditional institutions and undermined the local economy. Working in the mines had become an integral part of Swazi economic and social life.

Because wages in the modern sector were low, African migrant workers continued to rely on subsistence production in their home localities to make ends meet. In effect, they had to maintain one foot in the rural periphery and the other in the cash economy core, and had to migrate back and forth between them. The term *labor migrancy* has sometimes been used to describe this situation in which survival became dependent on combining wage labor and peasant subsistence production.

Political independence has altered but not lessened the influence of structural processes on migration. For example, modern development schemes involving irrigation and large-scale commercial agriculture displace small producers from their farms, contributing to increased population pressure in areas of land shortage. The deterioration of marginal environments used too intensively has contributed to widespread crop failures and hunger, while wars and civil unrest have forced millions to flee their homes as refugees.

The danger with purely structural explanations, however, is that the behavior of ordinary Africans may be seen as entirely determined by, and purely in the interest of, capitalism. People in socialist economies have also migrated for essentially the same reasons. Moreover, despite the importance of structural determinants, Africans have always made choices about how and where they earned their incomes, and their choices have not always been those with the greatest apparent benefits for capitalism.

Colonial and Postcolonial Labor Migration

Colonialism created a pattern of uneven development characterized by isolated nodes of modern economic activity, but with a universal requirement to participate in the cash economy. Consequently, it was inevitable that major migration streams would emerge between economic cores and their regional peripheries. Figure 12.2 shows the spatial differentiation of labor migration streams as it evolved during colonial times. Labor migrants were attracted to areas with developed cash economies, each of

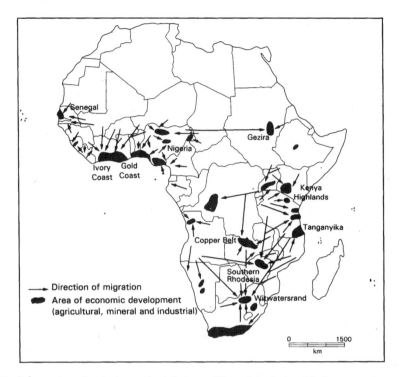

FIGURE 12.2. Labor migration in late colonial times. Note the relationship between the various labor reserve areas and the destinations—that is, islands of economic development. After R. M. Prothero. *Migrants and Malaria*. London: Longman, 1965, p. 42.

which had its own neglected periphery from which labor was drawn. The following were the major foci of the migrants' job search:

- South Africa, primarily the Witswatersrand mining region and major farming areas
- Central Africa, particularly areas of European agriculture in Zimbabwe and the copper belts of the present-day Democratic Republic of the Congo and Zambia
- East Africa, notably the former "White Highlands" of Kenya
- West Africa, especially the major cash crop areas of Ghana, Côte d'Ivoire, Senegal, and Nigeria

It is neither possible nor necessary to provide a detailed description of all the major movements of labor in Africa south of the Sahara. The general similarity of the patterns, processes, and meanings of migration in different parts of the continent is such that the discussion of these themes can be confined to a single

region—in this instance, migration to and from Ghana and Côte d'Ivoire. (The case of migration to and from South Africa is discussed separately in Vignette 12.2.)

In colonial times, hundreds of thousands of seasonal migrants were attracted to export crop zones in southern Ghana and southern Côte d'Ivoire. Most came from the more arid savanna states of the former French West Africa (Burkina Faso, Mali, and Niger; see Figure 12.3) and from the northern fringes of Ghana and Togo. In the early decades of colonial rule, both direct and indirect forms of coercion were employed to mobilize labor. Migration involved many hundreds of kilometers of travel, which often had to be undertaken on foot, owing to the poor development of transportation. In addition to the seasonal migrations of wage labor, permanent migrations of resettlement and agricultural colonization occurred. The best known of these colonial resettlement migrations is the expansive migrations are the Akwapim and Krobo peoples in Ghana to establish new cocoa plantations. At a local scale,

---

VIGNETTE 12.2. Myths and Realities in Cross-Border Migration into South Africa

During the era of apartheid rule, South Africa maintained a tightly regulated labor regime that severely restricted the movement of black South Africans, as well as that of people from neighboring countries. Workers from neighboring countries such as Lesotho, Malawi, and Mozambique were recruited in large numbers, and brought to South Africa on contract to work for a specific employer for a fixed period of time. Migrant workers had no option but to leave their families behind.

In 1997, South Africa abolished the Aliens Control Act, thus permitting an easier flow of people between Southern Africa and its neighbors in the South African Development Community (SADC). However, it is ironic that concerns about migration of fellow Africans from neighboring countries has increased greatly since the accession of the black majority government. Xenophobic reactions to African migrants circulated widely in South Africa, spurred by alarmist media reports and by statements from some politicians. A large-sample survey of residents of neighboring countries revealed major differences between popular perceptions and reality.

- *Myth 1: South Africa is being flooded by millions of migrants—estimates have ranged as high as 12 million—from other African countries, desperate to enter South Africa.* Half of those interviewed in neighboring countries had visited South Africa, but most were well established in their home countries, and few desired to live in South Africa.

- *Myth 2: Millions of the foreigners are illegal migrants who have entered clandestinely, sometimes taking great risks to do so.* Illegal migration is a problem, but it is far less pervasive than the South African news media have claimed. Of those surveyed, 89% possessed passports from their home countries, and the great majority had crossed the border legally.

- *Myth 3: South Africa's social services and job opportunities attract floods of illegal migrants.* Many migrants are attracted to South Africa by jobs and social services, but most people coming to South Africa are present on short-term trips to shop or visit friends.

- *Myth 4: Migration is very harmful to the economies and societies of the source countries.* Much of the academic literature has characterized migration as draining resources from the home country. However, the migrants surveyed considered migration to have been beneficial, both individually and for their home countries.

- *Myth 5: South Africa's neighbors and their citizens expect it to throw its doors open to migrants.* People in neighboring countries have a realistic view of their rights. They want to be able to enter South Africa easily, but acknowledge the right of South Africa to maintain border security and to limit the privileges accorded to noncitizens.

- *Myth 6: South Africa needs to act decisively to stem the rising tide of illegal migrants, before it is totally swamped with them.* Those most likely to seek entry into South Africa are young single men who are relatively well educated and have family members in South Africa. There is no sign of an imminent flood of illegals.

Based on D. McDonald et al., *Challenging Xenophobia: Myths and Realities about Cross-Border Migration in Southern Africa.* South African Migration Project, Migration Policy Series no. 7. Cape Town and Kingston, Ontario, Canada: Idasa/Queen's University, 1998. Available online at www.queensu.ca/samp/publications/policyseries/policy7.htm

---

many villages were relocated away from ancient defensive sites to downslope locations better suited to farming and with easier access to water.

After independence, the steady decline of the Ghanaian economy reduced the country's at-

tractiveness for migrants. Recurrent political tensions with Togo, which had been the largest source of migrants to Ghana, and the implementation of the Aliens Compliance Order of 1969, by which all foreign workers without a

FIGURE 12.3.Labor migrants from Niger, returning home to the farm after working in Côte d'Ivoire during the dry season. These migrants would likely be planning to return to Côte d'Ivoire in three or four months, after the farming season. Photo: author.

valid residence permit were expelled, also affected the size of Ghana's migrant population. The combined result was a net loss of about 400,000 foreign nationals between 1965 and 1975.

Migration to Côte d'Ivoire, especially from Mali and Burkina Faso, continued to increase after 1960. By the late 1960s, some 1.4 million people, about one-fifth of the Ivoirian population, was of foreign nationality. The proportion of immigrants was especially high in the former capital city of Abidjan (one-third of the population), and in the southern districts where export agriculture is concentrated. Table 12.1 demonstrates the overwhelming importance of Côte d'Ivoire as a destination, and of Burkina Faso as the primary source of migrants, in this regional migration system. In the early years of the new millennium, Côte d'Ivoire's migrant population continues to grow more rapidly than its indigenous population.

As Ghana's economic crisis deepened after 1970, growing numbers of Ghanaians left for other countries. Nigeria was very attractive because of the rapid growth of its economy, fueled by petroleum exports. The population of Ghanaians moving to Nigeria and elsewhere differed from the classic African labor migrant population because it contained large numbers of professionals, including teachers, pharmacists, engineers, nurses, and university lecturers. However, the increasing visibility of immigrants caused a backlash against them; the Nigerian government ordered all aliens without valid papers to leave in 1983 and again in 1985. An estimated total of 1.5 million non-Nigerians, including 700,000 Ghanaians, were expelled in 1983.

The expulsions of the mid-1980s only temporarily reversed the flow of Ghanaians to Nigeria. However, by the 1990s, the Ghanaian economy became much more robust, while at the same time economic stagnation and political violence made Nigeria a much less attractive destination. There is now a net flow of migrants into Ghana from Nigeria. Many Ghanaians still choose to leave, but their destinations are increasingly higher-income countries in Europe and North America.

Perhaps the single most important post-independence trend has been the increasing urban orientation of migration streams. Migrants still go to rural areas and work in agriculture, but over half of Côte d'Ivoire's immigrants now live in cities, compared to only one-third of the indigenous Ivoirian population. The influx to urban centers has continued despite the high cost of living and declining economic opportunities in

TABLE 12.1. Foreign Nationals (in Thousands) by Country of Nationality and Country of Enumeration, circa 1975

Country of nationality	Country of enumeration					
	Ghana	Côte d'Ivoire	Burkina Faso	Togo	Mali	Total
Ghana	—	43	17	30	NA	97
Côte d'Ivoire	18	—	44	NA	8	72
Burkina Faso	159	726	—	8	48	941
Togo	24	12	3	—	NA	259
Mali	13	349	22	NA	—	375
TOTAL	444	1,130	86	38	56	

After K. C. Zachariah and J. Conde, *Migration in West Africa: Demographic Aspects.* New York: Oxford University Press, 1981, p. 35.

most African cities since the imposition of structural adjustment programs, because cities are still seen to hold out the promise of success. Unlike in the colonial era, circulatory movements involving labor migrants now very commonly last for more than one dry season. Most migrants intend to stay in their new place of residence permanently, or at least for several years. They bring their families with them or marry in their new communities. Thus cities are growing rapidly, and an ever-growing proportion of their residents are becoming true urbanites.

In addition to migrations within Africa, a growing number of Africans, both skilled and unskilled workers, have made their way to Europe and North America. Certain countries and regions have predominated in these intercontinental labor migrations. There is a very large international Somali diaspora. Hundreds of thousands have migrated from Senegal and Mali to France and other European countries to seek work. Thousands of illegal migrants from this region are apprehended trying to enter Europe illegally.

The Brain Drain

There is a large and growing outflow of highly qualified Africans, bound especially for destinations in Europe, North America, Australia, and the Persian Gulf. Many African health care professionals, engineers, university teachers and researchers, and other skilled workers are drawn by the higher and more stable salaries, better working conditions and opportunities for professional development, and more attractive living conditions abroad. Others feel that their opportunities at home are limited by a range of factors, including ethnic and social barriers to advancement, political instability, crime, and the expectations of extended family members.

It is usually difficult to trace the international movement of trained professionals. The number of professionals leaving Africa for the United States increased by 1,000% between 1975 and 1985. It is estimated that 60% of highly educated people from Gambia and about one-third of those from Ghana and Sierra Leone are working abroad. France has attracted so many physicians and other professionals from its former African colonies that for certain countries of origin, more are working in France than at home.

One-fifth of all physicians practicing in the Canadian province of Saskatchewan gained their first medical degrees in South Africa. In rural Saskatchewan, 40% of all physicians were trained in Africa, while only 30% were trained in Canada. Saskatchewan rests in the middle of a "hierarchy of poaching," wherein Canadian doctors recruited to work in the United States and large Canadian cities are being replaced by doctors drawn from lower-wage countries such as South Africa. In turn, South Africa draws physicians from its neighbors to the north, such as Zimbabwe, Tanzania, and Ghana. The poorest countries have nowhere else to turn to replace their lost physicians. They find it very difficult to retain professional workers in poor, remote rural communities.

Refugees

Africa and southwest Asia are the two areas that have the largest concentrations of refugees in the world. In 2001, the United Nations High Commission for Refugees (UNHCR) registered 3.3 million refugees in Africa south of the Sahara. Both urban and rural areas in virtually all continental African countries south of the Sahara have been affected significantly as either origins or destinations of refugees since 1980 (see Figure 12.4). During the decade 1992–2001, 7 of the world's top 10 countries in terms of the largest refugee outflows, and 20 of the top 30, were located in Africa. Rwanda, from which 2,310,000 refugees fled in 1994 alone,

had more had twice as many refugees as any other country in the world for this period. Refugee outflows from five other African countries exceeded 400,000.

The numbers forced to flee are actually much larger than those given above; many international refugees are not registered with UNHCR, and others do not report to authorities in their countries of refuge. They may be able to blend unnoticed into the broader population of economically motivated migrants and are sometimes from the same cultural background as the local population. Moreover, available refugee data exclude the many displaced people who migrate to safer places inside their home countries. "Populations of concern" for the UNHCR—including not only refugees, but also recently returned refugees and internally displaced persons—total about 6.5 million in Africa south of the Sahara.

a

b

FIGURE 12.4. Refugee settlements, urban and rural. (a) A child plays soccer in a refugee camp at Lubango, Angola. (b) A camp in a remote part of the Ogaden region of Ethiopia houses 20,000 refugees seeking refuge from conflict in Somalia and drought. Photos: (a) CIDA (B. Paton), (b) CIDA (R. LeMoyne).

Table 12.2 provides data on refugee activity in the decade 1992–2001 for the 19 countries with the highest refugee numbers (i.e., where the combined total of incoming and outgoing refugees for the decade exceeded 200,000). In addition to the total number of fleeing (outflow) and incoming (inflow) refugees, data are provided on repatriated refugees because they often require help to become reestablished. Data are also provided on the total number of refugees living outside each country, as well as populations of concern to UNHCR living within each country. These figures are sometimes higher than the 1992–2001 numbers, because they include people who fled prior to 1992 but are still classified as refugees.

The African refugee problem is both pervasive and complex. Table 12.2 underscores the huge displacements of people that have occurred and continue to occur as a result of wars and civil unrest. Togo, for example, had a major but fairly short-lived refugee exodus in 1993, but has experienced little activity since. All of the 1.3 million refugees who had fled Mozambique between 1990 and 1993 returned by 1997, and there has been no subsequent outflow. Some countries have been exclusively places of refuge (e.g., Zambia, Guinea); others have been almost exclusively source areas (e.g., Eritrea, Somalia); and still others have had significant numbers both of those who have fled to other countries and those who have sought refuge (e.g., Sudan, Democratic Republic of the Congo).

Since the end of the Cold War, the geography of refugee spaces has changed. Increasing efforts have been made to establish "protected zones" within some countries of origin, and to attempt to limit refugees' movement into neighboring countries. For example, "safe haven" settlements

TABLE 12.2. African Countries with Major Refugee Activity (in Thousands), 1992–2001

| | Total numbers, 1992–2001 | | | As of 2001 | | During 2001 | |
	Outflow	Inflow	Repatriation (to . . .)[a]	Refugees (from . . .)[b]	Populations of concern	Outflow	Inflow
Angola	237	?	217	471	228	45	12
Burundi	675	317	601	554	126	16	2
Congo	108	106	96	24	122	0	
Congo (Dem. Rep.)	503	1,920	164	392	367	33	48
Côte d'Ivoire		287			128		8
Eritrea	101		195	333	36	1	
Ethiopia	81	195	208	59	162		10
Guinea		448			178		
Kenya		486		3	252		27
Liberia	420	177	518	245	253	12	
Mozambique	126		1,728	0	6		
Rwanda	2,475	26	3,178	85	58	7	5
Sierra Leone	536	7	350	179	103		4
Somalia	472		404	440	52	21	1
Sudan	367	98	82	490	354	35	
Tanzania		1,447			670		33
Togo	170	11	285	4	12		
Uganda	43	138	18	40	200	1	10
Zambia		145			285		35

Note. Blank fields: For "Total numbers, 1992–2001" columns, blank fields indicate that data were unavailable where the total was less than 5,000. For all "2001" columns, blank fields indicate that reported values were under 500.

[a]Repatriation (to . . .): Refugees who have returned from other countries; still receiving UNHCR support.

[b]Refugees (from . . .): Total number classified as refugees in 2001, originating from the country.

[c]Populations of concern: Total of persons registered with UNHCR in the country (refugees from other countries, returned refugees, internally displaced persons).

Data Source: United Nations High Commission for Refugees (UNHCR). **www.unhcr.ch/cgi_bin/texis/vtx/publ**

were established for Somali refugees along the Kenya–Somalia border, largely in response to Kenyan concern to limit the influx of refugees. However, these sites are commonly less secure than those in neighboring countries, with the result that inhabitants' mobility is very restricted.

Although the great majority of African refugees remain in Africa, a small proportion seek asylum in industrialized countries. Between 1992 and 2001, 148,000 refugees from Somalia, 104,000 from the Democratic Republic of the Congo, 77,000 from Nigeria, and 25,000–50,000 from each of seven other countries sought asylum in an industrialized country. Very few Africans have the resources needed to travel abroad to seek asylum even in peacetime, let alone during times of conflict.

Regional Variations

Southern Africa was formerly one of the most active regions of refugee activity, but since the end of South Africa's apartheid-era interventions in neighboring countries, this region has had the fewest refugees. Mozambique, which was Africa's largest source of refugees during the early 1990s, is now very stable. In turn, this has eased the burden on countries of refuge, especially Malawi and Tanzania.

Angola has been torn asunder by four decades of civil conflict. What started as the struggle to end Portuguese rule during the early 1960s and early 1970s continued as a bitter factional struggle fueled by South African intervention throughout the 1980s; it evolved into a struggle to control Angola's resource wealth, especially its diamonds, during the 1990s. The number of Angolans forced to seek refuge in neighboring countries ranged during this period from 200,000 to 500,000. Many more were displaced internally within the country.

In central Africa, the Democratic Republic of the Congo and Chad have been major sources of refugees on a continuing basis. Other countries, including the Central African Republic and the Congo, have been affected by refugee inflows from their neighbors, and at times have generated substantial refugee numbers themselves.

The "great lakes" region (i.e., east central Africa) has long been a major locus for refugee movements (see Figure 12.5 for patterns in the 1990s). The mass exodus of 2.3 million Rwandan refugees in 1994 was but the largest of many refugee surges in the region. Both Rwanda and Burundi have had several major exoduses of refugees since independence. Hundreds of thousands fled Uganda during its political crises of the 1970s and 1980s; smaller numbers continue to flee from areas in the north and west that remain unstable. During the late 1990s, conflicts in eastern Zaire (now the Democratic Republic of the Congo) led to large-scale refugee flight (see Figure 12.6). Most of the refugee burden has been borne by a handful of nations bordering the trouble spots. Tanzania, Uganda, Kenya, and the Democratic Republic of the Congo have all absorbed very large numbers of refugees from their unstable neighbors. During the mid-1990s, the former Zaire had 1.7 million refugees, most of them from Rwanda. Tanzania had 760,000 refugees in 2001, primarily from Burundi and the Democratic Republic of the Congo.

The Horn region has been a locus of long-standing political instability. The Ethiopia–Somalia conflict over the Ogaden region in 1977–1978, the civil war in Ethiopia that led to Eritrea's secession in 1993, and the Ethiopia–Eritrea border war of 2000–2001 all resulted in large-scale refugee migrations to Sudan and Kenya. Both famine and internal political repression, especially under the Marxist regime that came to power in Ethiopia in the mid-1970s, also forced very large numbers to flee. Between 350,000 and 800,000 Somalis have lived outside the country as refugees since the late 1980s, when Somalia collapsed into a state of anarchy. The number of refugees who have fled from conflict in the southern Sudan has remained very large since the early 1980s.

Until the late 1980s, West Africa had experienced relatively little international refugee movement. That changed dramatically with the outbreak of the Liberian civil war; some 750,000 (one-quarter of the population) fled in 1990 to Guinea, Sierra Leone, and Côte d'Ivoire, and many others were displaced internally. Subsequently, conflict in the region spread to Sierra

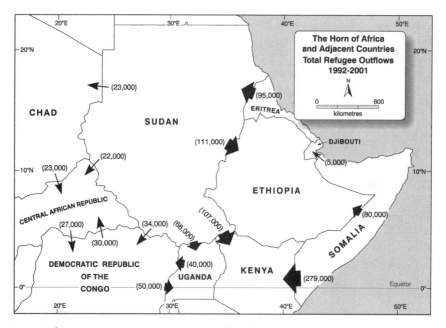

FIGURE 12.5. Refugee movements in east central Africa during the 1990s. Data source: U.S. Committee for Refugees. *World Refugee Survey, Africa.* www.refugees.org/world/articles/wrs02_Africa.cfm

FIGURE 12.6. Thousands of returning refugees trek home to Rwanda from Zaire (now the Democratic Republic of the Congo) in 1996. They had sought asylum there following the genocide of 1994, but were displaced once again when fighting broke out in the refugee camps. Photo: CIDA (R. LeMoyne).

Leone (1991) and Côte d'Ivoire (2002), in each case forcing many to take flight. West Africa has had other significant refugee movements, mostly of shorter duration, since 1990. The largest of these resulted from border skirmishes between Mauritania and Senegal, armed rebellion by Tuareg peoples in northern Niger and Mali, and political strife in Togo.

Refugee Impacts on Host Societies

Major influxes of refugees pose a dilemma for the countries to which they have fled. Most refugees arrive with neither material possessions nor any means to support themselves. Thus it is not enough simply to grant permission for refugees to stay: Their basic needs for food, drinking water, shelter, and health care cannot be ignored. Although the UNHCR and other voluntary agencies provide considerable assistance, much of the burden ultimately falls upon the countries of refuge.

For decades, Sudan has been a major destination for refugees fleeing from conflicts in neighboring countries—Uganda, Chad, Eritrea, Ethiopia, and the former Zaire. The number of officially enumerated refugees increased from 10,000 in the early 1960s to 400,000 in the early 1980s and to 1,200,000 in 1986. This number had declined to 490,000 by 2001. This refugee burden has been especially onerous for Sudan, already struggling with inadequate health care and education systems, recurrent economic crises, widespread drought and hunger, and unresolved religious and political tensions. Sudan discourages the spontaneous resettlement of refugees, directing them instead to government-organized farming settlements where they can grow crops and move toward greater self-reliance. These villages are meant to provide refuge only until repatriation can be carried out safely, not to facilitate permanent resettlement.

Sudan's acceptance of so many refugees, regardless of the large costs involved, is in marked contrast to the formidable barriers refugees encounter in many far wealthier countries. Unfortunately, other concurrent policies of the Sudanese government have helped to make southern Sudan one of Africa's largest sources of refugees. The southern Sudanese have fought for four decades to achieve the regional autonomy they consider essential to preserve their own cultures and religions within the Islamic state of Sudan. The prolonged, bitter campaign to quell this insurrection has cost many lives and has forced hundreds of thousands of southern Sudanese to flee their homes.

Beyond the statistical stories of millions of Africans forced to move are the individual human stories of personal, family, and community loss. The continuing, massive flow of refugees is not only a symptom of the diverse crises affecting Africa, but also a key cause of underdevelopment in both source and destination countries. Although effective short-term programs are needed to help poorer countries cope with refugee influxes and ensure that the basic needs of refugees are met, the longer-term priority must be to address the political, economic, and ecological disasters that continue to displace so many millions from their homes.

The Effects of Migration

Migration has diverse demographic effects both in areas of origin and destination that vary in relation to the type (e.g., labor or refugee), volume, and duration of migration. Migration alters not only the size of the population, but also its composition. Young adult males have been predominant in African labor migration streams, with the result that rural areas of heavy outmigration have a disproportionate number of females, while cities have more males than females. The male bias in migration streams developed in large part because of colonial policies that limited the mobility rights of those not holding jobs. Women left behind in the rural areas had to assume a greater share of the workload, in the absence of so many men. Male bias was particularly evident in migration streams to, and within, South Africa under the apartheid labor regime.

Whereas colonial labor migration consisted largely of short-term movements by single men, there was an increase in longer-term and permanent migration, and family migration increased in prevalence. Nevertheless, even where migration is apparently permanent, many urban residents maintain important social and economic ties with relatives in the countryside. Englund's study of migration in Malawi (see "Further Reading"), for example, emphasizes the simultaneous and overlapping roles of urban and rural spaces in migrants' lives. Even long-term residents of Lilongwe identify themselves in relation to their rural home communities. Most of these migrants farm, own houses, and regularly visit their rural homes. Nevertheless, living in the city provides essential income and in some cases an escape from problems back home. The hardships brought about by structural adjustment have reinforced, rather than weakened, the dual urban–rural identity of many Africans.

Migration is commonly said to benefit the economies of both sending and receiving areas. In reality, the benefits are very unevenly distributed. Recipient economies benefit from the availability of a ready supply of cheap labor. These benefits accrue primarily to employers; the arrival of a "reserve army" of would-be workers adversely affects the job security, wages, and working conditions of the established labor force.

Regions of outmigration may benefit from a somewhat reduced population pressure and from the infusion of remittances—money sent home to purchase necessities for family members, to buy consumer goods, or to start a business. Nigerians living abroad, for example, send some $1.3 billion in remittances, equivalent to 3.5% of Nigeria's gross domestic product. However, there are significant costs. Farming production in areas of rural outmigration often declines because of the loss of labor. Crops requiring high labor inputs may be replaced by less labor-intensive but inferior alternatives. For example, the substitution of cassava for yams and grain crops saves labor but adversely affects nutrition. Despite the constant infusion of remittances from migrants, most source areas have remained as poor as ever. Whatever remittances are received are poor compensations for the loss of the talents and energies of the young people who leave.

Further Reading

For discussions of theoretical issues related to African migration, see the following sources:

Curtin, P. *Why People Move: Migration in African History.* Waco, TX: Baylor University Press, 1998.

Gould, W. T. S., and R. M. Prothero. "Space and time in African population mobility." In L. A. Kosinski and R. M. Prothero, eds. *People on the Move.* pp. 39–49. London: Methuen, 1975.

Todaro, M. P. "A model of labour migration and urban unemployment in less developed countries." *American Economic Review,* vol. 99 (1969), pp. 138–148.

For studies on labor migration in various regions of Africa, see the following sources:

Arthur, J. A. "International labour migration patterns in West Africa." *African Studies Review,* vol. 34 (1991), pp. 65–89.

Crush, J. S. *The Struggle for Swazi Labour, 1890–1920.* Kingston, Ontario, Canada: McGill–Queen's University Press, 1987.

Crush, J. S., A. Jeeves, and D. Yudelman. *South Africa's Labour Empire: A History of Black Migrancy to the Gold Mines.* Boulder, CO: Westview Press, 1991.

McDonald, D., ed. *On Borders: Perspectives on International Migration in Southern Africa.* Cape Town and New York: Southern African Migration Project/St. Martin's Press, 2000.

Rain, D. *Eaters of the Dry Season: Labor Migration in the West African Sahel.* Boulder, CO: Westview Press, 2000.

Van Onselen, C. *Chibaro: African Mine Labour in Southern Rhodesia 1900–1933.* London: Pluto, 1976.

Whitman, J., ed. *Migrants, Citizens and the State in Southern Africa.* Basingstoke, UK: Palgrave Macmillan, 2000.

To learn more about refugees in Africa, see the following sources:

Bascom, J. *Losing Place: Refugee Populations and Rural Transformation in East Africa.* New York: Berghahn Books, 1998.

Hyndman, J. "A post-Cold War geography of forced migration in Kenya and Somalia." *Professional Geographer,* vol. 51 (1999), pp. 104-114.

"Internal Displacement in Africa." *Refugee Survey Quarterly*, special issue, vol. 18, no.1 (2000).

Rogge, J. *Too Many, Too Long: Sudan's Twenty-Year Refugee Dilemma.* Totowa, NJ: Rowman and Allanheld, 1985.

To explore the concept of migration as a linking of urban and rural spheres, see these sources:

Englund, H. "The village in the city, the city in the village: Migrants in Lilongwe." *Journal of Southern African Studies*, vol. 28 (2002), pp. 137–154.

Ferguson, J. *Expectations of Modernity: Myths and Meanings of Urban Life on the Zambian Copperbelt.* Berkeley: University of California Press, 1999.

Frayne, B., and W. Pendleton. "Migration in Namibia: Combining macro and micro approaches to research design and analysis." *International Migration Review*, vol. 35 (2001), pp. 1054–1085.

Simone, A.M. "The worlding of African cities." *African Studies Review*, vol. 44, no. 2 (2001), pp. 15–41.

The following studies discuss Africa's brain drain:

Commander, S., et al. *The Brain Drain: Curse or Boon? A Survey of the Literature.* www.nber.org/books/isit02/commander-et-al-CEPR-site 6-7.02.pdf

Harber, C. *Education, Democracy and Political Development in Africa.* Brighton, UK: University of Sussex, 1997.

McDonald, D., and J. Crush, eds. *Destinations Unknown: Perspectives on the Brain Drain.* Pretoria: African Institute and Southern African Migration Project, 2002.

Internet Sources

The following sites offer diverse resources related to migration, especially in southern Africa:

Idasa, South Africa and Queen's University, Canada. *Southern African Migration Project.* www.idasa.org.za or www.queensu.ca/samp (see "Migration Policy Series" publications).

International Organization for Migration (IOM). www.dfa.gov.za/for-relations/multilateral/iom.htm

The leading sites on refugees contain both detailed statistics and articles on current issues:

United Nations High Commission for Refugees (UNHCR). www.unhcr.ch/cgi_bin/texis/vtx/homere (see especially *State of the World's Refugees* [annual])

U.S. Committee for Refugees (see especially *World Refugee Survey* [annual]). www.refugees.org

Rural Economies

The majority of Africans live in rural areas and work as primary producers, relying on their skills as farmers, herders, fishers, and hunters to secure the necessities of life. The chapters in this section concentrate on primary production as the focus of rural economies. This topic provides many opportunities for examining such core themes as the importance of the environment for human sustenance, the cultural foundations upon which systems of production rest, and the significance for rural societies of colonial and postcolonial development initiatives.

Chapter 13 describes the major indigenous systems of food production, emphasizing the logic of these systems—the diverse strategies whereby sustenance is obtained from often difficult environments. These systems of production are an integral part of the African cultural endowment; each society has its own set of strategies and divisions of labor for the production of food. However, despite the resilience of indigenous systems of production, their health if not survival has been threatened in many areas by inappropriate policies, population growth, and environmental degradation.

Chapter 14 examines the legacy of a century of attempts to modernize African agriculture, with often disappointing consequences. The colonial policy of introducing cash crops was designed primarily to incorporate Africans into the commercial economy and to create a source of tropical products valued by Europeans, rather than to improve the lot of rural Africans. Many agrarian development strategies have been tried in recent decades, but the models for development have usually been foreign, costly, and unsustainable. Newer approaches that build upon existing systems of production and focus on small farmers provide some reason for optimism.

In Chapter 15, the subject is food security. The occurrence of critical food shortages has been very uneven in space, in time, and within the societies of affected regions. The chapter examines the distinction between chronic hunger, which occurs year after year in certain settings, and the acute food shortages of famines. The diverse explanations of hunger— environmental, demographic, technical, and political—are also considered. It is apparent that just as hunger is an urban as well as a rural phenomenon, its alleviation depends on changes not only in the countryside but also in urban Africa.

13

Indigenous Food Production Systems

Most Africans continue to earn their livelihood as primary producers—specifically, from raising livestock, fishing, hunting, and gathering. Indigenous systems of food production are predominant virtually everywhere on the continent. Research on African agriculture has repeatedly shown the sophistication of these indigenous production systems as means of extracting a living from often difficult and fragile environments. African primary producers simultaneously pursue several goals: maximizing yields, minimizing the risks associated with drought and other catastrophes, diversifying production for household self-sufficiency, and protecting the resource base. Growing recognition of the importance of environmental protection, the sustainability of production systems, and the need to save locally developed crop varieties (see Vignette 13.1) among other issues, is fostering an increased appreciation of indigenous systems of production.

The term *indigenous* is used in preference to the widely used *traditional*, which may seem to imply that these systems are rigid and unchanging. On the contrary, they have continued to evolve in response to diverse influences, including changing ecological circumstances, population growth, market opportunities, and

the introduction of new technologies. During the colonial era, African farmers found ways of incorporating new cash crops into their existing farming systems. More recently, such technologies as animal-drawn plows, chemical fertilizers, and improved varieties of seeds have been successfully incorporated into the existing system by many small-scale producers. On the other hand, plantation agriculture, European mechanized agriculture, large-scale irrigation, and cattle ranches are not merely introduced *technologies*, but entire *systems* of production.

Food Production

Indigenous food production methods may be grouped into three broad categories: (1) cultivating food crops; (2) raising livestock, including pastoralism; and (3) fishing, hunting, and gathering. Although it is possible, in theory, for primary producers to rely exclusively on one of these three sources of food, most rural Africans utilize all three. However, the relative importance of each varies greatly from culture to culture and among households in particular communities. Most agriculturalists supplement their farm-produced diets with livestock such

VIGNETTE 13.1. Seeds for Survival: Saving Ethiopia's Landrace Varieties

Ethiopia, one of the "culture hearths" in which agriculture first developed thousands of years ago, is a treasure trove of crop biodiversity. There are literally thousands of local varieties of such crops as sorghum, teff, barley, and coffee that have been maintained by farmers because of their particular qualities. Local varieties that have developed in a crop's region of origin are called *landraces*. The complex nature of environmental conditions in Ethiopia has contributed to the development of diverse varieties adapted to specific local conditions. So too has the country's very difficult terrain; many distinct varieties have evolved and been preserved in remote regions.

Seeds of Survival/Ethiopia (SOS/E) is an initiative supported by Canada's Unitarian Service Committee to study, preserve, enhance, and distribute Ethiopia's landrace seeds. It was initiated because of the threatened loss of many of these varieties. Some were lost as a result of the introduction of modern varieties into Ethiopia, while others disappeared during the drought, when some farmers were forced to eat their seed stocks to survive.

Such is the genetic diversity of Ethiopian agriculture that a study of sorghum grown in five communities found 60 distinct varieties, most of them grown by only a few individuals in one or two communities. Only five of the varieties were being grown in all of these communities.

The SOS/E project commenced with an inventory of varieties used by farmers in the study areas. Landraces were identified, their characteristics were catalogued, and seed samples were deposited in a national gene bank. These efforts were dependent on the active support and participation of farmers; it was their store of knowledge that was being tapped. Some farmers were paid to preserve and reproduce the most promising of the varieties. A second phase of the project involved research to develop local varieties with improved yields. As a result, yields were increased by up to one-third. The best of the landrace varieties were replicated and distributed to other farmers. This program enabled the reintroduction of landrace varieties into areas where modern seeds had previously replaced local varieties.

Ethiopian farmers responded enthusiastically to the program. They have a deep appreciation of the value of landrace seeds. Each variety has particular qualities—related to taste, nutritional value, drought resistance, soil tolerance, yield, storage quality, frost tolerance, pest resistance, and the feed value of stalks—that are attractive to farmers. Contrary to the received wisdom that modern, high-yielding varieties outperform traditional ones, farm yields from landrace varieties exceeded those from modern, high-yielding varieties by an average of 47%. These results were achieved without the application of chemical fertilizers.

The long-term health of our global food system depends on the maintenance of a large stock of landrace varieties. Indeed, modern biotechnology makes extensive use of traditional cultivars to develop new, specialized varieties of crops. This utilization of indigenous knowledge for profit raises important ethical questions about the ownership of such knowledge, developed over many generations, and profits derived indirectly from indigenous knowledge. Can we envisage a collaboration of farmers and scientists as equal partners in synthesizing indigenous and modern knowledge for the broader benefit of humankind?

Based on M. Worede. "SOS/E: Promoting farmers' seed—its conservation, enhancement, and effective utilization." Paper presented at SSC–Africa Workshop, Harare, Zimbabwe, 1998. (Available online at **www.cdr.dk/ sscafrica/wor4-et.htm**

as chickens and goats, and collect edibles from the wild. Many pastoralists cultivate crops, as well as hunt and gather. Systems of mixed farming, in which raising livestock and cultivating crops are integrated and of relatively equal importance, are widely practiced in certain cultures.

African primary producers also rely on the marketplace as a supplemental source of food for household use. The marketplace provides access to a variety of locally produced foodstuffs, as well as foods produced outside the community, including commercial products like flour and pasta.

African primary producers, to a varying extent, also market what they have produced. In addition to conventional "cash" crops like cocoa and cotton, foods of many kinds are sold: grains and other food crops, livestock and livestock products, fish, game, and other foods from the wild. Many farmers have shifted resources into the production of foodstuffs for resale, taking advantage of increased demand and rising prices in urban areas. These trends point to a blurring of boundaries between categories such as food crops and cash crops, or subsistence and commercial agriculture, around which discussions of African agriculture are often structured.

Fishing, Hunting, and Gathering

Hunting and gathering societies in Africa, as elsewhere, are on the verge of extinction. The !Kung San of the Kalahari and the Mbuti (Pygmies) of the Congo basin rain forest are societies where hunting and gathering have continued to be practiced widely. Studies of !Kung gathering strategies have shown that they obtain an extremely diverse and nutritious diet with relatively low labor inputs. Game, hunted by adult males, contributes about one-third of calories in the traditional !Kung diet, while vegetable foods, mostly collected by women and children, account for the remainder. Although the !Kung consume numerous plant foods, their staple is the highly nutritious mongongo nut. This diet has permitted the !Kung to live longer and healthier lives than most African farmers.

The !Kung hunting and gathering economy has become increasingly difficult to sustain, especially in Botswana. Pressures have come from the establishment of fenced cattle ranches in territories traditionally used by the !Kung, the creation of game reserves and official restrictions on hunting, and government programs to establish permanent !Kung settlements and encourage them to farm. This erosion of the traditional economy has been accompanied by an increase in previously uncommon health and social problems.

The marine resources of many parts of the African coast, as well as lakeshore and riverine environments, are an important source of food. For the Fante of the Ghanaian coast and numerous other societies, fishing and agriculture constitute the dual bases of economic activity. Fishing is a year-round activity involving the majority of the economically active population, but most families also farm. The work is usually divided according to gender, with men doing the fishing and women most of the farming, as well as the selling and preparing of fish.

For most Africans engaged in primary production, fishing, hunting, and gathering are secondary to the practice of agriculture and pastoralism. The kinds of edible flora and fauna that are harvested seem almost infinite. Such foodstuffs are often gathered seasonally and provide variety and nutritional balance, as well as dietary energy during times of seasonal hunger or crop failure. In many African societies, *bush meat* (meat from wild animals) and locally harvested fish are important sources of protein. The utilization of wild fauna and flora is discussed further in Chapter 24.

Farming Systems

African farms are small, typically 1–5 ha. Although the small size of farms sometimes reflects a scarcity of land in heavily populated areas, it more often reflects the limited labor and technological resources available in rural households. Unless hired labor or labor-saving technology is available, the size of farms will be limited to the area that a household can manage during the periods of heaviest labor demand—planting, weeding, and harvesting. Tradeoffs may be made, depending on the availability of land, between careful, labor-intensive cultivation of small farms and less intensive cultivation relying on lengthy fallows to restore soil fertility in sparsely populated areas.

In most African societies, women have primary responsibility for food production. In cultures where women do the farming, the involvement of men in food production tends to be limited to a few specific tasks, such as land preparation prior to planting. The active participation of children in food production not only enables the household to increase its total production of food, but also provides invaluable opportunities for socialization and instruction in farming techniques.

Various techniques of enhancing yields and protecting the environment, including crop rotation, intercropping, bush fallowing, and agroforestry, are widely used. *Intercropping* is the practice of planting two or more crops together in a field to maximize yield, reduce soil erosion, and take advantage of complementary nutrient requirements. For example, legumes (e.g., beans and peanuts), which convert atmospheric nitrogen into a form that other plants can use, are intercropped with nitrogen-dependent grain crops. In *crop rotation*, different combinations of crops are grown in each year of a cropping sequence. *Bush fallowing* is a method of restoring soil fertility by temporarily abandoning farmland so that it can be recolonized by natural vegetation. *Agroforestry* refers to farming systems in which there is a functional integration of planted and naturally occurring trees and shrubs with crop production.

African cultivators have developed several farming strategies, each of which involves its own set of adaptations to local social, economic, and environmental conditions. The following paragraphs examine the organizational structure and logic of the most important of these cropping systems: shifting cultivation, rotational bush fallow, and permanent cultivation. Nevertheless, it is important to emphasize that the actual implementation of these systems varies greatly in particular settings. It is also common for individual farmers to use different strategies in different fields—for example, cultivating fields close to home far more intensively than those located farther away.

Shifting Cultivation

In *shifting cultivation*, soil fertility and farm productivity are maintained by changing the location of cultivation. It involves clearing and burning natural vegetation to produce ash, then farming the cleared area for a few cropping seasons, and finally abandoning the plot for another to permit soil renewal. Because high land-to-population ratios are needed, it is a system

for sparsely populated regions. It is practiced in a variety of environments, including the tropical forest, moist woodland savanna, and dry parkland savanna. Cultivation strategies vary in different regions, reflecting both local ecological factors and culture (see Vignette 13.2, below, for one example).

Shifting cultivation is especially suited to the infertile laterite soils found in the tropical forest. The dense vegetation is nourished by decaying plant matter that covers the forest floor. Once the leaf-falling cycle that replenishes the litter layer is interrupted, soil fertility begins to decline. The cultivation cycle begins with the cutting and burning of vegetation, which releases nutrients from the vegetation into the soil. Higher yields of ash generally contribute to improved crop yields. Shifting cultivators plant several crops together in a way that mimics the diversity and layered structure of the natural ecosystem (see Figures 13.1 and 13.2). Legumes are intercropped to increase nutrient supply, reduce soil erosion and baking, and inhibit weed growth.

After a couple of years, when crop yields decline, the clearing and cropping sequence is repeated elsewhere. Plots are small and surrounded by forest, so the natural vegetation reestablishes itself soon. The stability of the system depends on the maintenance of fallows long enough to permit soil rejuvenation; these fallows average about 10–20 years in forest regions. The duration of cultivation may be longer in savanna regions, where nutrients are leached more slowly than in humid forest ecosystems. Crop rotation is used to prolong the farming cycle. Crops such as sorghum, sweet potatoes, and yams, which require nutrient-rich soils, may be planted initially and then replaced in the second or third year with less demanding crops such as cassava. Other crops such as beans, squash, maize, and bananas are intercropped with the major staples.

To the uninformed observer, shifting cultivation appears to be a primitive, inefficient, and disorganized system that needlessly destroys natural vegetation and is not amenable to modernization. However, agricultural scientists have come to see shifting cultivation as a sophisticated method of farming in marginal en-

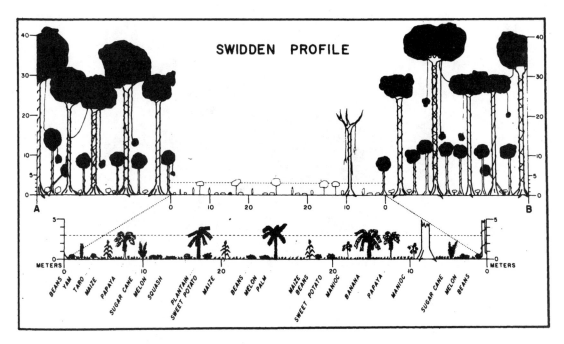

FIGURE 13.1. Cross-section of a shifting-cultivation farm. The diagram illustrates the diversity of intercropped species, mimicking the structure of the forest. Source: M. Hammond. *From Riches to Rags: A Critique of the Transformation of Shifting Cultivation Societies in Contact with Expanding Political Economies of Nation States.* Unpublished MA thesis, Michigan State University, 1977.

vironments: It uses almost no purchased inputs, and when done appropriately, it does not cause lasting ecological damage.

Rotational Bush Fallow

With increasing population density and shorter fallows, shifting-cultivation systems may have to give way to more intensive forms of cultivation, such as rotational bush fallow. Rotational bush fallow involves the development of a regular system of fallows in which farmland never reverts to natural forest or savanna vegetation. Strategies such as crop rotation and intercropping, greater use of animal manure, and intensive cultivation of floodplains and household gardens are used to compensate for the shorter fallows.

Intercropping and crop rotation are crucial to the success of rotational bush fallow. Unlike shifting cultivation, relatively few crops—typically two to four complementary crops—are grown together. Farmers apply a detailed knowledge of local environmental conditions

to select the best combinations and sequences of crops. For example, the varieties grown in well-drained and wet soils may be different. Some parts of the farm may support the most favored crops for three to five years before it is necessary to resort to cassava, whereas poorer soils nearby may be planted with cassava after only a single season. Vignette 13.2 (with Figure 13.3) discusses the use of termite mounds as preferred sites for farming—an example of farmers' utilization of local environmental knowledge.

In many rotational fallow systems, the soil is heaped into ridges or mounds to permit the concentration of fertile topsoil and ash around the roots of plants. Ridges and mounds also improve drainage and aeration, reduce soil erosion, and help to control weeds. Manure is commonly used to enhance fertility in areas where there are substantial livestock populations. Manure may be collected and spread on the farm, or pastoralists may be asked to graze their herds on the farm in the dry season and thereby to convert crop refuse into manure.

FIGURE 13.2. Shifting cultivation, southeastern Côte d'Ivoire. The random mixing of several different crops is characteristic of shifting-cultivation systems. Photo: author.

Figure 13.4 illustrates how soil nutrient supply is managed in a northern Nigerian community where short-fallow agriculture is practiced. Nutrients to support agriculture come from a variety of sources: Manure from livestock, natural nitrogen fixation as a result of intercropping, *harmattan* dust, and purchased inorganic fertilizers replace nutrients harvested as food and ensure the sustainability of the system. Farmers and herders benefit from living in close proximity to each other. Production from comparatively small but intensively farmed household gardens and floodplains becomes increasingly important as growing population densities put pressure on finite land resources. Heavy applications of manure and household refuse, together with careful cultivation techniques, ensure continual high yields. Economically valuable trees such as date palm and mango are deliberately cultivated to increase farm productivity.

Permanent Cultivation

Permanent cultivation as an indigenous system of production occurs in relatively few regions. In general, areas of permanent cultivation are notable for the intensity and sophistication of agricultural development, as indicated by their very high densities of agriculture-dependent population. Far from being confined to any particular ecological milieu, intensive, permanent cultivation is found in areas as diverse as the tropical-forest fringe in southern Nigeria and the rich volcanic soils in the high plateaus in southern Uganda, Rwanda, and Burundi. As noted earlier, women are primarily responsible for crop production; however, men are involved in agricultural tasks to a greater or lesser extent, depending on the culture and on the nature of the tasks (see Figure 13.5). The organization of production varies considerably in other respects, reflecting the diversity of local social, demographic, economic, and ecological conditions. To illustrate this diversity of permanent systems of cultivation, the organization of production among the Hausa and Igbo in Nigeria will be compared.

Permanent cultivation is characteristic of the *close-settled zones* of Nigerian Hausa cities, such as Kano, Zaria, and Sokoto. The Kano close-settled zone contains about 7 million people, with rural population densities of 250–500 per square kilometer. All arable land is cultivated every year and is enriched with heavy applications of animal manure, household refuse, and so-called "night soil" (human waste). Waste products to fertilize the soil are hauled to the countryside from the large cities and taken by farmers living in rural settlements to their

VIGNETTE 13.2. Farmers in Zambia: Making the Best of a Poor Resource Base

Very infertile soils—Oxisols—are prevalent in the *miombo* woodlands of central Africa. How, then, do farmers manage to function in this difficult region? Two of their strategies—using a special type of shifting cultivation, and farming on termite mounds—are outlined below.

Bemba farmers in northern Zambia use a cultivation method known as *citemene* to overcome the inhospitable growing conditions they encounter. Farmers commence a *citemene* cycle by cutting vegetation from an area five to ten times that of the farm and carrying the cut branches to a new farm plot, where the branches are spread out, dried, and then burnt. The burn temporarily decreases the soil's acidity and gives it a significant boost in phosphorous, calcium, potassium, and other nutrients. The elevated nutrient levels decline gradually over five years or so, after which a new farm is cleared and prepared.

Citemene is a controversial practice that the Zambian government tried to ban during the 1960s and 1970s. It was criticized as a primitive method that causes widespread deforestation. In fact, farmers usually harvest only branches, permitting the trees to grow back quickly. Cultural ecologists argue that *citemene* enables farmers to farm successfully in a zone where other farming systems generally fail. It is a sophisticated agricultural system that requires a subtle understanding of the impacts of fire on soil chemistry and fertility. However, because of the large areas cleared to fertilize each farm, the system is dependent on the maintenance of very low population densities.

Termitaria, structures that contain the nests of termites, are a common feature of African ecosystems, especially in moister savanna areas. Africa has many species of termites, each of which has a distinctive type of nest. Some termitaria are pillar-like and up to several meters

(cont.)

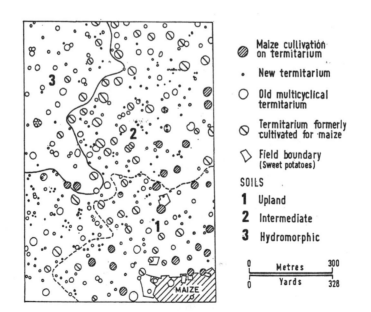

FIGURE 13.3. Termites and tillage in Zambia. Source: R. A. Pullan, "Biogeographical studies and agricultural development in Zambia," *Geography*, vol. 59 (1974), pp. 209–231. © 1974 by the Geographical Association. Reprinted by permission.

VIGNETTE 13.2. (*cont.*)

---

high, whereas others are large, relatively flat mounds up to 30 m across. The density of termitaria varies in relation to local environmental conditions; a Zambian study found an average of two to three mounds per hectare.

In the simple process of constructing its nests and sustaining colony life, the lowly termite enhances the fertility of soil, and thus contributes immeasurably to the well-being of millions of Africans. Termitaria often contain concentrations of fertile clays, brought up from many meters below the surface to construct the mound. They are rich in organic matter and nutrients derived from plant materials collected as food and from the cycle of life within the colony. In digging through heavy laterite soils, termites improve the soil's drainage and aeration.

African farmers recognize that termitaria, whether occupied or disused, have very fertile soils. In large fields, areas near termitaria stand out as having visibly healthier and higher-yielding crops. Figure 13.3 shows land use in a small study area with a large concentration of termitaria. The preferred use for these soils is maize, a crop requiring fertile soils to yield well. It is apparent that maize is being cultivated on many of the termitaria, and that numerous other termitaria had been used to grow corn in the past. Very little of the land between termitaria, and virtually none in the poorly drained lowland areas, is under cultivation.

Based on H. Brookfield. "Alternative ways to farm parsimonious soils." *Exploring Agrodiversity*, pp. 123–139. New York: Columbia University Press, 2001. R. A. Pullan. "Termite hills in Africa: Their characteristics and evolution." *Catena*, vol. 6 (1979), pp. 267–292.

---

nearby fields. The use of commercial fertilizers has increased in recent years.

Hausa farmers maximize yields from their small farms by intercropping beans and peanuts (legume crops) with millet and sorghum. Crop varieties are carefully matched to specific ecological conditions in each field. Chickens, goats, and sheep scavenge or are fed crop refuse and provide a source of dietary protein. Many types of foodstuffs and other useful products are

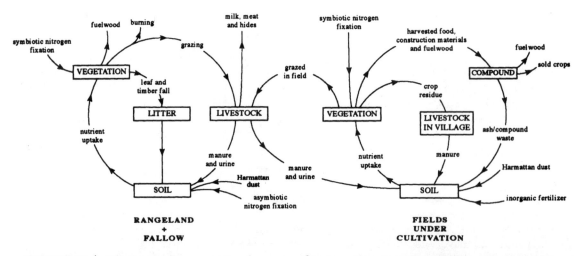

FIGURE 13.4. Nutrient cycling system in a dry savanna farming system in northern Nigeria. The diagram illustrates the complex ecological relationships that sustain this agrarian system. Source: F. Harris. "Nutrient management strategies of small-holder farmers in a short-fallow farming system in north-east Nigeria." *Geographical Journal*, vol. 165 (1999), pp. 275–285. © 1999 by *The Geographical Journal*. Reprinted by permission.

a   b

FIGURE 13.5. Gender and agriculture. (a) Women cultivators, Côte d'Ivoire. (b) Men preparing yam mounds, Ghana. Although women produce most of Africa's food, men are likely to be involved, even when women have primary responsibility for food crop production. The preparation of yam mounds is a good example of a male-dominated task; not only is the work heavy, but also the yam crop is a valuable source of cash—hence its attraction for men. Photos: (a) CIDA (R. Lemoyne), (b) CIDA (B. Paton).

obtained from flora growing in villages, on farms, and along farm boundaries. Farmers who have access to a floodplain and a source of water for irrigation will farm the floodplain very intensively, especially during the dry season, often to produce high-value vegetables for sale.

When land is scarce, maximizing returns from very small farms entails intensification of labor inputs. Farm labor use per cultivated hectare may be two to four times as high in the Hausa close-settled zones as in less densely populated regions where farms are larger. Off-farm earnings from the sale of livestock, trade, and other income-generating activities supplement households' returns from farming.

Permanent cultivation in the Igbo heartland of southeastern Nigeria applies many of the same principles and practices, but otherwise bears little resemblance to Hausa agriculture. The landscape of Igboland is dominated by the ubiquitous presence of the oil palm, once an important source of export earnings. The long rainy season enables crop production almost year-round—unlike Hausaland where most production is confined to the four- to five-month rainy season. The primary staples are root crops (yams, cocoyams, and cassava) rather than grains. The presence of tsetse severely limits possibilities for raising livestock; the Igbo consume considerably less animal protein than the Hausa.

Igboland has very high population densities, despite the prevalence of rather infertile and erosion-prone soils. Permanently farmed areas near settlements have the appearance of carefully tended gardens. Productivity is sustained at a high level through intercropping and the heavy application of manure, ashes, and crop refuse. In outlying fields, fertility is protected with short fallows during which soil-rebuilding legumes are grown.

Livestock Production

Although raising animals as a source of food and income is the work of many pastoral societies, livestock production is not monopolized by pastoralists. Most African farmers own poultry, and many own goats, sheep, donkeys, and cattle. This section focuses on pastoralist and mixed farming systems, where food production and the care of livestock are integrally linked. In the farming systems previously discussed, livestock are very commonly raised but are seldom a necessary element of the production system.

Pastoralism

Pastoralists utilize vast areas of savanna, semidesert, and desert primarily, if not exclusively, to raise livestock. Pastoralism is notable for its ability to utilize marginal environments that are too dry, too rocky, or too steeply inclined for successful farming. Cattle are the most important and most valued livestock for virtually all African pastoralists, except in desert and near-desert environments, where camels are kept. Most pastoralists also keep significant herds of goats and sheep.

The size and quality of family herds of cattle have traditionally been considered the preeminent measures of status and wealth. Gifts of cattle and sacrifices of cattle are reserved for ceremonial occasions like weddings and religious holidays. Cattle, especially females of breeding age, are sold with reluctance. Sheep and goats usually have less sociocultural significance, and thus are managed in a more utilitarian manner.

Cattle are the primary sustenance base for most pastoralist societies. Milk and yogurt are the preferred and most important foods for many pastoralists. Some East African groups consume blood taken from their cattle. Meat tends to be eaten quite rarely, except on ceremonial occasions. However, livestock are a form of "consumable capital" that can be sold or eaten in times of emergency. Cattle may also perform other utilitarian functions, such as pulling plows and carts or carrying loads.

Spatial mobility is fundamental to the pastoralist way of life (see Figure 13.6), for it permits an orderly utilization of available water and pasture throughout the year. The resources needed by pastoralists to sustain their herds are very unevenly distributed in time and space. Water and pasture are plentiful during the rainy season, but become increasingly scarce as the dry season progresses. Pastoralists gravitate during the dry season to moister localities, and they move as often as necessary to secure adequate forage. Seasonally varying risks such as livestock diseases also affect migration patterns. Many of the dry season's grazing grounds must be abandoned at the commencement of the rains, when tsetse become more numerous and widespread (thus increasing the risk of trypanosomiasis).

Pastoralism takes a variety of forms. So-called "true nomads," who move their herds in an apparently random fashion and maintain no permanent base camps, are very few in number. A more common pattern involves the maintenance of permanent settlements that are occupied throughout the year by women, children, and some men. Agriculture is often practiced around these settlements. Herders, together with the livestock, spend part of the year at the settlement and the rest on seasonally utilized pastures. Basic patterns of movement between seasonal pastures are generally well established. Certain lineages or families gravitate to places where they have an intimate knowledge of the environment, and where their access to wells and grazing lands has become customary as a

a   b

FIGURE 13.6. Pastoralists. (a) Pastoralists with a herd of sheep, Mali. (b) Pastoralist family breaking camp, Mali. Because they move frequently with their herds, nomadic pastoralists tend to have relatively few material possessions. Photos: CIDA (P. St. Jacques).

result of many years of use and cooperative interaction with local inhabitants.

The seasonal migrations of pastoralists take place within several distinct ecological frameworks (Figure 13.7). Many pastoralists move between upland pastures used during the rainy season and floodplains occupied during the dry season when pasture and water become scarce. In upland areas such as the Fouta Djallon and Abyssinian Highlands, seasonal mobility is between higher-altitude pasture and lowland pasture. In certain regions, mobility is governed by the seasonal oscillation of the boundary between tsetse-infested and safe zones; when fly density and range increase during the rainy season, pastoralists retreat to protect the health of their herds, as noted above. Another pattern, common in densely farmed areas, is organized in relation to the farming calendar. Camps are established near farming communities during the dry season, to take advantage of possibilities for trade and to utilize crop refuse as fodder. However, livestock are moved to sparsely occu-

pied areas during the rainy season, so that the animals do not damage crops.

In the marginal environments utilized by herders, environmental crises such as droughts may occur quite often. The survival of pastoralist families and their herds depends on a correct and timely "reading" of environmental signals. During bad years when normally utilized resources fail to materialize, longer, emergency migrations may be necessary to find suitable pasture and water. Decisions to deviate from established patterns of mobility are not made lightly; such migrations are risky because of the added stress of a longer journey for the livestock, as well as the herders' imprecise knowledge of unfamiliar territory and its resources. Movement into new territory also brings increased risk of conflict with farmers over land and the right to use it (Vignette 13.3).

Herd dispersal is a risk-minimizing strategy used by many pastoralists. Instead of keeping one very large herd, two or more smaller herds may be created. Herd dispersal gives some

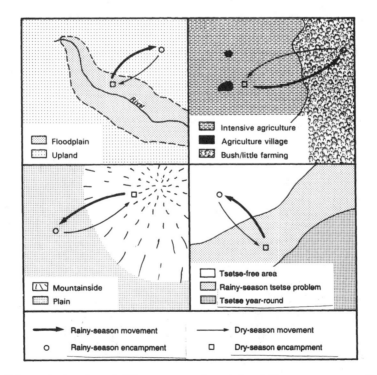

FIGURE 13.7. Schematic drawing of different types of spatial mobility among pastoralists. Mobility enables herders to use seasonally varying resources effectively, increasing returns and limiting ecological damage.

VIGNETTE 13.3. The Political Ecology of Pastoralist Societies under Pressure

Cultural ecologists have tended to emphasize how indigenous production systems involve an orderly allocation of resources and how they represent ideal strategies for the exploitation of particular environments. Reality is often more complicated. Scarce resources may be the focus of intense conflict between different ethnic groups or social classes. Thus environmental stress needs to be interpreted in relation to the broader context of political developments that affect decision making by households at a local level. This approach, which combines the concerns of ecology and political economy, is called *political ecology*.

An example is the escalation of conflict between indigenous Senufo farmers and recently arrived Fulani herders in northern Côte d'Ivoire. The Fulani moved into this region in the early 1970s from Mali and Burkina Faso, at the time of the Sahelian drought. The Fulani had previously avoided the area because of tsetse, which posed a substantial risk to the health of their animals. The Ivoirian government encouraged the settlement of Fulani as a means of strengthening the domestic livestock industry and reducing the country's dependence on imported meat.

The position of the Fulani in northern Côte d'Ivoire remains very tenuous. Herds must be moved very frequently because of the enhanced risk from tsetse. Moreover, as outsiders, they do not have traditional land use rights. The Senufo have strongly resisted the incursion of the Fulani into their territory, focusing on the financial losses they have suffered because of crop damage caused by Fulani-owned cattle and their inability to secure compensation for most of their losses. Conflicts over land and crop damage resulted in growing anti-Fulani sentiment, which culminated in several violent clashes in which numerous lives were lost. Many Fulani were forced to flee.

The Kereyu of the arid Awash region in the Ethiopian rift valley are other pastoralists under pressure due to a declining resource base. The creation of the Awash National Park deprived them of access to a large, traditionally important area of rangeland. The development of irrigated sugar farms also deprived them of critical grazing territory along the Awash River. Kereyu incursions into these now-restricted lands have brought about escalating conflict between the herders and farmers, as well as between the Kereyu and the Ethiopian government. The diminished resource base is especially critical in drought years, such as 2002, when many animals starved due to the lack of fodder. Kereyu communities near the road sold fuelwood and charcoal as a survival strategy—surely an example of longer-term pain for short-term gain in this fragile ecosystem.

Competition over disputed territory and scarce resources has fueled an increase in violent clashes between the Kereyu and neighboring pastoralist societies, which are similarly affected by drought and development projects. Because of the legacy of wars involving Ethiopia, Eritrea, and Somalia, weapons such as automatic Kalashnikov rifles are plentiful and cheap. These arms contribute to the escalation of violence. The Ethiopian government plans to establish permanent settlements for the Kereyu and their neighbors—an approach that the pastoralists utterly reject.

Sustainable solutions remain elusive for the complex issues of development affecting pastoralist societies. Approaches that pay insufficient attention to protecting the resource base needed by pastoralist communities, or that ignore the deep cultural attachments these societies have to their traditional way of life, have little chance of success.

Based on T. Bassett, "The political ecology of peasant–herder conflicts in the northern Ivory Coast," *Annals of the Association of American Geographers*, vol. 78 (1988), pp. 453–472. H. Wakota and D. G. Michael. *Awash Conservation and Development Project, CARE, Ethiopia*. Drylands Coordination Group, 1999. www.drylands-group.org/Report51999.pdf

protection against possible losses caused by drought, disease, and theft, and it reduces the rate at which the grazing resources in a particular location are exhausted.

Production techniques used by pastoralists have often been said to be uneconomic. Their cattle produce on average only 2–5 liters of milk per day, a small fraction of the production of cattle in modern dairy enterprises in the industrialized world. African pastoralists have also been accused of keeping excessively large herds, resulting in overgrazing, and are said to be very reluctant to market their livestock. However, careful analyses of pastoral economies have refuted these criticisms. Although the cattle produce little milk, they do so using scanty and unreliable resources; North American cattle would neither produce very much milk nor survive for long in most environments occupied by Africa's pastoralists. Moreover, studies of herd size have demonstrated that herds are seldom larger than the minimum needed to provide sustenance for a family, plus reasonable assurance against the possible impact of drought and disease.

For some pastoralists, growing crops has become a necessary adjustment because of the inadequate size of their herds. The loss of livestock during recent droughts of has forced many pastoralists into much greater reliance on agriculture. In Somalia, for example, many pastoral nomads turned to farming and fishing after almost all of their animals perished. Where pastoralist agriculture has developed as a strategy for economic diversification, there is often a class-based or gender-based division of labor between livestock rearing and cropping; agriculture is generally left for women or subordinate classes.

Pastoralists have come under increasing pressure because of the encroachment of agriculture and other forms of development into their traditional territories, ecological damage caused in part by increasing animal and human populations, and the devastation of drought and disease. They have been pressured by governments to "be more economical" in their behavior, and sometimes to abandon their migratory traditions. Despite such major pressures, pastoral societies are still resilient. The key to their survival has remained their sophisticated knowledge of animal husbandry and production systems that carefully and effectively utilize marginal environmental resources (see Figure 13.8).

Mixed Farming

Although most African farmers own some livestock, true mixed farming, in which there is a

FIGURE 13.8. Pastoralist's cattle grazing on crop refuse, Kaduna State, Nigeria. The photo illustrates one aspect of the symbiotic relationship between farmers and pastoralists: As the cattle graze, they deposit manure on the farmer's fields. Photo: author.

functional integration of raising livestock and growing crops, is uncommon. For example, farmers who own cattle or sheep, but do not have sufficient resources to grow fodder for the animals' use during the dry season, often entrust the care of their livestock to local pastoralists. The Serer of south central Senegal and the Amhara and Gurage of Ethiopia are examples of groups that do practice mixed farming.

The Serer own cattle and cultivate crops. Because of land scarcity, Serer agriculture involves heavy applications of animal manure to enhance yields. Cropped land and pasture are rotated in a systematic fashion. Cattle are also used extensively as draft animals to pull plows and carts.

Mixed farming is practiced in much of Ethiopia, the country with the largest number of livestock in Africa (see Figure 13.9). The Amhara are mixed farmers who fully integrate cropping and animal husbandry. The Gurage practice a distinctive and extremely intensive form of mixed farming in which enset is the main crop. Enset is a banana-like plant, the stem of which is used to make a starchy food. Manure from confined or communally pastured livestock is essential to the maintenance of this unique form of agriculture.

Indigenous forms of food production, as practiced by the Amhara and Gurage, still constitute the very heart of Africa's rural economies. Nevertheless, such systems are under pressure in many places, as a consequence of forces as diverse as environmental degradation, population pressure, the loss of a land base to competing uses, political instability and violence, and counterproductive government policies.

The long-standing conviction that indigenous systems of production are primitive, unchanging, and of little relevance to Africa's future—widespread among African government officials, aid workers, and agricultural scientists—is beginning to change. Newer research has provided a much better understanding of the logic and effectiveness of systems such as shifting cultivation and pastoralism, which, far from being primitive and rigid, continue to change in response to a variety of influences. The ideal rural development strategy is one that builds upon the strengths and potentialities of existing systems of production. Unfortunately, as the next chapter will show, this lesson has been learned very slowly.

Further Reading

Classic studies of hunting and gathering societies include the following sources:

FIGURE 13.9. Plowing with oxen, Tigre Province, Ethiopia. Two small stone barriers are visible behind the plowman; these structures have been built to trap rainwater and impede soil erosion. Photo: M. Peters.

Marshall. L. *The !Kung of Nyae Nyae.* Cambridge, MA: Harvard University Press, 1976.

Turnbull, C. *The Forest People.* New York: Simon and Schuster, 1968.

There are many studies that examine the logic of African indigenous farming systems in particular regional settings. These include the following:

Jones, W., and R. Egli. *Farming Systems in Africa: The Great Lakes Highlands of Zaire, Rwanda and Burundi.* Washington, DC: World Bank, 1984.

Mortimore, M., and W. M. Adams. *Working the Sahel: Environment and Society in Northern Nigeria.* London: Routledge, 1999.

Netting, R. M. *Hill Farmers of Nigeria: Cultural Ecology of the Kofyar of the Jos Plateau.* Seattle: University of Washington Press, 1968.

Scoones, I., ed. *Dynamics and Diversity: Soil, Fertility and Farming Livelihoods in Africa: Case Studies from Ethiopia, Mali, and Zimbabwe.* London: Earthscan, 2001.

The following sources consider aspects of the broader logic of indigenous farming systems, including innovation and soil conservation strategies:

Mook, J. L., ed. *Understanding Africa's Rural Household and Farming Systems.* Boulder, CO: Westview Press, 1986.

Reij, C., and C. Toulmin, eds. *Sustaining the Soil: Indigenous Soil and Water Conservation in Africa.* London: Earthscan, 1996.

Reij, C., and A. Waters-Bayer, eds. *Farmer Innovation in Africa: A Source of Inspiration for Agricultural Development.* London: Earthscan, 2002.

Richards, P. *Indigenous Agricultural Revolution: Ecology and Food Production in West Africa.* Boulder, CO: Westview Press, 1985.

Vaughn, M., and H. Moore. *Cutting Down Trees: Gender, Nutrition, and Agriculture in the Northern Province of Zambia.* Portsmouth, NH: Heinemann, 1993.

Issues pertaining to pastoral societies and economies are examined in the following sources:

Adamu, M., and A. H. M. Kirk-Greene, eds. *Pastoralists of the West African Savanna.* Manchester, UK: Manchester University Press, 1986.

Anderson, D. M., and V. Broch-Due, eds. *The Poor Are Not Us: Poverty and Pastoralism in East Africa.* Athens, OH: Ohio University Press, 2000.

Bassett, T. J. "Fulani herd movements." *Geographical Review*, vol. 76 (1986), pp. 233–248.

Dahl, G. *Suffering Grass: Subsistence and Society of Waso Borana.* Stockholm: Stockholm Studies in Social Anthropology, 1979.

Monod, T., ed. *Pastoralism in Tropical Africa.* London: Oxford University Press, 1975. (See Monod's introductory essay, pp. 99–183.)

Pratt, D. J., et al. *Investing in Pastoralism: Sustainable Natural Resource Use in Arid Africa and the Middle East.* Washington, DC: World Bank, 1997.

Smith, A. B. *Pastoralism in Africa: Origins and Development Ecology.* London: Hurst, 1992.

Internet Sources

The following are Web-based resources on African farming systems:

Brandt, S., et al., American Association for the Advancement of Science. *The Tree against Hunger: Enset-Based Agricultural Systems in Ethiopia.* www.aaas.org/international/africa/enset/studies.shtml

Consortium for International Crop Protection. *Indigenous Crop Protection Practices in Subsaharan East Africa: Their Status and Significance to Small Farmer IPM Programs in Developing Countries.* www.ippc.orst.edu/ipmafrica/kb/contents.html

Dixon, J., and A. Gulliver. *Farming Systems and Poverty: Improving Farmers' Livelihoods in a Changing World.* www.fao.org/DOCREP/003/Y1860E/y1860e00.htm

National Research Council on Science and Technology for International Development. *Lost Crops of Africa: Grains.* www.nap.edu/books/0309049903/html

Several organizations maintain websites dedicated to information on pastoralist issues:

Eldis. *Eldis Pastoralism Resource Guide.* www.eldis.org/pastoralism/

International Fund for Agricultural Development. *Livestock and Rangeland Knowledgebase.* www.ifad.org/lrkm/range/

Pastoral and Environmental Network in the Horn of Africa (PENHA). www.penhanetwork.org/page1.shtml

14

Agrarian Development and Change

Africa has more unexploited arable land than any other continent. Development planners have often expressed optimism about Africa's potential and have made many proposals over the years for the transformation of African agriculture and pastoralism. Attempts at major structural change, such as the establishment of plantations or state farms, have usually achieved poor results. The alternate approach, presently favored by many development agencies, focuses on the productive capacity of small farmers.

Colonial Effects on Indigenous Production Systems

Colonial rule brought many changes to African agriculture. Prior to the 1970s, colonial impacts on agrarian development were usually viewed in a positive light. Emphasis was given to the apparent benefits to Africa of the introduction of new crops, new market opportunities, and new technologies. Although coercion was often used to bring about desired changes, such measures were seen to be fully justified. Scholars from the dependency school dismissed this notion of colonial altruism; rather, the motivation behind colonial policies was seen as

blatant self-interest. These policies undermined the viability of indigenous systems by squeezing resources from small producers and upsetting ecological balances that, it was argued, Africans had maintained for centuries.

Large areas of land were expropriated and reserved for European farms, most notably in Kenya, Southern Rhodesia (now Zimbabwe), and South Africa. Africans expelled from their lands were relocated in confined reserves, where less fertile soils and necessarily shorter fallow periods caused declining productivity. The tax burden in settler-dominated colonies was placed squarely on the shoulders of Africans. Dual pricing systems were established that granted significantly higher prices for European crops and livestock.

Forest and game reserves were created, with often serious implications for farmers and pastoralists who had traditionally used these territories. Pastoralists lost vital seasonal pastures and had their normal migration routes severed. The increasingly dense vegetation and growing wildlife populations in these reserves created ideal conditions for the proliferation of tsetse, which in turn made the surrounding countryside increasingly dangerous for humans and livestock.

Rural labor was appropriated to serve the interests of the colonial state. Every colonial power relied at times on forced labor to recruit troops, porters, construction workers, and plantation laborers. For the peasantry, time lost from agriculture meant smaller harvests and often hunger. Other forms of labor appropriation—such as the "voluntary" seasonal migration of peasants to earn money for taxes—had similar implications for peasant agriculture, despite their more benign appearance.

Cash crops (see Figure 14.1) were promoted as a means of involving Africans in the commercial economy and ensuring a supply of tropical products for European industry. Some of these crops, including cotton, groundnuts, and oil palm, had been widely cultivated in precolonial Africa; others, such as tea (see Figure 14.2), were introduced for the first time. Africans frequently resisted the introduction of cash crops, because it diverted scarce resources away from food crop production. Some cash crops were more compatible than others with existing farming systems. For example, groundnuts could be intercropped with millet and actually increased grain yields by fixing atmospheric nitrogen in the soil, whereas cotton was viewed as a "soil robber" and poorly suited to intercropping.

When fairly attractive prices were offered for cash crops that were compatible with indigenous farming systems, African farmers could become enthusiastic adopters. Migrant farmers planted vast areas of unfarmed forest in the Gold Coast (now Ghana) with cocoa. However, the prices offered to African producers were seldom attractive, and cash cropping often clashed with food cropping. In such cases, peasant resistance to cash cropping made sense. The appropriation of profits through taxation and monopoly pricing was justified as a means of securing capital for development, but little of what was actually spent on development was of much benefit to rural areas.

Colonial agricultural agents attempted to introduce changes in indigenous agricultural practice that reflected European views of how farming should be done. For example, mixed farming integrating cropping and raising livestock was often promoted, but with little success, because few African farmers had sufficient land and wealth to engage in mixed farming. Monoculture was also promoted, despite evidence that traditional farming practices such as shifting cultivation and intercropping were actually superior. African farmers' disinclination to change seemed to confirm their backwardness. The myth of the unprogressive peasant helped to sustain the view that Africa's land and labor could be most effectively exploited by developing European-style farms and plantations.

Postcolonial Policies and the Neglect of Agriculture

During the 1960s, the expansion and diversification of cash cropping were seen as obvious vehicles to finance modern development. Farmers paid a "hidden tax" in the form of the low prices that they received from state agencies for cash crops. These prices were typically set well below prevailing world market prices (Table 14.1). But increased production often brought reduced prices in world markets, because the demand for products such as coffee and cocoa is relatively inelastic (i.e., consumption is unlikely to change much in response to changes in price). For other commodities, competition from other sources (e.g., U.S. mechanized peanut farm vs. African groundnut growers) or from substitute products (e.g., synthetic products used to replace sisal) depressed both demand and price. Finally, relatively little attention was paid to the possible role of locally important cash crops (see Figure 14.3).

Prior to the opening of the International Institute of Tropical Agriculture (IITA) in Ibadan, Nigeria, in 1968, food crops were virtually ignored in agricultural research and development. Little was known about the logic of indigenous production systems or the potential utility of improved food crop varieties and related innovations. When the need to actively encourage greater food production was finally recognized, many African nations were already suffering from recurrent food shortages.

The most common response to food deficits was to increase food imports, especially wheat and rice, which had become an important com-

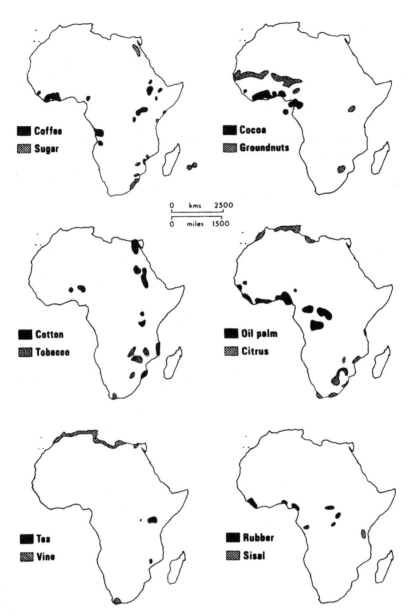

FIGURE 14.1. Major zones of cash crop production. Source: I. L. L. Griffiths. *An Atlas of African Affairs.* London: Methuen, 1984, p. 121. © 1984 by Methuen and Co. Reprinted by permission.

ponent of urban diets. Imported foods were often sold at less than the market prices of locally produced foods. Although food subsidies reduced the likelihood of unrest among the urban population, they also undercut indigenous food producers. Depressed prices for both food crops and cash crops created a massive disincentive for peasants to invest in their farms.

The overall structure of postindependence development has hastened the decline of agriculture. Schools and health clinics were constructed in many rural districts, but social services in urban areas remain far superior. The probability of migration was also increased by educational systems that seldom promoted agriculture or provided skills designed to foster rural development. Rather, the "hidden curriculum" tended to extol the values of modern ur-

FIGURE 14.2. Tea plantation, Tanzania. Although Africa produces only a small percentage of the world's tea, it is locally important in several East and central African countries. Photo: CIDA (D. Barbour).

ban life. It is little wonder that more and more youth aspired to an urban future and had little enthusiasm for agriculture or rural life.

Attempts to Restructure African Agriculture

Most African countries, eager to boost food production and export earnings, went through phases of importing modern agricultural machinery and establishing or promoting the development of large-scale farm enterprises. There were a few qualified successes, but these approaches typically had minimal effects on national production. In contrast, they typically

had adverse effects on indigenous production systems—often through alienating land normally used by peasants and herders, and through diverting developmental resources away from small-scale producers.

The following discussion outlines some characteristics and problems of four types of development initiatives to restructure agriculture: (1) state farms, (2) planned resettlement, (3) large-scale capitalist agriculture, and (4) production under contract.

State Farms

The first experiments with state farms occurred in the early 1960s in Ghana and Guinea. The

TABLE 14.1. The "Hidden Taxation" of Cash Crops in Selected Countries: 1976–1980 Producer Price as a Percentage of Fair Market Price[a]

	Cocoa	Coffee	Cotton	Groundnuts	Tobacco
Cameroon	45	36	79		
Malawi			75	59	28
Mali			44	43	
Sudan			60	67	
Togo	25	23	79		
Zambia				71	88

Note. Blank entries indicate that the country does not produce a significant amount of the product.
[a]World price minus transport and marketing costs.
Data source: World Bank. *Accelerated Development in Sub-Saharan Africa*. Washington, DC: World Bank, 1981 (Box D, p. 56).

FIGURE 14.3. Agricultural landscape, eastern highlands of Ethiopia. The most important cash crop in this region is *quat* (pronounced "chat"), visible as the clusters of bushes in the photo. Quat leaves have an intoxicating effect when they are chewed. Large quantities are exported to neighboring countries, especially Somalia, Djibouti, and Yemen. When we are thinking of cash crops, it is important to look beyond staples traditionally sold abroad and to consider a range of locally important crops. Photo: author.

collective farms of Eastern Europe provided the model for agrarian transformation. The experiments failed badly, largely because of poor management. Heavy investments in farm machinery were wasted because of parts shortages, poor maintenance, and adverse environmental conditions. Moreover, peasants were reluctant to commit themselves fully to collective work on state farms, preferring instead to farm their own land. Mozambique, Angola, and Ethiopia later tried to establish state farms as a part of a broader socialist transformation. In the former Portuguese colonies, farms formerly owned by white settlers were consolidated to form state enterprises. Mozambique invested heavily in mechanization, but was unable to mobilize adequate management. The results were so poor that most of the farms were abandoned by the mid-1980s.

Although state farms were most often associated with Marxist regimes, moderate socialist and capitalist regimes also experimented with them. For example, Tanzania established large-scale mechanized wheat farms with Canadian assistance (discussed in Chapter 25, Vignette 25.2). Zambia's experiment with state farms was unusual; each of several huge model farms was set up and managed as a foreign aid project by a different country (the United States, the Soviet Union, China, and Canada, among others). Zambia derived few benefits from this experiment, in which the quest for foreign aid took precedence over the development of a coherent national strategy of development built around an assessment of the country's own needs and resources.

Land Reform and Planned Resettlement

Planned resettlement has been initiated in several countries, but few of these initiatives achieved the anticipated benefits. One of the most common reasons for resettlement programs has been to accommodate people displaced by dams and other major projects. Resettlement has been used in several countries as a strategy for political and economic transformation. The settlement of Africans on farms reclaimed from European settlers in Kenya and Zimbabwe, the creation of *ujamaa* villages in Tanzania, and the removal of Ethiopians from drought-stricken plateaus to the sparsely populated periphery provide varied examples of politically informed resettlement.

One of the largest and most successful resettlement programs was undertaken in Kenya during the early 1960s. After the Mau Mau revolts, with independence imminent, many white farmers wished to leave the country. A series of resettlement projects was developed in which the government purchased European farms, subdivided them into smaller units, and sold them to Kenyans. The preparations involved careful evaluation of soil fertility and studies to determine optimum sizes for viable farms. The initial schemes focused on middle-class farmers who could afford to invest in relatively large farms of 6–20 ha. The resettlement program was later expanded to accommodate some 35,000 landless families in a new project, known as the Million-Acre Scheme. Colonial officials had little confidence in these poor, apparently inexperienced farmers. However, the performance of Kenya's resettlement schemes confounded the expert predictions. Total agricultural output increased 4% per year during the 1960s and early 1970s. Despite having inferior resources, smallholders were soon obtaining higher yields and higher profits than the middle-class owners. The extensive promotion of commercial farming innovations, including hybrid maize, various cash crops, and dairy cattle, was one of the keys to the success of the Kenyan program.

A similar approach to land reform was planned for Zimbabwe's transition to independence in 1980. However, few of the international funds pledged to purchase white-owned farms for redistribution to small farmers materialized. Most of the white farmers remained, and large-scale commercial farming continued to dominate the agricultural sector. Disparities in wealth and access to land were causes of resentment, and in 2000 land ownership became the country's paramount political issue. President Robert Mugabe declared that most of the land occupied by white farmers would be seized and allocated to landless Africans. This politically motivated land seizure resulted in violence as well as major economic and political instability in Zimbabwe, and was widely condemned internationally.

Tanzania attempted to achieve rural transformation by encouraging its scattered population to form new *ujamaa* villages. *Ujamaa* villagers were to devote a significant amount of their time to communal efforts, including work in collective village farms. The early promise of this scheme waned, especially when increasing levels of coercion were applied in what had begun as a voluntary program. Moreover, with its total economy in decline, Tanzania was less and less able to provide material support for rural development.

Ethiopia's socialist government began a program to resettle families from the densely populated heartland of the country to the sparsely populated western and southwestern peripheries. This program was greatly expanded at the time of the 1984–1985 drought; over half a million people were moved in just over one year. The government perceived resettlement as a cure-all for the famine and its deeper environmental roots, and as a way to lessen its dependence on foreign aid by opening up fertile, underutilized territory. However, the program was implemented hastily without careful planning and with inadequate resources to ensure an orderly and successful transition.

The claims of some international organizations that the Ethiopian government was forcing people to migrate, and that tens of thousands had perished because of inadequate planning, seem to have been overstated. Although many resettled persons (e.g., unemployed urban youths) were coerced, most migrated voluntarily, albeit because of their desperate circumstances during the famine. Many migrants later returned to their former homes, but most have remained behind and become successful farmers in their new environment. After 1986, the Ethiopian government curtailed the program, primarily because of the lack of resources to support successful resettlement.

Large-Scale Capitalist Agriculture

In most of Africa, large-scale capitalist agricultural projects, whether initiated by domestic or foreign companies, play a quite minor role. However, large-scale agriculture is well developed in some countries, such as Côte d'Ivoire, and is growing rapidly in others (see Figure 14.4). Nigeria allows transnational companies

a   b

FIGURE 14.4. Modern agriculture. (a) Modern dairy farm, near Lusaka, Zambia. (b) Large mechanized wheat farm, Eastern Rift Valley, Kenya. Modern wheat farms have been established in several countries, including Tanzania, Kenya, and Sudan, in an effort to reduce dependence on imported wheat. Photos: author.

to establish large estates to produce more of their own raw materials (e.g., cotton for textile mills and grain for breweries) and to grow food for the domestic market. This approach conforms to the long-standing view of those in power—namely, that future prosperity depends on modernizing agriculture, rather than on developing fully the potential of indigenous farming systems. According to this vision, more and more Africans will work as wage laborers rather than as independent producers. It implies that spatial patterns of development will become increasingly uneven. In most cases, it also implies that non-Africans will exert control over scarce agricultural resources.

In South Africa and Zimbabwe, agriculture was dominated by large commercial farms owned by white farmers. These farms produced a wide variety of commodities—livestock, grains, vegetables and fruits, and other specialty crops—using modern technologies and large amounts of black labor. Their output included high-value-added products, such as the wines produced in the Cape region. Legislation, particularly the South African Native Lands Act of 1913, and the expansion of white commercial farms during the 20th century left virtually no space for small-scale African farmers producing for the market. Africans who were not working on commercial farms were removed into tiny designated reserves, where

overcrowding and adverse environmental conditions limited agriculture to very basic subsistence production.

The structure of the South African agricultural sector has not changed substantially since the end of apartheid. Indeed, the removal of international sanctions has expanded market opportunities for commercial farmers to increase exports and make new investments in land and technology. Land reform, first promised by the African National Congress in its Freedom Charter of 1955, remains a major political challenge. The government has moved slowly and deliberately since 1994 to implement a land reform program to redress past injustice and alleviate rural poverty (Vignette 14.1). However, redistribution has fallen far short of the promised 30% of agricultural land within five years.

Production under Contract

During the 1980s and 1990s, increasing numbers of Africans became involved in the production under contract of higher-value horticultural export crops. These crops include fresh vegetables and fruit and fresh cut flowers, most of them bound for European markets. Kenya, Zimbabwe, South Africa, Zambia, Gambia, and Côte d'Ivoire have been among the major participants in this new trade. Although African producers have yet to overtake the leading

VIGNETTE 14.1. Land Reform Policies in Postapartheid South Africa

Centuries of white rule systematically dispossessed black South Africans of their land. Under apartheid, some 87% of the land was reserved for the white minority, while blacks had no rights to land outside of reserves occupying 13% of the land. Moreover, almost all of the higher-quality arable land was under white control.

Land reform to redress the injustices of dispossession under white rule has long been a fundamental commitment of the African National Congress, which it began to address through policy initiatives following its election in 1994. The new constitution of South Africa provided comprehensive protection for property rights. Land reform was to be financed by the government, with land to be acquired from willing sellers at fair market prices.

South Africa has initiated three parallel programs of land reform:

1. Land restitution provides a mechanism for the return of lands to individuals and groups dispossessed under the Native Lands Act of 1913, and as a result of subsequent racist policies. Some 67,000 claims were filed with the Land Claims Court and Commission; the resolution process has proven to be slow and difficult, especially in rural cases where up to 10,000 people may be involved in a single claim.
2. Tenure reform is designed to protect the property rights of groups with insecure tenure, including residents of the former reserve areas (where chiefs had exercised control over customary tenure), farm worker households, and periurban squatters.
3. Redistributive land reform is intended to provide opportunities for black households who were ineligible for the other programs to gain access to land. The program incorporated opportunities for diverse individual and communal tenure, and focused on the needs of poor households.

The South African government has worked hard to maintain a delicate balance in its approach to land reform. On the one hand, it recognizes the legitimate impatience of millions of poor South Africans hungering for justice and improved quality of life through land reform. On the other hand, it has remained dedicated to the rule of law—in the process, offering strong safeguards for the property rights of all landowners, including whites. This stance has been important in maintaining investor confidence and discouraging illegal occupations by landless groups. Their challenge also includes balancing land reform against other priority programs in such areas as housing, education, and health care.

Progress has been very slow. Less than 2% of land had been redistributed by the year 2003, far below the 30% five-year target established in 1994. Only 0.3% of the national budget is allocated to land reform initiatives. Nevertheless, the slower pace of the initial years has provided the state with opportunities to establish a strong administrative base and fine-tune its land reform policies. As a result, the pace of redistribution increased each year from 1994 until 2000.

Poverty reduction has been a fundamental goal of land redistribution in South Africa. The program has succeeded in giving priority to the poor. Most of the resettlement projects have achieved early economic success, with participating households achieving significant increases in income. Nevertheless, the great majority of South Africans who long for land and hope to benefit from land reform remain on the outside, looking in. How long will they remain patient? Although many South Africans are frustrated by the slowness of reforms, there have been relatively few cases of illegal occupations of land.

sources in Latin America (e.g., Mexico and Argentina) and Asia (e.g., China), Africa's market share is increasing rapidly. Kenya has become Europe's largest supplier of fresh horticultural produce, and South Africa is the world's third largest exporter of citrus fruits.

Production and marketing arrangements for higher-value horticultural goods differ from those associated with lower-value, bulky, less perishable "traditional" exports such as coffee. Assured product quality, flexible and reliable supply, and the speediest possible delivery to market are essential for highly perishable horticultural products. In contrast, traditional commodity exports required little or no special handling. In the horticultural sector, producer contracts govern the marketing chain that links growers to retailers, located continents apart. These contracts specify not only the quantity and timing of product delivery, but also production methods such as the use of pesticides.

Africa has several attractions as a source of produce for Europe. It offers diverse climate conditions, permitting the growth of most tropical and many temperate products. The climate permits year-round production wherever water supply is assured, while Southern Hemisphere locations offer counterseasonal supplies of produce. Cheap labor and land keep production prices low, and shipping time and costs may be lower because Africa is closer to Europe than some of its competitors are.

Produce exports offer new opportunities for producers in several African countries. Nevertheless, these opportunities come at a cost, because of the ways that importing firms control production and marketing arrangements (Vignette 14.2). Smaller producers often find it difficult to meet contract specifications, especially when these force them to give priority to the export crop over all other considerations. It has increased conflict over land and labor between male and female members of households. Heavy use of pesticides, especially for flower production, is a health threat that is especially serious for children and women. On the slopes of Mount Kenya, the rapid growth of horticultural production is resulting in water shortages and struggles over water rights.

Focus on the Small Farmer: The World Bank

In a 1973 speech, Robert McNamara, then president of the World Bank, signaled a major policy shift: He argued that the key to solving Third World poverty lay in raising the "low productivity of millions of small farms." The World Bank began to emphasize lending approaches such as integrated agricultural development projects (IADPs), which it claimed would assist the rural poor. Later, structural adjustment programs came to dominate the World Bank's agenda, including its approach to agrarian development.

Integrated Agricultural Development Projects

The IADPs were designed to promote a "package" of innovations and investments so as to attack possible barriers to progress. For example, it was reasoned that many of the benefits of improved yields would be lost if storage, transportation, and marketing systems remained deficient. The most important parts of the IADP package consisted of improved varieties of seeds and agricultural chemicals—fertilizers, herbicides, and pesticides—and a supply network to make these inputs accessible to farmers. Technological innovations, such as improved plows and small pumps for irrigation, were also promoted. Improved extension was seen as the key to convincing farmers to adopt the recommended practices. So-called "progressive farmers" were targeted for extension services, on the assumption that they would convince others to participate. IADP programs typically included components to improve local services, including investments in village water supplies, schools, and clinics to enhance productivity, and better roads to support more efficient marketing.

The integration of several dimensions of rural development was a vast improvement over previous approaches that focused narrowly on selected innovations (e.g., hybrid seeds) without considering other factors affecting their utility. The IADPs also won praise for attempt-

---

VIGNETTE 14.2. Contract Production of Horticultural Produce in Kenya: Opportunities and Challenges

In 1996, Kenya exported 85,000 metric tons of fresh horticultural produce—an increase of 58% over five years. Sales of cut flowers were increasing by 20% per year in the 1990s, while exports of fresh vegetables such as green beans and okra, and fresh fruits such as avocados, have grown steadily. This rapidly growing industry accounts for 10% of Kenya's export earnings, and it has become a major source of employment.

The growth of horticultural exports has been driven by the growing demand of consumers in high-income economies for fresh, healthy, ethically produced foods, and by the intensive efforts of supermarket chains to create new markets for fresh, exotic produce sold at premium prices. Supermarkets use their dominant position to dictate terms for would-be suppliers. Supply must be reliable, but also flexible (e.g., meeting significantly increased demand during holidays). Production methods (e.g., the use of pesticides) are monitored closely. Produce must conform to the chains' strict expectations for quality and uniformity. To ensure freshness, over 90% of Kenya's fresh produce exports are shipped by air. All possible means are utilized to increase the speed and efficiency of delivery from the farm to the supermarket, at the lowest possible price.

The strict demands of the produce trade favor large commercial producers over small farmers. Kenyan flower production, for example, is dominated by ten major growers concentrated near Lake Naivasha. Large growers have the resources of land, technology, labor, and credit to undertake large-scale, flexible production under contract for European importers. They also offer the assurance of traceability (i.e., the supermarket knows the source of the product and thus can verify production techniques and hold the grower responsible for supply or quality problems).

Small-scale producers have a secondary role in the supply chain, usually as subcontractors to larger producers for certain crops. These arrangements are tenuous; smallholders assume considerable risk when they commit scarce resources to producing a commodity that they may not be able to sell if their produce is judged to be of inferior quality, or if there is excess supply at the time of harvest. Exporters who purchase from small-scale growers often complain about the reliability of supply and quality of their produce. Over time, it is anticipated that most of the smaller producers will be squeezed out of the market.

Kenya's horticultural industry has achieved remarkable growth, but its long-term success remains in doubt. The supermarkets that dictate the terms of production and trade have no investments in Kenya, and they will shift quickly to other sources of supply if Kenya fails to maintain its comparative advantage with respect to price, quality, and assured supply. This system is noteworthy for the extent to which the chains are able to exert control over the minute details of production thousands of kilometers away, with significant implications for the development of agriculture in diverse African regions.

Based on H. Barrett et al. "Globalization and the changing networks of food supply: The importation of fresh horticultural produce from Kenya into the UK." *Transactions of the Institute of British Geographers*, vol. 24 (1999), pp. 159–174.

---

ing to integrate innovations into indigenous production systems. Supporters of the IADP approach pointed to significantly increased production in several projects. The gains were not always in the targeted areas, however. For example, the Funtua, Gusau, and Gombe IADPs in Nigeria did not attain their primary objective of increasing cotton production, although farmers did achieve a significant increase in their production of food crops.

Critics of the IADP approach focused primarily on the high cost and unevenly distributed benefits of these projects, and on the increased dependence on foreign inputs and capital. World Bank loans to finance IADPs resulted in increased indebtedness for a number of countries. With the decline in trade, increases in export crop revenues that were to pay for the loans often failed to materialize. Meanwhile, the cost of imported inputs, including seeds and chemicals essential to the IADP approach, continued to rise. The IADPs increased socioeconomic inequality as well—both between the project areas and other regions, and within the communities themselves. Local elites benefited from "progressive farmer" incentives to obtain IADP technology, but then used the profits to expand their holdings at the expense of poorer farmers.

Structural Adjustment and the Liberalization Agenda

The high cost of the IADP model limited its applicability to higher-potential regions in richer countries. For most of Africa, the IADP was never a realistic option. During the 1980s, the International Monetary Fund (IMF) and the World Bank initiated structural adjustment programs, with the stated objective of creating the conditions for sustainable economic growth. These programs were designed to unleash market forces through the removal of state subsidies and regulations that were considered impediments to investment and trade. It was anticipated that once the forces of supply and demand operated freely, rising prices would provide incentives for farmers to expand production. Market competition would ensure that farmers received the highest possible price for their goods, and that inputs would be sold widely and at the best possible price. Currency devaluation would protect farmers from cheap food imports and stimulate commodity exports, while reducing tariffs and taxation. Credit provision, extension, and other support formerly provided by the state would be transferred to the private sector; the state's role was mostly to avoid interference in the workings of the market.

Structural adjustment programs have helped to stimulate agricultural growth, but actual results have often fallen far short of expectations. These programs have often failed to take into account social and economic constraints that limit the ability of small-scale farmers to respond to market incentives, even when they wish to do so (Vignette 14.3). Most of them lack the financial resources to purchase inputs of land, labor, improved seed, and chemicals to increase their production. They do not have the means to support investments in tree crop production, where there would be little if anything to harvest for a number of years after the initial planting. The constraints of inadequate access to land, labor, and credit have been particularly serious for women farmers. As a result, progress has been too narrowly based; wealthier and male producers have benefited most, usually at the expense of the poorer and female farmers.

A second problem has come from market inelasticity, especially for traditional African commodities such as coffee and cocoa. World market prices have stagnated, wiping out anticipated gains for countries that had succeeded in increasing production. Supplying higher-value commodities (e.g., cut flowers and fresh produce for export) has provided farmers in several countries with new opportunities, but in the longer run these markets are likely to be quite unstable.

Ways Forward

"What kind of development and for the benefit of whom?" This question goes to the heart of debates about African agriculture. Will development be compatible with existing production systems, or will it have the effect of undermining them? Will development reduce rural poverty, or will innovations help to widen the gap between rich and poor? Will development be equally accessible and beneficial for men and women?

The ways forward must start with a genuine commitment by international financial institutions and governments to broad-based rural development. For example, many past initiatives

VIGNETTE 14.3. The Political Ecology of Failed Agrarian Reforms in Ghana

The Ghana Ministry of Agriculture, working closely with the World Bank, prepared a strategy to stimulate growth in the agriculture sector in 1990. The strategy was informed by the principles of Ghana's structural adjustment program, first implemented in 1983. It was predicted that annual growth of 4% in agricultural production would be achieved as a result of market incentives, such as higher producer prices, improved transportation, and the removal of other disincentives. It was anticipated that growth would come from a combination of areal expansion and intensified production from the increased use of inputs such as fertilizers. Growth was anticipated in diverse sectors—namely, in the increased production and marketing of food crops, the revival of cocoa and other traditional export crops, and the adoption of new cash crops (e.g., cashews).

Although production increased in each of these sectors, the 2% annual growth achieved in the early and mid-1990s was disappointing, and indeed did not keep pace with the rate of population growth in the country. When asked, farmers acknowledged that the improved market conditions were an attractive incentive to increase production. Why, then, were the results so poor?

Interviews of farmers in Berekum District in western Ghana revealed that social class and gender were significant factors determining whether a farmer increased his or her production. Whereas over 90% of farmers in the highest-income category had increased their production, less than 10% of the lowest-income group had done so. Male farmers were more likely to increase production than females, especially in the highest-income group. Women farmers face extra labor constraints because they have responsibility for domestic work, and thus have less time to farm.

The key to expansion was the farmers' ability to gain access to family and hired labor. Labor inputs have increased for Berekum farmers because of the recent proliferation of a noxious weed, *Chromolaena odorata,* which grows rapidly and chokes out the crop if it is not removed promptly. Weeding is an arduous task that accounts for almost half of the total labor requirements. Poorer farmers cannot afford to hire labor, and so must rely on household members for assistance. Women find it hard to find time for their own farming activities when faced with repeated requests for assistance from both their husbands and their matrilineal kin. The ability to "command" labor, whether family or hired, depends on one's social standing. Those who have displayed both generosity and humility—highly valued qualities in the society—find it much easier to obtain assistance than is the case for those whose social standing is poor. Hired workers are reluctant to work, much less to work carefully and hard, for someone who is poorly regarded in the community. Likewise, family members will be far from enthusiastic if the request for help is from someone who has neglected his or her family responsibilities.

Access to labor is not the only hurdle faced by poorer farmers and by women who wish to expand farming activities. Land is a scarce resource; those with wealth are in a much better position to purchase land or otherwise negotiate the secure tenure that is essential for investment in tree crops such as cocoa or cashews. Those lacking financial resources also find it hard to purchase fertilizers, herbicides, and other inputs for farming operations. Programs that seek to harness the development potential of small farmers are unlikely to succeed unless the full range of constraints facing the broad mass of producers is taken into account.

Based on L. Awanyo. "Labor, ecology, and a failed agenda of market incentives: The political ecology of agrarian reforms in Ghana." *Annals of the Association of American Geographers*, vol. 91 (2001), pp. 92–121.

have had the effect of increasing the marginalization of women, who outnumber men as farmers in most of Africa. As previously shown, current structural adjustment strategies have proven inadequate for the task. They have not only failed to achieve promised growth and diversification, but have actually increased rural poverty.

Greatly increased research on indigenous agriculture is needed to explore local practices in detail, focusing on elements that facilitate self-reliance and are ecologically sound. Agroforestry research provides an excellent example of the benefits of increasing our understanding of indigenous systems as a framework within which innovation—increased cultivation of income-producing trees, for example—may take place. Appropriate low-cost innovations can greatly improve the prospects for indigenous agriculture. For example, the development of *alley cropping* at the IITA has provided an ecologically sound, low-cost alternative to shifting cultivation. Alley cropping involves planting rows of leguminous shrubs that add nitrogen to the soil and reduce soil erosion (Figure 14.5). The use of alley cropping in marginal tropical forest ecosystems eliminates the need for lengthy fallow periods and the continual shifting of farm locations. It provides an excellent working model of a practical, sustainable agricultural system.

Progressive change also entails more self-reliant involvement of rural people in local development. However, the nature of official support for local improvements is crucial if local initiative is to be mobilized; peasant skepticism remains as a legacy of decades of heavy-handed government interventions. Participatory rural development provides an apt solution in which local communities make key decisions themselves about their needs and ways to achieve their objectives. For this strategy to be successful, the outside experts need to pass control over decision making to the community, and different groups—men and women, rich and poor—need to work together to create broad-based benefits.

Further Reading

The literature on colonial agricultural policies is large and diverse. See the following, for example:

Tosh, J. "The cash crop revolution in tropical

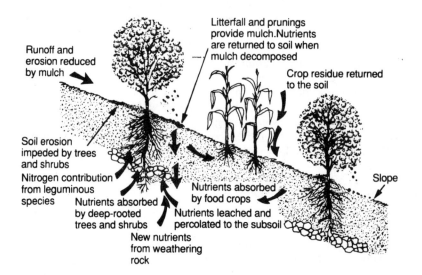

FIGURE 14.5. Alley cropping. The diagram shows some of the ecological dynamics that make alley cropping a sustainable farming system, even in marginal tropical forest environments. Source: B. T. Kang and L. Reynolds, eds. *Alley Cropping in the Humid and Sub-Humid Tropics*. Ottawa: International Development Research Centre (IDRC), p. 18. © 1989 by IDRC. Reprinted by permission.

Africa: A reappraisal," *African Affairs*, vol. 79 (1980), pp. 79–94.

The following sources provide excellent reviews of development policies and their impacts on the agricultural sector:

Barker, J. *The Politics of Agriculture in Tropical Africa*. Beverly Hills, CA: Sage, 1984.

Berry, S. *No Condition Is Permanent: The Social Dynamics of Agrarian Change in Sub-Saharan Africa*. Madison: University of Wisconsin Press, 1993.

Gibbon, P., K. J. Havnevik, and K. Hermele. *A Blighted Harvest: The World Bank and African Agriculture in the 1980s*. London: James Currey, 1993.

Richards, P. "Farming systems and agricultural change in West Africa." *Progress in Human Geography*, vol. 7 (1983), pp. 1–39.

Watts, M. "The agrarian crisis in Africa: Debating the crisis." *Progress in Human Geography*, vol. 13 (1989), pp. 1–41.

For examples of diverse experiences of agrarian change in specific countries, see these sources:

Hyden, G. *Beyond Ujamaa in Tanzania: Underdevelopment and an Uncaptured Peasantry*. Berkeley: University of California Press, 1980.

Leo, C. "The failure of the 'progressive farmer' in Kenya's Million Acre Scheme." *Journal of Modern African Studies*, vol. 16 (1978), pp. 619–638.

Pankhurst, A. *Resettlement and Famine in Ethiopia: The Villagers' Experience*. Manchester, UK: Manchester University Press, 1992.

Pingali, P., Y. Bigot, and H. Binswanger. *Agricultural Mechanization and the Evolution of Farming Systems in Sub-Saharan Africa*. Baltimore: Johns Hopkins University Press, 1987.

Schroeder, R. *Shady Practices: Agroforestry and Gender Politics in the Gambia*. Berkeley: University of California Press, 1999.

There is a growing literature on contract farming, especially in the horticultural sector. See the following:

Carney, J. "Struggles over crop rights and labor within contract farming households in a Gambian irrigated rice project." *Journal of Peasant Studies,* vol. 15 (1988), pp. 334–349.

Hughes, A. "Global commodity networks, ethical trade and governmentality: Organizing business responsibility in the Kenyan cut flower industry." *Transactions of the Institute of British Geographers*, vol. 26 (2001), pp. 390–406.

Lynch, K. "Commercial horticulture in rural Tanzania: An analysis of key influences." *Geoforum*, vol. 30 (1999), pp. 171–183.

Mather, C. "Agro-commodity chains, market power and territory: Re-regulating South African citrus exports in the 1990s." *Geoforum*, vol. 30 (1990), pp. 61–70.

For insights into the value of indigenous knowledge as a resource for agrarian development, see these sources:

Scoones, I., and J. Thompson, eds. *Beyond Farmer First: Rural People's Knowledge, Agricultural Research, and Extension Practice*. London: IT Publications, 1994.

Warren, D. M. *Using Indigenous Knowledge in Agricultural Development*. Washington, DC: World Bank, 1996.

Internet Sources

The following are websites maintained by particular commodity associations:

Kenya Flower Council. *A growing responsibility*. http://www.kenyaflowers.co.ke/

The Tea Board of Kenya. http://www.teaboard.or.ke

Africa is home to a number of major international agricultural research institutes:

International Institute of Tropical Agriculture (IITA). *About IITA*. http://www.iita.org

International Livestock Research Institute (ILRI). http://www.cgiar.org/ilri

West Africa Rice Development Association (WARDA). http://www.warda.cgiar.org/

15

Food Security

The intense coverage by the international news media of major famines in the Sahel, Ethiopia, Sudan, and southern Africa, among others, has marked Africa as a continent of hunger. Stories have emphasized the importance of ecological factors (drought and desertification) and political factors (instability and conflict) as explanations for these catastrophes. Such factors are very important, but provide only partial insight into Africa's food crisis. As Carl Eicher put it in 1982 (see "Further Reading"),

> The crisis stems from a seamless web of political, technical and structural constraints which are a product of colonial surplus extraction strategies, misguided development plans and priorities of African states since independence, and faulty advice from expatriate planning advisers. These complex, deep-rooted constraints can only be understood in historical perspective. (p. 157)

Food Insecurity in Africa

There are two main types of food insecurity confronting Africa: chronic undernutrition and abnormal food shortages (famine). Chronic undernutrition occurs when there are significant food deficits year after year. Famines accompany unusually large food deficits that occur because of crop failures or other catastrophes, often afflicting areas that normally have adequate food supplies.

What Is Food Insecurity?

Hunger is usually attributed to food production problems, but it is also the result of problems of food inaccessibility. It is the poor who go hungry and starve, because they lack the resources necessary to purchase adequate sustenance or to grow their own food. With the increase of market liberalization, hunger is increasingly prevalent among the poor, even at times when there is plenty of food in the marketplace. In short, food insecurity needs to be placed in socioeconomic context, in which the disparities in access to food are related to social structure and political–economic processes.

The research of Amartya Sen (see "Further Reading") has changed the way we conceptualize food scarcity and famine. Whereas famine was once considered a *geographical* problem afflicting particular regions, Sen argued that it is actually a *social* phenomenon in which different social classes and occupation groups vary in their ability to acquire food, regardless of its presence in the community. The term *entitlement* was used to characterize different households' varying levels of access to food. Household entitlements fluctuate in relation to such factors as the size of harvest, market prices for food, the value of saleable assets, dependence on casual employment, and social capital (including kinship and other networks of support that may be accessed in times of crisis).

Within households, men typically have first access to prepared food, with women and young children eating after the men. Men consume more food and more of the preferred foodstuffs such as meat, and thus are less likely to be undernourished than their wives and children. In spite of the fact that women farmers take primary responsibility for growing food in most African countries, men are likely to control most of the household income and assets, and thus to determine what is sold and purchased and for whose benefit. The effects of inequitable distribution within the household are especially evident in times of increased food scarcity.

The increasing prevalence of HIV/AIDS is adversely affecting total food production, as well as food entitlements in households affected by the disease. The impacts are diverse; they range from a decline in labor supply available for agricultural production, to the need to sell land and other productive assets to pay for food, to the precarious situation of children orphaned by the disease (see Vignette 15.1).

Our understanding of food security has changed from previous approaches that focused exclusively on staple foods such as grains, and on calorie intake. The Food and Agriculture Organization's (FAO's) new definition of food security was adopted at the World Food Summit in 1996. It incorporates not only assured access to sufficient food for all members of society, but also the nutritional adequacy and cultural acceptability of food:

'Food security' means that food is available at all times; that all persons have means of access to it; that it is nutritionally adequate in terms of quantity, quality and variety; and that it is acceptable within the given culture. (www.ausaid.gov.au//keyaid/rural.cfm)

Examples of Food-Insecure Households

The circumstances of food-insecure families vary greatly. Food insecurity is both a rural and an urban problem. The following brief sketches illustrate typical circumstances in which chronic undernutrition occurs.

A Female-Headed Household in Malawi

With only 0.4 ha of land to support her six children, this single mother struggles to make ends meet. She must sell part of her crop at harvest to repay debts, but then is forced to purchase food later at a much higher price. With trade liberalization, she cannot afford to pay for fertilizer and improved seed that could increase the size of her crop.

A Poor Family in Rural Lesotho

The small production of the family farm cannot support this family. Many families in Lesotho depend on remittances from members working in South Africa, but opportunities for foreign migrant laborers have become more restricted in recent years. Because this family does not have access to remittances, food obtained by participating in food-for-work projects supported with food aid is crucial for making ends meet.

An Urban Household in Liberia

After many years of civil war, the Liberian economy remains in tatters. The meager earnings of the eight people in this Monrovia-based household from jobs such as petty trading and hairdressing hardly cover the high cost of life in the city. The family eats only the cheapest staples, and often makes do with one meal a day. Rural-based relatives visit occasionally and bring gifts of food, providing a much-needed supplement to the family's diet.

National Data on Access to Food

Our focus now turns to national data, in an attempt to gain a macro-scale overview of food security issues. How prevalent is food insecurity in different countries? What are the trends in

VIGNETTE 15.1. How HIV/AIDS Affects Food Production

The HIV/AIDS epidemic is increasingly a major reason for the stagnation of agriculture, especially in the countries of southern and central Africa where HIV infection rates are highest. It affects agriculture in many ways, some of them less obvious than others. Virtually no facet of agricultural production and planning is immune from the effects of the epidemic.

Rural households affected by HIV/AIDS must produce their crops with fewer labor inputs. Not only is the labor of seriously ill or deceased individuals lost, but also other household members have less time to farm. They must care for ill members of the family, assume additional responsibilities within the household, and take time to attend AIDS-related funerals in the community. Those infected with HIV but still able to work are less productive as a result of the disease.

Reduced labor supply means that less land may be cultivated, or land may be prepared poorly for planting. Labor shortages may delay time-sensitive activities (planting, weeding, and harvesting), resulting in lower yields. The scarcity of labor may result in inadequate weeding and a decline in traditional methods of pest control. Farmers commonly switch to crops such as cassava that are easier to cultivate, but these substitute crops are likely to be less valuable or less nutritious. With lower yields and reduced incomes, rural households are forced to purchase more of their food, and have no money to hire labor or purchase inputs such as fertilizers.

The HIV/AIDS epidemic has also discouraged longer-range investments. Farmers are reluctant to engage in long-term soil conservation measures that do not yield immediate income, or that require too much labor. Similarly, there is increased reluctance to invest in tree crops that have short-term costs but only long-term profits. Households affected by HIV/AIDS may be forced to liquidate productive capital to cope with the disease. Land, livestock, and other assets may need to be sold to pay for medicines or for funerals, or these assets may be pledged as collateral for high-interest loans. Long-term decline in the productive capacity of the household, or even landlessness, are a consequence of these new expenses.

The highest rates of infection are among young adults, the most productive members of society. The decline in household fortunes means that children have less to eat and are often forced to leave school, because they must help support the family or because there is no money for school fees. Informal learning is also disrupted, because children are no longer able to learn farming and other skills from working with their parents. Children orphaned by AIDS may lose access to family lands, even if relatives in the community are able to take them in. Other orphans are forced to fend for themselves, often in cities.

People living with HIV/AIDS have increased nutritional requirements, both for calories and for protein. With declining crop yields, reduced household income, and increased prevalence of malnutrition, both the development of the disease and death from it are accelerated. Maternal malnutrition increases the risk of mother-to-infant transmission of the disease, particularly if a mother must supplement breastfeeding with other foods at an early age.

HIV/AIDS has devastated the agricultural extension services of many African countries. There are fewer extension agents to assist farmers, and there are significant losses of working time for agents who are coping with illness, whether their own or that of relatives.

Based on M. Haslwimmer. *AIDS and agriculture in Sub-Saharan Africa.* Rome: FAO, 1996. (Available online at www.fao.org)

food production? (Figure 15.1 illustrates one way of addressing this second question.) Is increased production providing larger stocks of food—at least at the national level—to address food insecurity problems? What are the trends in food imports, and to what extent does food aid compensate for shortfalls in production?

Nutritional Status

Table 15.1 provides estimates of the prevalence of undernourishment in selected countries. In the 39 countries of Africa south of the Sahara for which nutritional data are available, undernutrition rates range from 7% in Nigeria to 75% in Somalia. Over one-quarter of the population is undernourished in 26 countries. The data point to undernutrition as a widespread phenomenon, but one that varies greatly from country to country. These national data, in turn, mask major disparities in access to food within all countries.

Undernutrition is extremely prevalent among children under the age of five. Global estimates for Africa south of the Sahara show that approximately 31% of children under five are moderately to severely underweight, that 38% are moderately to severely stunted (much below the median height), and that 9% are moderately to severely wasted (much below the median weight for height). Undernutrition affects not only physical growth, but also the development of intellectual capacity. Undernourished children face a much-increased risk from a wide array of infectious diseases as well.

According to the Fifth World Food Survey of the FAO, the average dietary energy supply (DES) in Africa south of the Sahara was 2,070 kcal per capita (person) per day, which was 22% below the world average (2,660 kcal) and almost 40% less than the average for developed nations (3,390 kcal). However, DES represents only one of the components of a nutritious diet. The human body also requires many different

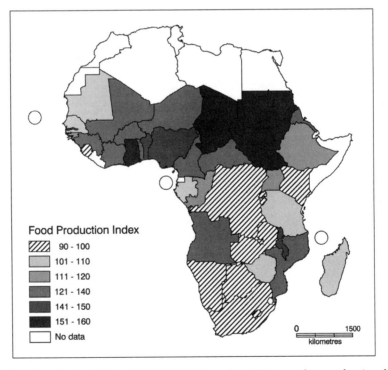

Food Production Index

▨	90 - 100
▧	101 - 110
▧	111 - 120
▨	121 - 140
▨	141 - 150
■	151 - 160
□	No data

FIGURE 15.1. Food production index, 1998–2000. The value 100 is equal to production during the baseline years of 1989–1991. Countries located north of the equator generally fared better than those in central and southern Africa. Data source: World Bank. *World Development Indicators 2002.* Washington, DC: World Bank, 2002.

TABLE 15.1. Food Consumption, Production, Trade, and Aid in Selected Countries

| Country | Undernour-ished people 1997–1999 (% popula-tion) | Food production index 1998–2000 (1989–1991 = 100) | | Food imports, 2000 | | Food imports per capita, 2000 (metric tons per 1,000 people/year) |
		Total	Net per capita	All imports (1,000 metric tons)	Food aid (1,000 metric tons)	
Angola	51	146	109	746	224	56.8
Benin	15	154	118	307	16	49.0
Botswana	23	93	76	446	0	289.4
Burkina Faso	24	136	109	210	19	18.2
Congo (Dem. Rep.)	64	88	65	360	53	7.1
Congo (Republic)	32	124	94	301	25	99.7
Côte d'Ivoire	16	137	110	1,244	11	77.7
Ethiopia	49	136	106	1,277	1,041	20.3
Gabon	9	114	90	170	1	138.4
Kenya	46	109	85	1,384	357	45.1
Lesotho	25	106	89	425	5	208.6
Mozambique	54	134	96	592	145	32.4
Namibia	33	111	88	512	4	291.4
Niger	41	137	102	357	32	32.4
Nigeria	7	153	119	5,059	1	44.5
Senegal	24	124	99	992	24	105.3
Tanzania	46	104	79	611	60	17.4
Zambia	47	101	80	50	33	4.8

Data sources: United Nations Development Program (UNDP). *Human Development Report 2002*. New York: Oxford University Press, 2002. Food and Agriculture Organization (FAO). *FAOSTAT Database*. http://apps.fao.org

minerals, vitamins, and proteins in varying quantities.

Trends in Food Production

Africa has lagged behind other major world regions in the production of food, especially since the 1960s, when the introduction of Green Revolution crops transformed agriculture in many parts of Asia and Latin America. The gap in per capita food production between Africa and other regions is especially great, reflecting Africa's very high rates of population growth. Indeed, per capita production of food fell by almost 20% between 1970 and 2000.

Figure 15.1, and the data in Table 15.1, show huge differences among African countries with respect to changes in food production between 1989–1991 and 1998–2000. Seven African countries had an absolute decline in production during this period, while an equal number achieved production increases of at least 50%. More significant are the data in Table 15.1 on production per capita, which show that many countries achieving modest increases

in total food production actually lost ground when population growth is taken into account. Per capita food production fell in 26 of 46 countries, most dramatically in the war-torn Democratic Republic of the Congo, where food production per capita in 1998–2000 was only 65% of the already low level of a decade earlier. There were also a few notable success stories, among them Nigeria, where a 19% increase in per capita food production was achieved during the 1990s.

Food Imports and Aid

Countries vary greatly in their use of imports, including food aid, as a means of meeting national food needs. Certain countries with some of the highest rates of undernutrition, together with declining or stagnating domestic food production, imported surprisingly little food in the late 1990s. The Democratic Republic of the Congo is a prime example; in spite of its catastrophic food situation, the country imported only 7 tons per 1,000 people per year—almost the lowest rate in Africa. Primary factors in this were the ongoing civil war and economic collapse, which made it impossible to purchase and distribute large amounts of food throughout the country. In contrast, Nigeria imported 5.1 million metric tons of food in 2000, in spite of very low rates of under-nutrition and robust agricultural growth during the 1990s. Cereals accounted for approximately 60% of Nigeria's food imports. These imports were used not only to meet the fundamental nutritional needs of the population, but also to satisfy demand for nontraditional foods—such as bread made from wheat, which has become an important part of the diet of many Nigerians (see Vignette 15.2 and Figure 15.2).

Food import data are difficult to interpret precisely. Several smaller countries, especially those with relatively uniform environments, import a large quantity of food per capita. These countries lack the advantage of ecological diversity possessed by larger countries such as Nigeria and South Africa, where interregional trade is an important factor in the national food economy. A large proportion of the food imports of these smaller, less diverse countries consists of nonstaple items. In the case of countries such as Botswana and Namibia, where per capita incomes are high by African standards, processed foods form a significant proportion of imports.

Poorer countries with food deficits, such as Ethiopia, Mozambique, and Tanzania, import large quantities of food (especially grains). About four-fifths of Ethiopia's food imports were obtained through food aid programs; the majority of these imports were distributed in regions with chronic food deficits through food-for-work programs, described in more detail later (see Vignette 15.3, below). Kenya is another major importer of food; its food aid shipments were second only to Ethiopia's in 2000. Yet Kenya was also a major food exporter—the largest supplier of fresh horticultural produce to the European market. The growth of export agriculture has been achieved at the expense of production for local needs. The distribution of food aid to helps to support export crop production by masking its impacts on the Kenyan poor.

Famine

Famines, or "abnormal food shortages" as they are called euphemistically by the FAO, can no longer be considered truly "abnormal" in the contemporary African context. From the late 1960s to the early 2000s, hardly a year has passed without reports of famine in some part of Africa. Crop failure and famine were particularly widespread in 1983–1984, 1991–1992, and 2002–2003. In each case, half of the countries of Africa south of the Sahara were designated by FAO as having abnormal food shortages (Figure 15.3), and special appeals were launched to provide these countries with emergency food aid.

One of the largest famine relief operations ever undertaken was launched in Ethiopia between December 1984 and December 1985. (Figure 15.4 shows the pattern of food aid distribution in Ethiopia during 1985 and the following year, 1986.) A total of 1,273,000 metric tons of international food aid provided basic rations for 7 million people. This massive relief effort involved 63 nongovernmental organizations (NGOs) from various parts of the world. The topography of Ethiopia was a major challenge; it was often impossible to get across canyons and up cliffs to places only a few kilometers away. The network of roads, railways, and airports was poorly developed and often unusable because of war-related damage and poor maintenance. The ongoing civil war in the north and east of the country provided the biggest challenge—not only because of insecurity, but also because of the government's unwillingness to involve rebel groups in food distribution in rebel-held areas. From a logistical standpoint, the 1984—1985 relief program was a success that saved many lives. Despite this short-term success, food insecurity has remained a constant challenge in Ethiopia.

Famines are often associated with an ecological crisis serious enough to drastically reduce harvests. Drought is the most common of these ecological crises; however, excessive rainfall, floods, swarms of locusts, volcanic eruptions, and plant diseases may also cause crops to fail. Famines are also associated with political strife that disrupts normal economic

VIGNETTE 15.2. The Wheat Trap

The rapid growth of food imports may encourage—and, in turn, be spurred on by—new patterns of consumer demand. The book *The Wheat Trap* is a study of Nigeria's growing reliance on imported wheat during the 1970s and 1980s. Similar processes of "taste transfer" to imported foods are found in virtually every country. The Nigerian experience illustrates how apparently simple, short-run policy decisions may have far-reaching implications that are very hard to reverse.

Nigeria's annual imports of wheat and flour increased 18-fold from 1960–1964 to 1980–1982. Wheat is not part of the traditional diet in most parts of Nigeria. However, by the 1980s, bread had become one of the most important foods of the masses in both urban and rural areas. Three factors that help to explain the growth of bread consumption are (1) its convenience, (2) its compatibility with local diets, and (3) its low cost compared to other staples.

The Nigerian government permitted wheat imports to increase during the 1970s because they helped to ensure that urban populations would have access to cheap food. However, the high cost of imported wheat provided the justification for massive investment in irrigation schemes to grow wheat. Once this commitment to domestic production had been made, continuing wheat imports could be treated as an interim measure, pending the availability of local supplies. This vision paid scant regard to the high cost and technical difficulty of growing wheat under irrigation in a tropical setting. A provisional estimate put the true cost of growing 1 ha of irrigated wheat at $2,000 per year, equivalent to a government subsidy of about $1,400 for every metric ton of wheat produced.

Beset with rapidly increasing debts and stagnant export revenues, Nigeria resolved in 1985 to reduce wheat imports and to encourage consumption of indigenous staples in its place. Then wheat imports were banned in 1987, and farmers were called upon to increase their production of wheat to enable Nigeria to become self-sufficient.

In trying to replace wheat imports with indigenous grains and local wheat production, Nigeria faced a number of hurdles. Several flour mills and thousands of bakeries had been established to produce bread for the Nigerian market; the milling companies and bakers formed a

FIGURE 15.2. Billboard promoting bread consumption, Nigeria. Advertising is one of the factors accounting for the rapid growth of bread consumption and the growth of Nigeria's "wheat trap." Photo: author.

VIGNETTE 15.2. *(cont.)*

---

potent lobby arguing for unrestricted imports of wheat. As well, consumers demanded access to cheap bread, especially because of the high cost of locally grown staples. Wheat imports had also contributed to the decline in domestic agriculture, which meant that food scarcity and higher food prices resulted from the sudden reduction in imports.

Nigerian wheat production did increase about fivefold, from an average of 25 million metric tons in the early 1980s to 139 million metric tons in 1987. However, this increased domestic supply fell far short of demand; late 1980s wheat production represented no more than about one-tenth of wheat consumption prior to the imposition of the import ban. Nigerian wheat production declined from its 1987 peak in subsequent years. Many farmers who had experimented with wheat cultivation were disappointed with their returns, or found their land poorly suited to the crop.

The ban on wheat imports was lifted in 1990, and consumption again grew. In 2000, Nigeria imported 2,230,000 metric tons of wheat, approximately twice the level of imports in the early 1980s. The domestic crop now accounts for only about 5% of the country's wheat consumption.

Based primarily on G. Andrae and B. Beckman. *The Wheat Trap*. London: Zed Books, 1985.

---

activity. Beyond these immediate causes, a host of contributory factors may be identified that help to account not only for famine, but for its differential impacts within affected communities. Each episode of famine increases susceptibility to subsequent crises, particularly among the poor. Food stores are exhausted, debts are incurred, and land may have to be sold to acquire food. Smaller food reserves, coupled with debt and poverty, mean that

what was manageable hardship in the past may result in famine today.

In drought-prone regions, where periodic famines have occurred since time immemorial, people rely on many strategies to cope with food shortages. Stored grain and so-called "famine foods" (edibles that are consumed when other, preferred foods are unavailable) may provide some relief from hunger. Temporary interregional or rural–urban migration may provide

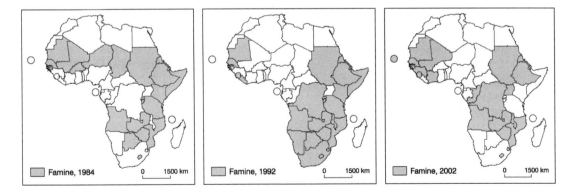

FIGURE 15.3. Countries experiencing abnormal food shortages, 1984, 1992, and 2002. Abnormal food shortages were reported in 37 countries in at least one of the three years; they occurred in all three years in 12 countries. Data sources: Food and Agriculture Organization (FAO). *World Food Report 1985*. Rome: FAO, 1985. FAO. *Food Outlook, 1992*. Rome: FAO, 1992. World Food Programme. *Reliefweb*. www.reliefweb.int

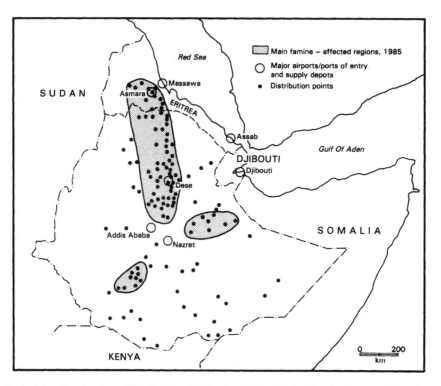

FIGURE 15.4. The distribution of food aid in Ethiopia, 1985–1986. The challenge of distributing a massive amount of food aid in extremely difficult circumstances necessitated diverse and innovative strategies. After K. Jansson, M. Harris, and A. Penrose. *The Ethiopian Famine*. London: Zed Books, 1987.

a source of income to purchase food. In very serious crises, personal property or land may be sold, or members of the household may be forced to move away permanently. As Figure 15.5 shows, these strategies may be viewed as points on a hierarchy of responses, ranging from those that are relatively easy to those with increasingly drastic and irreversible consequences.

The research of Michael Watts in northern Nigeria (see Figure 15.5 and "Further Reading") helps us to see how historical changes in political economy have changed the meaning of food shortages. He describes the workings of the 19th-century "moral economy," in which the ruling class received grain from peasants as tax but was obliged in turn to provide relief in times of crisis. The moral economy provided a degree of security for the poor. During the colonial era, heavy taxation, cash crop promotion, and the erosion of traditional values irreparably damaged the moral economy. Consequently, African rural communities today are less able to cope with crop failures; they must increasingly

rely on external aid, as well as such drastic responses as outmigration, to survive when food shortages occur.

International famine relief provided through multinational channels such as the World Food Programme, and disaster assistance from development agencies and NGOs in donor countries, have come to play a large role in meeting emergency food needs. Yet the history of famine relief shows that international aid has been unreliable at best; it is affected significantly by factors such as the nature and extent of media coverage, the international standing of a country's political leadership, the country's perceived strategic importance, and other issues that "compete" for donors' attention. The massive international response to the 1984–1985 Ethiopian famine is in stark contrast to the repeated lack of international response to food crises in southern Sudan.

Ethiopia and Malawi were among several countries that established national emergency grain reserves during the 1990s to serve as a first line of defense during times of hunger. It is ironic that Malawi's attempt to create an emergency response system of its own was undermined by the International Monetary Fund (IMF), which directed the Strategic Grain Reserve to be sold, insisting that it was trade-distorting and too expensive to maintain. When famine struck in 2002, traders who had purchased the food stores made windfall profits, and many Malawians living in poverty went hungry because they could not afford to purchase food.

Chronic Undernutrition

Africa's most serious problem of food insecurity is not drought or mass starvation. Rather, it is the pervasive growth of undernutrition. Fam-

ilies, communities, regions, and nations in all parts of the continent are affected by chronically inadequate food supplies. As a result, food imports and food aid have progressed from being short-term relief to becoming an integral part of national food budgets in most African countries. Strategies such as food-for-work programs have been implemented widely in food deficit regions. Rations of food aid are distributed to those in need, in return for their labor on local conservation and development projects (see Vignette 15.3 and Figure 15.6).

The patterns of undernutrition in society and space, and the factors that help to account for the growth of the crisis, are equally diverse. However, several generalizations may be made about chronic undernutrition in contemporary Africa:

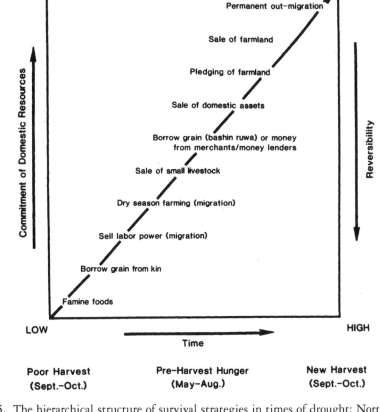

FIGURE 15.5. The hierarchical structure of survival strategies in times of drought: Northern Nigeria. As the effect of famine deepens, survival strategies shift, typically involving increased commitments of domestic resources and difficulty in reversing the strategies. Source: M. Watts. *Silent Violence: Food, Famine, and Peasantry in Northern Nigeria.* Berkeley: University of California Press, 1983, p. 436. © 1983 by The Regents of the University of California. Reprinted by permission.

VIGNETTE 15.3. What Makes for a Successful Food-for-Work Program?

Most Westerners associate food aid in Ethiopia with the large-scale famine relief programs that occurred in the mid-1980s. These programs—which involved widespread distribution of food to large groups of famished people in temporary camps, and made extensive use of air drops to deliver food in remote areas—provided dramatic media images that have shaped popular perceptions of food aid.

Ethiopia is a country with chronic food deficits, exacerbated by high risks of drought and crop failure. Food aid has continued to be distributed each year in Ethiopia in areas of particular need, through programs that remain largely unknown in the Western world. Yet these programs have achieved startling results—not only in helping to alleviate hunger, but also in transforming the Ethiopian environment through a variety of public work projects.

Food-for-work programs involve distributing food aid, presumably to those in greatest need, as a payment for participating in communal work projects. Cash-for-work programs operate in a similar way, except that recipients receive cash with which to purchase food for their families. These programs are widely supported by both donors and recipients. Food as payment for work is less likely to undermine the self-esteem of recipients than food handouts. Projects that involve road construction or reforestation offer the prospect of long-term benefits for the entire community.

Food-for-work programs help to alleviate environmental conditions such as deforestation, soil erosion, and rainwater loss that contribute to food insecurity. One of the most popular projects has been the construction of terraces on hillsides, to reduce erosion, trap rainwater, and accumulate soil so that cultivation or reforestation may take place. Hundreds of hillsides across Ethiopia have been terraced from top to bottom during the past two decades through food-for-work programs (see Figure 15.6).

One of the major challenges of food-for-work programs is to ensure that food is made available to those with the greatest need. Most of these programs have been initiated close to major roads, neglecting less accessible regions in need. Such spatial biases reflect the reality of Ethiopia's difficult terrain and poor network of roads. Within communities, it is important to recognize the limitations of work-linked food distribution; some households, such as those headed by women, are less able to participate in the designated projects.

Food-for-work programs may also draw household labor away from other productive tasks,

a   b

FIGURE 15.6. Two faces of food aid in Ethiopia. (a) Distribution of famine emergency food at the time of the 1985–1986 famine. (b) The work to terrace and partially reforest this hillside in Tigray Province was done by community groups which were compensated through a food-for-work program. Photos: (a) CIDA (D. Barbour); (b) M. Peters.

VIGNETTE 15.3. *(cont.)*

---

including farm preparation and weeding. In such a case, it may lessen the household's capacity to grow its own food. The price paid for work is likewise important—large enough to serve as an incentive, but not so large that it is more lucrative than other employment. If too much food is introduced into the marketplace, food aid programs have the potential to distort the market for locally produced food. Under these circumstances, local farmers are penalized with lower prices for their crops.

The most successful food-for-work programs provide peasants with a real voice in the planning, implementation, and management of community projects. Without a strong sense of community ownership, the initial work is likely to be done carelessly, and there is little enthusiasm for maintaining the project after its completion. Although results have varied, Ethiopia's food-for-work projects have generally been successful.

Based on L. Humphrey. *Food-for-Work in Ethiopia: Challenging the Scope of Project Evaluations.* IDS Working Paper no. 81. Brighton, UK: University of Sussex, Institute of Development Studies, 1999.

- Chronic undernutrition is widespread in both urban and rural settings.
- It is often seasonal in nature.
- It is not only, or even primarily, a production problem.
- It is clearly linked to increasing levels of poverty and dependency.
- The crisis has continued to deepen in the majority of countries since the 1970s, when global attention focused initially on food insecurity in Africa.

No countries and no ecological regions are totally free from chronic undernutrition. Contrary to popular belief, it is not just a crisis of arid and semiarid environments. Major food exporters such as Kenya and South Africa, and countries with abundant rainfall such as the two Congos, are all affected by serious and widespread undernutrition. The wide distribution of food insecurity in diverse settings forces us to look beyond drought and other environmental conditions to recognize hunger's social and political determinants.

Often the apparent abundance of food for sale in markets masks the extent of urban hunger. This food tends to be extremely expensive in relation to the earning power of the urban poor; poverty, rather than food production failures, is what prevents the urban poor from obtaining adequate sustenance. A variety of government policies, on issues as diverse as minimum wages, food imports, and the maintenance of the transportation infrastructure, affect not only the volume of food in urban markets but also its cost.

Undernutrition often occurs seasonally, especially in rural areas with a semiarid climate. Food production in such areas is virtually confined to a short rainy season. Although food is relatively plentiful at harvest time, and people gain weight then, the months before a new harvest are often characterized by food scarcity and hunger. Studies in several countries have shown that body weights and general health status decline because of inadequate dietary energy. Because these food deficits are most acute during the farming season, when good health and plenty of energy are needed to undertake a heavy onerous workload, the deficits may lead to reduced farm productivity.

It is hard to overestimate the importance of poverty as an explanation for chronic undernutrition. As Sen has argued, hunger will be manifest in the midst of apparent food surpluses if part of the community lacks the means to grow or purchase its own food. Chronic indebtedness is often the key to hunger in rural Africa. Loans obtained to buy food, to acquire seed and fertilizer, or to obtain medical care for ill relatives are usually payable at harvest time, when market prices are lowest. Thus poor families sell their crops cheaply and then purchase food expensively at a later date. Deeply indebted households may be forced to sell the means of production, especially their land.

Food Insecurity Debates

Clearly, human factors such as poverty and disease, as well as physical conditions such as the occurrence of drought, are determinants of hunger. But what are the *underlying causes* of food insecurity in Africa? This problem has long been the focus of intense debate in both academic and policy-making circles. Some of the key debates are outlined briefly in this section.

Population Growth and Environmental Decline?

Whereas per capita food production has declined in recent years, the total amount of food produced increased in all but seven African countries during the 1990s. The implication of these data *seems* clear: If Africa's population was not growing so rapidly, its food problems would dissipate. Although few would dispute the value of a slower rate of population growth, many rapidly growing poorer countries, both in Africa and in other parts of the world, have achieved notable improvements in per capita food production. Moreover, labor shortages in rural areas have at times been found to be a factor in declining production. In short, population growth by itself provides little insight into food scarcity.

According to the neo-Malthusian perspective (see Chapter 11), one of the most important ways in which population growth undermines food security is by causing degradation of the environmental resource base. By linking population growth to the degradation of vegetation, soil, and water in all African ecosystems, the neo-Malthusian approach has had a powerful influence on public perceptions and policy development. However, while environmental degradation has indeed taken place in association with the expansion of agriculture, research in many countries and in diverse ecosystems is challenging received wisdom about the ubiquity of environmental degradation and the role of humans in ecological change. Indeed, studies in different parts of the continent—on the forest–savanna boundary in southwestern Guinea and in Machakos District in the Kenyan Highlands, for example—have shown how human manipulation of the environment has *improved* its capacity to produce food and support an increasing population. These studies offer support for Boserup's hypothesis that population growth stimulates development.

A Failure of the Peasantry?

The African peasantry represents an easy target; after all, the peasants have failed to produce enough food to make hunger a thing of the past. According to this view, indigenous producers should be criticized for their reluctance to adopt new farming methods and for their use of practices that allegedly harm the environment. The solution is seen to lie in transforming agriculture with modern technology and large-scale production systems. Although this perspective began to wane with the new emphasis in the 1970s on small producers, it continues to be enunciated—particularly by those who stand to profit from the promotion of large-scale, highly technological agriculture. The influence of this approach is still to be seen in the support for capitalist investment in large grain and livestock enterprises in many African countries.

There is no shortage of evidence concerning the willingness of smallholders to respond to market opportunities. In recent decades, this has often involved switching from production of commodities such as cocoa and coffee for export markets, to production of staple foods not only for household use but also for sale in local and urban markets. Research on indigenous agricultural knowledge has also demonstrated the continuing importance of these systems, both for the environment and for food production.

Inappropriate Government Policies?

Many international development agencies have placed the blame for food deficits on the policies pursued over the years by Africa's governments. This view is often identified with the World Bank, and especially with its pivotal 1981 report *Accelerated Development in Sub-Saharan Africa: An Agenda for Action*. Pricing policies have been cited as a key issue. When states import a substantial quantity of food and fix its price below the market value for locally grown staples, they undermine domestic food production. Agriculture is also discouraged when state-run produce-marketing agencies pay farmers less than a fair market price for their crops. Moreover, governments have been criticized for inadequate maintenance of transportation networks, thus hampering both the distribution of production inputs to farmers and the flow of food to consumers. Finally, a whole range of "urban biases" in development policy is seen as an inducement for young people to move to the city.

The IMF and the World Bank have waged war for two decades against state policies they have considered counterproductive. Currency devaluation, the establishment of free market pricing, cutbacks in government spending, and other policies associated with structural adjustment programs have been imposed. Increased domestic food production and major reductions in food insecurity were among the promised benefits of these reforms. Although the international financial institutions often claim successes in certain countries—Ghana and Uganda, for example—in arguing that structural adjustment is the answer to food insecurity, the larger picture is discouraging. The continuing food crisis after two decades of structural adjustment raises the question of why the IMF and World Bank should not accept major responsibility for the widespread failure of their imposed policies. Indeed, there is evidence that food insecurity is increasing in a number of countries where these institutions' interventions were most pervasive.

Commodification of Agriculture and Increased Dependency?

Underdevelopment theorists have put special emphasis on the role of colonialism and of capitalist development in the postcolonial era as the causes of food insecurity in Africa. *Commoditization* leads to increased dependency on purchased foods and reduced self-sufficiency. It may accompany increased cash crop production (and thus the diversion of resources into nonsubsistence production), the incorporation of local farmers into large-scale production systems, and/or the increased dependence of the poor on selling their labor. The key issue is the development of gross disparities in the distribution of land and other resources that occur with the commoditization of food systems, leaving a large proportion of the population increasingly vulnerable to the effects of economic recession, crop failures, and food that is too expensive for them to purchase.

The question of appropriate agrarian policies has proved to be more difficult for critics of commoditization. Liberals have tended to advocate small-scale, self-reliant rural development that uses local resources whenever possible to achieve greater equity. Many community-based projects, developed through participatory methods and building upon indigenous knowledge, have achieved notable local success. Nevertheless, because such projects are necessarily small-scale and carefully adapted to local circumstances, they cannot be readily mass-produced.

At the dawn of the new millennium, genetically engineered crops are being touted as another "magic bullet" to end the shortfall of food in Africa. Although some genetically engineered crops may offer benefits for Africa, it seems likely that the same factors that limited the impact of the Green Revolution will apply once again. Is this technology compatible with common practices such as intercropping? Will these new seeds, and the chemical inputs that generally are required to grow them, be accessible and affordable to small-scale farmers? Or will this latest innovation only serve to increase the gap in food entitlements between rich and poor?

Far more effective would be the restoration of peace and security in areas troubled by persistent conflict. Most of the countries with the greatest food deficits and highest rates of undernutrition—Somalia, the Democratic Republic of the Congo, Ethiopia, Angola, Mozambique, and Sierra Leone—share a common history of bitter, protracted wars that undermined the productive capacity of the country. In war-ravaged countries, recovery has usually been extremely slow, not least because of the lack of substantial international aid for rehabilitation.

The policy response to food problems has been too narrowly conceived and too often divorced from local realities. That food insecurity has not only persisted but also become steadily more pervasive, despite the attempts to implement myriad remedial strategies, points to the complexity of the challenge. Simplistic explanations and the simplistic solutions that follow them should not be trusted.

Further Reading

The following sources provide discussions of theoretical issues concerning the interpretation of Africa's food crises:

Bryant, C, ed. *Poverty, Policy and Food Security in Southern Africa.* Boulder, CO: Lynne Rienner, 1988.

Bryceson, D. F. "Nutrition and the commoditization of food in sub-Saharan Africa." *Social Science and Medicine,* vol. 28 (1989), pp. 425–440.

Devereux, S., and S. Maxwell, eds. *Food Security in Sub-Saharan Africa.* London: Intermediate Technology Development Group, 2001.

Eicher, C. K. "Facing up to Africa's food crisis." *Foreign Affairs,* vol. 61 (1982), pp. 151–174.

Lawrence, P., ed. *World Recession and the Food Crisis in Africa.* Boulder, CO: Westview Press, 1986.

Raikes, P. *Modernizing Hunger: Famine, Food Surplus and Farm Policy in the EEC and Africa.* London: James Currey, 1988.

Watts, M., and H. Bohle. "Hunger, famine, and the space of vulnerability." *Geojournal,* vol. 30 (1993), pp. 117–125.

The following sources provide insights into Amartya Sen's entitlement theory as a framework for understanding hunger:

DeWaal, A. "A reassessment of entitlement theory in the light of recent famines in Africa." *Development and Change,* vol. 21 (1990), pp. 469–490.

Dreze, J., and A. Sen. *Hunger and Public Action.* Oxford: Clarendon Press, 1989.

Sen, A. *Poverty and Famines: An Essay on Entitlement and Deprivation.* Oxford: Clarendon Press, 1981.

Seasonal aspects of hunger are discussed in the following source:

Chambers, R., R. Longhurst, and A. Pacey. *Seasonal Dimensions to Rural Poverty.* London: Frances Pinter, 1981.

For studies of the international response to African famines, see these sources:

DeWaal, A. *Famine Crimes: Politics and the Disaster Relief Industry in Africa.* Bloomington: Indiana University Press, 1998.

Jansson, K., M. Harris, and A. Penrose. *The Ethiopian Famine.* London: Zed Books, 1987.

Pushpanath, K. "Disaster without memory: Oxfam's drought programme in Zambia." In D. Eade, ed. *Development and Social Action.* Oxford: Oxfam GB, 1999, pp. 120–132.

Salih, M. A. R., ed. *Inducing Food Insecurity: Perspectives on Food Policies in Eastern and Southern Africa.* Uppsala, Sweden: Scandinavian Institute of African Studies, 1994.

Internet Sources

The following are among the many websites that deal with hunger and food aid issues:

Alan Shawn Feinstein International Famine Center, Tufts Nutrition. www.famine.tufts.edu

Famine Early Warning Systems (FEWS). www.fews.net

Food Aid Management. www.foodaidmanagement.org

Food and Agriculture Organization (FAO) of the United Nations: Helping to Build a World without Hunger. www.fao.org (provides access to the FAOSTAT database)

Food Insecurity and Vulnerability Information and Mapping Systems (FIVIMS). www.fivims.org

United Nations Standing Committee on Nutrition. *Africa Nutrition Database Initiative Website.* www.africanutrition.net

U.S. Agency for International Development (USAID). *Food for work.* www.usaid.gov/hum_response/crg/module_2.html

World Food Programme. www.wfp.org

Urban Economies and Societies

Although most Africans continue to live in rural areas, it is in urban Africa that the most rapid growth is taking place. The residents of Africa's cities now account for over one-third of the total population. This rapid growth in the population of cities since independence has been accompanied by a comparable increase in the size and diversity of the cities' economic, cultural, and political roles. The interpretation of this growth has continued to be the subject of debate. Some see the cities as dynamic centers of growth and innovation, taking the lead and showing the way for rural and small-town Africa. Others are critical of the "urban bias" that they see pervading most development theory and practice, and these critics see the cities as parasitic centers that drain the countryside of development capital.

Chapter 16 surveys the changing structure of African cities, from the precolonial cities that existed in several parts of the continent to the cities of colonial Africa to the African cities of today. Looking at African urbanization in historical perspective highlights the dramatic changes in the size, functions, and social and spatial organization of the cities that have occurred and continue to occur.

The economic geography of African cities is discussed in Chapter 17. The chapter is structured around a comparison of the organization and significance of formal and informal economic activity. At the same time, it is important not to look at urban economic activity in isolation from that of rural areas. Not only are rural–urban economic linkages of increasing importance as urbanites continue their struggle to make ends meet during very difficult economic times, but also urban dwellers rely to a considerable extent on farming (the quintessential "rural" occupation) to survive.

Chapter 18 looks at housing in the urban environment and, in particular, the policy options used by governments to address the acute problems of housing that have inevitably occurred in conjunction with rapid growth. However, in the provision of housing, as with most other amenities, the efforts of government pale in comparison to the self-help efforts of countless millions of urban Africans who have constructed homes for themselves. Terms such as "squatter settlements," which are often used to designate these self-housed communities, hardly do justice to the dynamism, ingenuity, and self-reliance that these places represent.

16

The Evolution of Urban Structure

This chapter examines the changing size, spatial organization, and functions of African cities in precolonial, colonial, and postcolonial times. Although urbanization has been a feature of the 20th century (and continues into the 21st) in most of Africa, the continent has had a long history of urban development. African cities are diverse, resulting from the melding over many centuries of indigenous African, Arabic/Muslim, and European influences. Thus standard North American theories of urban development and models of urban form must be used with caution in the African context.

The Precolonial City

Although most Africans in precolonial times lived in rural communities, there was a widespread and dynamic tradition of urban development (Figure 16.1). Urbanization first appeared in Africa south of the Sahara some 3,000 years ago in present-day Sudan. Most precolonial urban development, however, occurred between the 10th and 19th centuries A.D. Although Meroë, Great Zimbabwe, and many other early African cities exist today only as architectural ruins, some have survived into the 21th century

and remain important today. Ibadan, Kano, Kumasi, and Mombasa are all examples of modern cities that were established and flourished before the colonial era.

Most precolonial African cities were political capitals, the headquarters of powerful emperors or chiefs holding dominion over neighboring territories. The religious shrine and the ruler's palace in such a city were typically situated close to each other—a symbol of the spiritual base of political power. As political centers, subject to attack from competing powers, early cities were often surrounded by a wall or located on an easily defended site. Residents of outlying villages could seek refuge in the walled city in times of danger. Precolonial cities also had important economic functions. Some were located near sources of vital raw materials; the first settlers in Kano, for example, exploited local deposits of iron ore. Manufacturing activities such as weaving, iron making, and metal casting were an important source of wealth when the products were sold in local and regional markets, or internationally.

Precolonial African cities tended to be structured "organically" around the palace and shrine as primary focal points (Figure 16.2). Particular occupational groups (e.g., black-

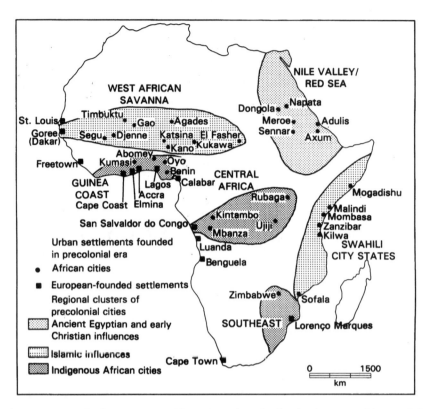

FIGURE 16.1. Precolonial urbanization in Africa. Precolonial urbanization was widespread. It reflected a variety of indigenous and external influences, specific to each of several regions of urban development.

smiths, weavers, traders) were generally clustered in their own wards. Members of royal families and their associates tended to be concentrated near the locus of power—the palace. In some of these cities, the huge size and high walls of the royal compound literally created a city within a city.

The development of cities in the middle section of the Nile Valley dates back 3,000 years to the founding of Napata and Meroë. The forms and functions of these cities, most notably the construction of ceremonial temples and pyramids, reflected their close ties to Egypt. The Nile Valley later experienced distinct Christian (6th–15th centuries) and Islamic (post-15th century) phases of urbanization. The large, sprawling cities of Christian Nubia were replaced by compact cities of Islamic design, the most important of which were Sennar and Khartoum.

A separate tradition of urbanization evolved in what is now Eritrea and Ethiopia, through cultural contact and synthesis between local African and migrant Arab populations. The port of Adulis, which became a very important trading center, was among the earliest of these cities. By the 1st century A.D., the city of Axum had been established as the political capital of the powerful state of the same name. The legacy of Axum's 4th-century conversion to Christianity includes the development of several cities as centers of trade, Christian worship, and rule. The founding of Addis Ababa in 1886 as a new capital marked the final phase of two millenia of premodern Ethiopian urban development.

Africa's most remarkable example of indigenous urban development occurred along some 1,500 km of the East African coast, between Mogadishu in Somalia and Sofala in present-day Mozambique. Some 40 major cities and hundreds of smaller settlements were established between the 8th and 19th centuries A.D. Some of these cities, such as Kilwa, Mombasa, and Mogadishu, have been occupied continuously for a thousand years or more, while others were abandoned at various stages. The reasons for abandonment are likely to have included environmental factors (e.g., poor water supplies or disease), conquest, and competition from other settlements having locational advantages for trade.

Newer archaeological evidence and a reexamination of written and oral historical records have led to a reinterpretation of the development of these cities. They were once thought to have been founded by migrants from Arabia and the Persian Gulf, but are now known to have been established by Africans. They developed as trading *entrepôts*, serving their local hinterlands and developing longer-distance connections into the interior, along the East African coast, and to several parts of Asia. Opportunities for trade attracted Arab and Persian merchants, many of whom settled in the larger cities. Using proceeds from a flourishing trade in gold, ivory, and many other goods, the urban elites created planned urban landscapes, containing elaborate mosques, palaces, tombs, homes, and other buildings. These buildings, constructed from locally quarried stone and wood, provide clear evidence of the culture's sophisticated architectural knowledge and craftsmanship. The majority of homes, and all of the buildings in many smaller towns, were constructed from less durable materials

such as mud. The language (kiSwahili) and culture that developed in the coastal communities reflected centuries of the synthesis of elements brought by Arabian, Persian, and African newcomers with the locally dominant African cultures.

Islamic cities were established in the West African savanna empires of Ghana, Mali, Songhai, and Borno, and in the Hausa states of Nigeria, starting in the 9th century. Most of these cities were located near the southern end of major trans-Saharan trading routes and developed in conjunction with the growth of economic linkages between West and North Africa. Their layout and architecture show a very strong North African Islamic influence. Many had separate districts inhabited by traders from North Africa. Manufacturing was concentrated in quarters occupied by specific types of artisans, while trade was centered within marketplaces. Mosques, the focal point of religious life, were located in the heart of the city. The most important of these cities were dynamic and sophisticated. For example, Timbuktu in the 15th and 16th centuries was a world-class city with impressive architecture, a bustling trading economy, and a renowned international university.

The writings of Heinrich Barth, a German explorer who visited Kano twice during the

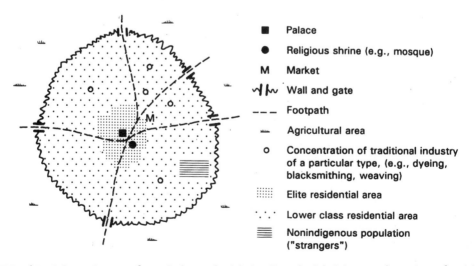

FIGURE 16.2. Schematic map of a typical precolonial city. Precolonial cities served a variety of social, economic, and political functions.

mid-19th century, provide a graphic description of the vitality of that city. Kano's economy was based on both manufacturing and commerce. The principal industry—the weaving and dyeing of cotton cloth—was undertaken during the dry season in the city and throughout the rich farming region nearby. Kano cloth was worn throughout the savanna of West Africa and across the Sahara in North Africa. Barth noted that the influx of foreign traders during the dry season caused a doubling of Kano's normal population of 30,000.

Indigenous (non-Islamic) African cities evolved in several regions, including Zimbabwe, the Kongo kingdom in the present-day Democratic Republic of the Congo, and the forest zone of West Africa. These cities, having developed in diverse cultural and ecological settings, were more varied in form and function than were the Islamic urban centers.

The precolonial urbanization process was particularly well developed in the Yoruba kingdoms of southern Nigeria and in the neighboring kingdom of Benin. The earliest Yoruba city, Ife, dates back at least to the 10th century. Several Yoruba cities were founded as military camps. Craft industries such as weaving, carving, and metalworking developed, with most towns having their own specializations, but the economies of Yoruba cities remained primarily agricultural. By the late 19th century, three Yoruba cities (Ibadan, Oyo, and Abeokuta) had populations of approximately 100,000, and 20 other urban settlements contained 10,000 to 60,000 people. A number of European visitors to Benin, starting in the 15th century, described it as a very impressive city, with broad and straight avenues lined with neatly arranged and solidly constructed houses. The king's palace, which reportedly occupied as large an area as the Dutch town of Haarlem, contained numerous dwellings and huge galleries where bronze plaques portraying the history of Benin were on display. Merchants from Benin conducted large-scale trade in ironwork, weapons, ivory, and other goods throughout western Africa.

Europeans arriving for the first time found cities in many parts of Africa. In other regions there was no urban development, even where conditions seemed to favor urbanization. For example, the Igbo of southeastern Nigeria did not build cities, despite their proximity to other urbanized societies, their very high population densities, and relatively advanced social and economic institutions. It has been hypothesized that the absence of cities is related to the egalitarian structure of Igbo society: They had no chiefs, and therefore had no reason to build cities to house and glorify an aristocracy.

Starting in the late 15th century, Europeans established themselves at isolated points along the coast, the intention being to secure and protect regional trading rights and to provide refueling stations for ships en route to India. Fortifications were built at many of these sites, and small settlements grew up around them. Modern cities such as Dakar, Accra, and Luanda grew rapidly during the colonial era at sites that Europeans had occupied for centuries. Other centers, such as Libreville and Calabar, emerged later in the precolonial era in conjunction with the growth in commodity trade and early efforts to liberate slaves.

The Colonial City

Although a large proportion of Africa's cities and towns originated in colonial times, only one-quarter of today's largest, most important cities had not been established prior to the colonial conquest. These previously existing settlements, of both African and European origins, grew as a result of expanded trade and new administrative functions. Elsewhere, particularly in the interior and in areas where there was little or no previous tradition of urbanization, Europeans established cities for the first time. Figure 16.3 is a schematic map of a typical colonial African city.

Although most colonial cities in Africa performed a variety of economic and political functions, it is often possible to identify an initial one. Lusaka and Abidjan are among several cities established as colonial administrative centers. Other, generally smaller cities developed as provincial and district administrative centers in each country. The administrative capitals usually contained military garrisons, which rep-

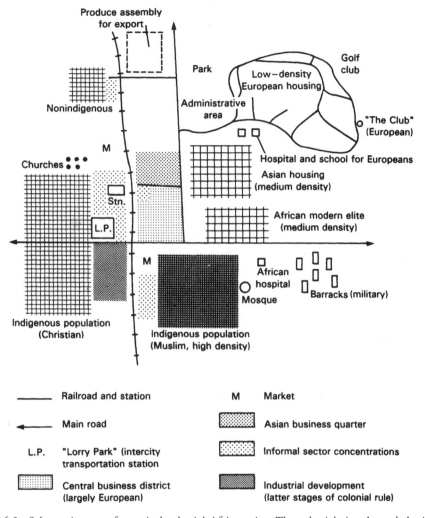

FIGURE 16.3. Schematic map of a typical colonial African city. The colonial city showed the influence of urban planning: a grid pattern of streets and a spatial separation of urban functions and population subgroups.

resented the ultimate basis of colonial authority.

The exploitation of African resources was possible only because of the development of a transportation system. Port cities such as Douala, Matadi, and Beira were founded where good harbor sites were available. Railways were constructed to move minerals and farm produce from the interior to the ports for shipment overseas. Cities located along the rail lines served as bulking points where produce from the hinterland could be gathered for export. In major mineral areas such as the copper belt of

Zambia and the Democratic Republic of the Congo, cities were developed to house and provide services for mine employees. Thus many of Africa's colonial cities were created for economic reasons; administrative functions were clearly of secondary importance in these centers.

Some of the cities established by Africans were rejected by the new colonial rulers as administrative centers, or were bypassed by the new transportation routes and thereafter declined into obscurity. Kukawa and Abomey, the capitals of the kingdoms of Borno and Dahomey, respectively, were cities that declined under colonialism. However, other precolonial cities gained new economic and governmental functions and prospered. In the British colonies, the forms of precolonial cities were left essentially intact; new colonial cities were grafted onto them. Kano is the classic example of a dual city, incorporating both a bustling modern sector and an ancient walled Islamic city (see Vignette 16.1, with Figures 16.4 and 16.5). In contrast, precolonial cities in French colonies were redesigned substantially in the European image. Most of the old cities were demolished and rebuilt with straight avenues, public squares, and other typically European features.

Urbanization during the colonial era reflected a concerted attempt to control the form, size, and function of cities, and to achieve orderly development through the application of European town-planning principles. The grid pattern of streets, the central business district, and the architecture of public buildings were all reminiscent of Europe.

Other attributes of colonial cities reflected the particular concerns of Europeans living in Africa. Cities were deliberately and rigidly segregated on racial grounds. European, Asian, and African living and working spaces were placed in separate parts of the city. Segregation was justified on grounds of security and public health (see Vignette 16.2 with Figure 16.6). Efforts were also made to segregate Africans of different religious and ethnic origins, allegedly to reduce the likelihood of conflict. In reality, segregation permitted the colonialists to exert greater control by "dividing and ruling" groups that were potential sources of discontent and resistance. Economic functions were similarly divided among (1) a central business district catering to Europeans; (2) a so-called "second-class" business district occupied by Asian merchants, who were "middlemen" linking the African and European economic sectors; and (3) marketplaces for African traders.

Salisbury (now Harare), Lusaka, and Nairobi differed from most other colonial cities in that they were designed to cater to the needs of European settler populations. The urban structure and architecture of these cities were very similar to those in Europe. There were not many signs (especially in the central business district and in European residential districts) that these cities were located in Africa. Their European character was protected by means of legislation strictly controlling rural–urban migration, and thus limiting both the number of African residents and their location in the city. Figure 16.7 shows Lusaka, as well as a smaller town in present-day Eritrea with colonial origins.

Because of the heavy emphasis given to export-oriented agriculture in African economic development, it is hardly surprising that colonial authorities did not encourage migration to urban centers. Colonial urban economies were dominated by European commercial companies and Asian (primarily Lebanese and Indian) merchants. As a rule, rather limited opportunities were left for African traders. The public sector also provided few job opportunities for Africans during colonial times.

Although the history of South African urbanization reflects the unique character of its racially mixed population and its dynamic capitalist economy, there are strong parallels between South African and colonial African urban development. The implementation of apartheid as an official policy after 1950 was the culmination of decades of policies like those of colonial Africa that attempted to limit the migration of Africans to the city and intensified segregation within the city (as described in Vignette 16.2). In Cape Town and other cities, the quest for a strictly segregated city involved uprooting vast numbers of people from their homes and sending them to locations designated by the state. The dismantling of apartheid has enabled some Africans (mostly wealthier and middle-class people) to move to areas formerly reserved for other groups, but the vast majority cannot afford housing outside the townships or the massive squatter settlements that have mushroomed in recent years.

The Contemporary African City

As of the early 1960s, Africa south of the Sahara had few substantial cities. Of the 25 countries gaining independence between 1960 and 1964, only 3 had a city with at least 300,000 people: Nigeria (Lagos and Ibadan), the present-day

VIGNETTE 16.1. Kano: Quintessential African City

The urban landscapes of Kano, Nigeria, reflect over 1,000 years of development and change. The early history of the city is quite well known, in part because of the detailed reports of several Arab and European visitors over the centuries and in part because of the Kano Chronicle, a unique written record of the political and economic history of the city under 42 kings between the 9th and 19th centuries.

Kano grew at the base of Dala Hill, where iron was being smelted as early as the 7th century. The old city is surrounded by a wall—first constructed between 1095 and 1134, and subsequently enlarged three times. Kano, and particularly Kurmi Market in the heart of the city, became the center of a booming trade in textiles, slaves, ivory, and other goods. Kano trade networks extended across the Sahara to North Africa, south into the forest zone, and across the savannas from present-day Sudan to the Atlantic. By the 19th century, Kano had become probably the most important manufacturing and trading center of black Africa; it was a cosmopolitan city of 30,000 permanent residents.

Under colonial rule, the traditional trade patterns of Kano were severely disrupted, but the construction of the railway to Lagos created new opportunities, and Kano became the center of a regional economy based on exporting groundnuts and importing manufactured goods. The structure of the old city was not significantly changed during the colonial era. However, a new city was established outside the walls, containing a modern central business district; administration and transportation facilities; and new residential areas for Europeans, migrants from southern Nigeria, and newcomers from other parts of the north.

The city continued to expand after independence, and particularly after the establishment of Kano State in 1967. Its population surpassed 1 million inhabitants during the 1980s, and exceeded 2.3 million by the year 2000. The economy has diversified considerably beyond the traditional economic base in trade. Manufacturing has become increasingly important, as have education and administration. The form of the city has continued to evolve. Newer residential areas increasingly reflect class divisions in the society, rather than ethnicity. The emergence of squatter settlements, occupied for the most part by the poor, is a new phenomenon. Elsewhere, middle- and upper-middle-class suburbs have been constructed to house civil servants, professionals, and businesspeople. With the proliferation of cars, traffic has become a major problem. Nevertheless, away from the few main thoroughfares, the old city remains a place apart; most of it is accessible only along a circuitous maze of footpaths.

The accompanying photographs (Figure 16.4) show a variety of Kano landscapes. Their locations are indicated by letters on the map (Figure 16.5) corresponding to those for each photograph.

a

b

FIGURE 16.4. Photographs of Kano, Nigeria. (a) Kofar Kansakali, the best preserved of the gates of the 20-km-long old city wall, dates from the 14th century. (b) The Central Mosque, along with the Emir's Palace and Kurmi Market, are the most important focal points in the old city.

(cont.)

VIGNETTE 16.1. *(cont.)*

c

d

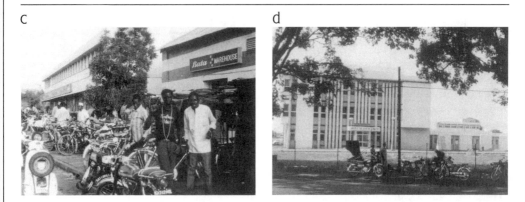

(c) In the central business district, the informal sector (in this case, bicycle repairs) and the formal sector (here, the multinational Bata Shoe Company) exist side by side. (d) Kano State Ministry of Home Affairs, an example of government-related functions that have been among the most important "growth industries" since the 1967 founding of Kano State.

e

f

(e) Gwagwarwa, a working-class suburb established about 1920, is inhabited mainly by northern Nigerian Muslims. (f) Sabon Gari is a neighborhood established in colonial times to house southern, largely Christian Nigerians. The building shown here combines commercial and residential space.

g

h

(g) Kurna, a lower-class squatter settlement that grew up during the 1970s and 1980s, is located directly under the main flight path to the airport. (h) An upper middle-class residence in the newer suburb of Gyadigyadi. Photos: author.

(cont.)

VIGNETTE 16.1. (*cont.*)

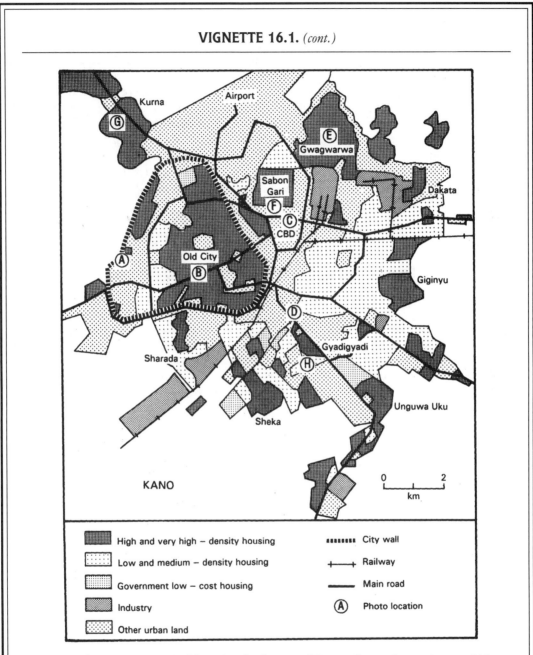

FIGURE 16.5. Kano, Nigeria. The urban landscapes of Kano reflect a thousand years of history and the effect of burgeoning growth in recent years. After H. A. C. Main, "Housing problems and squatting solutions in metropolitan Kano." In R. B. Potter and A. T. Salan, eds. *Cities and Development in the Third World*. London: Mansell, 1990, pp. 12-31.

Democratic Republic of the Congo (Kinshasa), and Senegal (Dakar). The population of the largest city was under 100,000 in 11 countries, and in 4 countries the largest city had fewer than 30,000 people. In all but 2 of these countries, Cameroon and Mauritania, the administrative capital was the largest city. Lesser urban centers were generally much smaller than the capital city, except where these urban centers were the sites of major economic activity (e.g., mining or shipping).

The pace of urban population growth quickened after the attainment of independence. By the year 2000, about 240 million people, representing over one-third of the population, lived in African cities south of the Sahara. The urban population had been only 15% in 1970 and 23% in 1980. The rate of growth of large cities during the mid-1990s was approximately 4–7% in most countries, roughly twice the rate of natural increase of most national populations. These exceptional rates of urban growth reflect the widespread reality of economic stagnation in the countryside and the common perception

VIGNETTE 16.2. Sanitation and Urban Planning

The spatial form of African cities still shows the influence of policies that sought to protect the health of whites from the allegedly unhealthy Africans. These measures, which grew out of the discovery of the relation between ill health and crowded, filthy environments, took very different forms in Europe and in the colonies. Massive investment in piped water and sewage systems did much to improve public health in European cities. In Africa, racial segregation brought health benefits to few.

In Cape Town, South Africa, epidemics of bubonic plague at the turn of the 20th century provided a pretext for the government to forcibly relocate nonwhites from their inner-city homes to remote, barren sites outside the city. Thus apartheid in this formerly integrated city began with the labeling of nonwhites as a threat to the health of whites.

In Freetown, Sierra Leone, the British decided to segregate African and European residents to protect the health of colonial officers, in the wake of the discovery in 1898 that malaria was spread by the bite of the *Anopheles* mosquito. Ronald Ross, who had made this discovery, organized an expedition to Freetown (reputedly the most malarious place in the British Empire) to advise the government on malaria control. Ross made a number of policy recommendations, starting with the elimination of mosquito-breeding sites. However, the Colonial Office chose a different approach—namely, the segregation of Europeans away from mosquito-breeding sites and from the African neighborhoods where most of these sites were to be found. A new European settlement, the Hill Station, was constructed overlooking the city.

The alternatives rejected by the Colonial Office, such as insect elimination and swamp drainage, would have contributed to better health for both Africans and Europeans. The choice of segregation as a strategy meant that only Europeans would benefit.

On closer examination, the racist underpinnings of the segregation policy become evident. Africans were blamed for being too lazy and primitive to maintain sanitary conditions in the town. Also cited as a justification for segregation was the alleged cost of providing "civilization" to Africans. But the implementation of segregation was not permitted to interfere with the basic comforts of European life; officers were permitted to have resident domestic servants for their Hill Station homes.

Sanitation regulations are no longer enforced with a view to the maintenance of racial segregation. Nevertheless, the distinctive appearance of the old European and African neighborhoods, sometimes still separated by open space—the former *cordon sanitaire,* or "sanitary barrier," where no development was permitted (see Figure 16.6)—are reminders of how and why early colonial cities were designed.

(cont.)

VIGNETTE 16.2. (*cont.*)

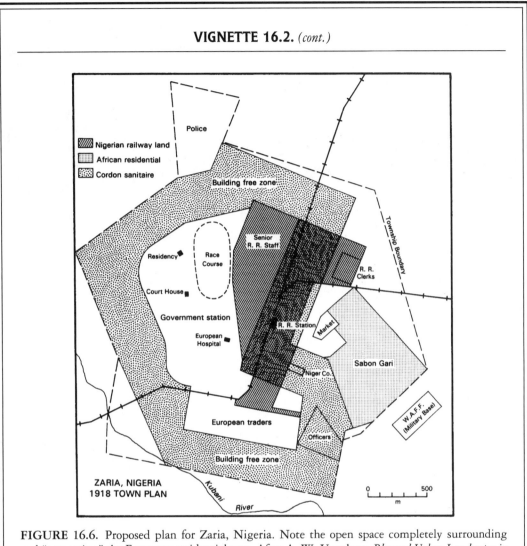

FIGURE 16.6. Proposed plan for Zaria, Nigeria. Note the open space completely surrounding and "protecting" the European residential area. After A. W. Urquhart. *Planned Urban Landscapes in Northern Nigeria.* Zaria, Nigeria: Ahmadu Bello University Press, 1977, p. 39.

Based on S. Frenkel and J. Western. "Pretext or prophylaxis?: Racial segregation and malarial mosquitoes in a British tropical colony: Sierra Leone." *Annals of the Association of American Geographers,* vol. 78 (1988), pp. 211–228. M. W. Swanson "The sanitation syndrome: Bubonic plague and urban native policy in the Cape Colony, 1900–1909." *Journal of African History,* vol. 18 (1977), pp. 387–410.

that fame and fortune are attainable in the city. Monrovia, Liberia and Maputo, Mozambique are among several cities located in countries experiencing protracted instability and violence where above-average rates of urban population growth have occurred. Large cities may offer relative safety and stability during conflict, as well as access to goods and services unavailable in the countryside.

The rapidity of urbanization is reflected in patterns of growth of specific cities (Figure 16.8). Lagos, which had only 0.76 million people in 1960, had grown to 4.45 million by 1980 and had an estimated population of 12.5 million in

a b

FIGURE 16.7. Two cities established in colonial times. (a) Central business district, Lusaka, Zambia. A well-developed central business district and railway (visible on the left of the photo) are both typical elements of colonial-era cities. (b) Keren, Eritrea. Two elements that show the colonial origins of this small town are its layout, radiating from a central plaza, and the Italian influences visible in the architecture of some of the buildings. Not surprisingly, the cities of different colonial powers varied somewhat in layout and appearance. Photos: (a) author; (b) M. Peters.

2000. Kinshasa had grown from 0.45 million in 1960 to 2.24 million in 1980; its estimated population in 2000 was 4.35 million. Johannesburg was the only city over 1 million in 1960; by 2000, there were some 30 "million-plus" cities in Africa south of the Sahara.

One of the consequences of very rapid growth has been the acute shortage of affordable housing. Many poor families can afford to occupy only one rented room, while others have responded by building their own dwellings in squatter settlements where they have no formal tenure. Surveys suggest that between half and four-fifths of the residents of large national cities live in slums and squatter settlements. The rapid growth of cities has put immense pressure on urban infrastructure, including schools, water and sewage systems, and transportation. As a result, urban residents—and, in particular, people living in densely populated squatter set-

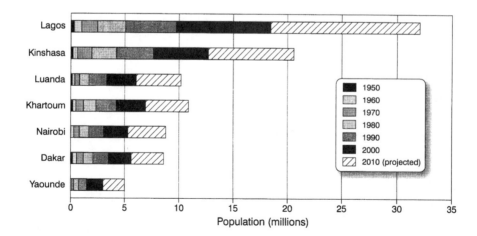

FIGURE 16.8. Past and projected growth of selected cities. Many African cities had some reduction in the rate of population growth during the 1990s, due in large part to the impacts of structural adjustment. Data source: *Compendium of Human Settlement Statistics.* New York: United Nations, 2001.

tlements on the edge of a city—expend much time and money trying to get access to services that they need.

The modern African city looks very different from its colonial predecessor. Apart from the very obvious difference of size, there are important structural and functional differences. Formal racial segregation, one of the fundamental principles of colonial urban design, has been superseded by informal divisions reflecting the class structure of society. African elites have moved into higher-class districts of the type formerly occupied only by whites. Class-based segregation also extends to the middle and lower classes, who tend to occupy separate neighborhoods.

The central business districts of the leading cities in the more prosperous countries have large stores, skyscrapers, and traffic jams reminiscent of cities in industrialized countries. The names of transnational corporations—for example, Toshiba, Mitsubishi, Sheraton, and Hilton—are evident, while the names of colonial trading companies such as the United Africa Company (UAC) and the Compagnie Française d'Afrique Occidentale (CFAO) have become quite obscure. African entrepreneurs have developed sizable companies by taking advantage of expanding business opportunities and, in many cases, laws restricting expatriate involvement in certain types of activities. The streets of the central business district also contain many petty traders plying their wares. This type of informal street trade in the central business district had been strictly controlled, and usually prohibited, during the colonial era. Petty trading and other informal-sector enterprises are certainly not confined to the central business district. Small-scale businesses making available an infinite variety of goods and services are found along major thoroughfares, in markets, and almost everywhere in the poorer neighborhoods, where these informal-sector jobs are crucial for making ends meet.

Colonial industrialization was very rudimentary, mostly involving the primary processing of local products for export. Now the outskirts of larger African cities have substantial industrial estates where a variety of goods are manufactured, primarily for domestic markets. These estates—along with national buildings, modern sports stadiums, and universities—are not merely business ventures and places of employment, but also symbols of national development and hope for the future.

The symbolic importance of modern urbanization is especially important in a handful of new, planned cities established since independence. The port city of Tema, developed as a focus for the anticipated industrial development of Ghana in the early 1960s, was the first of these postcolonial planned cities. New capital cities have been established in Malawi (Lilongwe), Côte d'Ivoire (Yamoussoukro), and Nigeria (Abuja). These cities have taken their inspiration from Brazil's futuristic capital on the frontier, Brasilia. In each case, their locations (away from the old centers of power), their planned designs, and their focal buildings were intended to evoke an image of hope for a brighter future (see Figure 16.9).

There is another perspective on the symbolism of new, modern capital cities, well expressed in this excerpt from a letter to the editor of a Nigerian newspaper:

> Today there seems to be a growing distance between Abuja and the rest of Nigeria. To a stranger, Abuja paints a picture of a rich and economically buoyant Nigeria. The towering story buildings and exotic structures are beautiful. The massive construction works going on all over the city give an impression of a serious nation. The orderly flow of traffic and the good roads network create the image of a peaceful nation. The truth is that Abuja is an artificial impression of the true state of the nation. So, nearby Nyanya makes one understand that Abuja is detached from the reality that is Nigeria. Nyanya and its likes are but true slums, call it squalor in splendour. (Isa Sanusi, *Weekly Trust* [Kaduna], November 22, 2002)

Although it is important to look for generalizations that can be made about the cities of Africa and their postcolonial development, the limitations of generalized statements must be stressed. Many of the leading cities in African nations have developed into bustling metropolitan centers, but other cities bear the stamp of persistent economic and political crises that

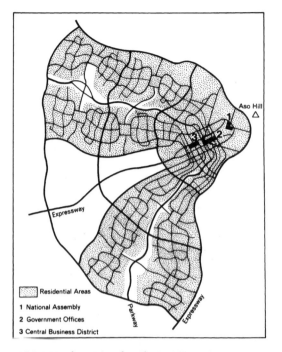

FIGURE 16.9. Plan for Abuja: Nigeria's new capital, established during the 1980s. Source: W. T. W. Morgan. *Nigeria.* Harlow, England: Longman, 1983, p. 155. © 1983 by W. T. W. Morgan. Reprinted by permission.

have stifled modern urban development, though not the growth of urban population. The contrast in appearance and economic vitality between economically stagnating cities like Freetown, Mogadishu, Luanda, and Kitwe, and entrepreneurial hives of activity like Nairobi, Abidjan, Lagos, and Douala, is much greater than it was at the time of independence. Between these two extremes are cities like Lusaka, where relative prosperity and rapid urban development in the early independent years have since given way to decline and stagnation. In other cases, cities such as Accra and Kampala that fell on hard times during much of the 1970s and 1980s are again prospering.

The discussion in this chapter has focused primarily on national cities, almost all of which are capitals. The postcolonial transformation has been most dramatic in these cities. Although the development of smaller cities has been much less spectacular than that of national cities, many of these secondary urban centers have also experienced considerable growth since

the 1960s. Several states have attempted to encourage development in these smaller centers through administrative decentralization, transportation development, and the locating of public-sector facilities and enterprises away from the national capital. Still, with very few exceptions, the more cosmopolitan and better-serviced capital cities continue to attract much more investment and many more migrants than the secondary urban centers do.

Though most African societies remain predominantly rural, cities have grown explosively, especially in the three decades since independence. Urban centers have become the primary hub not only of political affairs, but also of change in national economies and societies. This concentration of wealth and power ensures that Africa's leading cities will continue to grow in size and importance within their respective countries.

It has been noted that many residents of contemporary African cities struggle to make ends meet and to secure basic amenities. Governments must also grapple with these same challenges—that is, striving to achieve economic growth and stability, and ensuring that housing and other amenities are available. Planning for the future is the easiest part; the real challenge is to find ways of bringing about orderly and progressive development when resources are so scarce and when urban residents simply choose to get on with meeting their own needs, paying no heed to utopian official plans. The next chapter examines in greater detail individual and official responses to problems of urban economic development.

Further Reading

Reference sources on African urbanization include the following sources:

O'Connor, A. *Urbanization in Africa: An Annotated Bibliography.* Boston: G. K. Hall, 1982.
Stren, R., and C. Letemendia. *Coping with Rapid Urban Growth in Africa: An Annotated Bibliography.* Montreal: McGill University, 1986.

Various aspects of the development of African cities are discussed in the following sources:

Briggs, J., and I. Yeboah. "Structural adjustment and the contemporary African city." *Area*, vol. 33 (2001), pp. 18–26.

Coquery-Vidrovitch, C. "The process of urbanization in Africa (from the origins to the beginning of independence)." *African Studies Review*, vol. 34 (1991), pp. 1–98.

Hanna, W. J., and J. L. Hanna. *Urban Dynamics in Black Africa: An Interdisciplinary Approach*, 2nd ed. New York: Aldine, 1981.

Mabogunje, A. L. "Urban planning and the postcolonial state in Africa." *African Studies Review*, vol. 33 (1990), pp. 121–203.

O'Connor, A. *The African City*. London: Hutchinson, 1983.

Rakodi, C., ed. *The Urban Challenge in Africa: Growth and Management of Its Large Cities*. New York: United Nations, 1997.

Stren, R., and R. White, eds. *African Cities in Crisis: Managing Rapid Urban Growth*. Boulder, CO: Westview Press, 1989.

Swilling, M., R. Humphries, and K. Shubane, eds. *Apartheid City in Transition*. Cape Town, South Africa: Oxford University Press, 1991.

Precolonial urbanization is discussed in the following sources:

Nast, H. "Islam, gender, and slavery in West Africa circa 1500: A spatial archaeology of the Kano Palace, northern Nigeria." *Annals of the Association of American Geographers*, vol. 86 (1996), pp. 44–77.

Winters, C. "Traditional urbanism in the north-central Sudan." *Annals of Association of American Geographers*, vol. 67 (1977), pp. 500–520.

The effect of colonialism on African urban development is discussed in the following sources:

King, A. *Colonial Urban Development*. London: Routledge and Kegan Paul, 1976.

Myers, G. *Verandahs of Power: Colonialism and Space in Urban Africa*. Syracuse, NY: Syracuse University Press, 2003.

Southall, A. "The impact upon urban development in Africa." In V. Turner, ed., *Colonialism in Africa, Profiles of Change: Society and Colonial Rule*, pp. 216–255. Cambridge, UK: Cambridge University Press, 1971.

Each of the following case studies examines a particular African city:

Lloyd, P. C., A. L. Mabogunje, and B. Awe. *The City of Ibadan*. Cambridge, UK: Cambridge University Press, 1967.

Skinner, E. P. *African Urban Life: The Transformation of Ouagadougou*. Princeton, NJ: Princeton University Press, 1974.

Western, J. *Outcast Cape Town*, reprint ed. Berkeley: University of California Press, 1997.

Internet Sites

International Network for Urban Development. http://www.inta-aivn.org

Brinkhoff, T. *City Population. The Principal Cities and Agglomerations in Africa.* http://www.citypopulation.de/Africa.html

United Nations-Habitat. *United Nations Human Settlements Programme.* http://www.unchs.org

Some of Africa's cities have municipal websites. See the following, for example:

Bienvenue Ville de Dakar. http://www.dakarville.sn
Cape Town. http://www.capetown.gov.za
Durban. http://durban.kzn.org.za/durban

17

Urban Economies

Chapter 16 discussed the explosive growth that has been characteristic of African urban centers, especially since 1960. The focus now turns to how urban Africans make their living and to the broader characteristics of the urban economies within which Africans work. Whereas most of the policy initiatives by governments were once directed at the development of large-scale projects in the formal economic sector, the focus has increasingly shifted to the informal activities of small-scale traders, artisans, and the like, which account for the majority of economic activity.

The Structure of Urban Economies

Economists speak of four basic types of economic activity:

- Primary production, which involves the harvesting of products of the environment by farmers, fishers, forestry workers, and miners
- Secondary production or manufacturing, which transforms products into more valuable and useful forms
- Tertiary activities, entailing the provision of services such as health care, entertainment, and the sale of goods to consumers

- Quaternary activities, including government and other administrative services

Although *primary production* is usually associated with rural economies, it also plays an important role in many African cities. Johannesburg and Kimberley in South Africa, Ndola in Zambia, and Lubumbashi in the Democratic Republic of the Congo are examples of urban centers that started as mining towns and where mining remains important. Primary producers engaged in farming, fishing, and forestry are to be found in virtually all African cities. In certain regions, including indigenous Yoruba cities like Ibadan and Oyo, some farmers migrate between permanent residences in the city and seasonally occupied farmsteads.

Small market gardens (see Figure 17.1) are an integral part of the landscape within and adjacent to most cities. These small farms contribute to family food supplies and take advantage of a large demand for fresh produce. Only quite recently have social scientists begun to recognize the scale and importance of this activity, which puts to use countless nooks and crannies inside household compounds, beside roads and streams, and wherever else vacant land is to be found. A study of urban agriculture in Nairobi

FIGURE 17.1. Urban agriculture, Kampala, Uganda. Produce grown by urban dwellers on small parcels of land within the city accounts for a significant proportion of urban food supplies. Photo: author.

(Vignette 17.1, with Table 17.1) demonstrates the value of urban agriculture to the well-being of families and the economy as a whole.

Secondary production ranges in sophistication from simple handicrafts to modern manufactures. Although small-scale handicraft production is prevalent in both urban and rural settings, large-scale manufacturing is essentially an urban-based activity. In most countries, some 65–90% of manufacturing industry is located in the primate city, with virtually all the rest being situated in smaller, regional centers.

Certain *tertiary activities* are heavily concentrated in cities. For example, interregional transportation services, as well as import–export and wholesale trade, are clustered in large port cities where transportation routes converge. Financial services such as banking and insurance are largely urban-based, as are the corporate headquarters of larger companies. Africa's large cities also have a quite diverse range of retail, entertainment, and other services, catering in particular to those with above-average incomes.

Quaternary activities are perhaps the most urban-focused of all. African governments tend to be highly centralized in countries' national capitals, and to a lesser extent in regional or state capitals. So too are *parastatal* corporations—companies established by governments to en-

gage in particular business ventures. The public sector, which oversees the establishment and management of schools, hospitals, and other social institutions, is a major employer of urban workers.

Formal and Informal Economies

Until the early 1970s, studies of urban economies in Africa focused almost exclusively on modern economic activity undertaken in factories and by larger-scale commercial enterprises. Meanwhile, the "subsistence" economic activities of the poor were virtually ignored. Later, it was recognized that many urban residents obtained few benefits from standard modernization strategies. The poor constructed their own houses, manufactured a wide variety of products (often from recycled materials), and earned a living as self-employed sellers of an infinite variety of goods and services. In some African cities, as many as 90% of urban workers are active in this "informal economy."

The urban informal economy consists of small-scale, labor-intensive enterprises, heavily reliant on family capital, skills, and labor. To attempt precise definitions—exactly how small and how dependent on family resources, for example—would probably be a futile exercise.

VIGNETTE 17.1. Making Ends Meet: The Significance of Urban Agriculture

Most earlier studies of the urbanization process assumed that migrants from rural areas abandon agriculture when they move to the city. More recent studies, however, are finding that urban agriculture is both prevalent and economically significant. Table 17.1 (based on Freeman [see footnote below], p. 129) shows something of the importance of farming in three of Kenya's largest cities.

TABLE 17.1. Urban Agriculture in Three Kenyan Cities

	Kisumu	Mombasa	Nairobi
HH[a] with food-growing land in city	28%	15%	11%
HH growing food on urban or rural land	70%	55%	65%
HH selling urban crops	18%	21%	21%
Median quantity produced	90 kg	125 kg	29 kg
HH keeping livestock	55%	47%	51%

[a]HH, households.

Evidence of the activities of urban farmers is to be seen in virtually all parts of Nairobi, even in the heart of the central business district. These vegetable gardens, or *shambas,* are on average about 2,300 m^2 in size, ranging from 5 m^2 to over 6 ha. Private residential land is used by about one-third of cultivators; most others farm roadsides, riversides, and other public spaces. Private land, when available, is more secure from theft, but the alluvial soil found in river floodplains is likely to be more fertile.

Two-thirds of urban cultivators in Nairobi are female, but otherwise they vary greatly in income, age, education, employment status, and duration of urban residence. Close to half had no job at the time of Freeman's survey.

Many different crops are grown on urban *shambas.* The most important is maize, produced on over half of all farms. Beans, potatoes, cocoyams, various types of vegetables, and bananas are also grown by many cultivators. Although most urban farmers (70%) consume all that they grow, there is also a significant market component to agriculture in Nairobi. It was found that 13% of those surveyed sold half or more of their crop.

Urban agriculture contributes to the local and national economy and to the social well-being of the urban poor. Freeman has observed that farming benefits the community economically in five ways: by (1) contributing to urban productivity (with an estimated aggregate production for all Kenyan cities of 25 million kg); (2) generating employment, not only on the plots themselves, but also through the sale of agricultural inputs and produce; (3) giving women entrepreneurs entry into the informal sector; (4) exploiting a vacant niche in the urban economy, focused on the food needs of the urban poor; and (5) valorizing vacant and often derelict urban land.

Urban farming is a practical, self-reliant, and sustainable response to people's pressing needs for an adequate diet, a job, and extra income. For Nairobi's poor, the *shambas* represent, according to Freeman, "gardens of hope, not wastelands of despair" (p. 121).

Based on D. B. Freeman. *A City of Farmers: Informal Agriculture in the Open Spaces of Nairobi, Kenya.* Montreal: McGill–Queen's University Press, 1991.

Rather, definitions of the informal economy have emphasized ways in which formal and informal economies differ (Table 17.2).

Although the contrasts between formal and informal economies are evident, it is important to remember that they are part of a total urban economy and are neither as clearly differentiated nor as functionally separate as Table 17.2 seems to imply. All cities contain enterprises such as medium-sized, locally owned retail outlets and industries that fall between the "definitely formal" and "definitely informal" sectors. There are also many forms of interaction between the formal and informal economies that are not reflected in this dualism. Labor moves back and forth between the two sectors, transferring skills as well as raw materials and finished products from one to the other.

The Formal Economy: Manufacturing

Africa south of the Sahara remains the least industrialized part of the world, accounting for only 1% of global manufacturing value added. Colonial policy retarded the development of an industrial sector; manufacturing was quite simply incompatible with Africa's assigned role of supplying raw materials to the industrialized world. The only notable exceptions were a few raw-material-processing enterprises, such as cotton gins, vegetable-oil mills, and mineral concentrators.

Newly independent governments invested heavily in industrial development in the 1960s and 1970s. They gave generous incentives to industrialists, formed partnerships with transnational companies, and established industrial parastatal corporations. The case for rapid industrialization was stated by Kwame Nkrumah in *Africa Must Unite* (New York: Praeger, 1963): "We have here, in Africa, everything necessary to become a powerful, modern, industrialized continent. . . . Every time we import goods that we could manufacture if all the conditions were available, we are continuing our economic dependence and delaying our industrial growth" (p. 112). In socialist Mozambique and Tanzania, almost 90% of industrial projects were undertaken with government participation, prior to the imposition of structural adjustment.

The record of state involvement in manufacturing has been mixed. Incentives and partnership arrangements helped at times to overcome the reluctance of transnational corporations to undertake risky investments. However, hundreds of industrial "white elephants" provide clear evidence that sound economic reasoning was often bypassed. Rates of industrial growth declined during the 1970s and 1980s, reflecting the very serious effect of rising petroleum prices, falling export revenues, and growing debts. The decline in industrial production became acute during the early years of structural adjustment, when many factories closed because they could not compete without subsidies in an unregulated market, and because the costs for imported inputs were increasing as a result of massive currency devaluations. "Deindustrial-

TABLE 17.2. Attributes of Formal and Informal Economies

Informal sector	Formal sector
Ease of entry	Difficult entry
Predominantly indigenous inputs	Overseas inputs
Predominantly family property	Corporate property
Small scale of activity	Large scale of activity
Labor-intensive activity	Capital-intensive activity
Adapted technology	Imported technology
Skills from outside school system	Formally acquired (often expatriate) skills
Unregulated/competitive market	Protected markets (e.g., tariffs, quotas, licenses)

ization" was particularly severe in several countries, including Ghana, Benin, and Tanzania, where as little as 30% of installed industrial capacity was being used in the late 1980s and early 1990s. In countries where structural adjustment did little to stimulate economic recovery, the majority of industrial capacity remains idle.

Table 17.3 shows that the downturn was especially severe in low-income countries, where there was no growth in output between 1975 and 1980, and (with the exception of semiarid countries) where there was significant decline between 1980 and 1987. The most significant decline since 1980 has been among oil exporters; the collapse of oil prices was instrumental in bringing about a decline of 2.5% per year between 1980 amd 1987 in industrial output, compared to an annual growth of 11.5% during the boom years of the late 1970s. Industrial crisis is manifested in the abandonment and underutilization of industrial capacity.

During the 1990s, a period of considerable growth in the world economy, African industrial production continued to stagnate. Its growth of 1.6% per year was half that of Latin America and one-quarter the rate achieved in South Asia. South Africa and Nigeria each had annual growth rates of 1% during the 1990s. Côte d'Ivoire, with 5.1% annual growth, was among the few countries where the industrial sector prospered during the 1990s.

Industry in underdeveloped economies tends to be concentrated in the largest urban centers, where energy supplies, transportation, communications, and other infrastructural resources are most accessible. The largest cities also contain better-educated and more experienced workers, as well as the most affluent consumers. The extreme concentration of Senegal's manufacturing in the capital city of Dakar (Vignette 17.2) is typical of industrial development in Africa. South Africa and Nigeria are the only countries in which industry is fairly well dispersed.

The composition of Senegal's manufacturing is typical of the African pattern. Manufacturing in most countries consists almost exclusively of enterprises that process exports and produce substitutes for imports. The first category includes ore concentrators, vegetable-oil mills, fruit canneries, sawmills, and other facilities preparing primary goods for export. Import substitution industries make various consumer goods for local markets, using either local or imported raw materials.

Capital-goods industries, producing inputs utilized in other types of production, are poorly developed except in South Africa. Though Africa has significant deposits of coal and iron ore, only South Africa has a functioning, integrated iron-and-steel complex. Nigeria's steel complex, begun in 1976 at Ajaokuta, has been plagued by technical difficulties and fiscal crisis; over 25 years after its inauguration, the plant is still not in operation. Ghana, Cameroon, and Guinea have major aluminum smelters (see Figure 17.2), but their products are almost all exported as unfinished aluminum rolling stock. Although several countries have petroleum refineries producing fuels and basic petrochemicals, only South Africa has a well-developed chemical industry.

Africa south of the Sahara has not yet attracted much export-oriented manufacturing

TABLE 17.3. Growth of African Manufacturing Output, 1960–1987 (Percentage per Year at Constant 1970 Prices)

	1960–1965	1965–1970	1970–1975	1975–1980	1980–1987
Low-income countries					
Semiarid	4.8	9.7	1.8	0.4	2.9
Other	9.3	8.2	2.4	−0.2	−2.7
Middle-income countries					
Oil importers	7.5	7.6	7.7	4.2	NA
Oil exporters	3.7	15.9	6.6	11.5	−2.5
Total for sub-Saharan Africa	7.3	9.3	5.3	4.4	−1.2

Data sources: W. F. Steel and J. W. Evans. *Industrialization in Sub-Saharan Africa: Strategies and Performance.* Washington, DC: World Bank, 1984, p. 33. World Bank. *From Crisis to Sustainable Growth: A Long-Term Prospective Study.* Washington, DC: World Bank, 1989.

VIGNETTE 17.2. Industrial Development: A Case Study of Senegal

By the standards of Africa south of the Sahara, Senegal is a moderately industrialized country. It lags well behind Nigeria, Côte d'Ivoire, Kenya, and Zimbabwe in volume and diversity of industrial production, but it is more industrialized than most African countries, including some more populous ones like Tanzania and Ethiopia. Manufacturing accounted for 13% of Senegal's gross domestic product in 1999 but provided jobs for only 40,000 employees, representing only 1% of the economically active population.

Over 85% of Senegal's manufacturing activity is located in Dakar and its suburbs. Most of the remainder is established within 150 km of Dakar in the provincial towns of Thiès, Kaolack, and Diourbel. The very few industries found elsewhere in the country are generally linked to sources of raw materials; the sugar factory at the Richard Toll irrigation project is an example. The concentration of industry in Dakar reflects not only the size and skill level of its workforce, but also the better transportation and communication systems, and other services that are readily available there.

Senegal's industries process local raw materials for export and produce goods that substitute for imports in the domestic market. Food processing, which accounts for about 40% of industrial production, includes groundnut-oil mills, fish canneries, and a sugar refinery (export processing), as well as breweries, flour mills, and confectionaries (import substitution). Textiles, metal products, and chemicals and allied products each account for over 10% of total industrial production. Although some local raw materials are used in these industries—cotton for textiles and phosphates for fertilizers, for example—most industries depend on imported raw materials such as unrefined petroleum or imported parts for local assembly.

Senegal possesses a good variety of basic industries producing goods for local consumption, but there are few capital-goods industries. The exceptions are plants that refine petroleum, make fertilizer, produce cement, build ships, and assemble agricultural machinery. Senegal's domestic market is presently neither large enough nor wealthy enough to support a more sophisticated industrial base. The country benefits from being able to export industrial goods to its less developed neighbors, such as Mauritania, Mali, and Gambia. The government has emphasized the development of industries to process and add value to its own raw materials, such as phosphates, fish, and groundnuts; varying proportions of these products are exported to other parts of the world. A free-trade zone has been established in Dakar to attract foreign industry geared to export production, but so far it has had little success.

Senegal's industrial production has stagnated since 1986, primarily due to the impact of structural adjustment policies on less competitive industries. For example, textile production declined in the mid-1990s because of shortages of cotton; much of the cotton crop was sold abroad to reap higher prices following the devaluation of the currency. The textile and shoe industries are among several that have been adversely affected by cheap imports. Under structural adjustment, the government was forced to divest its interests in many parastatal companies. Between 1990 and 1999, the value of food-related manufacturing declined from 60 to 44%, while the chemical sector increased threefold to 26% of the value of manufactures.

Senegal's manufacturing sector is of considerable importance to the domestic economy but nevertheless remains small, even within the regional context of West Africa. The size and composition of the industrial sector are unlikely to change significantly in the foreseeable future.

investment, unlike Thailand, Mexico, and many other Third World countries. A number of African countries (including Senegal, Togo, Liberia, Tanzania, and Sudan) have established free ports, where tariffs and other regulations are relaxed for companies producing for export. However, none has had more than limited success attracting exporting industries. Africa has no great advantages of location, labor costs, or skills, compared to competing locations in

FIGURE 17.2. Inside an aluminum smelter, Douala, Cameroon. Regions with an abundance of hydroelectric power have some comparative advantage in producing aluminum for export. Photo: CIDA (R. Lemoyne).

Latin America and Asia. The continent's reputation for political instability and inadequate infrastructures have been major disincentives to large-scale foreign investment.

Industrialization has brought together workers from diverse backgrounds who share common concerns about wages, working conditions, and job security. One result has been the growth of a working-class consciousness that transcends ethnic, regional, and religious cleavages. Working-class solidarity typically becomes stronger as a result of union membership and of collective actions (e.g., strikes and protests against government policies that are harmful to labor). Nevertheless, the participation of workers in national affairs is constrained by restrictions imposed on unions in the majority of African states.

The Informal Economy

Whereas the emerging industrial sector has received careful attention and many favors from African governments, the informal sector has continued to function in relative obscurity. As previously mentioned, the informal economy consists of small-scale enterprises producing and selling a wide variety of goods and services,

using comparatively simple technology and resources mobilized primarily from within family units.

The urban informal economy resembles, but is distinct from, the traditional craft industries and petty trading found throughout rural Africa. As a rule, craft production in rural areas is done seasonally by people who depend primarily on farming for their livelihood. Most of those employed in the urban informal economy are, by necessity or by choice, almost totally dependent on nonagricultural income. There are, however, tangible connections. For example, migrants from the countryside often enter the urban informal economy by employing or adapting their rural skills.

The urban informal economy is important not only as a means of subsistence, but also for its enormous contribution to the larger economies of cities and nations. The majority of the African urban work force is employed in the informal economy. In Dakar, 56.5% of workers are in informal-sector jobs, 26.5% are in the formal sector, and 17% are unemployed. This sector is especially important for women. In Benin, 97% of women and 83% of men in the nonagricultural labor force have informal-sector jobs. A very significant proportion of the manufactured goods utilized by Africans is produced by informal-sector workers. These items provide lower-cost alternatives to the products of large-scale industries; in the smaller, less industrialized countries, these items constitute the majority of manufactured goods produced domestically.

It is sometimes forgotten that informal economic activity is not a new phenomenon, in either African or non-African cities. The structure of the informal economy continues to evolve in response to changes in the larger economy and society (see Figure 17.3). As certain opportunities wane, others emerge and are exploited by innovative entrepreneurs.

Programs initiated by states and by international agencies, notably the International Labour Organization (ILO), have helped to foster the development of the urban informal economy through relaxation of legal restrictions affecting the informal sector, improved access to credit, and technical assistance. Still, the view persists in many countries that the spontaneous development of the urban informal economy is incompatible with sound planning. Attempts to regulate the informal sector have had little success, since the first "law" of the informal economy is that opportunities must be exploited where and when they occur, regardless of whether officialdom approves.

Three brief case studies of informal-sector activities are given below. These examples should illustrate both the diversity of the informal sector and the ingenuity and adaptability of its entrepreneurs. They also touch upon several aspects of the interdependence of formal and informal economies.

Leather Workers in Dakar

Leather workers produce a variety of goods, including shoes, handbags, and amulets that are worn as protective "medicine." Most traditional leather workers are artisans who work individually and sell their products directly to consumers. Some work cooperatively in larger workshops and sell their goods to local traders.

Adaptability is crucial to survival in the informal economy; entrepreneurs must be able to adjust, for example, to changes in raw-material supply and customer demand. In the past, Dakar shoemakers worked with leather purchased from Mauritanian women who tanned hides in the traditional manner. After the Senegalese government granted the transnational Bata Shoe Company a virtual monopoly to purchase leather, local leather workers lost their source of raw materials. The local shoemakers solved this problem by using plastic from Europe. However, a tiny number of importing firms control the trade in plastic, leaving the local shoemakers more vulnerable to interruptions in supply and increases in price. These shoemakers have retained a large share of the local market, but the plastic goods that they now make bring a lower price than those made of leather.

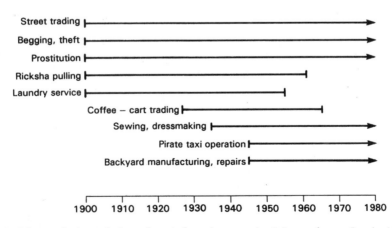

FIGURE 17.3. The evolution of the urban informal sector in Johannesburg, South Africa. After C. Rogerson. "Johannesburg's informal sector: Historical continuity and change." *African Urban Quarterly*, vol. 1 (1986), pp. 139–151.

Secondhand Clothing Sellers in Lusaka

Merchants who sell *salaula*—secondhand clothing—are a key link in a commodity chain that starts with the donation of secondhand clothing in Europe and North America. The majority of this clothing, after it has been sorted and bundled, is shipped to wholesalers in countries such as Zambia. The wholesalers in turn sell to *salaula* retailers, who operate within the informal sector. Lusaka's two largest markets between them had some 2,500 *salaula* merchants in 1995; better-established dealers have stalls, while many newcomers are located in open-air areas. The number of traders increased during the 1990s, when Zambia experienced tough economic times and many workers lost their jobs.

To succeed as a secondhand clothing merchant requires a lot of business acumen and considerable luck. Traders must decide not only on where to sell, but also what to sell and how to market it. They purchase clothing from wholesalers by the bale; if they get a poor-quality bale, their profits evaporate! Larger traders can afford to purchase new stock on a continuing basis, so are in a better position to keep their sales high. Many specialize in particular types of clothing. Items that are not readily saleable because of poor quality or an unpopular style are sent to tailors, who refashion them into new, saleable items.

One of the keys to the thriving *salaula* market in Zambia is the popularity of secondhand clothes among consumers. The eclectic mixture of styles from around the world provides consumers with vast choice at a reasonable cost. Wearing *salaula* offers the poor both dignity and opportunities for self-expression.

The Entertainment Industry in the Mathare Valley, Nairobi

Women are often attracted to informal-sector businesses that operate out of the home and that therefore can be combined with child care and other domestic work. Some 90% of the women in the Mathare Valley, a squatter community in Nairobi, work in informal-sector jobs. Many women brew local beer from maize in their rooms or houses. Although home brewing is illegal, it is at-tractive as a fairly lucrative business requiring low capital inputs, readily available raw materials, and skills that are already known to many women. The beer is sold from tiny informal bars located on the same premises. Independent women may also supplement their incomes through sexual liaisons with men whom they meet as customers.

Making Ends Meet in the City

Julius Nyerere, former president of Tanzania, commented on the difficulties of urban dwellers in a 1968 speech: "The vast majority even of our town dwellers live extremely poorly, and in most cases they are worse off . . . than the people in rural areas could be. An unskilled worker . . . earns wages which are hardly sufficient to enable a family to eat a proper diet and live in a decent house."

The prospects for most urban dwellers are much grimmer in the new millenium than they were when Nyerere made this statement. For most urban Africans, life is a constant struggle to make ends meet. Whereas rural Africans are able to produce most of their own food, urban Africans must rely primarily on earned income to survive. They are vulnerable to market fluctuations and have no real protection when crop failures, changes in government policy, or hoarding drive up prices. As shown in the case study described in Vignette 17.1, many urbanites produce some of their own food. However, for the majority, including most of those who are especially vulnerable, lack of access to land and other agricultural inputs means that farming is not a viable option.

The most vulnerable residents of the city are those who are cut off from the possibilities of support from their extended families. There has been a massive increase in the number of street children in recent years, including many who have been orphaned by the AIDS epidemic or separated from their families as a result of civil conflict. Street kids earn meager from odd jobs and begging, but mostly depend on their wits to survive.

In most African countries, the International

Monetary Fund (IMF) has imposed severe economic-restraint programs, forcing governments to reduce public-sector employment, to remove subsidies that had ensured the availability of relatively cheap imported food, and to devalue the currency. Previous government programs to provide affordable housing were eliminated, and user fees were implemented for a wide range of services, including water and sanitation, health care, and schooling. Licensing fees were imposed for many types of small-scale enterprises. The impact of these fees was especially serious for the urban poor, who were already struggling to make ends meet. Middle-class Africans have also suffered a significant decline in their standard of living under structural adjustment.

How do people make ends meet? They do so by severely curtailing consumption, by finding ways to supplement their incomes, and by making use of social networks. Few families can afford to rely on a single income. Workers in the formal sector commonly supplement their earn-

ings with part-time informal-sector jobs, such as taxi driving or petty trading (see Figure 17.4a). Housewives contribute to the family income by keeping gardens, trading, preparing food for sale, and minding other people's children. In poorer families, the children work, often as messengers or street hawkers (see Figure 17.4b).

The cultivation of social networks is crucial for survival. Relatives, friends, neighbors, and colleagues at work may provide invaluable information about when and where housing, employment opportunities, and scarce commodities may be found. Social networks are also a possible source of help in times of need.

For many African households, straddling the urban–rural divide is fundamental to their survival. Close linkages are maintained between urban and rural household members. Urban residents remit small sums of money when possible, and provide temporary lodging for visitors and new migrants to the city; in turn, they receive occasional gifts of firewood and food, as

a b

FIGURE 17.4. The informal sector at work. (a) Cassette seller, Kano, Nigeria. Locally (illegally) duplicated tapes are sold by this vendor, who seeks out customers by traveling from place to place on his bicycle. He advertises his wares by playing music on the tape recorder. (b) Children selling peanuts and cigarettes, Asmara, Eritrea. Children are heavily represented in the informal sector. Their earnings, even if they are small, are often crucial for the family budget. Photos: (a) author; (b) M. Peters.

well as assistance in times of difficulty. Whereas in the past this flow of resources was usually seen primarily as the city dwellers' subsidizing the country people, the net flow of resources is often reversed: Many of the urban poor could not survive without support from their relatives in the countryside.

The survival strategies of poorer urban households depend not only on a two-way flow of resources, but also of a movement of household members between countryside and city. Those most likely to remain in the city are young adults, especially younger men. Young children may be sent by their urban-based parents to stay with grandparents or other relatives in the countryside—in part because of the difficulty of raising children in the city, especially for those able to afford only a room or two. Older household members, who are no longer able to contribute to household income, may retire from the city to the village. Similarly, people who have been stricken with debilitating illnesses (especially AIDS) typically return to stay with, and receive care from, rural relatives.

Although life for the urban poor is very hard indeed, it is quite wrong to see the poor as merely passive victims. Community associations provide a means to achieve modest improvements in living conditions by working together to conduct community clean-ups, to organize activities for children, to petition governments to address their most urgent needs, and to initiate microcredit programs.

In order to survive, the poor need to be creative and daring—open to all possibilities at all times. These qualities are clearly evident in the informal-sector artisans, traders, and service providers at work in every conceivable location—not just large markets such as the one depicted in Figure 17.5, but everywhere. Though this ever-changing, ever-adapting informal economy may be an urban planner's worst nightmare, it is also a hopeful symbol of African self-reliance and ingenuity.

Life in the African city has become increasingly difficult in recent years, due to government cutbacks and the deteriorating economic situation. Many formal-sector jobs in manufacturing and civil service have been lost, and competition in the informal sector is fierce. As the cost of food, shelter, transport, and other necessities continues to increase, it becomes harder and harder to make ends meet. Nevertheless, despite this gloomy picture, the city's "bright lights" still shine for many Africans. The city still exemplifies opportunity—the possibility of achieving something better than what can normally be expected in the countryside.

FIGURE 17.5. "Roche Santeirs," located near the harbor in Luanda, Angola, is reputed to be the largest market in Africa south of the Sahara. Photo: CIDA (B. Paton).

Further Reading

Aspects of Africa's experience with industrialization are discussed in these sources:

Mytelka, L. "The unfulfilled promise of African industrialization." *African Studies Review,* vol. 32 (1988), pp. 1–72.

Steel, W. F., and J. W. Evans. *Industrialization in Sub-Saharan Africa: Strategies and Performance.* Washington, DC: World Bank, 1984.

The following sources focus on the emergence of an African working class and consider the significance of this development:

Lubeck, P. *Islam and Urban Labour in Northern Nigeria: The Making of a Muslim Working Class.* Cambridge, UK: Cambridge University Press, 1986.

Sandbrook, R., and R. Cohen, eds. *The Development of an African Working Class.* Toronto: University of Toronto Press, 1975.

The following sources are recommended as introductions to the urban informal economy:

Bromley, R., ed. *Planning for Small Enterprises in Third World Cities.* Oxford: Pergamon Press, 1985.

Gerry, C. "Petty production and capitalist production in Dakar: The crisis of the self-employed." *World Development,* vol. 6 (1978), pp. 1147–1160.

Hansen, K. T. *Salaula: The World of Secondhand Clothing and Zambia.* Chicago: University of Chicago Press, 2000.

Sethuraman, S. V., ed. *The Urban Informal Sector in Developing Countries: Employment, Poverty and Environment.* Oxford: Pergamon Press, 1985.

Tripp, A. M. *Changing the Rules: The Politics of Liberalization and the Urban Informal Economy in Tanzania.* Berkeley: University of California Press, 1997.

Many studies have looked at urban poverty and its effects, as well as at poverty alleviation strategies:

Ferguson, J. *Expectations of Modernity: Myths and Meanings of Life on the Zambian Copperbelt.* Berkeley: University of California Press, 1999.

Jones, S., and N. Nelson, eds. *Urban Poverty in Africa: From Understanding to Alleviation.* London: Intermediate Technology, 1999.

Rakodi, C., and T. Lloyd-Jones. *Urban Livelihoods: A People-Centred Approach to Reducing Poverty.* London: Earthscan, 2002.

Yeboah, I. "Structural adjustment programs and emerging urban form in Accra, Ghana." *Africa Today,* vol. 47 (2000), pp. 61–89.

To learn more about urban agriculture, see these sources:

Egziabhor, A., et al. *Cities Feeding People: An Examination of Urban Agriculture in East Africa.* Ottawa: International Development Research Centre, 1994.

Koc, M., R. MacRae, J. Mougeot, and J. Welsh, eds. *For Hunger Proof Cities: Sustainable Urban Food Systems.* Ottawa: International Development Research Centre, 1999.

Smith, O. B., ed. *Urban Agriculture in West Africa.* Ottawa: International Development Research Centre, 1999.

Internet Sources

For general information on urban development and on specific African cities, see these websites:

Eldis: the Gateway to Development Information. *Urban Development.* www.ids.ac.uk/eldis/urban/urban.htm

France: Ministry of Foreign Affairs. *The Dynamics of Urbanization in Africa.* www.france.diplomatie.fr/cooperation/developp/dossiers/d_urban/index.gb.html

For information on urban economic activity, see the following:

City Farmer. *Africa and Urban Agriculture.* www.cityfarmer.org/subafrica.html

International Labour Organization (ILO). *Focusing on the Informal Economy.* www.ilo.org/public/english/employment/infeco/publ.htm

Mbendi. Information on Africa. www.mbendi.co.za (formal sector economy)

Women in Informal Employment: Globalizing and Organizing (WIEGO). www.wiego.org

18

Housing Africa's Urban Population

The urban population of Africa south of the Sahara has undergone a 12-fold increase over the past half century, growing from 19.8 million in 1950 to 240 million in 2000. Although the rate of urban growth has slowed somewhat in recent years, it is still very rapid. It is projected that the urban population will have grown to over 400 million by 2010. Explosive growth exerts tremendous pressure on urban infrastructures. It means more demands for health care and education. It means more vehicles on the roads, as well as more passengers crowding onto public transportation. It means an increased demand for water, and thus the probability that water shortages will affect both domestic consumers and industry. It means a scarcity of reasonably priced housing. Where are newcomers to the city, most of whom are poor, to find shelter?

Rapid urban growth affects different parts of the city in distinct ways. In the older central districts, land use tends to become more intensive. One- and two-story houses and shops are replaced with multistory commercial and residential buildings. On the periphery, services have to be expanded to cater to the needs of emerging suburbs. Government officials and urban residents face the same challenges—namely, coping with contemporary crises of strained infrastructure, and trying to predict and plan for future needs.

Housing and Social Class

The geography of social class intersects closely with the geography of housing. The size and style of houses, along with their location within the city, are powerful indicators of wealth and status (see Vignette 18.1 and Figure 18.1). Even if the residential spaces occupied by the rich and the poor are not far removed from each other, rich and poor seldom trespass on each other's turf and rarely live side by side.

The housing situations of Africa's elite, the middle class, the working poor, and the unemployed populations vary in many obvious ways:

- Location within the city (residential segregation)
- The appearance (design) of houses
- Building materials used to construct houses
- Space (both interior and exterior) and privacy
- Amenities in the house—piped water, toilets, kitchens
- Quality and quantity of nearby community services

268

- Occupants' control of the property (owners vs. tenants)
- Extent of government support and subsidization

Elite Housing

During the colonial era, the government constructed housing for expatriate administrators and military officers on large lots in estates well separated from African townships. The houses were European in design and very spacious. However, elite neighborhoods have long since ceased to be the exclusive preserves of European government officials. African civil servants, professionals, and entrepreneurs predominate, although some expatriate businesspeople, diplomats, and contract employees are still present, especially in national capitals. Architectural styles have changed, but the houses are fundamentally the same: They are very spacious, essentially European in style, and situated in pleasant, landscaped surroundings. High security walls and guarded entrances are common and speak of the contemporary elites' wariness of the possibly restive masses.

Until the 1980s, most African governments allocated a disproportionate share of scarce resources to providing housing for higher officials. With the advent of structural adjustment, governments were forced to divest themselves of most of their housing stock. Policies were implemented to encourage the development of an active market economy for housing. Today, the great majority of the housing in upper-income neighborhoods is privately owned.

Middle-Class Housing

The term *middle class* in Africa is used loosely to refer to middle-level civil servants such as teachers, clerks, soldiers, and police officers. Middle-income employees of companies, and middle-level independent professionals and businesspeople, may also be included. Colonial governments provided basic housing for many of its African employees. In the newly established colonial cities, especially "European" cities like Salisbury and Lusaka, companies took an active role in providing housing for their African employees. Much less of this type of housing was erected in established precolonial cities like Kano.

In the postcolonial era, the proportions of middle-level civil servants and employees living in government and company housing estates declined greatly. The main exceptions were police and military personnel, many of whom are still housed in barracks. The middle classes tended to be the main beneficiaries of so-called "low-cost" public-housing projects that were developed by many governments from the 1960s to the 1980s, as a response to the urban housing crisis. However, even before the 1980s (when most of this public housing was privatized), the majority of the middle classes looked after their own housing needs.

Structural adjustment policies have brought significant changes to the housing options available to middle-class families. A generation ago, most middle-class families were able eventually to purchase their own homes. Many benefited from low-interest loans from government or company sources, and they tended to receive favored treatment when states sold building lots to the public. Middle-class homes ranged in size from modest two- to four-room single-family homes, to larger dwellings with space for members of the extended family and rooms for rent to supplement the household income. Now the elimination of most government subsidies, together with major declines in real wage levels, has left most of the middle class struggling to survive and with much-reduced prospects for home ownership.

Housing for the Working Poor

Housing the poor was a very low priority in colonial times. Basic housing was provided for the servants of Europeans. Many of the large mining companies built barracks to house their employees, but often not the employees' families. This type of housing was clearly designed to benefit expatriate employers rather than African employees.

The housing available to the poor in most African cities still provides little more than the minimal requirements for shelter. Tenants pay a substantial portion of their incomes to rent this rudimentary housing. Figure 18.2 shows

VIGNETTE 18.1. Residential Landscapes in Lusaka

Like all colonial and postcolonial cities, the residential areas of Lusaka are structured according to social class. The dominant elites have their own elite neighborhoods, separated spatially from those of the less well off. At the other end of the scale, the poor provide their own housing, often in squatter settlements with little in the way of modern amenities.

The photographs in Figure 18.1 show typical housing in different parts of Lusaka. They are accompanied by small maps showing the layout of the same or similar neighborhoods, and giving some indication of the density of development. The letters on the larger map show the approximate locations of the small, detailed layout maps.

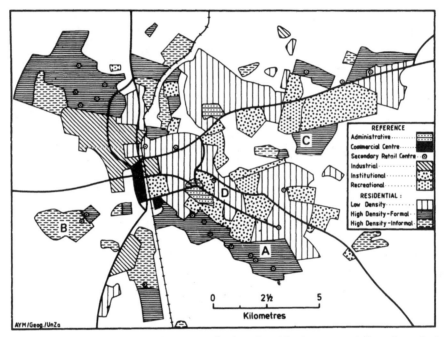

FIGURE 18.1. Social class and housing in Lusaka, Zambia. The inset maps below show the layout of various residential districts and highlight the magnitude of difference in house size and density in relation to social class. (Scale of inset maps: 1 cm = 100 m.)

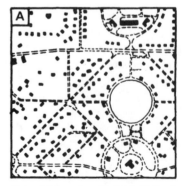

(a) Housing constructed during the colonial era for African workers.

Colonial housing in Old Chilenje.

(cont.)

VIGNETTE 18.1. *(cont.)*

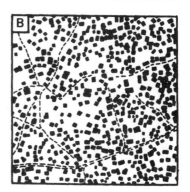

(b) Postindependence squatter settlement, occupied mostly by lower-class residents.

Squatter settlement at Kanyama.

(c) A newly-developed site-and-service scheme provides opportunities for lower middle-class people to build houses which serve their own needs.

Site-and-service center at Mtendere.

(d) Hidden behind a wall and surrounded by gardens, the opulent home of either a wealthy Zambian or an expatriate.

Colonial low-density European housing at Ridgeway.

Source (city map and inset maps): G. Williams. *The Peugeot Guide to Lusaka*. Lusaka: Zambian Geographical Association, 1983. © 1983 by G. Williams. Reprinted by permission. Photos: (a) and (b) G. Williams; (c) and (d) author.

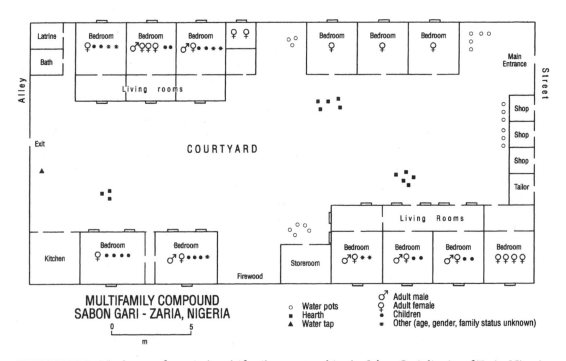

FIGURE 18.2. The layout of a typical multifamily compound in the Sabon Gari district of Zaria, Nigeria. Sabon Gari ("new town") was established during colonial times as a residential area for migrants from southern Nigeria. After A. W. Urquhart. *Planned Urban Landscapes in Northern Nigeria.* Zaria, Nigeria: Ahmadu Bello University Press, 1977, p. 49.

the layout of a multifamily dwelling in a working-class migrant suburb of Zaria, Nigeria.

The accommodations of the poor of Nairobi are fairly typical of the shelter available to Africa's urban poor—either a one-room house made of scrap materials, or a single rented room in a larger dwelling, with kitchen and bath shared by several families. An extensive housing survey conducted in low-income districts of Nairobi found that the great majority of respondents were tenants rather than owner-occupiers. Over 90% of dwellings had only one room, and very few homes had individual kitchens and toilets. Few of the houses in the newer low-income suburbs were constructed of permanent building materials. However, even rudimentary housing of this sort was very costly in relation to the earnings of the urban poor.

The poor must make do with very rudimentary public services. Local schools and health clinics, if available at all, are likely to be very overcrowded and understaffed. The inadequacy of the water supply and of disposal systems for sewage and refuse poses a threat to health. Wide-

spread cholera epidemics in the early 1970s provided dramatic evidence of this health threat, but less dramatic illnesses such as gastroenteritis and dysentery regularly cause still greater levels of sickness and death. The extremely high population densities found in many poorer districts increase the concentration of pollution and facilitate the spread of epidemic disease.

Governments have done little to alleviate the housing difficulties faced by the urban poor. In most large cities, affordable, legally registered building plots are very scarce. There are many regulations governing plot development, such as rules stating that homes must be constructed of permanent materials and finished within a specified period. The poor have often responded by building their own houses on land they do not legally own.

Housing for the Unemployed

The cities of Africa south of the Sahara contain significant populations of desperately poor unemployed people. Many of them find shelter with relatives. Others may stay in house entranceways and courtyards, often doing casual work for the house owners. Still others are forced to sleep on the street, in markets, or in other public spaces.

The majority of the unemployed population consists of men. Some are unmarried, while others have left their families at home in the countryside. Many families also lack basic housing. Drought, armed conflict, and other crises that force people to leave their rural homes are likely to increase the number of homeless families in cities (Figure 18.3). A growing proportion of the homeless urban population consists of street children, many of them orphaned by AIDS or war. The unemployed are very poorly served—not only because they have no housing of their own, but also because they have extremely poor access to amenities and social services. Many live in environments that are literally threatening to their health.

The Cultural Meanings of Housing

The quality of housing cannot be evaluated by using only utilitarian indices, such as number of rooms, amenities, and building materials. Appropriate housing not only offers shelter, but also facilitates the expression of the community's cultural values. The cultural dimension is clearly evident in the design of houses in rural communities and often in older, indigenous neighborhoods in cities.

For an example of how cultural values are reflected in house design, we may consider the houses of Islamic societies in countries such as Mali and Sudan. Islam permits polygynous marriages and calls for the seclusion of wives (*purdah*). Seclusion is accomplished with windowless walls surrounding the compound, as much to exclude men as to keep women within. Separate rooms are constructed for each wife around a central courtyard. The communal courtyard space and shared cooking facilities symbolize the common identity of household members and the importance of cooperation.

Housing in the typical modern African city, particularly the housing normally available to the poor, is often ill designed in relation to cultural values. Single-room apartments and houses provide little privacy from other family members and neighbors. Outside space for food preparation and socializing is cramped, if it is

FIGURE 18.3. Shelters occupied by refugees, Addis Ababa, Ethiopia. As "internal refugees," these Ethiopians displaced from Eritrea, when it was still part of Ethiopia, were not recognized as refugees by the United Nations High Commission for Refugees (UNHCR) and thus were not entitled to receive assistance. Photo: M. Peters.

available at all. The high cost of adequate housing may force poor men to leave their children or wives in the countryside with relatives. Housing problems may also serve as a disincentive for people to have the culturally preferred number of children.

African public-housing projects were often criticized for ignoring cultural values. Even the spatial arrangement and directional orientation of houses might convey unintended cultural meanings. The clearest examples were the all-male company barracks built for migrant workers in South Africa. Workers were forced to leave their families behind; "normal" family life was possible only during brief visits every few months or after retirement from the mines. The socialist government of Zanzibar commenced a large-scale project in the 1960s to construct high-rise apartment complexes, based on the model of Eastern European cities. These dwellings were designed with little regard for the cultural expectations and practical needs of Zanzibari households. Even the luxurious houses that many governments provided for senior civil servants were often ill suited for traditional lifestyles. Architects and planners need to become better educated about cultural values in order to design better houses and communities.

Despite the importance of cultural values, the poor are pragmatic about their housing. For many of the poor, more spacious and culturally appropriate housing is a lower priority than other needs, including longer-term "investments" such as school fees for their children. Moreover, those able to set aside some money do not necessarily invest it in urban housing; many urban Africans with strong ties to their rural communities of origin prefer to construct village homes. In doing so, they reaffirm the importance of kinship and also provide themselves with some security for their retirement years.

Access to housing is determined not only by social class, but also by gender. Women have considerably less access to home ownership than men. This reflects their lower income levels and lack of access to borrowed capital, as well as customary factors such as inheritance laws and customary land tenure systems that have privileged men. Nevertheless, some women—the

numbers vary from society to society—have been able to purchase homes of their own. These homes offer income-earning potential, as well as invaluable security and status, for their owners (Vignette 18.2).

Housing Policy Alternatives

The conventional wisdom about housing policy has changed radically. Prior to 1980, many governments endeavored to develop public-housing estates, in theory to help the poor. Unfortunately, these provided no real answer to the urban housing crisis, in part because high costs have prevented the construction of enough units to make a difference. So-called "low-income" housing was usually priced beyond the means of the majority of the population. In some projects, undemocratic allocation procedures favored the relatives of those in power and excluded those without connections.

The housing crisis was worsened by the shortage of affordable plots where private-sector housing could be built. The shortage of plots was more acute in southern, central, and East African cities, where colonial policies had established strict planning guidelines to limit migration and urban growth, than in older West African cities, where legal access to plots could be obtained through both traditional and governmental channels.

One response by Africans denied access to affordable, registered building plots has been to erect houses on someone else's land in contravention of planning regulations. These illegal communities—*squatter settlements*—have mushroomed, particularly in marginal areas (e.g., floodplains, rocky hillsides, or forest reserves within or just beyond urban boundaries). Sometimes landowners may illegally subdivide and sell their holdings.

Officials have tended to view squatter settlements as unhealthy eyesores that subvert the entire planning process. Until the 1980s, the common response was to bulldoze them—often with no warning—as a means of reaffirming "the rule of law." However, despite official disapproval and harsh penalties for building illegally, squatter settlements continued to prolif-

VIGNETTE 18.2. Claiming Space: Gender, Housing and Mobility in Niger

Housing has diverse cultural meanings, which may shift in relation to changes in the social and political environment. For Muslim women in Maradi, a city in south central Niger, the ability to own or control housing is one of the strategies used to maintain their social position in society. Whereas most urban women in adjacent regions of Nigeria are confined to purdah, women in Niger have traditionally been much freer to move about. However, men who hope to establish a more rigid Muslim society increasingly seek to limit women's mobility rights.

Women and men in Maradi occupy quite separate spheres. Whereas men have privileged access to public spaces, women's spaces are primarily internal ones, situated behind compound walls and hidden from public view. However, most women are mistresses only of their own sleeping rooms, which they make their own by decorating them with their dowry and other possessions. Other spaces within the compound are shared with other household members, often including cowives.

In Maradi, marriage is a fragile institution; most women have several marriages during their lifetimes. Between marriages, a woman traditionally returns to stay with her kin, or goes to live in a "house of women" as a prostitute. As such, women are essentially "perpetual migrants," moving into and out of spaces that are not their own.

Married women in Maradi have begun to acquire houses of their own. Some have done so through inheritance, usually following the death of their husbands. Others have purchased houses as investments, sometimes using pension money from husbands who had worked in the public service, or from their savings accrued from trade profits. These houses may be rented, thus providing their owners with a steady source of income. An owner might also use a house as a base for a business venture.

A divorced or widowed woman who owns a house has it to fall back upon as a place to live, as a place to earn an income, and as a place to entertain friends and kin. Home ownership enables her to have greater choice about remarriage; indeed, remaining single without working as a prostitute becomes a viable option. The possibility of providing a home for her older children, and to employ them in a home-based business, helps her to retain close ties to family. Moreover, home ownership in itself is a significant badge of social status in Hausa society.

There has been an increase in the observance of Islamic customs of wife seclusion and veiling in Maradi, especially among wealthy traders and religious leaders. Today, all Maradi women wear a veil when entering public spaces. However, whereas veiling is commonly seen in the West as a practice that limits women's freedom, the perspective of women in Maradi is quite different. The veil gives women a sense of social and religious respectability, and it makes it possible for them to travel about quite freely and enter public spaces usually marked as male. The freedom of movement conferred by the veil is especially important for women who are unmarried.

This case study illustrates the power of human agency in shaping urban space. These women—who from afar may seem to have few options in the patriarchal religious society of present-day Maradi—have used the purchase of real estate, along with the adoption of the veil, to consolidate their social status and enhance their opportunities.

Based on B. M. Cooper, "Gender, movement, and history: social and spatial transformations in 20th century Maradi, Niger." *Environment and Planning D: Society and Space*, vol. 15 (1997), pp. 195–221.

erate. The bulldozers only served to provoke public outrage and to deepen the housing crisis by destroying homes without providing alternate shelter.

The adaptive and dynamic nature of squatter settlements has gradually come to be recognized. The initial shacks built of scrap materials gradually disappear and are replaced by more elaborate dwellings made of permanent building materials. After a few years, there may be little apparent difference between planned suburbs and squatter communities. This upgrading process typically takes place in several stages as homeowners put aside savings and become more confident about their tenure. Some governments and aid agencies have begun to assist in this process by providing low-interest loans to finance home improvements and by installing services for community use.

In South Africa, policies designed to restrict the growth of African urban populations and to effect strict residential segregation by race were intensified under apartheid. As Vignette 18.3 (with Figure 18.4) shows, there are no easy, effective policy options to alleviate the acute disparities in housing availability and quality that are part of the continuing legacy of apartheid.

Urban Upgrading Projects

The reevaluation of squatter housing such as that depicted in Figure 18.5 has led to the development of *urban upgrading projects*, sometimes called *site-and-service schemes*, in which the government begins to develop a new suburb by surveying plots, constructing roads, and installing electricity, water, and other basic services. After obtaining a plot, prospective homeowners are free to build houses appropriate to their needs. Since the costs to the state are much lower than with public-housing projects, site-and-service schemes may be developed on a large scale. However, the demand for serviced lots has exceeded the supply, and the costs of buying a plot and building a house remain beyond the means of many of the working poor and certainly the unemployed.

In addition to creating new site-and-service suburbs, governments in several countries have implemented programs to upgrade existing squatter settlements. Upgrading schemes typically involve the development of improved public infrastructures and incentives for individual householders. These incentives may include granting legal title to squatter properties and providing low-interest loans for renovation work. Improvements are often accomplished by mobilizing local voluntary labor to work on roads, drainage systems, and schools, with materials supplied by the organizing development agency.

Upgrading programs have contributed substantially to better living conditions in many of the communities where the programs have been launched. Still, comprehensive upgrading schemes can be quite expensive, and only infrequently have they been considered a high priority by governments. Not uncommonly, these schemes show up as demonstration projects in a nation's capital city, to give the impression that something is being done.

An assessment of urban upgrading projects in several countries, undertaken in 2002, found a mixed record. In nations where governments had initiated comprehensive schemes, including both land tenure reform and the provision of several services, the pace of development was slow and cost overruns were common. In countries where the schemes were kept quite basic, by contrast, it was possible to address the needs of a larger number of people. In Burkina Faso, for example, the government developed a massive number of building plots that were made available at nominal prices; as a result, it was able to increase the proportion of planned settlements from 29% to 73% in a decade.

Although urban upgrading might seem like a simple concept that could be replicated at will, the reality has been quite different. Because of variations in the amount of available land and varying systems of traditional land tenure, such projects must be planned individually in relation to local circumstances. As well, women and men have quite different housing needs and constraints. Upgrading projects have tended to be most successful when the community has participated actively in project design and implementation.

Market Reforms

In urban areas, as elsewhere, structural adjustment policies sponsored by the International Monetary Fund (IMF) have focused on market reforms. Governments were forced to curtail their involvement in the housing market. Public-sector housing projects were privatized, and

VIGNETTE 18.3 Khayelitsha: Housing Policy and the Struggle for Shelter in a South African City

Under apartheid, Cape Town, like other South African cities, was divided into zones for allocation to each racial group. So-called "coloreds" (people of mixed race ancestry) who composed over half of Cape Town's population, were allocated only 27% of the land. The fate of residents of District Six, a long-established inner-city neighborhood, was typical of that of tens of thousands of "coloreds." They were uprooted from their homes and sent to an isolated, desolate plain 20 km from downtown. The new community had virtually no housing or services. Not only squatters, but also people with legal tenure, were evicted from their homes.

A massive housing crisis emerged as a result of forced removals, rapid population growth, and a grossly inadequate housing policy. The government eventually responded to this crisis with a number of remedial measures, one of which was the establishment in 1983 of the new town of Khayelitsha to house Africans legally residing in Cape Town. The town was to grow in a gradual and orderly fashion, with families moving in as new public housing was built. However, Africans protested being forced to move into 14.4-m^2 "houses" in a distant suburb with few services, virtually no possibilities for employment, and no rights to land ownership.

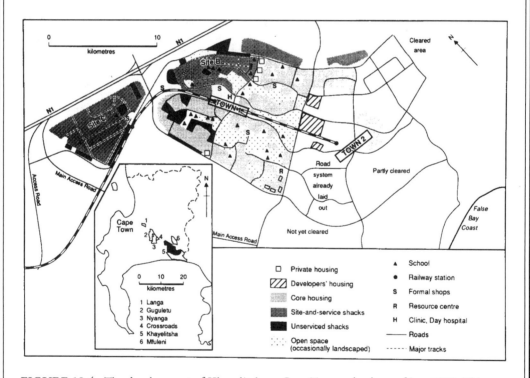

FIGURE 18.4. The development of Khayelitsha, a Cape Town suburb, as of late 1990. This rapidly growing community is primarily a squatter settlement, but has some private and site- and-service scheme development. Postapartheid municipal and national governments have invested resources to upgrade housing and services in Khayelitsha. Source: G. P. Cook. "Cape Town." In A. Lemon, ed. *Homes Apart: South Africa's Segregated Cities*. London: Paul Chapman, 1991, p. 31. © 1991 by Paul Chapman. Reprinted by permission.

(cont.)

VIGNETTE 18.3. (cont.)

By 1985, the government had relented, dropping the idea of moving all Africans to Khayelitsha and adding site-and-service plots where people could construct their own homes.

A survey in 1988 estimated that between 110,000 and 189,000 people were living in Khayelitsha. By late 1990 (see Figure 18.4), the number of persons in Khayelitsha had swollen to 450,000, only 14% of whom were in formal housing. A further 54% occupied serviced and partly serviced sites. The remainder consisted of squatters, who illegally occupied whatever space they could find. The squatter areas received no services of any kind. The rapid growth of Khayelitsha, despite its remoteness, unattractive appearance, lack of jobs, and very inadequate services, was a reflection of the virtual absence of viable alternatives under apartheid for the vast majority of Africans in Cape Town.

Khayelitsha continued to grow rapidly following the demise of apartheid. By 2002, it had a population estimated to be as high as 900,000. The government selected it as one of eight nodes to be given preferential funding for housing and urban development initiatives. One program offers government subsidies to collectives of 200 households for the construction of new homes. Homes are constructed by private companies, which in turn employ members of the community. Initiatives to privatize garbage collection and other services have been much more controversial. Many of the township's poor have had services cut because they could not afford to pay the user fees, and some neighborhoods have been neglected. Illegal dumping of refuse has increased.

To the casual observer, Khayelitsha appears to be a haphazard maze of crudely constructed dwellings. With close to 60% unemployment and infant mortality three times as high as that for Cape Town, Khayelitsha is clearly a community facing daunting challenges. However, residents of Khayelitsha have a positive image of their community; relatively few say they would like to live elsewhere. Individual residents struggle to find innovative ways of making ends meet. Several local nongovernmental organizations work to improve the quality of life. Finally, all levels of government are playing an active, supportive role in the search for sustainable solutions to the township's problems.

Based primarily on G. P. Cook. "Khayelitsha: New settlement forms in the Cape Peninsula." In D. M. Smith, ed., *The Apartheid City and Beyond: Urbanization and Social Change in South Africa,* London: Routledge, 1992, pp. 125–135.

housing subsidy programs were severely limited. Other reforms sought to stimulate private investment in housing.

In Ghana, privatization has spurred growth in the home construction business, as well as greater private investment in housing. However, housing has become prohibitively expensive for all but the richest Ghanaians—the result of rapidly increasing prices for construction materials and high interest rates. For a senior public servant, a one-bedroom house costs the equivalent of almost 10 years' salary, six times as much proportionally as in 1980. Middle-class families who formerly aspired to home ownership can now only afford to rent, and

poorer families can only afford the most basic shelter.

Market reforms have also extended to urban services, such as public transportation, water supply, and waste disposal. In theory, the privatization of these services transfers the cost of service provision from the public to the private sector and promises to increase their efficiency. In practice, the costs of the services have often become prohibitively high, and their quality varies considerably. The quality of privatized services tends to be positively related to a neighborhood's ability to pay. The overall impact of market reforms in housing and urban services has been to make it extremely difficult

FIGURE 18.5. Crossroads, Cape Town, South Africa. The housing of Crossroads is typical of the squatter settlements that mushroomed on the peripheries of South African cities in the late apartheid era. Photo: D. Bowen.

for the poor, and for many middle-class Africans as well, to make ends meet.

The housing crisis in African cities mirrors a wider crisis of access to basic needs. The poor are severely disadvantaged in their quest for adequate water, transportation, health, education, and other services (see Figure 18.6). Although "basic needs" strategies of urban development have helped to improve service accessibility, the scarcity of resources, market reforms, and the self-interest of ruling classes have limited their effect. As always, the poor must rely on their ingenuity in finding ways of providing for their own basic needs and making ends meet.

FIGURE 18.6. Uncollected garbage in a residential area, Kano, Nigeria. Inadequate sanitation, often a characteristic of poorer districts in Africa's largest cities, is a health hazard and an aesthetic eyesore. Photo: author.

Further Reading

For an intriguing study of the cultural meanings of housing in one society, see this source:

Donley-Reid, L. "A structuring structure: The Swahili house." In S. Kent, ed. *Domestic Architecture and the Use of Space*, pp. 114–126. Cambridge, UK: Cambridge University Press, 1990.

The following sources provide overviews of housing problems in contemporary African cities:

Amis, P., and P. Lloyd, eds. *Housing Africa's Urban Poor.* Manchester, England: Manchester University Press, 1990. (See in particular the Introduction and Part 1.)

Morrison, M., and P. Gutkind, eds. *Housing in Third World Cities.* Syracuse, NY: Maxwell School of Citizenship and Public Affairs, Syracuse University, 1982. (See Chapters 1, 2, 6, and 8.)

Policies to address housing crises by developing site-and-service schemes and upgrading squatter settlements are discussed in the following sources:

Burgess, R. "Some common misconceptions about self-help housing policies in less developed countries." *African Urban Quarterly,* vol. 2 (1987), pp. 365–378.

Gilbert, A. "Self-help housing and state intervention: Illustrative reflections on the petty commodity production debate." In D. Drakanis-Smith, ed., *Urbanization in the Developing World,* pp. 175–194. London: Croom Helm, 1986.

Gulyani, S., and G. Connors. *Urban Upgrading in Africa: A Summary of Rapid Assessments in Ten Countries.* Washington, DC: World Bank, 2002. (available online at www.worldbank.org/urban/upgrading/docs/afr-assess/countrysummary)

Laquian, A. A. *Basic Housing: Policies for Urban Sites, Services and Shelter in Developing Countries.* Ottawa: International Development Research Centre, 1983.

Obudho, R. A., and C. C. Mhlanga, eds. *Slum and Squatter Settlements in Sub-Saharan Africa.* New York: Praeger, 1988.

Much has been written about the experiences with housing policy in particular African countries. Examples include the following sources:

Bamberger, M., B. Sanyal, and N. Valverde. *Evaluation of Sites and Services Projects: The Experience.* *from Lusaka, Zambia.* Washington, DC: World Bank, 1982.

Grant, M. "Moving and coping: Women tenants in Gweru, Zimbabwe." In K. Sheldon, ed. *Courtyards, Markets, City Streets,* pp. 169–189. Boulder, CO: Westview Press, 1996.

Grootaert, C., and J. L. Dubois. *The Demand for Housing in the Ivory Coast.* Washington, DC: World Bank, 1986.

Kariuki, P. W. "Street children and their families in Nairobi." In S. Jones and N. Nelson, eds. *Urban Poverty in Africa,* pp. 138–148. London: Intermediate Technology, 1999.

Konadu-Adyemang, K. "Structural adjustment programs and housing affordability in Accra, Ghana." *Canadian Geographer,* vol. 45 (2001), pp. 528–544.

Lemon, A., ed. *Homes Apart: South Africa's Segregated Cities.* London: Paul Chapman, 1991.

Smith, D. M., ed. *The Apartheid City and Beyond: Urbanization and Social Change in South Africa.* London: Routledge, 1992. (Part 2 and Part 5 both have a number of chapters on housing.)

Internet Sources

These sites are excellent comprehensive sites dealing with urban renewal issues, and providing data on urban settlements:

United Nations-Habitat. *United Nations Human Settlements Programme.* www.unchs.org (see especially *Human Settlements Statistics; Global Urban Indicators Database*)

World Bank. *Urban Development: Upgrading Urban Communities.* http://worldbank.org/urban/upgrading

The following sites focus on the provision of municipal services:

Municipal Services Project. http://qsilver.queensu.ca/~mspadmin

Network for Water and Sanitation (NETWAS). www.netwasgroup.com

United Nations-Habitat. *Urban Water and Sanitation.* www.unhabitat.org/hd/hdv8n2

Social Geography

The discussion in the following three chapters may be situated within the welfare school of social geography. This approach, originally conceptualized by David Smith as the "study of who gets what where," examines social and spatial variations in welfare. However, the social-welfare approach goes beyond mere description to search for the underlying causes of social injustice, whether historical, economic, cultural, or political. It is a policy-oriented approach in which the ultimate objective is to show how more equitable future states might be attained.

Chapter 19 explores the social geography of African women. The substantial and diverse economic roles played by African women are finally being recognized. Although development agencies have responded by supporting women's development projects, there continues to be disagreement about whether this aid is a harbinger of changing attitudes toward women or is mere tokenism. This chapter also considers how the role of women is defined within the context of African culture, and what the implications are for African women when increasing access to education and other changes in a modernizing society subject traditions like early marriage and female circumcision to greater scrutiny.

Chapter 20 looks at issues related to children's welfare. It considers the cultural and social context within which children are raised. The chapter looks in some detail at children's opportunities for education. Education is very unevenly developed, in large part as a result of colonial policies. The educational system continues to be characterized by large rural–urban, gender, and class imbalances. The final part of Chapter 20 looks at children at risk—from diseases such as HIV/AIDS, from deep poverty, and in some countries from being abducted to become child soldiers or forced laborers.

In Chapter 21, the focus shifts to health, particularly to the policy debates that have surrounded the development of health services in Africa. In recent years, as the African economic crisis has deepened, it has become increasingly necessary for governments to reexamine how health care is provided and financed; several approaches have been used to address these problems. Finally, the chapter examines how the rapid spread of HIV/AIDS has created a number of new challenges to health care systems.

19

African Women and Development

Until the 1970s, scholars and development planners paid very little attention to the social, economic, and political roles of women in Africa. Implicit in this silence were the assumptions that what women did was not very important, and that their interests were adequately served by conventional "gender-neutral" development strategies. The quantity, quality, and diversity of feminist scholarship on gender and development have since increased vastly, leaving no room for doubt about the fundamental significance of gender relations for development.

At the same time, the voices of African women, speaking both individually and as members of women's organizations, have begun to be heard. One of their greatest contributions has been to insist that the values, opinions, and objectives of African women be respected and used as the basis for development initiatives.

This chapter examines distinctive features of the social geography of African women, looking in particular at the effect on these women of conventional approaches to development. This is not to imply that there is a single, unified geography of African women. On the contrary, their situations and strategies are extremely diverse, reflecting differences in social class, ethnicity, religion, and place of residence.

It has been said that African women have suffered under "two colonialisms": that of Europe and that of men. European officials and missionaries brought with them their own notions of the proper social and economic roles for women. The Victorian ideal was that women should stay at home and concentrate on child rearing and domestic labor. If the African reality clashed with these values—as it usually did—then measures should be taken to change the Africans! Hence, for example, taxes were seen as an incentive for men to assume their "proper" role as farmers. European policies did not challenge the strong traditions of patriarchal control of women in African society; indeed, such policies often strengthened those traditions.

The end of colonial rule seldom brought substantive changes to either of the "two colonialisms." Women have continued to suffer disadvantage and discrimination in many facets of their personal lives, in options accessible to them in their communities, and in their treatment by their nation-states. At the same time, it must be emphasized that although African women are disadvantaged in many ways, they

283

should not be seen as merely passive victims either of male domination or of larger economic forces. Women play vital roles in development, and they struggle in various ways to exercise greater control over their own destinies and to maximize opportunities for themselves and their families.

The economic contribution of African women has tended to be forgotten and ignored—in part because so much female economic activity is classed as housework or subsistence, and hence is not recorded in such macro-scale measures of economic activity as the gross national product. In most African societies, it is appropriate to portray women as family "breadwinners." Not only do they grow most of the food, but also women, not men, are the ones who take ultimate responsibility for seeing that basic needs are met. In Africa, women work longer hours than men; rural women work longer than urban women; and women of the poorest countries work the longest hours simply to provide the basic needs of food, shelter, and clothing for their families.

The Double-Double Workload of African Women

Many studies of women's work have stressed the importance of women's double workload: in reproduction (the domestic sphere) and in production (paid labor in the public sphere). The term *double workload* refers not only to the dual focus of women's work but also to the amount of work performed by women. The double-workload construct, the validity of which is now increasingly questioned by those interested in industrial societies, has always been of uncertain value for the analysis of women's work, in the Third World. As a general rule, African women work longer and harder, do more types of work, and struggle against more formidable barriers than women in industrial societies— hence the reference in the heading above to their "double-double" workload. Much of the work of African women cannot easily be categorized as either reproductive or productive, in the usual sense of these terms. Moreover, the nature and productive significance of women's work vary greatly both among societies and within specific societies. In the discussion that

follows, five broad categories of women's work (the first two of which are illustrated in Figure 19.1) are considered:

- Child rearing, the preparation of meals, and other types of domestic labor centered primarily in the home
- Household subsistence activities that are usually done by women and performed outside the home, such as farming, collecting fuelwood, and fetching water
- Money-earning activities centered in the informal economy, including the preparation of goods for sale, the provision of services, and retail trade
- Employment in the formal economy—in factories, as civil servants, and so on
- Voluntary activities in civil society, as members of social and political groups

Only a small minority of women are actively involved in all five of these types of work. As a rule, there are significant rural–urban and class distinctions. Rural women have few if any opportunities to work in the formal economy, and the various subsistence tasks centered outside the home are much more likely to affect rural than urban women. As for social class, the poorer the household, the greater the likelihood that women will be forced to undertake many types of work.

The strength of women's economies is important, not least because of the precarious situation of older women who have been widowed or divorced. With virtually no social security and relatively poor prospects for remarriage, unmarried older women must find ways of supporting themselves. Some receive help from family and friends; others engage in petty trade (Figure 19.2); and those who are most destitute beg.

Domestic Labor Centered Primarily in the Home

Women have responsibility for work that is particularly associated with the home. These tasks include, among other things, preparing and serving meals, keeping house, and caring for children and for those who are sick. As a rule,

FIGURE 19.1. "A woman's work is never done": The daily tasks of women in Botswana. (a) Bringing home water from the village borehole. (b) Fetching firewood for use as fuel for cooking and warmth. (c) Working together on the farm. (d) Winnowing grain. (e) Preparing food using a mortar and pestle. (f) Caring for children. Photos: (a) R. Dixey; (b) to (f) CIDA (B. Paton).

men do not participate in domestic work, including child rearing; such tasks are considered the exclusive domain of women. Although men in some societies are obliged by custom to provide their wives with sufficient money to cover the costs of food and day-to-day household costs, in other societies—particularly in cases where women farm—men's obligations regarding household maintenance are more ambiguous. It is a common complaint that men

FIGURE 19.2. Older women selling produce at a market, in Jigawa State, Nigeria. Older widowed or divorced women often live in extreme poverty, dependent on meager income from trade to survive. Photo: author.

do not take sufficient responsibility for providing for their families, but rather treat the income that they earn as theirs to spend.

Most Africans would undoubtedly assert that bearing and rearing children constitute women's primary role. This role involves a tremendous commitment of time and energy. African women give birth an average of five to seven times, and may breastfeed each of their babies for 18–30 months. Thus many African women spend perhaps 15–20 years of their lives bearing or breastfeeding children. This is not to imply that caring for an infant brings much relief from other types of work; quite the opposite is true. Carrying their babies with them wherever they go, women continue to cook meals, fetch water, work in the fields, carry goods to market, and do whatever else needs to be done.

The burden of combining child care and other work tends to be more severe in poorer monogamous households than in polygynous households, especially where there are no older children and few relatives living nearby to lend assistance. In polygynous households, there are opportunities for cowives to share the burden of domestic work, including child care.

Many African cultures traditionally practice polygyny and consider it the preferred form of marriage. Islam sanctions polygyny, but Christian denominations based in Europe and North America have always strongly condemned it. The preference for polygynous marriages, especially among men, relates primarily to its value as a symbol of status and wealth. Because the costs of marrying and supporting more than one wife are increasingly beyond the means of men with modest incomes, polygyny may well become less common, but its image as a status symbol is unlikely to diminish.

Household Subsistence Activities outside the Home

In addition to their home-centered activities, African women do the majority of household subsistence work consisting of tasks outside the home. According to an International Labour Organization (ILO) study by Ruth Dixon-Mueller (see "Further Reading"), African women contribute 70% of the total hours spent in agricultural production, as well as doing 50% of the care of livestock, 80% of the collection of wood for fuel, and 90% of water collection for household use.

Women compose the majority of agricultural workers in Africa and do most of the farm work. Rural women in Gambia, for example, spend an average of 159 days working on the farm. In Uganda, most women—approximately 70% in many districts—reported that their husbands seldom or never assisted them in their work on the farm. In one common gender division of labor, which arose with the emergence of a cash economy during colonial rule, women assumed sole responsibility for food crop production, while men concentrated on cash crops. The sale of cash crops was essentially equivalent to the sale of labor; both were methods of earning money to pay taxes and obtain consumer goods. As men focused increasingly on work to generate income, they left to women a larger proportion of nonmonetarized work, such as growing food for household consumption. Colonial strategies that converted more productive land to cash crop production forced women to travel farther and work harder to produce the same amount of food on soils that were becoming less fertile (see Vignette 19.1).

Although the majority of food production is undertaken by rural women, this gender division of labor is not universal. In some Islamic cultures, where most women stay in purdah, the only women actively involved in agriculture are older widows or women from the poorest, most labor-deficient households, where survival has to take precedence over preferred religious practice.

Women in rural areas lacking easy access to water must spend hours each day carrying heavy water jugs from distant streams and shallow wells. The work is onerous, and time spent carrying water must be taken from other tasks. Compromises have to be made, an example of which is the substitution of relatively easily cultivated but less nutritious crops like cassava for other, more demanding food crops. Collecting fuelwood for cooking, heating, and income-producing activities (e.g., smoking fish) is another time-consuming process, almost entirely performed by women and children. In the forest zone, women spend an average of 45–60 minutes per day collecting wood. In savanna and semidesert environments, especially where population densities are quite high, considerably more time may be needed to collect sufficient fuel. The expansion of agriculture, the cutting of fuelwood for sale in urban markets, and (in some cases) government restrictions on tree felling are forcing women to spend longer hours and to walk farther to obtain enough wood for household needs.

Labor migration has generally increased the burden of work for rural women. With many and sometimes most of the able adult men absent, women have been forced to do additional, traditionally "male" tasks (e.g., clearing farmland). The situation of rural women was especially precarious in the former homelands of South Africa, since South Africa until the mid-1990s prevented the wives and families of black migrant workers from accompanying their husbands to the city.

Money-Earning Activities Centered in the Informal Economy

In Ghana, as in several other parts of Africa, to enter a marketplace is to witness the economic power of women. Most of the traders are women. Moreover, many of the goods that women sell have also been produced, processed, and transported to market by women. For example, in Ghana, every stage in the processing and marketing of fish except the actual fishing is likely to have been done exclusively by women. Although the majority of economically active women operate small, marginal enterprises from their homes, a small number of women have become wealthy entrepreneurs controlling major trading, contracting, and manufacturing enterprises. The market women of Togo, Ghana, and southern Nigeria are legendary for their economic savvy and wealth (see Vignette 19.2 for a Togolese example).

Even in Islamic societies practicing strict purdah, women prepare handicrafts and foods for sale and participate in trade. In the city of Kano, where very few married women of childbearing age ever venture outside the house except on ceremonial occasions or for medical care, 90% of women are involved in income-producing activities. Women's economic activity accounts for up to one-third of the total economy of Kano. This "hidden trade" by secluded women relies largely on using children as messengers and sellers. These often forgotten women's economic activities may be eroded by state licensing requirements and even by otherwise progressive social

VIGNETTE 19.1. Changes in the Gender Division of Labor under Colonialism

How colonialism changed the lives of rural women depended on a variety of factors, such as types of export crops introduced, extent of labor migration, the control of land, and cultural traditions about the division of labor. What follows is a brief comparison of changes in four groups: the Luo and Kikuyu of Kenya, the Mandinka of Gambia, and the Ewe of Ghana.

Among the Kikuyu and Luo, men cleared the land, and women were responsible for the planting, weeding, and harvesting of food crops. As male labor migration became more common and longer-term, men often failed to contribute their traditional share of labor, forcing women to work harder. Because men thought that providing food was women's responsibility, women had to find ways of earning their own cash income to buy supplemental food. Luo women became more involved in trading, while Kikuyu women began to modernize farming practices and relied on traditional forms of cooperative craft production. Women lost some of their best land when it was taken over for settler agriculture, and they came under further pressure with the 1955 land tenure reform that created individual ownership and made it possible for absent husbands or sons to sell farms without women's consent.

During the colonial era, Ewe men increasingly shifted land and labor from yam cultivation to cocoa production; they also engaged in labor migration. As men laid claim to the most fertile land for their cocoa farms, women found it harder to get access to land to grow food. Women lost their traditional rights to use lands that had been common to those of a particular lineage, and had to get permission from individual men to use fallow areas. Because of the shortage of male labor for farm clearance, women also had to use land that was less densely forested and thus easier to clear; however, this land tended to be the least fertile. They increasingly turned to growing cassava and to petty trading to compensate for the shortfall in yam production. While women's workload increased, family nutrition suffered because of cassava's low protein content.

Among the Mandinka of Gambia, women traditionally concentrated on swamp rice cultivation, while men were responsible for upland crops like millet and sorghum. Mandinka men migrated infrequently during the colonial era; instead, they intensified their cultivation of groundnuts as a cash crop. Competition for land was less of a problem, since men and women made use of different types of land to grow their crops. The workload of Gambian women increased, however, because they had to grow more rice to compensate for the declining production of men's staples, which necessitated using more distant and less attractive swamplands.

Specific effects and adaptive strategies varied, but in each case described above, women had to find ways of producing more food with less support from their husbands and often with decreased access to land. Moreover, women continued to have responsibility for such tasks as child rearing, food preparation, and fuelwood and water collection. The double workload of women was redoubled.

Based on E. Trenchard. "Rural women's work in sub-Saharan Africa." In J. Momsen and J. Townsend, eds., *Geography of Gender in the Third World* (pp. 153–172). Albany: State University of New York Press, 1987.

legislation; making school attendance compulsory in Nigeria, for example, deprived many secluded women of the labor they needed to carry on their economic activities.

The economic circumstances of most urban households have changed for the worse under structural adjustment, due to declining real incomes and the increased cost of living in the city. As a result, more and more women are starting small, informal-sector businesses of their own, most notably in countries where urban women have until recently been less active in the economy than men. In cities such as Dar es Salaam, the earnings of women and children

VIGNETTE 19.2. The "Nanas-Benz" of Togo

A 1987 article describes a very powerful group of businesswomen in Togo:

Who are the 'Nanas-Benz'? These venerable ladies are Togolese cloth merchants. Relatively elderly—some as old as 75—and often illiterate, they are nevertheless canny businesswomen who certainly know their numbers. 'Nanas-Benz' is a combination of two words: nana, meaning 'established woman' or 'woman of means' and Benz from their passion for the Mercedes-Benz.

The major feature distinguishing the Nanas from other female cloth merchants is the sheer volume of their businesses. The most successful have been known to turn over the equivalent of $600,000 per month. The location of a Nana's business—which may be an important stall in the Grand Marché of Lomé, Togo's capital, or a well-known boutique in the street of a neighboring quartier—contributes as much to her prestige as to her sales volume. Status also depends on buying power, ability to negotiate, and size of clientele. The Nanas-Benz act as agents between the importers and a wide-ranging clientele in West Africa. About 85 per cent of the cloth delivered to Lomé is sold to buyers from other West African countries. Togo's low rate of customs duties allows buyers from neighboring nations to purchase more cheaply in Lomé than in their own countries, even after transportation costs are taken into account.

The Nanas select their fabrics from samples created by European designers, who sometimes follow their suggestions. Batik cloth arrives on container-ships. The bolts are addressed to the particular Nana and distributed on the importer's premises. Boat arrivals are irregular: up to eight months may elapse between order and delivery, so the Nanas must rely on intuition and foresight in anticipating the tastes of their clients and trends of fashion.

At the top of Togo's cloth trade are the wholesale dealers and major retailers—that is, the Nanas and a few others. All of them take in retail clerks or assistants, who are not paid but are given room and board. Often they come from the same village as the owner, and become trusted associates—sometimes they even take over the business. At the bottom of the scale are pedlars who criss-cross the market, selling fabric by the piece. They pay their employers only after making a sale, so the relationship is based on mutual trust.

All successful 'Nanas-Benz' make sure their children attend university. Girls study economics, management, and administration; boys become architects, teachers, or bankers. These women consider it important to have female children, partly to inherit the business. If no daughter exists, they often leave the enterprise to a niece or cousin. In this way, a kind of family matriarchy is created within a social system which is both patriarchal and polygynous.

On Lomé's Grand Marché, that wonderful gathering-place for women, the only men seen behind the stalls are butchers. And any other men here are merely clients, strollers, or relatives of the lady merchants.

Source for quote: M. Crouillère. "The 'Nanas-Benz.'" *Development*, Spring 1987, pp. 7–8.

have become essential for survival in more and more households.

The economic activities of women are characterized by the development of mutual-support mechanisms. One example is the revolving-credit system, in which a group of women pool a specified amount of money on a weekly or monthly basis; members in turn collect the total sum and use it for projects of their choice. Such institutions help to create networks of mutual support, and they can be used as the basis for organizing cooperatives to undertake projects beyond the scope of individual members.

Employment in the Formal Economy

Women in urban areas, particularly those with education, are beginning to face new economic challenges. Many of these women work in the formal sector as teachers, health service personnel, factory workers, and the like. Their concerns include such workplace issues as promotion, job security, scheduling, and maternity leave, as well as such public concerns as the reliability of public transportation systems and the quality of education. Most of these formal-sector jobs are considered very desirable, although many do not command especially high rates of pay.

Although the number of women employed in the formal sector is small in relation to the total population of women in Africa, these women are becoming more influential. However, as a result of economic reforms under structural adjustment, the number of women in formal-sector employment has declined. Women have borne a disproportionate share of job layoffs. Other women have chosen to move into the informal economy, where they anticipate being able to make more money.

Involvement in Civil Society

Women have had to struggle to make their voices heard. Nevertheless, they play a very important role in politics and civil society. In the 1997 Kenyan election, Charity Ngilu's campaign for the presidency raised the political profile of women. In several countries, women's voting power has been instrumental in electing the governing party. Politicians have become increasingly aware of the power of women's political voice.

Women have also taken the lead in establishing nongovernmental organizations (NGOs) to address their needs. In some cases, these NGOs are essentially modern versions of the revolving-credit schemes discussed above, designed to facilitate greater economic opportunities for women. In other cases, women's NGOs are playing a critical role in providing social services—for example, offering support for persons with HIV/AIDS.

Social and Political Issues

Each society has its unwritten rules—guidelines that are neither so rigid that change is impossible, nor so flexible that "unacceptable" behavior will go unchallenged. The direction of change reflects struggles between conservative factions determined to protect cultural tradition and radicals striving for a new order. Tradition remains a very powerful force limiting the choices open to African women. Strict adherence to culturally defined rules of behavior is seen as both the individual's and the society's best defense against misfortune. The defense of tradition also provides the ultimate rationale for the exercise of power by chiefs, elders, priests, and other "traditional" leaders.

Those arguing for greater choice for women must also confront a generally ambivalent attitude from most men who consider themselves progressive. Many men are preoccupied with other issues and do not really wish to change fundamentally the subservient status of women, both within the family and in the broader society. Moreover, African women are divided about the extent to which tradition should give way to greater choice; many influential women argue that the best guarantees for women's fulfillment lie in careful adherence to culturally defined norms.

The debate about women's choice has been particularly heated on the issue of female circumcision. In many African cultures, some form of surgical alteration of females' genitalia is performed as a rite of passage. The surgical procedures used vary, and the practice is not universal (Figure 19.3). Many Western feminists have strongly condemned female circumcision, pejoratively (from the African perspective) calling it "female genital mutilation." African women are divided about whether these practices should be retained, modified, or abolished, not least because the underlying cultural meanings of the surgery are complex. Most African women are united, however, in condemning outsiders' intervention in a matter that they wish to see settled in Africa, by Africans, and in accordance with African values.

Controversy also surrounds the issue of female choice in marriage. Customarily, most African cultures prefer to have their girls marry early—at or even before puberty. Such customary marriages are not very compatible with female education at the postprimary level. Marriages are traditionally arranged by the bride's

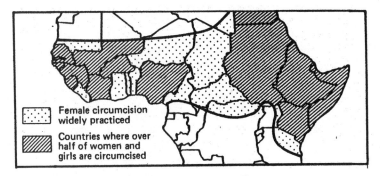

FIGURE 19.3. The practice of female circumcision. After J. Seager and A. Olson. *Women in the World: An International Atlas.* London: Pan Books, 1986.

and groom's families, with the bride herself having little or no say about her future husband. Young women are increasingly assertive about their right to reject arranged marriages that they dislike. "Bride wealth" payments are customarily given to the parents of the bride. Parents are inclined to demand whatever the market will bear, thus forcing poorer families to go into debt and to delay marriage for years until bride wealth is accumulated. Where revolutionary governments were established, laws to abolish forced marriages and the payment of bride wealth received early assent, but also suffered from widespread public resistance. Marriage customs are evolving gradually, and girls—especially those from better-educated, middle-class, urban families—are likely to have a greater choice about marriage than was the case in the past.

Providing universal opportunities for education is very important for women's progress. Literacy gives access to an infinite range of written material, and vocational and post-primary education opens new employment opportunities to women. There has been real progress in expanding educational opportunities for women. Females accounted for 46% of students in African primary schools in 1999, compared to only 32% in 1960. At the secondary level, they composed 42% of students, up from 26% in 1960. Still, much remains to be accomplished. Within both the African family and the modern state, the tradition of attaching higher priority to the education and the post-education employment of boys than girls dies hard.

The declines in household income under structural adjustment have had paradoxical effects on the status of urban women. More and more women have gone into business as a means of earning income, which they use for purposes including personal needs, children's school fees, collateral for future business ventures, and the cost of basic household necessities. As a result, they are working harder than ever to make ends meet. However, as women entrepreneurs gain experience and confidence, and as they accumulate capital of their own, they are in a better position to negotiate from a position of strength within the household. Men in patriarchal African societies have been ambivalent about their wives' success. On the one hand, many feel threatened by the growing economic independence of women, and some seek to limit the nature and scope of their wives' businesses. On the other hand, many men recognize their own inability to meet all household costs from their own earnings.

The economic and social policies that benefit women will inevitably be shaped within the political arena. Thus, if women are to advance, they must succeed in having a meaningful influence in politics. The evolution of a more active political role for African women has been assisted indirectly by developments in the international women's movement. The United Nations Decade for Women and associated conferences—especially the concluding World Conference held in 1985 in Nairobi—served as a catalyst for the growth of national women's organizations in several countries, the discussion of women's concerns, and the exchange of ideas about programs to benefit women.

Achieving real political influence in present-day Africa will not be easy, given the male domination of the media, politics, the civil service, and the military. Women have made the greatest strides toward political equality in states such as Eritrea and Guinea–Bissau, where independence was won through armed struggle. Progress in these states evolved from the recognition that anti-imperial struggles could not succeed without the full commitment of women, and that full commitment was impossible as long as women suffered under "two colonialisms." However, most states provide few opportunities and little encouragement for female political activism. Although many countries have appointed female cabinet ministers, only a few have made serious attempts to involve women in the political process. In Ghana, President Jerry Rawlings's election victories in 1992 and 1996 owed much to his prolonged efforts to mobilize women's support, particularly in rural areas, by emphasizing the benefits to women of water supply and other rural development projects.

Political power and influence are not entirely new to African women. History records several examples of women who were influential leaders, such as Queen Amina of Zaria (Nigeria). In certain matriarchal societies, such as the Ashanti of Ghana, women play an important role in political affairs. The Queen Mother holds a position of great honor in Ashanti society and heads an organization of women who participate in the selection of new kings. During the era of colonial rule, African women often took the lead in protesting taxes and other government measures that they considered to be unfair.

Women and Development

In 1970, Ester Boserup published a groundbreaking book, *Women's Role in Economic Development* (see "Further Reading"). Boserup's analysis began with a survey of the crucial economic role played by women in the Third World—a role, she noted, that was usually discounted and forgotten because of the exclusion of domestic and subsistence work from economic statistics. She showed how governments and development agencies had ignored the crucial work of women, both during and after the colonial era. Moreover, she argued, the "development" process had contributed to a decline in the situations of many women.

The colonial promotion of cash cropping directed at men excluded women from the income so derived, and often displaced them from the best and most accessible farmland. Following independence, women's agriculture was further marginalized when development policies and programs put even more emphasis on cash crops to increase foreign exchange earnings. This reflected not only a bias toward export crop production, but also a belief that food production would continue to "look after itself." It was also assumed that males as heads of households controlled whatever happened on the farm. Thus women continued to work almost invisibly, using traditional, labor-intensive technology to produce food, mainly for household consumption.

When growing food shortages and rising prices for food finally focused attention on food production in the 1970s and 1980s, the solutions proposed—irrigation schemes, large-scale farming, and land tenure reform, for example—often favored men instead of women. Land, traditionally a communal resource in most parts of Africa, was converted to private title in a number of countries, in part to permit its use as collateral for loans. Men have often obtained title to land normally farmed by women, in the process eroding women's customary rights of use.

Boserup did not critically examine the dominant model of development; her point was that women needed to be included in the development process. The publication of her book stimulated interest for the first time in projects specifically for women. Development agencies established "women-in-development" programs that typically attempted to improve domestic skills and methods of crafts production. These types of projects were attractive because they did not challenge the economic domain of men, and thus were politically and culturally "safe." A few of these projects proved quite successful (see Figure 19.4).

FIGURE 19.4. Women in the Oodi Weaving Cooperative, Botswana. Oodi tapestries are internationally renowned for their beauty. The Oodi cooperative was a notable success among the early "women-and-development" aid projects. Photo: CIDA (B. Paton).

The limitations of disjointed, token "women-in-development" projects soon became evident. Domestic work and crafts generate comparatively little money relative to the work effort and tend to reinforce the stereotypical view of "women's work." Moreover, many women lack the time, energy, and financial means to take full advantage of development programs intended to be of benefit to them (Vignette 19.3). More ambitious projects for women's development also failed when they were not designed with enough attention to local social realities. A case in point is the Yagoua agricultural project in Cameroon, which attempted to give women assistance to produce rice. In Cameroon, women traditionally produce sorghum and men produce the more valuable crop, rice. The project failed to meet its objectives because planners had failed to realize that, unlike sorghum income, rice income was controlled by men even when women produced the crop.

More recent approaches have emphasized the inclusion of women in integrated community development projects. Although women may be less visibly "ghettoized" in these integrated projects, questions remain about whether women's voices are truly heard and whether women reap their fair share of benefits. In most cases, men continue to dominate and control what happens at the local level, and women continue to do more than their share of the work.

It would be wrong to conclude that development programs cannot be of substantial benefit to women (see Figure 19.5). For example, the provision of safe, accessible water not only contributes to improved family health, but may also reduce the time that women spend hauling water—thus freeing them for other things. Maternal and child health programs that address women's concerns about the birth and survival of children also contribute to the well-being of women.

Since the mid-1980s, the number of women's organizations, operating at all levels from the local community to the international, has grown significantly. These organizations are helping to initiate a feminist transformation in the way in which women's development is done. At the international and national levels, the analysis is often explicitly feminist—focusing on gender relations as the key to understanding women's oppression and as the necessary starting point for progressive, liberating development. Where overt political activity is possible, feminist strategies increasingly center on ensuring that the political voice of women is strong and effective. Even in local communities, women are adopting strategies that are at least implicitly feminist. Women not only are forming their own local organizations to address needs that they themselves have identified as important, but also are insisting that women

VIGNETTE 19.3. Gender and Poverty:
Why Poverty Alleviation Programs May Fail

Western nations have pledged, as a goal for the new millennium, to end absolute poverty by the year 2015. This program includes a range of specific targets such as universal primary education, gender equality, universal primary health care, and a two-thirds reduction in mortality among children under five. These are important goals, especially for women in extremely poor countries such as Ethiopia. The reality, however, is that various constraints limit the probability of success. These include macro-scale problems related to aid conditionalities and the lack of sufficient financial and human resources, as well as micro-scale problems related to the reality of women's lives.

Households in rural Ethiopia spend 90% or more of their meager resources on food, but still have inadequate diets. Recurrent droughts have resulted in lost harvests and the death of livestock—traditionally the source of resources to deal with emergencies. As more households fall into deep poverty, traditional forms of reciprocity in the community become less effective. Female-headed households are especially vulnerable.

Various forms of gender inequality affect the daily lives of women. They are left with a disproportionate share of the workload—ranging from domestic duties, to the collection of water and firewood, to agricultural labor. Men acknowledge women's burden of work, but do little to relieve that burden. Moreover, women in the poorest households eat a smaller share of food than men; it is said that women and girls are able to last longer without food.

Education and health care facilities in rural Ethiopia are scattered and rudimentary. The spatial inaccessibility of services combines with other factors—women's disproportionate workload, and cultural attitudes that the needs of females in the household are less important than those of males—to perpetuate gender inequity in access to basic health care and education. User fees for services introduced as part of structural adjustment are formidable barriers for the poor, for whom prayer is a necessary substitute for medical care.

Programs with worthy objectives such as the elimination of poverty have little chance of success if they do not address fundamental economic and cultural barriers that help to perpetuate the marginalization of the poor, and especially of poor women and girls.

Based on F. von Massow. " 'We are forgotten on earth': International development targets, poverty, and gender in Ethiopia." In C. Sweetman, ed. *Gender in the 21st Century*, pp. 45–54. Oxford: Oxfam GB, 2000.

remain in control of these initiatives and that women reap whatever benefits can be derived.

As pressure on African societies increases because of population growth and environmental degradation, and as governments become less and less able to respond to urgent development needs because of declining revenues and increasing debt, the hardships that confront the majority of African women continue to grow. Metaphorically, the geographic landscape for African women is a rocky terrain, but African women are not to be pitied as passive victims. Rather, they are proud, innovative, and assertive within the constraints imposed by their so-

cieties, and they are increasingly determined to shape their own destinies.

Further Reading

For insightful and provocative feminist overviews of the "world of women," see the following sources:

Hodgson, D. L., and S. McCurdy, eds. *'Wicked' Women and the Reconfiguration of Gender in Africa.* Portsmouth, NH: Heinemann, 2001.

Seager, J. *Women in the World: An International Atlas,* 3rd ed. London: Myriad Editions, 2003.

a   b

FIGURE 19.5. Social development projects designed to assist rural women. (a) Women learning to sew, Jigawa State, Nigeria. At the end of their six-month course, the participants are allowed to keep the sewing machines and given one year to pay for them. (b) A Zambian volunteer working on a Canada–Mozambique aid project teaches improved cooking techniques to women in Nampula, Mozambique. Photos: (a) R. Maconachie; (b) CIDA (B. Paton).

Stichter, S. B., and J. Parpart, eds. *Patriarchy and Class: African Women in the Home and the Workforce.* Boulder, CO: Westview Press, 1988.

United Nations Development Fund for Women (UNIFEM). *The World's Women, 2000.* New York: United Nations, 2001.

Economic issues associated with "women and development" are discussed in these sources:

Boserup, E. *Women's Role in Economic Development.* London. Allen and Unwin, 1970.

Dixon-Mueller, R. *Women's Work in Third World Agriculture.* Geneva, Switzerland: International Labour Organization (ILO), 1985.

Creighton, C., and C. K. Omari. *Gender, Family and Work in Tanzania.* Aldershot, UK: Ashgate, 2001.

Ferguson, J. *Expectations of Modernity: Myths and Meanings of Urban Life on the Zambian Copperbelt.* Berkeley: University of California Press, 1999.

Parpart, J. L., and K. A. Staudt. *Women and the State in Africa.* Boulder, CO: Lynne Rienner, 1989.

Plewes, B., and R. Stuart. "Women and development revisited: The case for a gender and development approach." In J. Swift and B. Tomlinson, eds. *Conflicts of Interest,* pp. 107–132. Toronto: Between the Lines, 1991.

Stamp, P. *Technology, Gender and Power in Africa.* Ottawa: International Development Research Centre, 1989.

Tripp, A. M. *Changing the Rules: The Politics of Liberalization and the Urban Informal Economy in Tanzania.* Berkeley: University of California Press, 1997.

Selected political and social issues concerning African women are discussed in the following:

Baylis, C., and J. Bujra, eds. *AIDS, Sexuality and Gender in Africa: The Struggle Continues.* London: University College London Press, 2001.

Charleton, S. E. M., V. Everett, and K. Staudt, eds. *Women, the State and Development.* Albany: State University New York Press, 1989.

Coquery-Vidrovitch, C. *African Women: A Modern History.* Boulder, CO: Westview Press, 1997.

Gujit, I., and M. K. Shah, eds. *The Myth of Community: Gender Issues in Participatory Development.* London: Intermediate Technology, 1998.

Sheldon, K., ed. *Courtyards, Markets, City Streets: Urban Women in Africa.* Boulder, CO: Westview Press, 1996.

The lives of African women are explored in a diverse body of nonfictional and fictional literature. As a starting point, see the following sources:

Emecheta, B. *Second Class Citizen.* London: Fontana, 1974. (Other novels by the same author include *Double Yoke, The Bride Price,* and *The Slave Girl.*)

Hansen, K. T. *Keeping House in Lusaka.* New York: Columbia University Press, 1997.

Smith, M. F. *Baba of Karo: A Woman of the Muslim Hausa.* New Haven, CT: Yale University Press, 1954.

Schipper, M. "Women and literature in Africa." In M. Schipper, ed., *Unheard Words: Women and Literature in Africa, the Arab World, Asia, the Caribbean and Latin America,* pp. 22–58. London: Allison and Busby, 1984.

Internet Sources

The Internet is a rich resource for learning about African women. Start with the United Nations sites, and that for Stanford University, which offers comprehensive linkages to many other sites:

United Nations Economic Commission for Africa (UNECA) *Empowering Women.* http://www.un.org/Depts/eca/women.htm

United Nations Development Fund for Women (UNIFEM). http://www.unifem.org

WomenWatch: UN Gateway on the Advancement and Empowerment of Women. http://www.un.org/womenwatch

Stanford University. *Africa South of the Sahara: African Women.* http://www-sul.stanford.edu/depts/ssrg/africa/women.html

An increasing number of African women's NGO networks maintain websites. For example, see the following:

The African Women's Development and Communication Network (FEMNET). http://www.femnet.or.ke

Enda Synfev [Senegal]. *Synergy, Gender and Development.* http://www.enda.sn/synfev/synfev.htm

Gender Learning Network (GLN) [Kenya]. http://www.arcc.or.ke/gln

Kabissa [Nigeria]. *Niger Delta Women for Justice.* http://www.ndwj.kabissa.org

Tanzania Gender Networking Programme (TGNP). http://www.tgnp.co.tz

Women's Net [South Africa]. http://www.womensnet.org.za

Women of Uganda Network (WOUGNET). http://wougnet.org

Zambia National Women's Lobby Group. http://www.womenslobby.org.zm

Other sources on issues of concern to African women include these:

Female Genital Multilation Education and Networking Project. http://www.fgmnetwork.org

Women in Development Network (WIDNET). http://www.focusintl.com/widnet.htm

20

Children in Africa:
Prospects for the Next Generation

Whereas children are often seen in Western societies as a costly burden for families and for society as a whole, children in African societies are valued greatly. Attitudes toward fertility are beginning to change, but children are still seen as a blessing. To be childless is generally considered a tragedy; it is almost inconceivable for couples to choose to remain childless.

Nevertheless, childhood in African societies is not always idyllic. Children are generally expected to help with the work of the household, and for many, long hours of work mean that there is no time for going to school. Moreover, social changes associated with modernization, urbanization, and the spread of the HIV/AIDS epidemic are putting an increasing proportion of African children at serious risk.

At Home and in the Community

Infants establish a strong with their mothers during the first years of their lives. African babies are traditionally breastfed for the first 18–30 months of their lives. They are with their mothers constantly, typically carried on the mothers' backs. African fathers tend to have a more distant and formal relationship with their children. Responsibility for child care, as well as for routine daily decision making about children's needs, rests with mothers.

The rearing of African children is the responsibility not only of parents, but also of the extended family and the community as a whole. From a relatively early age, especially in rural areas, children are free to explore outside their homes with older siblings or on their own. Adults in the community participate actively in watching out for the safety of children at play, and disciplining them for inappropriate behavior when necessary.

The institutions of family and household are conceived more broadly in Africa than is the case in Western societies, and this is reflected in the rearing of children. Child fostering is not uncommon in African societies. Among the Hausa, for example, the first-born child is normally sent to live with a relative or someone else in the community. Such traditions not only share the responsibility of child rearing, but also represent a

297

form of fertility sharing; a child may be sent to stay with a childless couple, for example.

Child fostering has remained an important part of the strategies used by households to maintain a presence in both urban and rural settings. Families living in the city often send their young children to stay with grandparents or other relatives in the countryside. This arrangement helps to relieve pressure on crowded living conditions and extremely tight family budgets. The countryside is also seen as a safe place for children to learn about their cultural roots and go to school.

Children in rural communities entertain themselves with a variety of locally produced toys and games (Figure 20.1). Toys include simple "cars," sometimes just large wire hoops that can be rolled along the street with a stick. Games using marbles, seeds, or stones are commonplace; the most ubiquitous is a type of "board game" that is played by "sowing" seeds or stones sequentially in rows of small holes. The game has many names, among them *wari* (Asante), *ayo* (Yoruba), and *giuthi* (Kikuyu). Like their counterparts in most other parts of the world, African boys are enthusiastic soccer fans and players, although they are often forced to play with a homemade ball.

Each society has its own songs and musical games that are sung by groups of children as a form of self-entertainment. These songs, along with riddles and stories, help to develop children's mental and oral dexterity, and help them to learn valuable social lessons.

At Work

Life for African children—except for those whose parents are urban-based, well off, and modern in their thinking—involves much more than play and going to school. The majority of African children contribute in tangible ways to the household economy, either by performing household or farm chores or by earning money to help support the family.

Children growing up in rural Africa assist their families in many ways (see Figure 20.2). Preadolescent girls help to care for their younger siblings. Children, especially girls, commonly fetch water or firewood for the household. This task may involve hours of work carrying heavy containers filled with water from distant pumps, wells, or streams, or gathering wood outside the village. In many herding societies, boys as young as eight years of age are assigned responsibility for looking after herds of goats, sheep, or cattle. In agricultural communities, boys help their parents to plant, weed, and harvest crops.

The role of children in the labor force presents a dilemma for governments and for development workers. On the one hand, children's rights to go to school and to be protected from being exploited as labor are enshrined in the United Nations Universal Declaration of Human Rights, and in international protocols developed by the International Labour Organization (ILO). When children cannot attend school because they are working, their future prospects

a   b

FIGURE 20.1. Children at play. (a) A game of football, Koubri, Burkina Faso. (b) A wire hoop, rolled endlessly down the road, is a toy for these Zambian children. Photos: (a) R. Maconachie, (b) CIDA.

FIGURE 20.2. Children at work in Mankessim, Ghana. These children selling tomatoes in the market are also caring for an infant sibling. Many poorer households rely significantly on children's labor to make ends meet. Photo: CIDA (P. St.-Jacques).

are jeopardized. Children are also robbed of years of opportunities for intellectual growth through play and socialization with their friends. Even children who start school may be taken out when they are a bit older, with negative consequences for their futures.

On the other hand, children's help may be critical to the viability of poorer households. Their help enables farming households that are short of labor to grow just enough food to survive. The money children earn as hawkers of foods and other products may be necessary for poor urban households to pay for the bare necessities of food and shelter. When the Universal Primary Education (UPE) program (to be discussed later in this chapter) was introduced in Nigeria, it had major adverse effects on the earnings of Muslim women, who had long used

the help of children as messengers and vendors to conduct businesses while they themselves remained confined in purdah.

In addition to the role of children in contributing their labor to the household is the growing phenomenon of forced child labor. Child labor is considered briefly in the final section of this chapter.

At School

For most African children, education is not simply a process of passing from grade to grade through school. Education incorporates various forms of traditional and non-Western learning, as well as modern schooling. What children learn in the formal education system is also far from certain. Is schooling conducted in a child's native tongue, or in a foreign (quite probably European) language? Do children learn to read and write, or is the class essentially untaught because the teacher has not been paid, is sick, or has died from AIDS? Will a child be able to complete several years of schooling before being forced to work to help support the family?

Indigenous Education

One of the greatest myths about African development is that education was a European gift. Although formal Western education was introduced in colonial times, it is also true that education, both formal and informal, existed long before the arrival of Europeans. African societies relied on oral communication to convey traditions, beliefs, values, and knowledge from one generation to the next. Some societies incorporate formal instruction in the rituals, beliefs, and traditions of the community into the ceremonial initiation into adulthood. Forms of indigenous education are still widely practiced, and they are still the basic means by which African values and knowledge are passed to children.

The most widespread form of education is the learning that occurs informally as children observe and explore the environment in which they live. They do so at play and while assisting their parents and elder siblings with work on the farm or in the household (see Vignette 20.1).

VIGNETTE 20.1. Learning to Farm: The Socialization of Children in Rural Sudan

What children learn about the environment and about using it as a basis for sustenance is crucial for the survival of rural economies and societies from generation to generation.

In the village of Howa in eastern Sudan, as in most traditional African communities, learning occurs in conjunction with work and play. Children are active participants in the activities of the household. They help their parents with such agricultural tasks as clearing the farm, planting, weeding, and harvesting. They play an important role in the care of livestock. They are often made responsible for collecting firewood and obtaining water for domestic use. They are sent to gather wild foods, such as fruits and green vegetables, that constitute an important part of rural diets.

Work of this sort provides unlimited opportunities for learning about the environment. Children learn by observing their parents and older siblings as they work with them in the fields. They learn by asking questions, and they learn from mistakes; if, for example, they mistakenly weed out a useful plant, their parents will explain what should have been done.

Children in rural Sudan also learn through play. One example is when boys trap small birds to be roasted and eaten, and sometimes to be sold for a few pennies. The line between work and play isn't always well defined; trapping birds for fun may help to supplement the family diet and may contribute to the control of crop-destroying pests.

Changes in the economy and the ecology brought about by development initiatives may change the nature and significance of traditional learning. In Howa, the development of a large irrigation scheme has brought about an increase in the work of children. For example, collecting firewood takes longer because it is less readily available close to the village. The labor requirements of irrigated cotton production are greater than those for crops grown previously. Households need more income to survive in the increasingly commercial local economy, and children must contribute their share.

The irrigation scheme has increased economic disparities within the community and has resulted in the widespread alienation of land. As a result, most Howa children are unlikely to be able to have their own farms in the future. Many will work as farm laborers, and others will seek employment in the city.

There is a growing discontinuity between what the children of Howa have learned and the opportunities they will have as adults. Modern, formal education often appears to be more relevant preparation for this new reality. However, the growing workload of children in the local economy has become a significant barrier to school enrollment; enrollment is not increasing as might have been expected, given the socioeconomic changes that have taken place.

Based on C. Katz, "Sow what you know: The struggle for social reproduction in rural Sudan." *Annals of the Association of American Geographers*, vol. 81 (1991), pp. 488–514.

Working alongside elders provides countless opportunities to learn, both by example and by instruction.

In areas where Islam was important, formal Koranic schools were established (see Figure 20.3a). Children were (and still are) instructed in the Arabic language; they studied the Koran and read other books on Islamic belief and practice. The more gifted students could pursue higher Islamic studies, specializing in areas such as Islamic law or training to become religious leaders and teachers, and occupying positions of great prestige in society.

Education in Colonial Africa

Despite the lofty rhetoric of the "civilizing mission," education was not a high priority for the

a   b

FIGURE 20.3. Portraits of education, Niger. There is a vast difference in the curricula, formal and informal, of these two schools. (a) A Koranic school, where traditional Islamic education, conducted in the Arabic language, focuses on the study of the Koran and other religious texts. (b) Scene in a modern primary school in rural Niger. Photos: CIDA (R. Lemoyne).

colonial rulers of Africa. They recognized the need to train a cadre of local people who would be literate in the European language, capable of doing numerical calculations, and familiar with European ways of doing things. Graduates could assume lower-ranking positions within the colony—for example, as teachers, nurses, drivers, mechanics, or assistants in government offices. On the other hand, it was feared that educated Africans could become a focus of political dissent, questioning the authority of both colonial rulers and compliant traditional leaders. Many believed that too much schooling would "spoil" Africans, leaving them divorced from their own culture but not part of the modern culture.

The approach taken in most British colonies differed from that used in the French and Portuguese colonies. The French and Portuguese facilitated the creation of a very small class of "assimilated" Africans, who were given special privileges by virtue of an elite education that sometimes included a period of study in Europe. It was hoped that this assimilated class would identify with the colonial power. The educational system in British colonies was somewhat less elitist and did not aim explicitly to create a class of officially "assimilated" Africans.

Missionary societies did not have the same ambivalence as the colonial states did about education. Schooling was identified as a powerful tool for spreading the Christian gospel. Schools, especially boarding schools, provided an atmosphere in which children could be taught Christian beliefs and attitudes on marriage (monogamy), lifestyle (a rejection of traditional beliefs and practices), and work ("discipline"). In short, establishing schools was seen by missionary societies as a means to an end, rather than an end in itself.

Education in Contemporary Africa: Who Gets What Where?

The view that modern education (see Figure 20.3b) is *the* key to unlocking Africa's potential is widespread and deeply held. It prevails within development agencies such as the World Bank, as well as among national political leaders. In many African families, scraping together money for children's school fees takes precedence over all but the most crucial of other needs.

The results of efforts to invest in education have been impressive, particularly when the achievements in education are compared to those in other sectors. Between 1960 and 1999–2000, the number of children in primary schools in Africa south of the Sahara increased sixfold, from 13 million to 80 million. To make these gains possible, African countries have invested vast sums in physical infrastructure (i.e., buildings, books, and supplies) and human capital (i.e., teachers and administrators).

The state of educational development varies considerably from country to country (Figure 20.4). For example, fewer than one-third of adults are illiterate in 11 countries, while more than two-thirds are illiterate in another 13 countries. The enrollment in primary schools ranges from fewer than 30% of school-age children in Niger, Ethiopia, Somalia, and Chad to 100% in South Africa.

Table 20.1 provides an overview of children's access to education in selected countries. Girls' educational opportunities lag behind those of boys in most African countries; only in South Africa and Namibia are girls as likely as boys to attend primary school. In several poorer countries, such as Mozambique and Mali, girls are more likely than boys to drop out of school early (before reaching grade 5). Although national wealth is usually a good predictor of enrollment levels, this is not always the case. For instance, although Côte d'Ivoire is a much wealthier country than Tanzania, a significantly lower percentage of its children attend primary

school, and more of them drop out before grade 5.

Limited educational opportunities for girls reflect patriarchal attitudes entrenched in both African and European (colonial) society. Because girls were generally seen as destined to bear children and keep house, female education had a low priority. Moreover, African families have tended to see education as an investment, wherein a graduate could secure a well-paying job and contribute to the support of the extended family. Because young women take their training and earning power with them when they marry, girls' education has often been seen as a poor investment. Most ordinary African families cannot afford the considerable cost of educating every child beyond the primary level.

These national data do not tell the whole story, however. They do not indicate levels of progress; in Mali, for example, the net primary enrollment almost doubled from 20% to 38% between 1980 and 1998. The data also do not identify the extent of disparities in children's access to schooling *within* particular countries in relation to social class or region. Finally, na-

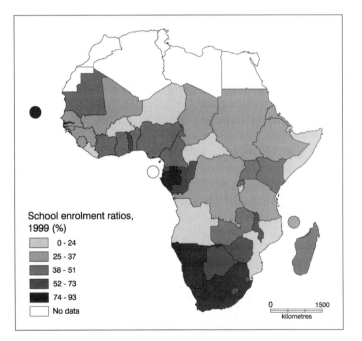

School enrolment ratios, 1999 (%)

- 0 - 24
- 25 - 37
- 38 - 51
- 52 - 73
- 74 - 93
- No data

0   1500
kilometres

FIGURE 20.4. Primary school enrollment ratios, 2000. The data show the proportion of the school aged population attending school. Data source: United Nations Development Programme (UNDP). *Human Development Report, 2002.* New York: Oxford University Press, 2002.

TABLE 20.1. Children's Access to Education in Selected African Countries

Country	Net primary enrollment, 1998		% of students reaching grade 5		Public expenditure on education (% of GDP[a])
	Female	Female as % of male	Male	Female	
Burkina Faso	28	68	74	77	3.6
Côte d'Ivoire	50	81	77	71	5.0
Ethiopia	30	73	51	50	4.0
Mali	34	70	92	70	2.2
Mozambique	46	61	52	39	—
Namibia	90	108	78	82	9.1
Senegal	70	84	89	85	3.7
South Africa	100	100	72	79	7.6
Tanzania	88	94	78	84	—
Zambia	72	98	—	—	2.2

[a]GDP, gross domestic product.

Data sources: World Bank. *World Development Report 2001.* New York: Oxford University Press, 2001. United Nations Development Programme (UNDP). *Human Development Report 2002.* New York: Oxford University Press, 2002.

tional data do not address the important subject of the quality and relevance of the education children receive.

Modern education as a strategy for social mobility has become much less attractive for the poor in recent years. Most African countries have many unemployed teachers and other educated professionals. The probability of being appointed to a public-sector job is increasingly dependent on wealth and social contacts in circles that count. Education today is more an instrument for the perpetuation of class cleavages—a means for the ruling classes to consolidate their position—than an opportunity for social mobility.

Education means much more than learning the three R's. In addition to the formal curriculum, children learn a "hidden curriculum" of ideas and values that are conveyed implicitly. One of the most important elements of the hidden curriculum is the message that indigenous African culture and knowledge is not important—that everything worth knowing comes from Europe. Education has been called the "enemy of the farm," since so many graduates of rural schools seem determined to pursue their dreams in the city, regardless of how scarce jobs may be there.

Four decades after the demise of colonialism, European languages remain preeminent in African education, even in the first year of primary school. Children learn in an indigenous language in early primary classes in 13 of 15 countries formerly ruled by Great Britain, but in only 4 of 15 former French colonies and in none of the former Portuguese territories. Although the great number of indigenous languages spoken and the lack of published materials in most local dialects are barriers to the use of many African languages in schools, the most important barrier has been a reluctance to fundamentally rethink priorities in education.

The powerful words of Frantz Fanon, in his book *Black Skin, White Masks* (New York: Grove Press, 1967), focus on why the language of instruction is a potent symbol of what education represents:

> Every colonized people—in other words, every people in whose soul an inferiority complex has been created by the death and burial of its local cultural originality—finds itself face to face with the language of the civilizing nation; that is with the culture of the mother country. The colonized is elevated above his jungle status in proportion to his adoption of the mother country's cultural standards. He becomes whiter as he renounces his blackness. (p. 18)

Julius Nyerere, Tanzania's first president, wrote an essay entitled "Education for Self-Reliance" (see "Further Reading"), in which he provided an insightful critique of conventional education and proposed an alternate approach that gave priority to local needs and indigenous wisdom. Nyerere argued that children needed to learn skills relevant to the lives that the majority of them would lead in rural communities. He urged schools to develop a sense of pride, independence, and self-reliance in their students. His ideas were widely discussed, but outside Tanzania there were very few attempts to implement them.

In addition to the content of African education, its quality has been questioned. The quality of education has been jeopardized by the economic crises affecting virtually every country. There is a crucial shortage of books and supplies; expenditure on educational materials at the primary level averages only 60 cents per pupil per year throughout the continent. Many school buildings are closed because there is no money for essential repairs. Because of very poor salaries, teachers have to juggle their teaching duties with other income-producing activities in order to make ends meet. The effects of educational cutbacks have been uneven. The wealthy purchase books for their children, patronize private schools, and hire personal tutors. The poor, particularly the rural poor, get whatever remains.

Significant spatial variations in access to education first emerged during colonial times and still persist. As a rule, urban areas have received proportionally more investment in education than rural areas have; this has been particularly true at the secondary and tertiary levels. For rural students, not having a nearby school may mean either having to bear additional costs for room and board in a town, or being unable to attend school at all.

Nigeria provides an illustration of how regional inequities in educational development came into existence. In southern Nigeria, where missionaries became very active at an early date, 36,000 students were enrolled as of 1912. By 1957, 2,343,000 students were enrolled in 13,473 primary schools in southern Nigeria. In northern Nigeria—home to over half of Nigeria's population—there were fewer than 1,000 students in primary schools in 1912 and only 185,000 in 1957. Colonial authorities virtually excluded missionaries from the Islamic north-

ern part of the region. The government ran the few schools established in Muslim areas.

The extent of the consequent inequities is revealed in enrollment data from 1972, when Nigeria launched its UPE program (mentioned earlier). Three states in northern Nigeria had fewer than 10% of primary-school-age children in school, while three southern states had well over 80% enrolled (Figure 20.5). The northern states have moved decisively to close this educational gap, but it will take generations to do so. In the meantime, these disparities have contributed substantially to the ongoing social, political, and economic tensions within Nigeria.

At Risk

The happy faces of children at play leave the impression of childhood as a carefree time. Yet childhood is a difficult phase of life for Africa's children, exemplified by the fact that many newborns fail to survive to adulthood. Tens of millions of African children live in deep poverty, and suffer hunger and material deprivation as a result. Moreover, children who are living in war zones, or whose parents have been infected by HIV/AIDS, cannot take the security of home and family for granted.

Children's Morbidity and Mortality

Africa's children face a major challenge from birth—the challenge of surviving their first years of life. Infectious and parasitic diseases (e.g., malaria, whooping cough, measles, and dysentery) are the prime reasons for the death of 92 of every 1,000 children born within the first year of life. Some 15% of children fail to reach the age of five.

The expansion of immunization programs during the 1980s brought a significant reduction in infant and child mortality. However, the recent rise of AIDS as a cause of infant and child deaths has negated most of the gains achieved through vaccination campaigns. The under-five mortality rate in Zimbabwe increased from 108 per 1,000 in 1980 to 125 in 1998, primarily because of the HIV/AIDS epidemic. AIDS has had an indirect effect as well

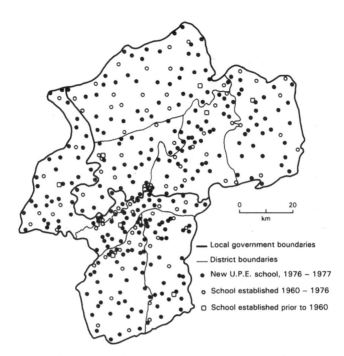

FIGURE 20.5. The Universal Primary Education (UPE) program in Hadejia Emirate, northern Nigeria. UPE brought about a fourfold increase in the number of schools in only two years. After R. Stock. "The rise and fall of Universal Primary Education in northern Nigeria." *Tijdschrift voor Economische en Sociale Geografie*, vol. 76 (1985), pp. 277–287.

on children's health and survival through its adverse impact on food production, since children are the most vulnerable group in times of food shortage.

The health risks that children face vary greatly from country to country, as well as regionally within countries. Under-five mortality rates range from 250 per 1,000 children in Niger to 83 in South Africa. Rates are typically considerably higher in rural than in urban settings within each country. Children in rural areas are less likely to have the benefits of safe water supplies, are less likely to have been vaccinated, and are less likely to have ready access to medical care when they become ill.

Children as Orphans

The HIV/AIDS epidemic has given rise to a huge number of orphaned children, especially in eastern, central, and southern Africa. This is hardly surprising, given that over 30% of the adult population is infected with HIV in certain countries. It is estimated that 15–25% of all children in a dozen countries will be orphans by the year 2010. In Zimbabwe alone, it is estimated that 900,000 children will be living without their mothers by 2005. This is in a country with a population of only 12 million.

Orphaned children, especially those orphaned because of AIDS, face an uncertain future. The most fortunate will be adopted by relatives, but they may be sick with AIDS themselves, or may be overwhelmed by the responsibility of caring for more children with very limited resources. An adopting relative may be an impoverished elderly grandparent. Urban-based families may have no nearby relatives who could care for their children. The ability of the community as a whole to care for orphaned children, as would have been the case in the past, has been severely compromised. In the worst-hit countries, there are simply too many children in need. Nevertheless, efforts have been made to address these children's needs. In Zimbabwe, land has been set aside in many villages for the use of orphaned children, and commercial farmers have supported initiatives to care for the children of deceased farm workers. Important though these measures have been, they look after only a portion of children in need.

Many orphans gravitate to the city, where they survive as best they can on the street (Vignette 20.2). These children use their wits to survive. They are doubly stigmatized—as street children, and as offspring of parents who died from a dreaded disease. One of the growing threats faced by children, especially those orphaned by AIDS, is sexual assault. This troubling phenomenon is symptomatic of the broader social breakdown occurring in countries where there is an exceptionally high rate of HIV infection.

Forced Labor by Children

A large and growing number of African children are victims of forced-labor schemes. These children are "recruited" are in varying ways; many are kidnapped. The children of poor families are also at risk; parents may succumb to the false promises of recruiters that their children will have paying jobs and will be able to remit money to their families. Children are sometimes transported across international boundaries, as in the case of children from Mali and Burkina Faso taken to work on farms in Côte d'Ivoire.

Many child workers are subjected to physical abuse. They are commonly fed bare-minimum diets and are forced to work for little, if any, pay. The majority of the victims of forced child labor are domestic servants or farm laborers. The ILO, human rights organizations, and many governments are working to abolish child labor, but achieving this goal is a difficult challenge.

Children and War

Over a dozen sub-Saharan African countries have been involved in major international conflicts or civil wars during the past decade. More limited conflicts have occurred in several other countries. Countless African children have been traumatized by their experiences of war. These wars have resulted, directly or indirectly, in the deaths of hundreds of thousands of children. Millions have been forced to flee their homes as refugees; women and children form a disproportionately large segment of refugee populations. In southern Sudan, groups of young boys have fled into the bush—in many cases, eventually reaching sanctuary in neighboring countries—out of fear of slave raiding by northern Sudanese.

The use of children, some under the age of ten, as combatants in wars is a most disturbing phenomenon that has become increasingly common in recent years. In northern Uganda, the rebel Lord's Resistance Army kidnapped thousands of children. The boys were trained as fighters, and the girls were used as servants and sex slaves. During the continuing Liberian civil war, all sides have made extensive use of child soldiers. In Sierra Leone, child soldiers formed a large proportion of the Revolutionary United Front forces. There has also been widespread and deliberate use of child soldiers in the civil wars of Angola and the Democratic Republic of the Congo.

Seizing children by force is an easy way for unscrupulous rebel—and in some cases government—forces to gain recruits. Because of their youth, child soldiers are easily manipulated psychologically to ensure their loyalty. They quickly come to believe that returning to their homes and families is no longer an option. They are also very effective as soldiers, because they tend to fight without either fear or remorse. There is also a sense that they are viewed and treated as more expendable than adults.

Efforts made in several countries to free and rehabilitate child soldiers have had rather discouraging results. Even if it is possible to find the parents of freed child soldiers, both the children and the parents often find it equally difficult to reunite and to repair the trauma of war. Many former child soldiers are psychologically scarred, and find it difficult to function in society or concentrate on their schooling (Figure 20.6). As a result, many of them end up on the street—or, if fighting has continued, return to their lives as soldiers.

Most African countries have signed the 1998

VIGNETTE 20.2. Who Are the Street Kids of Nairobi?

In recent years, there has been a major increase in the number of street children in all of Africa's major cities. Kariuki, in her study of Nairobi street kids, notes that they fall into a number of distinct categories. Children *in* the streets are there because their parents have sent them to earn money on the streets; most return home at night. Children *of* the streets have made their own decisions to wander the streets because of poverty and other difficulties at home. Then there are orphans and abandoned children—both boys and girls—who have no option but to live and support themselves on the streets.

When asked, "Who are street children?", the children themselves identified several characteristics: street children eat waste (spoiled food from dustbins); they sleep wherever they can because they have no parents; or they beg on the streets but go home at night.

Street children survive by whatever means they have at their disposal. Some sort through garbage from shopping centers in more affluent areas, in search of food, paper, and other items that can be used or sold. Some congregate near the Kenyatta Market to carry people's loads, while others work in teams to wash cars. Some beg, but begging is not a preferred occupation. Most children say they would much prefer to work for their sustenance, rather than beg. Some street kids survive as thieves or pickpockets.

Whatever their occupations, street kids are normally treated badly by police and the public, who typically label them all as thieves. This public reaction seems to be part of a larger phenomenon of anger directed at the poor. The community support systems that once offered a safety net for those who were poor and hungry no longer function well.

To explain the phenomenon of street children, social scientists have pointed to the breakdown of traditional values and traditional family structures as part of the modernization process. Family crises are exacerbated by deep poverty, which makes it impossible for many parents to offer their children a secure home environment that meets even basic needs. Some researchers paint a more disturbing picture of the home environments from which street children have come, claiming that many of these children have been the victims of physical, emotional, and sexual abuse in dysfunctional families.

Street children in Nairobi survive in squalor. They are exposed to many dangers, especially disease and violence. Despite their stigmatization and the precarious circumstances in which they live, street kids are resilient and resourceful. Authorities and the general public must develop greater empathy with these children, and develop innovative programs to address their needs for protection, education, vocational training, shelter, and nourishment.

Based on P. Kariuki. "Street children and their families in Nairobi." In S. Jones and N. Nelson, eds. *Urban Poverty in Africa: From Understanding to Alleviation*, pp. 138–148. London: Intermediate Technology, 1999.

international treaty barring the recruitment and use of child soldiers. Several countries in which young soldiers had previously been taken into the national armed forces have tightened their recruitment policies and established training programs to rehabilitate these youth. However, the enforcement of international law among rebel organizations is a much greater challenge. As long as civil wars persist on the African continent, the forcible recruitment of child soldiers is likely to continue.

What Future for Africa's Children?

African societies value children highly. Nevertheless, the crises of underdevelopment epitomized by growing poverty, the HIV/AIDS epidemic, and wars in many parts of the continent have created an extremely precarious environment for more and more of Africa's children. Creating a better future for Africa's next generation involves better child health and educational programs, but it especially involves the

FIGURE 20.6. Children at risk: A former child soldier living at the Gulu Center for Traumatized Children in northern Uganda. Many former child soldiers are unable to return to live with their families. Photo: CIDA (P. Bennett).

establishment of peaceful and secure environments in which children can grow.

Further Reading

For statistics on African children, their welfare, and their education, see the following:

United Nations Children's Fund (UNICEF). *State of the World's Children.* New York: UNICEF (published annually).

United Nations Development Programme (UNDP). *Human Development Report.* New York: UNDP (published annually).

United Nations Educational, Scientific and Cultural Organization (UNESCO). *Statistical Yearbook* and *World Education Report.* New York: UNESCO (published annually).

Colonial educational policies are discussed in the following sources:

Mugomba, A. T., and M. Nyaggah. *Independence without Freedom: The Political Economy of Education in Southern Africa.* Santa Barbara, CA: ABC-Clio, 1980.

Tibenderana, P. Z. "The emirs and the spread of Western education in northern Nigeria, 1910–1946." *Journal of African History,* vol. 24 (1983), pp. 517–534.

The relationship between education and development is discussed in the following sources:

Freire, P. *Pedagogy of the Oppressed.* New York: Seabury, 1970.

Nwomonoh, J. *Education and Development in Africa.* San Francisco: International Scholars, 1998.

Nyerere, J. "Education for self-reliance." In J. Nyerere, *Ujamaa: Essays on Socialism,* pp. 267–290. London: Oxford University Press, 1968.

Pampallis, J., and E. Motala, eds. *The State, Education and Equity in Post-Apartheid South Africa: The Impact of State Policies.* Aldershot, UK: Ashgate, 2002.

Simmons, J. "Education for development, reconsidered." *World Development,* vol. 7 (1979), pp. 1005–1016.

World Bank. *Education in Sub-Saharan Africa: Policies for Adjustment, Revitalization and Expansion.* Washington, DC: World Bank, 1988.

The following Human Rights Watch publications offer case studies of African child soldiers:

Ehrenreich, R. *The Scars of Death: Children Abducted by the Lord's Resistance Army in Uganda.* New York: Human Rights Watch, 1997.

Fleischman, J. and L. Whitman, eds. *Easy Prey: Child Soldiers in Liberia.* New York: Human Rights Watch, 1995.

Rone, J. *Children in Sudan: Slaves, Street Children*

and Child Soldiers. New York: Human Rights Watch, 1995.

For studies of HIV/AIDS and its impact on children, see these sources:

Cornia, G. A., ed. *AIDS, Public Policy and Child Well-Being.* Florence, Italy: United Nations Children's Fund–International Child Development Centre (UNICEF–ICDC), 2002.

Desmond, C., and J. Gow, eds. *Impacts and Interventions: The HIV/AIDS Epidemic and the Children of South Africa.* Durban, South Africa: University of Natal Press, 2002.

Guest, E. *Children of AIDS: Africa's Orphan Crisis.* London: Pluto Press, 2001.

To learn more about Kenya's street children, see the following source:

Kilbride, P., C. Sucha, and N. E. Njeru. *Street Children in Kenya: Voices of Children in Search of a Childhood.* Westport, CT: Bergin and Garvey, 2000.

Internet Sources

Organizations with broad-based programs related to children's well-being include the following:

Human Rights Watch. *Children's Rights.* www.hrw.org/campaigns/crp

Ohio University. *Institute for the African Child.* www.ohiou.edu/afrchild/IAC.htm

UNICEF. *Why Children Must Be Heard.* www.unicef.org

For information on issues related to child labor, please see these sources:

Anti-Slavery: Today's Fight for Tomorrow's Freedom. Child Labour. www.antislavery.org/homepage/antislavery/childlabour.htm

Global March against Child Labour. *Child Labour in Africa.* www.globalmarch.org/cl-around-the-world/barometre-africa1.php3

International Labour Organization (ILO). *International Programme on the Elimination of Child Labour (IPEC).* www.ilo.org/public/english/standards/ipec/about

Several organizations are dedicated to helping Africa's street children. See the following, for example:

Street Child Africa. www.streetchildafrica.org
Street Kids International. www.streetkids.org

The following websites concern the use of child soldiers in Africa:

Dridi, B. A. *Child-Soldier Rehabilitation Programs in Africa—Effective Antidotes to War?* www.ossrea.net/announcements/dridi.pdf

Center for Defence Information. *CDI's Children and Armed Conflict Project.* www.cdi.org/atp/childsoldiers

Soldier Child. www.soldierchild.org

www.childsoldiers.org: Giving Voice to Children Affected by War. www.childsoldiers.org (see especially *Child Soldiers Global Report 2001*)

21

Social Policy: The Health Sector

In Africa, as in all other parts of the world, there are gross inequalities of risk and opportunity. This chapter looks at the provision of health care, particularly the kinds of health-policy responses that have come from governments. The choice of strategy—or rather, in many countries, the lack of a coherent policy—has serious implications for public health, especially the health of comparatively vulnerable groups such as women, children, and the poor.

The chapter starts by identifying several long-standing controversies about how African health care systems should be developed. Debates on these issues began during the colonial era; continued during the first two decades of independence, when health care systems were being expanded; and have extended into the present, when cost control has become a top priority. To be effective, health policies have to look beyond the provision of services to address a full range of factors—environmental, economic, social, and political—that profoundly influence levels of risk.

Health Policy Issues

The specific issues of greatest concern to the state and the public have varied at different times, as have the approaches favored by medical and social scientists and by policy makers. Nevertheless, the key debates for the most part have involved the same health policy issues. Four of these controversies are outlined below.

Health versus Other Priorities

Although few indeed have ever disputed the importance of having a healthy population, the question of the appropriate commitment of resources by the state to health promotion has long been a matter of controversy. Health may be important, but so too are many other things. Those who argue that health should be given priority point out that improvements in health bring other benefits, such as greater productivity, but investment in education, for example, is also an investment in future development.

The role of the state has remained very important in Africa because of the uneven and usually weak development of the private sector as a source of health-related initiatives. In making budget allocations, states signal where their priorities lie. The unevenness of governments' commitment to health is reflected in major variations in health care expenditure, especially in comparison to other sectors, such as military spending.

Under structural adjustment, African governments have been pushed to reduce their commitment of resources to publicly funded health care. There has been an increased emphasis not only on the privatization of service delivery, but also on user fees as a demand-regulating and cost-recovering mechanism for health care, water, and other services. As for international development assistance, many donors have treated health as a second-order priority.

General versus Targeted Programs

The allocation of health expenditures, and particularly the best ways to obtain desired benefits at an acceptable cost, have long been subjects of controversy. According to one school of thought, the best results may be obtained with programs to control or even eliminate certain diseases with carefully structured mass campaigns. The successful eradication of smallpox more than 30 years ago was the most successful of these campaigns. Critics of the targeted-disease strategy argue that far broader and longer-lasting benefits can be obtained from general programs that address the full range of health problems. The HIV/AIDS epidemic has rekindled this debate. Although the AIDS epidemic is clearly a medical emergency that has overwhelmed many health care systems, targeting HIV/AIDS creates the danger that other health problems, such as malaria, dysentery, and maternal mortality, will receive diminishing attention.

Closely related to this debate about whether programs should be general or more narrowly focused is the question of whether health care resources should be distributed as uniformly as possible, or whether selected areas (usually cities) should be given priority. Unlike the colonial states, which provided health care for a few select groups and areas, all governments now pay lip service to universal access to health care as a goal. However, the actual distribution of resources often tells a different story.

Preventative versus Curative Medicine

The gap between theory and practice is frequently evident in the commitment to preventative health interventions, such as the provision of safe water, mass vaccination of children, and health education. Often governments that talk about the importance of prevention allocate the great majority of resources to health care facilities that focus on curative medicine. On the other hand, because the demand for curative services seems limitless, it is always difficult to find ways of freeing more resources for prevention without appearing to jeopardize curative care. In the struggle against the HIV/AIDS epidemic, clear priority has been given to health education over treatment. Although few indeed would question the necessity of effective preventative programs, controversy has erupted over the low priority given to chemotherapy for HIV/AIDS in Africa (see Vignette 21.1).

The Necessity of Modern Technologies and Advanced Training

Controversy also surrounds the subjects of modern technology and advanced medical training. At least in theory, African countries have access to a full array of modern medical technology, ranging from CT scanners to the latest pharmaceuticals. It is often argued, especially by the medical establishment, that these technologies are needed if teaching and specialist hospitals are to function effectively. The counterargument is that expensive advanced modern technologies not only fail to address the major health issues of African nations, but also drain scarce resources from more pressing needs.

The debate about technology extends to the training of health care personnel. Should physicians and other medical professionals continue to be trained to Western standards in conventional medical schools, or should resources be reallocated into programs to train large numbers of paraprofessional health workers? Africa faces an acute shortage of health care professionals to offer care, provide health education on a day-to-day basis, and undertake medical research. This shortage of personnel is especially acute in the poorest countries and in rural areas. A substantial proportion of Africa's health care professionals—trained at great expense to the nation—have migrated to Europe, North America and other high-income destinations

VIGNETTE 21.1. Global Apartheid and Response to the AIDS Crisis in Africa

Is it simply random chance that four-fifths of the 3 million deaths each year worldwide from AIDS occur in Africa? Why is the response to HIV/AIDS, in terms of the commitment of resources to stem the epidemic and alleviate individual suffering, so dramatically different in Africa than in Europe and North America? Is "the enjoyment of the highest attainable standard of health" not included as one of the fundamental human rights in the 1948 Universal Declaration of Human Rights? Comparisons are now being drawn between apartheid under the former white regime in South Africa and the differential response worldwide to HIV/AIDS. In their article, Booker and Minter define *global apartheid* as

> . . . an international system of minority rule whose attributes include: differential access to basic human rights; wealth and power structured by race and place; structural racism, embedded in global economic processes, political institutions and cultural assumptions; and the international practice of double standards that assume inferior rights to be appropriate for certain "others," defined by location, origin, race, or gender. (p. 11)

In wealthy countries, the development of antiretroviral drugs has enabled people infected with HIV virus to halt, or at least slow, the development of AIDS. These same drugs are available to only an extremely tiny minority in Africa. The basic antibiotic and other drugs available to Africans are ineffective in slowing the course of the disease or alleviating patients' suffering. Hospitals are overwhelmed with the needs of sick and dying patients, but most Africans die quietly at home with palliative care from relatives. Multinational pharmaceutical companies that hold patents on drugs to fight HIV/AIDS, and have profited enormously as a result, have strongly resisted calls to make drugs available in developing countries at an affordable price. Indeed, when South Africa passed legislation to make it easier for its people to obtain antiretroviral drugs, the pharmaceutical companies filed a lawsuit to block its implementation.

Although the fight against HIV/AIDS has received much rhetorical support from the world's wealthiest countries, the commitment of financial resources has been small. Moreover, financial support to expand access to effective drugs has not been a priority in development aid programs. The industrialized world has turned its back on Africa and its people with AIDS, saying in effect that their lives are not worth the cost of the medications.

Antiretroviral drugs are not a cure; making these drugs widely available in Africa will not end the devastation of AIDS, but offering drug therapy can make a difference. Providing drugs to pregnant women who are HIV-positive greatly reduces the likelihood that their babies will be infected. The availability of drugs that offer extended survival and less suffering will encourage people to be tested for the virus; when no treatment is available, people are reluctant to be tested because a positive test is literally a death sentence. Widespread, early testing is crucial to slowing the spread of the disease through public education campaigns and precautionary behavior by HIV-positive individuals.

Apart from these practical reasons is the fundamental matter of human rights. What possible justification can be given for practicing a system of global apartheid, in which corporate profits trump the lives of those considered insignificant?

Based on (and quotation from) S. Booker and W. Minter. "Global apartheid." *The Nation*, July 9, 2001, pp. 11–17.

for professional and personal reasons, exacerbating the crisis of human resources in Africa. Governments have grappled with this challenge, but have had little success in stemming the medical brain drain.

Health Care: The Colonial Legacy

The primary objective of colonial health policy was quite clear: It was designed to protect the health of Europeans in the "white man's grave." Hospitals were constructed in the major cities to care exclusively for white patients. Public health measures—ranging from the strict segregation of African and European residential areas (see Chapter 16, Vignette 16.2), to drainage as a means of reducing insect populations—were designed to protect Europeans from the threat supposedly posed by "unhealthy" Africans. Basic health care was often made available to Africans employed by colonial governments or European companies. These services were concentrated in such areas of economic growth as mining enclaves or export cropping zones, where health care services could increase worker productivity. In poorer rural areas, colonial regimes gave scant regard to health care.

In much of the continent, the only health care was provided by Christian missionaries, who linked proselytizing to medical care. Missionary medicine was very unevenly distributed; few missionaries went to Muslim areas, for example. In areas of indirect rule, such as northern Nigeria, a few simple dispensaries were constructed by African local governments and financed by tax levies. At the end of the colonial era, most of rural Africa had no effective access to Western health care. Moreover, the few existing facilities were rudimentary—in most cases, small clinics staffed by a dispenser trained to give basic first aid and vaccinations (see Figure 21.1).

Since time immemorial, African societies have had their own healers charged with responsibility for the health of the community. Far from being nurtured by colonial governments, indigenous medicine was seen either as irrelevant or as primitive and dangerous. In many colonies, certain kinds of healing rituals were banned, because such ceremonies were thought to be potential foci of anticolonial resistance.

Colonial governments launched campaigns against specific diseases—in particular, yellow fever, sleeping sickness, leprosy, malaria, and yaws. Although a few of these campaigns achieved some success, most were paralyzed by

FIGURE 21.1. Village dispensary, Jigawa State, Nigeria. For most rural Africans, access to modern health care means a walk of several kilometers to a small clinic staffed by two or three primary health care workers. Photo: author.

their uneven implementation, the crudeness of available technologies, and Africans' widespread skepticism about colonial public health interventions. It was often assumed that the government had ulterior motives in mind—increased taxation, for example—when it implemented public health measures. Moreover, some of the problems that these campaigns were designed to alleviate were themselves the products of ill-conceived colonial development policies. For example, the rising incidence of sleeping sickness resulted in no small part from the establishment of resettlement schemes in high-risk environments, the creation of forest reserves that became potential disease "hot spots," and the increased mobility of infected people and disease vectors along newly built transportation routes.

The health care systems that the colonialists bequeathed to newly independent governments were fragmented in structure, biased in favor of urban areas, oriented to curative instead of preventative medicine, and organized to serve colonial interests. The colonial neglect of health care for Africans was exemplified by the situation in the present-day Democratic Republic of the Congo, which at the time of its independence in 1960 did not have a single indigenous graduate in medicine in the country.

The 1960s and 1970s: Growth (and Development?)

Following independence, African governments moved to upgrade health services; however, they had to contend with the scarcity of resources to build new facilities, train health workers, and initiate new policies. Compared to the increasingly difficult years that followed, the 1960s and 1970s were times when African governments were able to strive for growth and development in the health sector. The approach adopted by most countries reflected the still-predominant faith in modernization theory and the uncritical acceptance of conventional models of health care delivery.

Foreign aid donors gave financial support for the construction of modern "disease palace" teaching hospitals that reproduced the technological medicine of industrial countries, but paid scant attention to ordinary though absolutely crucial health problems, such as intestinal parasites, malaria, and malnutrition. These hospitals proved to be so expensive to staff and operate that little was left for the health care needs of the population. Such inequities in the distribution of resources conform to the prediction of the "inverse care law" (Figure 21.2), which states that there tends to be an inverse relation between the availability of medical care and the needs of those served.

Most African countries attempted to enhance health services in the previously neglected rural areas. Kenya developed a *basic health services* (BHS) scheme, with a network of spatially accessible health centers and clinics catering to the needs of local and regional populations. The BHS approach envisaged the creation of a hierarchy of services—ranging from rural health clinics where primary (basic, first-stage) health care would be provided, to tertiary centers for more seriously ill patients referred from nearby secondary clinics. However, BHS brought no

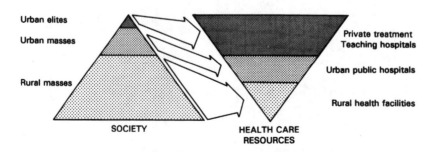

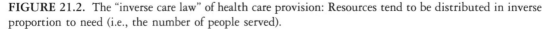

FIGURE 21.2. The "inverse care law" of health care provision: Resources tend to be distributed in inverse proportion to need (i.e., the number of people served).

substantive changes in the old biases favoring urban centers.

The Kenyan BHS model was emulated in several other countries. In practice, it was far too expensive to be widely implemented, even in relatively wealthy Nigeria. Nigeria's ambitious BHS plan proposed to cover the country with local networks of 20 clinics for basic care, four health centers for more advanced care, and one comprehensive health center for every 150,000 people. However, the actual cost of 8 million naira per unit was 16 times the budgeted estimate. In Kano State, only one-tenth of the planned facilities were constructed. Seven years after Nigeria launched its BHS plan, there were only 30 doctors to serve 8.7 million people in the rural districts of Kano State.

In 1978, the Alma Ata Conference of the World Health Organization (WHO) ended with a declaration that "health for all" should be a reality by the year 2000. *Primary health care* (PHC) was identified as the approach most likely to make this goal attainable. Many of the ideas associated with the PHC model came from the experiences of China. A notable example was the Chinese program of training "barefoot doctors"—ordinary citizens with basic medical training who were given responsibility for community public health and the treatment of common illnesses. The PHC strategy emphasized preventative health care and health education, and stressed the importance of addressing diverse factors (such as clean water, universal education, and food production) that affect well-being.

The PHC approach has the potential to change the state of health and health care quite radically. When local health workers are selected by the community and supported both materially and morally by it, the health workers become accountable to the community. However, although lip service is often paid to this ideal, governments and aid agencies seldom have any real commitment to empowerment at the grassroots; among other things, it is widely feared that empowerment is an invitation to dissent. In countries such as Kenya and Côte d'Ivoire with a strongly entrenched capitalist ruling class, urban elites continue to obtain good-quality care from private physicians, from

teaching hospitals, and even from overseas clinics, while many rural areas still have no modern health care at all.

The health care systems of Guinea–Bissau and Mozambique were developed differently. During the prolonged struggle against the Portuguese, the liberation movements organized social services in the areas they controlled. The PHC model implemented during the struggle continued to be applied after victory, with emphasis given to preventative programs such as immunization and sanitation and to the training of village health workers. In the early 1980s, the commitment of Guinea–Bissau and Mozambique to health as a basic right and first-order priority was reflected in their ranking as the two African countries allocating the highest proportion of government expenditures to health. Unfortunately, these countries have more recently had to reduce expenditures on health to address other pressing economic and political concerns.

Post-1980: The Challenge of Doing More with Less

Since 1980, African governments have faced growing economic pressures resulting from rising debt loads and declining terms of trade. These problems have forced them to reexamine their development priorities and strategies. Meanwhile, the growth and redistribution of population and the emergence of new sources of demand, most notably from the spread of HIV/AIDS, have placed increased pressure on health care systems. In short, governments have had to find ways of doing more with fewer resources.

New Strategies for Organizing and Financing Health Care

The deteriorating economic situation has left governments with very little money for the maintenance of existing health infrastructures, or for the purchase of drugs and other supplies. Cutbacks in public-sector spending, often made at the behest of the International Monetary Fund (IMF), have resulted in large staff layoffs and significant salary reductions.

Africa's 42 poorest countries reduced health care spending by 50% during the 1980s. In Nigeria, for example, per capita expenditure on health care declined by 75% between 1980 and 1987. As a result, hospitals often become, in the words of Nigeria's former President Ibrahim Babangida, "mere consulting clinics" without drugs and sometimes without medical staff. Many facilities were closed because of staff shortages or the lack of funds for needed repairs. Cutbacks that forced health care workers to take second jobs to make ends meet resulted in increased absenteeism and poorer care. Many doctors left the public service, and some took advantage of their international-standard training to emigrate.

The PHC model, like the BHS approach that preceded it, was criticized as being too expensive for the majority of African countries. An alternate approach, *selective primary health care* (SPHC), was promoted by the major international agencies as being more cost-effective. The SPHC approach involves the prioritization of programs to address a few targeted diseases that are the most important causes of mortality. There is heavy reliance on low-cost interventions such as child immunization, oral rehydration therapy (the use of a simple solution of salt and water to counteract dehydration in infants with diarrhea), and the provision of clean water to reduce mortality from the targeted diseases.

The SHPC approach has been very popular among the major development agencies, because it focuses attention on manageable parts of the larger health issue; is target-oriented (i.e., programs can be evaluated in relation to a numerical goal, such as an $x\%$ increase in immunization); and puts its faith in a set of technological "magic bullets." However, SPHC has been widely criticized for (among other things) its failure to address the root causes of ill health in Africa, as well as a top-down, technological approach that does not empower communities to tackle their own health problems.

Selective is the operative word in evaluating SPHC, which has involved the *selective* application of certain technologies to a few health problems and has brought about *selective* progress—reductions in infant mortality, for example. However, its piecemeal approach pro-vides no basis for a long-term, comprehensive strategy to improve the health of Africans. Moreover, where inadequate provision has been made for maintaining SPHC gains, the benefits have been quite temporary. For example, Tanzania's program of developing rural water supplies was undermined by the lack of funds for maintenance, with the result that an estimated half of all wells became inoperative.

Governments have been urged to introduce fees to recover some of the costs of providing health care. These fees have been introduced in virtually all countries where treatment was formerly free. Patients are expected to purchase their own drugs and supplies, and the families of inpatients are often expected to provide meals and routine care. Although lip service is paid to alleviating the effects of user fees on the poor, the reality is that fees are, for many, formidable barriers to obtaining care (see Vignette 21.2). Cost recovery has also been pursued through community revolving drug schemes, in which drugs are provided for sale at near cost, and the funds received are used to purchase replacement supplies. Mozambique was one of the first countries to initiate a basic-drugs purchasing program as a means of reducing the high cost of imported pharmaceuticals.

The "Discovery" of Ethnomedicine

Ethnomedical practitioners exist in every African society and continue to provide the bulk of health care for most Africans (see Figure 21.3). Several distinct healing traditions, including herbalism, midwifery, bone setting, and spiritual healing, are practiced in each African culture. In stable rural communities, where medical knowledge is passed from one generation of healers to the next, the reputations of healing families are established through their accomplishments over many decades. Traditional healers are popular, not only because they are so much more numerous and accessible than modern health care facilities, but also because people trust and value the cures that these healers provide.

Until recently, African ethnomedicine was dismissed as mere superstition, particularly in

---

VIGNETTE 21.2. Health Care User Fees: What Benefits? What Costs?

The effect of prolonged economic crisis and rising debt in Africa south of the Sahara has been an acute shortage of money for health care programs. Governments have been unable to maintain existing services, much less to start new programs. For many Africans, health care means going to a hospital where there are no drugs and quite possibly no doctor. Moreover, the true extent of such cutbacks, hidden by inflation and currency devaluation, is often greater than it seems.

The World Bank and other development agencies have promoted user fees (so-called "cost recovery") as a strategy for generating revenues for the health sector. Users of the health care system are charged a small fee that can be used to purchase drugs and finance new services. It is argued that patients will readily accept such fees, especially if the quality of care is improved as a result. Fees are also justified as a means of discouraging unnecessary visits and thus bringing about more efficiency. The World Bank acknowledges that fees may have a negative effect on the ability of the poor to obtain care; it proposes that very poor families be exempted from fees, and that certain services be offered free of charge.

The imposition of user fees has major implications for the affordability of health care—and, consequently, major effects on utilization. When Swaziland imposed modest user fees, attendance at state health facilities declined by 32%, and the use of mission-run clinics fell by 10%. Visits for childhood vaccination and for the treatment of sexually transmitted diseases fell by 17% and 21%, respectively. Studies in Mali, Côte d'Ivoire, and elsewhere have also shown that user fees cause people to delay treatment or to do without it, and that higher disease and death rates occur as a result. Programs to exempt the poor from paying fees have proven ineffective, in part because confusion about who qualifies for exemption keeps many poor people from seeking assistance.

The health benefits from user fees have been very small indeed. The monies collected generally contribute no more than 5–15% of national health budgets—an amount that can hardly justify either (1) the allocation of scarce resources to collect the fees, or (2) the health, social, and economic costs of excluding so many of the poor from access to basic services. Charges that effectively exclude much of the population impose significant costs not only on individuals but also on the society as a whole when they impede a timely and effective treatment of illness.

Since 2000, the World Bank's position on user fees has become more ambiguous. The bank now acknowledges that user fees have had some negative impacts, and promotes free primary care. However, there is continuing support for the general concept of user payment for services.

The solution to the funding crisis in health lies not in collecting small user fees from patients, but in focusing on the root causes of the crisis—particularly the effect of debt repayment on the ability of African governments to deliver even rudimentary health care services. Debt forgiveness is not the entire answer, but it is an obvious starting point.

Based primary on A. L. Colgan *Hazardous to Health: The World Bank and IMF in Africa.* Washington, DC: Africa Action, 2002. (www.africaaction.org/action/sap0204.htm)

---

the scientific community. Scientific research has now begun to prove the efficacy of many herbal medicines and to better understand the importance of the healer–patient relationship for social-psychological healing. The WHO and several African states, responding to the problem of shortages in the formal health care system as well as to the new scientific interest in ethnomedicine, have shown considerable interest in programs that involve healers in national health care systems. The WHO has established a series of objectives for its traditional medicine strategy for 2002–2005. The WHO initiative is a major advance, although its strategy puts

FIGURE 21.3. Traditional healer and patient, Botswana. Ethnomedicine accounts for a substantial proportion of the health care obtained by Africans. Photo: R. Dixey.

more emphasis on *regulating* traditional medicine than on understanding its utilization in the context of culture. The WHO proposes to do the following:

- Develop national policies leading to a recognition of traditional medicine
- Facilitate dialogue between healers and other providers of health care
- Protect and preserve indigenous medical knowledge and medicinal plants
- Establish regulatory systems for herbal medicines
- Undertake research to ascertain the effectiveness and safety of herbal medicines, and formulate guidelines for their use
- Educate health providers and consumers about the use of traditional medicines

There are various potential avenues for involving healers—ranging from programs to upgrade the skills of village midwives, to initiatives in which spiritual healers assist graduate psychiatrists in the treatment of mentally ill patients. Such programs are still relatively rare and small-scale, but they have opened the door for further integration in the future. They represent an important attempt to look beyond Western technology to indigenous knowledge

and resources in the search for more cost-effective and culturally relevant solutions to Africa's development needs.

New Pressures on the Health Care System: The Challenge of HIV/AIDS

As of 2003, an estimated 30 million Africans were HIV-positive, with about 5 million new infections occurring annually. Some 70% of worldwide HIV/AIDS cases, and 77% of AIDS deaths, occur in Africa. As noted in Chapter 11, rates of infection vary greatly from country to country, and between high-risk and lower-risk populations in both rural and urban areas.

AIDS has usually been viewed as a medical and public health problem, but the disease also has an important social dimension—namely, the "ripple effect" that it may have on the health of individuals, families, and communities living in the shadow of the disease. Caring for a patient with AIDS can deal a crippling blow to an African family. Medicine, when it is available at all, is prohibitively expensive. Even the cost of repeatedly washing and replacing soiled clothing may represent a major cost when family budgets are stretched to the limit. Moreover, income may be adversely affected, particularly if the patient is a wage earner or if a household member has to give up a job to stay home and provide care. When AIDS causes expenses to increase dramatically or income to be reduced, the health of other family members is put at risk. With less money available for food, nutritional status declines, and opportunistic infections like pneumonia and tuberculosis become more prevalent. HIV/AIDS also interacts symbiotically with famine, as noted in Chapter 15: Families coping with AIDS are able to produce less food, and the progression of AIDS is hastened in people whose diet is inadequate.

The most dramatic health effects occur among children who have been orphaned as a result of AIDS (see Chapter 20). When extended families and whole communities have been ravaged, traditional mechanisms for the care of orphans—elder siblings, or aunts and uncles assuming a parental role—may break down. Elderly grandparents, who are no longer able to do strenuous work, may be the only available guardians. In severely affected countries, especially in the cities, many "AIDS orphans" survive without family support.

Governments have responded to HIV/AIDS very differently. Uganda has been quite open in acknowledging its HIV/AIDS problem and welcoming external assistance in research and program development. This openness has not been without a cost, however, since Uganda was subjected to prolonged international scrutiny as a "global AIDS capital." However, the Ugandan government's open and determined program of public education has yielded impressive results; new infections are occurring at a much-reduced rate. Conversely, in other countries where governments have failed to address the HIV/AIDS issue effectively, the epidemic has grown explosively.

The experience of HIV/AIDS-related public health programs in Uganda and elsewhere has demonstrated the value of innovatively using drama, film, and peer testimonies to speak frankly to youth about risks and responsibilities in language that is real to them. Effective health education is not simply a matter of presenting information; rather it entails breaking the silence and stigma that have surrounded the disease, facilitating dialogue, and encouraging affirmative behaviors that reduce risks of infection.

The HIV/AIDS epidemic presents several major challenges to the orderly development of the health care system. Because of its rapid spread, very scarce resources in the health care system may be diverted from other needs to the fight against HIV/AIDS. In countries such as Botswana, patients with AIDS are occupying most hospital beds. Because of the high profile of AIDS, development aid in the health sector is often channeled into AIDS programs. Experienced researchers and specialists focus on AIDS to the detriment of other health needs. Finally, AIDS threatens to rob Africa of scarce medical expertise because doctors and other health care workers are susceptible to HIV infection themselves, not least because of the nature of their work. Although there seems to be an increasing recognition of the need to view AIDS holistically, the policy response in most countries remains much too narrowly focused on treating it as a *biomedical* rather than a *social* disease.

International publicity has often seemed to imply incorrectly that the fight against HIV/AIDS is in the hands of outsiders, with Africans as mere passive recipients. In reality, African health care workers, nongovernmental organizations (NGOs), and communities have remained in the forefront in the struggle against AIDS. With the epidemic having an overwhelming impact on the formal health care systems, the burden of care has fallen largely upon families and community NGOs such as Uganda's The AIDS Support Organization (TASO).

Toward an Integrated Approach to Planning for Health

The WHO (see "Internet Sources") has defined health as "a state of complete physical, mental and social well-being and not merely the absence of disease or infirmity." This is a utopian definition, not just for Africa, but for the world at large, although it helps to direct our attention beyond curative medicine to a much wider view of health. Ideas about how to improve health conditions in Africa have changed radically, especially since the late 1970s. Greater attention is now paid to measures such as clean water and sanitation, improved diet, and immunization (see Figure 21.4), and on targeted expenditure on the most essential drugs.

Despite the WHO's call for "health for all the world's people by the year 2000," the reality has been very different: health for some, and good health for the privileged few. Scarcity of resources is a real problem, but one that may be overcome with adequate commitment to social justice and sustained support at a modest level. Unfortunately, the connections between health and productivity usually receive too little attention. People who are ill cannot work effectively, for example. Expenditure on health should therefore be seen as a cost-effective investment, not as a burden. In fact, it should be unnecessary to resort to economic arguments; effective access to basic health care is a fundamental human right that every state has a duty to make possible.

There continue to be large disparities in the provision of health services. The percentage of the population with access to health services varies from under 30% in Malawi, Niger, and several other countries, to over 75% in 12 countries, including Guinea, Lesotho, and Senegal. Although these data must be interpreted cautiously because of country-to-country variations in the definition of *accessibility*, the same cannot be said for the percentage of children who are immunized (Figure 21.5). The gap between the highest and lowest rates of vaccination coverage is as high as that for access to health care.

The priority given to the provision of safe water says much about a government's commitment to its people. The absence of safe water means that people are at risk for a host of infectious and parasitic diseases. When water is not readily available, sanitation inevitably declines. Yet the construction of a borehole to provide accessible and unpolluted water for a community costs only a few thousand dollars. Similarly, it costs only pennies to immunize a child against several life-threatening diseases.

Medical professionals, interacting with elite interests, have often impeded real progress toward the achievement of health for all. They have argued for the preservation of expensive curative medicine based in hospitals, and against a redistribution of resources into programs that are designed to give the most benefit to the most people. The results have been predictable: underfunded and ineffective token programs, and policies that do not address the root causes of ill health. The few states—Mozambique and Zimbabwe, for example—that have attempted to redistribute resources out of hospitals and into the community have met with strong opposition from medical interests.

Life expectancy and the quality of life are only partly matters of chance. They are not primarily reflections of the hostility of the natural

a   b

FIGURE 21.4. Health interventions in Mozambique. (a) A health care worker prepares to vaccinate young children at a health clinic. (b) A village well provides an accessible, safe, and dependable supply of water— one of the cheapest and most effective defenses against a host of diseases. Photos: CIDA (B. Paton).

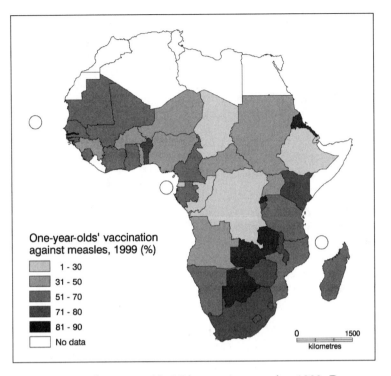

FIGURE 21.5. Immunization of one-year-old children against measles, 1999. Data source: World Bank. *World Development Indicators 2002.* Washington, DC: World Bank, 2002.

environment, as many once believed. Life expectancy and the quality of life are significantly affected by the choices made by governments—their priorities for development in general and for health policy in particular. The question "Who gets what where—and why?" goes to the heart of how to assess states' policies toward and on behalf of their people.

Health involves much more than the provision of curative medical care. Good health policy cannot be formulated apart from policies to improve nutrition, sanitation, housing, education, and general well-being. Good health policy is politically grounded in a firm commitment to socially just development and to the empowerment of communities to work for their own improvement. Such a commitment is an essential ingredient, but good intentions are hard to put into practice when sufficient resources are unavailable. Thus the future development of health depends as much on the general improvement of Africa's economic health as on specific policies in the health sector.

Further Reading

On the relation between underdevelopment and health, see the following sources:

Stock, R. "Health care for some: A Nigerian study of who gets what where and why." *International Journal of Health Services,* vol. 15 (1985), pp. 469–484.

Turshen, M. *The Political Economy of Health in Tanzania.* New Brunswick, NJ: Rutgers University Press, 1985.

There is a large literature on primary health care in theory and in practice. See, for example, the following sources:

Gish, O. "The political economy of primary health care and 'health by the people': An historical exploration." *Social Science and Medicine,* vol. 13C (1979), pp. 203–211.

Green, R. H. "Politics, power and poverty: Health for all in 2000 in the Third World?" *Social Science and Medicine,* vol. 32 (1991), pp. 745–755.

Lafond, A. *Sustaining Primary Health Care.* London: Earthscan, 1995.

United Nations Children's Fund (UNICEF). *The Bamako Initiative: Rebuilding Health Systems.* New York: UNICEF, 1995.

World Bank. *Better Health in Africa: Experience and Lessons Learned.* Washington, DC: World Bank, 1994.

The effects of user fees, and of structural adjustment more generally, on access to health care are discussed in the following sources:

Anyinam, C. "The social cost of the IMF's adjustment programs for poverty: The case of health care in Ghana." *International Journal of Health Services,* vol. 19 (1989), pp. 531–547.

Epprecht. M. "Investing in amnesia, or fantasy and forgetfulness in the World Bank's approach to health care reform in sub-Saharan Africa." *Journal of Developing Areas*, vol. 31 (1997), pp. 337–356.

Turshen, M. *Privatizing Health Services in Africa.* New Brunswick, NJ: Rutgers University Press, 1999.

Yoder, R. A. "Are people willing and able to pay for health services?" *Social Science and Medicine,* vol. 29 (1989), pp. 35–42.

There are numerous studies of African traditional medicine. The following are especially recommended. The Katz et al. and Somé volumes focus on spiritual aspects of African medicine.

Anyinam, C. "Availability, accessibility, acceptability and adaptability: Four attributes of African ethnomedicine." *Social Science and Medicine,* vol. 25 (1987), pp. 803–812.

Good, C. M. *Ethnomedical Systems in Africa.* New York: Guilford Press, 1987.

Janzen, J. M. *The Quest for Therapy: Medical Pluralism in Lower Zaire.* Berkeley: University of California Press, 1978.

Katz, R., M. Biesele, and V. St. Denis. *Healing Makes Our Hearts Happy: Spirituality and Cultural Transformation among the Kalahari Jul'hoansi.* Rochester, VT: Inner Traditions, 1997.

Somé, M. P. *The Healing Wisdom of Africa.* New York: Penguin-Putnam, 1998. "Traditional medicine: Growing needs and potential." Geneva, Switzerland: WHO, 2002. (Available online at www.who.int/medicines/organization/trm/wpe10.jpg)

The impacts of HIV/AIDS on health and the health care system are discussed in many studies, including the following:

Cornia, G. A., M. Patel, and F. Zagonari. "The impact of HIV/AIDS on the health system and child health." In G. A. Cornia, ed. *AIDS, Public Policy and Child Well-Being.* Florence, Italy: United Nations Children's Fund–Internation Child Development Centre (UNICEF–ICDC), 2002.

Hope, K. R. *AIDS and Development in Africa: A Social Science Perspective.* Binghamton, NY: Haworth Press, 1999.

Internet Sources

The following are websites for major governmental organizations dealing with international health:

U.S. Department of Health and Human Services. *Centers for Disease Control and Prevention.* www.cdc.gov

Joint United Nations Programme on HIV/AIDS (UNAIDS). www.unaids.org

United Nations Children's Fund (UNICEF). www.unicef.org (see UNICEF's annual reports, *The Progress of Nations*, and *The State of the World's Children*)

World Health Organization (WHO). www.who.int

Policy and Strategy on Traditional Medicine (WHO). www.who.int/medicines/organization/trm/orgtrmmain.shtml

The following are websites for a selection of international health NGOs:

African Medical and Research Foundation (AMREF). www.amref.org

International Committee of the Red Cross. www.icrc.org/eng

Malaria Foundation International. www.malaria.org

Doctors without Borders/Médecins sans Frontières (AMREF). www.msf.org

Panos Institute. *Panos: Making Sense of Global Change.* www.panos.org.uk

Resources

African economies remain firmly based on primary production. Most African workers are engaged in farming, mining, forestry, and fishing. The primary sector is the direct source of sustenance for the great majority of export earnings. Moreover, other elements of the economy—purchasers of primary products, and suppliers of goods and services to primary producers—are closely linked to the primary sector. Thus the spatial distribution of natural resources, the quantity and quality of these resources, and the people and groups controlling their development are important determinants of economic strengths and of political and social stability.

Chapter 22 focuses on the development of Africa's rich but unequally distributed mineral and mineral fuel resources. Although Africa is an important world producer of several minerals, African governments remain hostages to the realities of foreign ownership and the uncertainties of the international market for minerals. Mineral exploitation has brought relative prosperity to several countries, but it has been a dependent prosperity with an uncertain future.

Chapter 23 looks at the rivers, lakes, wetlands, and groundwater that constitute Africa's aquatic resource base. Water is vital for irrigated agriculture, and is exploited as a source of protein (inland fisheries) and hydroelectricity. Though water is plentiful in some parts of Africa, it is extremely scarce elsewhere. Where water is scarce, there are particular dangers of international struggles over access to water, and of damage to the resource base as a result of ill-advised development projects.

Chapter 24 examines the diverse uses of flora and fauna in the development of local, regional, and national economies. Vegetation is important as a source not only of export earnings (e.g., in the logging *and wood-processing industries), but also of food, traditional medicines, building materials, and fuel. The role of wildlife in providing a basis for tourism is well known, but fauna are also important as a source of food and other economically useful products. This resource base is increasingly threatened, both directly and indirectly, by environmental degradation and unsustainable rates of harvest. Community-based management offers an alternative to the usual bureaucratic decision-making process, and as such may help to preserve the resource base.*

22

Mineral Resources

Gold and diamonds, more than any other substances, symbolize opulence and solid, inflation-proof wealth. Yet, though about one-quarter of the world's gold and over half of the world's diamonds are mined in Africa south of the Sahara, the continent is known for its poverty rather than its prosperity. Gold and diamonds glitter for few Africans.

This chapter examines why Africa's mineral and energy resources have so seldom stimulated real economic development. It points to stark contrasts in the organization and significance of mining in precolonial times and after the colonial conquest. The mining industry in Africa, as elsewhere, is controlled by companies that operate globally and make decisions reflecting corporate self-interest rather than Africa's welfare.

Mining in Precolonial Africa

Inherent in the myth of "darkest Africa" was the denial of indigenous development; European capital and expertise were said to have been essential for the development of African resources. Yet Africans have exploited the minerals of the earth's crust for thousands of years. Iron, gold, copper, and tin were mined in sig-

nificant quantities in the precolonial era and were used for making a variety of utilitarian and ceremonial objects. These minerals were very important components of interregional and even international trade. Gold from the headwaters of the Niger River and from present-day Ghana was reaching North Africa and Europe as early as the 8th century A.D., and it brought extraordinary wealth and fame to the empires of Ghana and Mali. Similarly, gold from Zimbabwe was traded as far afield as the Persian Gulf, India, and China.

The technology of iron making reached Nok in central Nigeria between the 7th and 4th century B.C., apparently having spread from Carthage via Saharan Berber nomads. In Nubia (in present-day Sudan), iron making had begun in the 5th century B.C., and was crucial to the kingdom's prosperity. Archeological evidence points to the independent invention of iron making in the lake district of East Africa prior to the 3rd century B.C. and perhaps as early as the 7th century B.C. Knowledge of iron making spread gradually from these initial sites to many other parts of Africa.

Copper, primarily from the copper belt of present-day Zambia and the Democratic Republic of the Congo, and tin, primarily from central Nigeria and the eastern Democratic Republic of the Congo, were mined and used in bronze casting to make ceremonial and utilitarian objects. (Figure 22.1 illustrates traditional tin mining.) The intricate bronze art of Ife and Benin demonstrates the vitality of these precolonial African metalworking industries.

Salt was one of the most important commodities of precolonial African trade; thousands of caravans moved annually from Saharan producing centers such as Bilma into the savanna and forest regions, where salt sources were unknown. Salt was so valuable that it was used as a medium of exchange. The Saharan mines produced many types of mineral salts used as medicine, food additives, and raw materials for tanning and dyeing. The salt industry continues to function much as it did in precolonial times, though it has declined in importance. In many coastal areas, such as the lagoons of Ghana, salt has long been extracted by evaporation from seawater for local use and as a valued trading commodity.

Development of the Modern Mining Industry

Although Europeans had long been aware that Africa contained rich stores of minerals, these resources remained firmly in African control prior to the colonial conquest. The rulers and merchants of the ancient empire of Ghana had managed to prevent Arab traders and visitors from learning the source of their vast supplies of gold. Likewise, early European explorers found Africans extremely reluctant to show them locations from which strategic and valuable minerals were obtained.

Following the discovery of diamonds and gold in South Africa, interest in the mineral potential of Africa rose to a fever pitch. Thousands of fortune seekers lured by stories of huge diamonds rushed to the Kimberley area in 1870 and 1871. Small mining operators soon gave way to capitalist mining companies, which in turn fought each other for an increased share of mineral property and market share. The ultimate victor was Cecil Rhodes's company, De Beers Consolidated Mines. By 1890, De Beers had sufficient control of the world diamond market to be able to set prices by restricting supply.

Gold was found in the Witwatersrand in Transvaal in the mid-1880s. The deposits were huge, but special technology was needed to exploit them because of their depth and the difficulty of extracting gold from the local bedrock. The investments needed to develop these gold deposits came from large companies, including those involved in diamond mining. The hope of finding even greater mineral wealth spurred

FIGURE 22.1. Traditional mining of alluvial tin deposits, Jos Plateau, Nigeria. Photo: author.

Rhodes to expand northward. By means of deception and military force, territory in present-day Zimbabwe and Zambia was seized for the British South African Company and was named Rhodesia.

A similar pattern of occupation and mineral exploitation took place in Katanga, the region of what was then the Belgian Congo adjacent to the Rhodesian copper belt. During the 1920s copper boom, the Belgian consortium Union Minière emerged as the third largest copper producer in the world. Other companies were established to exploit Congolese diamond and tin deposits. Where mineral deposits were found elsewhere in Africa, the pattern of exploitation was similar, although usually on a smaller scale. Concession mining companies invested in diamond mines in Angola, South-West Africa, and Sierra Leone; tin mines in Nigeria; and gold and manganese mines in the Gold Coast. Investment and production fluctuated in response to world demand, collapsing during the depression of the 1930s, but growing rapidly during and after World War II.

In territories where mining development occurred, there were diverse economic, social, and political effects. Railroads were constructed from mineral deposits to ports or previously existing lines, and new towns were established to house the mine workforce. The new urban, wage-based economy of the mining communities was suddenly inserted into the midst of agricultural subsistence economies. Although coercion was often needed in the initial stages to recruit labor for the mines, the attraction of wage employment and the stagnation of traditional economies made labor recruitment increasingly simple. Except in South Africa, permanent mine workers gradually replaced seasonal workers who returned home to farm for several months each year.

During the 1950s, there was a new wave of investment in mining. Increased prospecting resulted in the discovery of new ore bodies, both in the established mining regions and in other parts of the continent. Many new mines were opened in South Africa, Southern and Northern Rhodesia, and the Belgian Congo. Major iron ore mines were established in Liberia, Mauritania, and Angola; the world's largest deposits of bauxite were tapped in Guinea; and

Gabon became an important producer of uranium and manganese. Perhaps the most important development was the discovery of petroleum in the Niger Delta of Nigeria, and along the western coast in Angola, Congo, Gabon, and Cameroon. The subsequent improvement of technology for offshore drilling facilitated exploitation of the region's offshore petroleum and natural gas reserves.

Relatively few major mining developments have been initiated since independence. The exceptions have included new uranium mines in Niger and Namibia, and new copper–nickel and diamond mines established in Botswana (see Vignette 22.1, with Figure 22.2 and 22.3). Since 1995, new petroleum fields have been developed offshore in Equatorial Guinea and in southern Sudan. Although the proportional contribution of mining to the African economy doubled from 5 to 10% during the 1960s, the 1970s brought mixed fortunes. Petroleum producers profited greatly from the successful manipulation of marketed supply and prices by the Organization of Petroleum Exporting Countries (OPEC), but producers of most other minerals were badly hurt by global recession and reduced demand for industrial raw materials. Production declined, and marginal mines were closed. During the 1980s, with the weakening of OPEC, petroleum producers suffered a similar slackening of demand and price collapse that lasted until the late 1990s.

Especially since the early 1980s, there has been a major shift from the dominance of large mining corporations to the growing role of smaller, venture capital companies involved in exploration and development. Some of these companies have shown more willingness to work in riskier investment situations, such as in politically insecure countries (e.g., Sierra Leone and Democratic Republic of the Congo). Some of these ventures have been very profitable, but some have also generated considerable negative publicity for the companies involved.

Who Produces What Where?

Figure 22.4 illustrates which minerals were mined in which sub-Saharan African countries as of 2001.

VIGNETTE 22.1. Botswana's Mineral-Based Economy: Success or Illusion of Success?

When Botswana first became independent in 1966, it was considered one of the poorest countries in Africa. Its population of 500,000 was scattered over an area of 660 million km². Exports in 1968 amounted to only $10 million. The domestic economy was based on cattle herding and remittances from migrant workers in the mines of South Africa.

Shortly after independence, several important mineral discoveries were made (Figure 22.2). Large, productive diamond deposits were found at Jwaneng and Orapa. Botswana now ranks second in the world in the volume of diamonds produced, and it leads in the production of gem-quality diamonds. A major copper–nickel deposit was discovered in 1966 at Selebi-Pikwe near the Zimbabwean border (Figure 22.3). The mining of this deposit commenced in 1973. Coal production in 1999 amounted to almost 800,000 metric tons. Smaller mining operations have opened more recently to exploit gemstone deposits near Gaborone, and soda ash at Sua Pan.

The development of the mining industry has brought such startling growth to Botswana that it has been characterized as one of Africa's greatest economic success stories. The 2000 gross national income per capita was $3,300, twice the 1989 figure and the second highest in Africa south of the Sahara. The value of exports grew from $22 million in 1970 to $1.4 billion in 1988 and $2.6 billion in 1998. Minerals account for about 80% of export earnings, most from the sale of diamonds. With rapidly growing export earnings, Botswana's balance of payments situation remained favorable until the mid-1990s, but the value of imports now exceeds export earnings.

Samatar states that Botswana has been a true success story—not only because of economic growth, but also because mineral-generated wealth was used wisely to create broad-based development. Strong, effective leadership from the state was instrumental in achieving economic progress. Since independence, Botswana has remained a free society with a democratically elected government, despite the coercive presence until 1994 of the South African apartheid regime on its borders. Botswana's civil service is competent and operates without the undue political interference that characterizes the public service of many African countries. Even though Botswana's economy is based on the exploitation of nonrenewable resources, strong leadership, a clear sense of vision, and prudent investments – social as well as economic – have set the stage for a sustainable future.

Curry argues that rapid economic growth has not brought prosperity to all of Botswana. A small elite class of officials, managers, and businesspeople has become extremely wealthy. About one-third of the population is supported in reasonably well-paying jobs in the mines, the public service, and the military, but the other two-thirds live in dire poverty. The gap between the rich and poor

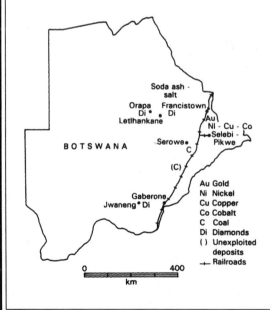

FIGURE 22.2. Mining in Botswana. After U.S. Bureau of Mines. *Minerals in Africa: 1989 International Review*. Washington, DC: U.S. Department of the Interior, 1989, p. 26.

VIGNETTE 22.1. *(cont.)*

FIGURE 22.3. Open-pit copper mine, Selebi-Pikwe, Botswana. Photo: CIDA (C. McNeill).

has increased markedly. Botswana's elites have invested heavily in cattle ranching. As the rich were accumulating an increasing proportion of the land and cattle, 80% of the rural population was being relegated to less productive lands. Many poor rural families have been forced to seek urban employment to make ends meet.

Despite the growth of cattle ranching, Botswana's food production declined by 7% during the 1990s. Even in years when harvests are excellent, Botswana falls far short of self-sufficiency. Although the decline in food production is in large part attributable to drought, the effect of increasing societal inequalities of access to scarce economic resources in rural areas has been even more significant. Food imports have increased markedly, reaching 446 metric tons in 2000. Food imports are but one aspect of Botswana's growing dependence on South Africa. The role of South African capital in the mining sector has continued to grow. Botswana also depends increasingly on energy imports from South Africa to fuel its commercial economy. The demise of apartheid has permitted Botswana to develop closer, more cooperative relations with its powerful neighbor to the south.

Based on R. L. Curry. "Poverty and mass unemployment in mineral-rich Botswana." *American Journal of Economics and Sociology*, vol. 46 (1987), pp. 71–87. A. I. Samatar. *An African Miracle: State and Class Leadership and Colonial Legacy in Botswana Development*. Portsmouth, NH: Heinemann, 1999.

Metals

The metals most commonly associated with Africa are gold and copper. South Africa, principally the Transvaal region (known as the Rand), has been the world's largest source of gold for a century. Most of the gold occurs at great depths in conglomerate formations known as *reefs*, which are often quite expensive and technically difficult to mine. Ghana is Africa's second-ranked gold producer. Smaller amounts of gold are produced in several countries, including Zimbabwe and the Democratic Republic of the Congo. The quest for new sources of African gold has brought many exploration companies to Africa in recent years. Mali has been the focus of

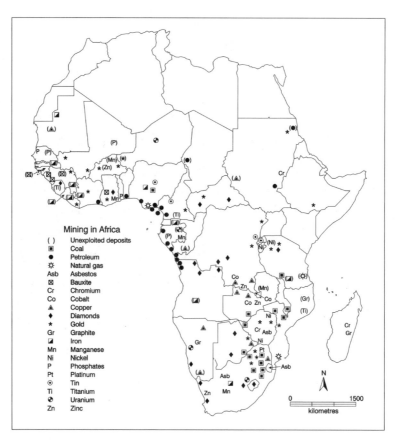

FIGURE 22.4. Mineral exploitation. Many of the mining operations shown on the map have operated on a small scale, often intermittently. Primary data source: U.S. Bureau of Mines. *Minerals in Africa: 1990 International Review*. Washington, DC: U.S. Department of the Interior, 1992. Updated to 2001.

considerable attention, as companies seek to pinpoint the source of the gold that brought fame and fortune to ancient Mali.

The copper belt of Zambia and adjoining areas of the Democratic Republic of the Congo is, after the South African Rand, Africa's most famous metal-producing area. Other metals (especially cobalt and platinum) are mined in the same region, often as by-products of copper production. Copper and associated metals are also mined in Botswana, South Africa, and Zimbabwe. South Africa is the world's leading source for manganese and chromium, both important ferro-alloy minerals. Gabon and Ghana are other major producers of manganese.

Africa has many deposits of iron ore, but few are of a sufficiently high grade to be viable sources of supply for steel mills thousands of kilometers away in Europe and Asia. Mauritania remains an exporter of iron, but formerly

mined deposits in Angola, Liberia, Guinea, and Sierra Leone are now inactive because of high costs and the effects of civil wars. South Africa is Africa's largest iron ore producer; ore from mines in the Northern Cape Province supplies the domestic steel industry and is also exported.

Bauxite is widely dispersed in Africa, but only in Guinea is it mined on a large-scale basis. Ghana and Mozambique are minor sources of bauxite, which is shipped as a concentrate to aluminum smelters in Ghana and Cameroon, as well as abroad.

Industrial Minerals

Diamonds are far and away the most important of the industrial minerals produced in Africa south of the Sahara. Included are both high-value gem diamonds and smaller industrial diamonds used in the manufacture of cutting tools and abrasives. Although the most productive diamond fields are in southern and south central Africa (principally in South Africa, Botswana, Namibia, Angola, and the Democratic Republic of the Congo), diamonds are also found in East Africa (principally in Tanzania), west central Africa (primarily in the Central African Republic), and West Africa (where Ghana and Sierra Leone are the largest of several producers). Other precious and semiprecious gemstones come from Madagascar, Botswana, and several other countries.

Phosphates, destined primarily for use in fertilizer production, are mined in substantial quantities in South Africa, Togo, and Senegal; the majority of production is exported. The other major industrial mineral entering the world market from Africa is asbestos, mined in Zimbabwe, Swaziland, and South Africa. Many countries produce cement from limestone, as well as sand and crushed stone for the local construction industry. Clays are widely used by local artisans to make pottery and bricks, and are used in some countries to manufacture bricks and ceramic products on a commercial basis.

Mineral Fuels

Commercially viable deposits of hydrocarbons—petroleum, natural gas, and coal—are found in only a quarter of the countries of Africa south of the Sahara. Except for small deposits in southeastern Nigeria, Africa's exploited coal deposits are all in the south central and southeastern parts of the continent. South Africa is the only major coal producer, with 6% of world production. Large deposits occur in thick seams in the Natal and Transvaal regions. Zimbabwe and Mozambique have substantial deposits of good-quality coal that are mined commercially, while Zambia and Tanzania have minor coal-mining industries based on small, low-quality deposits.

Petroleum production is concentrated in Nigeria (with about two-thirds of production) and other countries of west central Africa, principally Angola, Congo, Gabon, Cameroon, and Equatorial Guinea. Equatorial Guinea commenced production from offshore wells in 1996, and within four years had become Af-

rica's fourth-ranked source of petroleum. Oilfields in south central Sudan have been brought into production in recent years. Exploration and initial drilling for petroleum have taken place in several areas along the West African coast from Nigeria to Senegal, in the Lake Chad basin, and along the East African coast in Tanzania and Mozambique. Petroleum benefits the producing countries primarily as a source of export income, rather than as a major energy resource for indigenous industrial development. Natural gas occurs with the petroleum, but most of it is flared because of high development costs and competition from cheap Russian, North American, and North Sea gas in European markets.

Africa's Minerals and the World Economy

Africa contributes substantially to the global supply of minerals, with a total production valued at over $50 billion per year. Except for South Africa, which has a large and diverse industrial sector, most African production is exported. Nigeria and South Africa together account for two-thirds of the total value of mineral production. Angola is the third largest exporter by value, followed by Botswana, Gabon, and Equatorial Guinea. A further five countries had sales between $0.5 billion and $1.0 billion in 1999 (Table 22.1).

Africa is an important source of several minerals and is the dominant source of supply for a few mineral commodities (Table 22.2). It produces half or more of the world's chromium, diamonds, and platinum, and 20% or more of the world's cobalt, gold, manganese, uranium, and titanium. Bauxite, phosphates, copper, and petroleum are also produced in large quantities. Note that although petroleum accounts for some 60% of the mineral exports of Africa south of the Sahara, the largest African source, Nigeria, ranks only 11th among the world's petroleum producers.

With few exceptions, Africa has lost market share to other parts of the world, especially since 1990. Between 1989 and 1999, its share of world copper production fell from 12% to less than 4%. Whereas Africa had over half of

TABLE 22.1. Major Exporters of Minerals, 1989 and 1999

Country	1989 mineral exports Value (U.S.$, billion)	1989 mineral exports Share of exports (%)	1999 mineral exports Value (U.S.$, billion)	1999 mineral exports Share of exports (%)	Leading product
1. Nigeria	7.5	90	20.4	99	Petroleum
2. South Africa	11.5	65	12.6	28	Gold
3. Angola	2.2	95	5.6	90	Petroleum
4. Botswana	1.5	>90	2.4	85	Diamonds
5. Gabon	1.5	79	1.9	80	Petroleum
6. Equatorial Guinea	—		1.3	95	Petroleum
7. Cameroon	0.9	48	1.0	47	Petroleum
8. Congo	0.8	76	0.9	50	Petroleum
9. Zambia	1.3	92	0.7	80	Copper
10. Namibia	0.8	75	0.7	43	Diamonds
11. Zimbabwe	0.8	45	0.7	27	Gold

Primary data source: U.S. Bureau of Mines. *Mineral Industries of Africa* (1989 and 2001 editions). Washington, DC: U.S. Department of the Interior, 1989, 2001.

the world's gold production in 1981, it now has one-quarter. South Africa's gold mines and Zambia's copper mines are examples of prominent producing areas now in decline because the higher-quality reserves are rapidly diminishing. Since the demise of the Soviet Union, Western mining companies have been very active in Russia and the other successor states; Africa's share of world trade in several minerals has declined as the share of the former socialist states has increased. As well, political instability and violence has played an important role in the demise of commercial mining in several countries, including Liberia, Sierra Leone, and Angola.

Africa's mineral wealth has had geopolitical repercussions on several occasions. In the 1960s, Western mining companies supported the attempted secession of the mineral-rich Katanga Province from what was then Congo—Kinshasa, which had a leftist government at the time. Later, troops from Belgium and France twice intervened to support Mobutu's repressive dictatorship when revolts in what became Zaire threatened to disrupt mineral production and dislodge him from power. Prior to the collapse of the former Soviet Union, Africa was the main alternate source to the Soviet Union for chromium, manganese, gold, platinum, and cobalt. The perceived strategic value of South Africa as a source of these minerals contributed to the reluctance of the Western alliance to impose strong sanctions against South Africa for its apartheid policies. Direct and indirect inter-

TABLE 22.2. Africa in the World Mineral Economy, 1999

Mineral	% from Africa 1981	% from Africa 1989	% from Africa 1999	Leading producers among world's top five (% of world supply, 1999)
Diamonds	72	50	54	Botswana (17%); Dem. Rep. Congo (16%); South Africa (9%)
Cobalt	66	76	39	Dem. Rep. Congo (23%); Zambia (16%)
Gold	54	35	25	South Africa (19%)
Platinum	>40	49	46	South Africa (46%)
Manganese	23	33	31	South Africa (16%); Gabon (11%)
Uranium	33	31	24	Niger (10%)
Bauxite	16	20	13	Guinea (12%)
Chromium	38	38	54	South Africa (48%); Zimbabwe (5%)

Primary data source: U.S. Bureau of Mines. *Minerals Yearbook*. Washington, DC: U.S. Department of the Interior, 2001.

ventions in civil wars in Sierra Leone, Chad, and Nigeria have been related to the strategic importance and value of mineral resources—petroleum in Nigeria, uranium in Chad, and diamonds in Sierra Leone.

Mining and Underdevelopment

Minerals provide substantial export revenues in about half of the countries of Africa south of the Sahara. In 12 countries, more than half of export earnings are from the sale of minerals. Nigeria, Angola, and Equatorial Guinea are countries where at least 90% of exports consist of minerals. However, despite the importance of African minerals in the world economy, African nations and their citizens have not become wealthy as a result of mining development. They remain heavily dependent on exporting primary commodities (see Figure 22.5), and are therefore vulnerable to changes in world mineral markets.

South Africa is an exception, in that mineral industries have clearly been instrumental in its growth as an industrial power. Several factors distinguish South Africa from other countries on the continent. Its mineral resources are diverse and vast, and this fact has historically given it considerable leverage in world markets, especially for diamonds, gold, and chromium. Indigenous capital, rather than colonial or transnational companies, took the lead in the development of the South African mining industry. These mining companies later used their profits to diversify into other industrial enterprises. The profitability of the mining industry under apartheid was also attributable to government regulations that strictly controlled wages and working conditions for African miners.

South Africa has taken advantage of its position as Africa's only country with major coal reserves, as well as the low cost of exploiting these deposits. Coal is used in South Africa's iron and steel, mining, and transportation industries, to generate thermal electricity and manufacture synthetic petroleum. Because South Africa has no petroleum fields, the apartheid state was highly committed to its petroleum-from-coal industry, which made South Africa less susceptible to possible international economic sanctions.

Even during the years when economic sanctions were in place, South African mining capital managed to maintain strong regional and global influence in sectors of the world mineral economy. The best example was the virtual monopoly of De Beers in the production and trade of diamonds. Even Angola and Tanzania, two of

FIGURE 22.5. Loading bauxite for export, Guinea. Most African minerals receive only primary processing before being exported. Photo: CIDA (P. Chiasson).

the most fiercely antiapartheid countries, had to market their diamond production through De Beers on terms set by the company. It was only in 2000, after increasing negative publicity about illicit so-called "blood diamonds" (Vignette 22.2), that De Beers relinquished its monopoly and stated that it would only purchase legitimately mined stones.

Except in South Africa, most mining regions have remained essentially enclave economies

---

VIGNETTE 22.2. Lethal Gems: Diamonds and Conflict in Africa

How are insurgencies financed? Where do rebel groups obtain funds for arms and ammunition, as well as other requirements for their campaigns? During the Cold War, countries on both sides provided covert support to groups seeking to overthrow governments they disliked. South Africa worked tirelessly to destabilize regimes opposed to it, especially in Angola and Mozambique. But with the end of the Cold War and of the apartheid regime in South Africa, insurgent groups were forced to become more self-reliant. In a number of African countries, diamonds became a guerrilla's best friend.

Diamonds were especially attractive because of their high value and small size, making them comparatively easy to smuggle from one country to another. Moreover, it was almost impossible to verify the source of diamonds, uncut or cut, sold on the world market. De Beers had long maintained a global purchasing and marketing monopoly that promoted diamonds generically (i.e., without reference to their origin). There were no safeguards to ensure the legitimacy of gems offered for sale in Antwerp, Belgium, and the other major diamond exchanges worldwide.

In Angola, the National Union for the Total Independence of Angola (UNITA) rebels, who had fought the national government since independence in 1975, gained control of the diamond-mining areas close to what was then the border with Zaire in 1992. Gems from these mines were smuggled out and sold or exchanged for weapons. This illicit trade provided the means for merchants and regional military commanders to amass huge personal fortunes. Diamonds fueled an escalation in what was already an unspeakably brutal civil war.

In 1991, the civil war in Liberia spilled over into Sierra Leone. A group called the Revolutionary United Front (RUF) seized territory in the southeast and vowed to overthrow the government. The RUF waged a campaign of terror against Sierra Leone's citizens that continued for almost a decade. The sale of gem-quality diamonds from the Koidu area not only financed the RUF's insurgency in Sierra Leone, but also fueled the war in Liberia. Almost all of the RUF-mined diamonds were marketed via Liberia. In 1998, 2.6 million carats of diamonds from Liberia were sold in Antwerp, but only 150,000 carats were mined in that country.

Publicity campaigns by organizations such as Human Rights Watch Africa increased international awareness about conflict diamonds. Separate reports commissioned by the United Nations and the World Bank further documented the linkage between the illicit trade in diamonds and the misery of civil wars. The negative publicity began to have a significant effect on retail sales – who could tell which gems were legitimate and which were so-called "blood diamonds" or "conflict diamonds"? Such diamonds only represented only 4% of total world sales, but their image tarnished the entire industry.

The solution proposed by the United Nations and accepted by De Beers in July 2000 was to develop a system to document the source of all diamonds and to exclude all conflict diamonds. De Beers had little choice but to accept this new system, which effectively ended its purchasing monopoly, in order to restore consumer confidence.

Although the certification program did not result in a total elimination of conflict diamonds from world markets, it made the illicit trade considerably more difficult and less profitable. As of 2004, the campaign against conflict diamonds appears to have contributed to the ending of civil wars in Sierra Leone and Angola.

without strong linkages to other sectors. Their products are destined for overseas markets, and machinery and other inputs are virtually all imported. Relatively little value is added between the mine and the port; ore concentrates and (in some cases) refined minerals, not manufactured products, are exported. The most important benefit has been the employment of unskilled and semiskilled labor. However, increasing mechanization is now reducing the job-creating potential of mining development.

In countries highly dependent on mineral exports, the mining industry has tended to distort national priorities; the needs of the mining sector that "pays the bills" take precedence over other needs. Government neglect of agriculture and the lure of mine wages stimulate outmigration and economic decline in rural areas. In some cases, such as the Niger Delta, pollution from the mining development has destabilized rural economies by polluting water sources and farmland. It is a vicious circle; agricultural decline ensures even greater dependence on mineral export revenue to finance development projects and pay for imported food.

The world market for minerals is extremely unstable and is characterized by major fluctuations in demand and price. These fluctuations are illustrated in Table 22.3 with data on copper exports from Zambia in the years 1968 to 2000. Copper production declined from 20 to 2% of the world's total during this 30-year period. The declining size of high-grade reserves, coupled with the high cost of production in Zambian mines (in 1989, $1.22 per pound of refined copper in Zambia, versus an average of $0.63 in all major producing countries), indicates that further declines in Zambia's copper industry are to be expected. The stability of the Zambian mining industry is also challenged by the rise of HIV/AIDS, which is having a devastating impact on its workforce (Vignette 22.3).

The value of copper exports has fluctuated from year to year and has shown no general upward trend. Meanwhile, the price index for the cost of imports increased sixfold between 1968 and 1991. Therefore, apparent increases in the value of imports actually translate into far fewer goods. The growing imbalance in trade was offset in the short run by increasing levels of debt. However, with falling government revenue and the rising cost of debt servicing, Zambia had no option but to implement deep austerity measures prescribed by the International Monetary Fund (IMF). Currency devaluation and the reduction of basic services such as health care have significantly reduced the standard of living for most Zambians.

Zambia is not alone among African mineral producers in experiencing a long-term decline in the standard of living. Trends over time in human development index (HDI) values tell a disturbing story. Zambia ranked 153rd among 172 nations in the 2000 HDI ratings; this rank is 13 places lower than its position in 1990. During the same period, South Africa's HDI ranking dropped by 19 places, Botswana's by 18 places, and Nigeria's by 4 positions. The world's lowest HDI ratings are for Sierra Leone and Niger; both countries have export economies dominated by minerals. Clearly, mineral wealth has not been translated into continuing improvements in the quality of life for the great majority of citizens. Rather, most export earnings pay for mining industry inputs, increased food imports, and corporate benefits in the form of repatriated profits and wealth for local elites.

TABLE 22.3. Trends in the Value and Cost Structure of Zambian Trade

	1968	1978	1988	2000
Share of world copper exports (%)	20	11	5	2
Exports (U.S.$, million)	1,363	1,382	966	789
Price index for mineral exports (1980 = 100)	56	63	118	
Imports (U.S.$, million)	456	628	848	1,008
Price index for all imports (1980 = 100)	19	69	110	
External debt (U.S.$, billion)		2,585	6,498	5,600

Data sources: World Bank. *World Tables* (1989–1990, 1993, and 2001 editions). Washington, DC: World Bank, 1990, 1993, 2001.

VIGNETTE 22.3. Migratory Labor, HIV/AIDS, and the Mining Industry in Southern Africa

The mining industry's biggest challenge at the beginning of the new millennium does not involve the uncertainties of the world market for minerals, or technical matters of getting ore from the ground. Rather, the mining industry in southern Africa is increasingly preoccupied with the impact of HIV/AIDS on its workforce, and thus on the profitability of the sector. With prevalence rates among working-age adults estimated at 20% in South Africa and over 30% in Botswana, Zimbabwe, Lesotho, and Swaziland, AIDS is having a pervasive impact in the region.

The early stages of the spread of HIV/AIDS in southern Africa were closely tied to the migrant labor system. For most of the 20th century, South African mines relied on workers hired on limited-term contracts from the homelands and from neighboring countries. Migrant workers were not allowed to bring their families with them. With prolonged separation from their families and money to spend, it is not surprising that migrant miners commonly purchased the services of commercial sex workers. Periodic home visits provided opportunities for the spread of HIV.

In the early stages of the epidemic, during the late 1980s and early 1990s, South Africans targeted migrant workers as a public health threat. Fears that miners would infect South Africans with HIV led to the banning of Malawian workers, because of Malawi's comparatively high rate of infectivity at the time. However, the social construction of AIDS as a migrant worker's disease offered a convenient pretext to replace foreigners with South African miners at a time of declining employment in the industry. It also served to deflect discussion away from social and economic issues in South Africa itself that made it vulnerable to the HIV/AIDS epidemic.

The economic impact of HIV/AIDS on the mining sector is pervasive and huge. As skilled workers are lost to the disease, it is increasingly difficult to find suitable replacements for them. It is estimated that there will be a fivefold increase in training costs to replace lost workers. HIV/AIDS has become the most important factor in increased absenteeism. Workers' absence is attributable not only to their own illnesses, but also to the need to care for sick relatives or to attend the funerals of those who have died from AIDS. Companies face increased costs for worker benefits, such as for medicines and coffins. A study of the Botswana Diamond Valuing Company estimated the annual cost of AIDS to the company at $237 per employee—a figure especially high in the African context.

The reduced productivity of the mining sector adversely affects government revenues and hence the ability to fund government programs. It is also certain to affect the ability of mining in the region to attract and retain international investment.

If there is a silver lining to this very dark cloud, it may be that the current crisis has made the mining companies acutely aware of the value of healthy, productive workers in an industry with a dark history of treating its workers as a readily disposable commodity. Many companies have become active participants in community programs promoting healthier lifestyles.

Further Reading

For recent data on the production, trade, and use of minerals, see the following sources:

United Nations Conference on Trade and Development (UNCTAD). *Handbook of World Mineral Trade Statistics, 1995–2000.* New York:

UNCTAD, 2002. (Available online at www.unctad.org/en/pub/ps1itcdcomd37.en.htm)

U.S. Bureau of Mines. *Minerals Yearbook. Area Reports: International. Vol. III: Africa and the Middle East.* Washington, DC: U.S. Department of the Interior, published annually. (Also available online at www.mineral.usgs.gov/minerals/pubs/myb.html)

The following sources provide useful overviews of the mining industry in Africa:

Fozzard, P. M. "Mining development in sub-Saharan Africa: Investment and its relationship to the enabling environment." *Natural Resources Forum*, vol. 14 (1990), pp. 97–105.

Labys, W. C. *The Mineral Trade Potential of the Least Developed African Countries*. New York: United Nations Industrial Development Organization (UNIDO), 1985.

Lanning, G. *Africa Undermined: Mining Companies and the Underdevelopment of Africa*. Harmondsworth, UK: Penguin, 1979.

Ogunbadejo, O. *The International Politics of Africa's Strategic Minerals*. London: Frances Pinter, 1985.

There are many case studies of the mining industry in particular settings. Examples include the following:

Crush, J. "Scripting the compound: Power and space in the South African mining industry." *Environment and Planning D: Society and Space*, vol. 12 (1994), pp. 301–324.

Cunningham, S. *The Copper Industry in Zambia: Foreign Mining Companies in a Developing Country*. London: Praeger, 1981.

Freund, B. *Capital and Labour in the Nigerian Tin Mines*. Atlantic Highlands, NJ: Humanities, 1981.

Greenlaugh, P. *West African Diamonds: An Economic History*. Manchester, UK: Manchester University Press, 1985.

Hirsch, J. L. *Sierra Leone: Diamonds and the Struggle for Democracy*. Boulder, CO: Lynne Rienner, 2001.

Kanfer, S. *The Last Empire: DeBeers, Diamonds, and the World*. New York: Farrar Straus Giroux, 1993.

Khan, S. A. *Nigeria: The Political Economy of Oil*. Oxford: Oxford University Press, 1994.

Okunta, I., and O. Douglas. *Where Vultures Feast: Shell, Human Rights, and Oil in the Niger Delta*. San Francisco: Sierra Club, 2001.

Internet Sources

The following sites offer excellent information on geology and mining in southern Africa:

Chamber of Mines of South Africa: News, Data and Policy Information in the South African Mining Industry. www.bullion.org.za

Geological Society of South Africa. www.gssa.org.za

Geological Survey of Botswana. www.gov.bw/government/geology.htm

Geological Survey of Namibia: Earth Sciences for Namibia's sustainable development. www.gsn.gov.na

Zimbabwe Geological Survey. www.geosurvey.co.zw

Most of the mining conglomerates have websites; see, for example, the Anglo-American site:

Anglo-American plc: A Powerful World of Resources. www.angloamerican.co.uk

The issue of conflict diamonds is addressed in many sites. The second site offers the diamond industry's perspective:

Amnesty International. *The True Cost of Diamonds*. http://web.amnesty.org/pages/ec_kimberley_process

Conflict Diamonds: Analyses, Actions, Solutions. www.conflictdiamonds.com

United Nations. *Conflict Diamonds: Sanctions and War*. www.un.org/peace/africa/Diamond.html

United Nations Security Council. *Diamonds in Conflict*. www.globalpolicy.igc.org/security/issues/diamond

23

Water Resources

Water, a resource that is fundamental for any type of development, is very unevenly distributed across the face of Africa. In vast areas within and adjacent to the Sahara and Kalahari, the scarcity of water is the defining constraint that severely limits opportunities for humans. Elsewhere, especially in equatorial regions, water is extremely abundant; on occasion, the abundance of water becomes a development constraint. How then do Africans evaluate and utilize this resource base? And what are the major challenges that confront Africa's water users and development planners?

The Resource Base

This section describes Africa's rivers, lakes, wetlands, and groundwater, which collectively comprise the freshwater resource base of the continent. The natural sources of freshwater that Africans use vary over space and during the year, both *reflecting* and *affecting* regional ecology and patterns of human utilization.

Rivers

Several of the world's largest rivers are located in Africa (Table 23.1). Apart from this common characteristic, the rivers of Africa are extremely diverse. The Nile is the world's longest river, but its discharge rate is low. In contrast, the Congo is second only to the Amazon in its rate of discharge; 38% of the total runoff from the African continent is carried by the Congo.

Variation in the hydrology of Africa's rivers reflects the regional diversity of climates, as well as such factors as vegetation cover, soils, and slope characteristics. The difference in discharge rates of the Nile and Congo illustrates these disparities. Whereas the Nile flows for most of its length through arid regions, most of the Congo basin has equatorial climates with high, year-round rainfall. Many African rivers—for example, the Zambezi, Orange, and Niger—are noteworthy for major seasonal variations in their flow. Such variations are typical of rivers that drain savanna regions, and reflect the seasonal nature of precipitation in tropical wet-and-dry environments.

Flooding is a typical feature of river catchments where precipitation is seasonal and quite heavy. Such flooding may be destructive or beneficial. The huge floods that affected the Zambezi Valley in 2000, destroying property and displacing millions of people, were a major disaster for Mozambique. In the headwaters of river systems, especially where slope, soil, and

TABLE 23.1. Largest Rivers in Africa South of the Sahara

River	Length km	Length Rank[a]	Drainage basin km²	Drainage basin Rank[a]	Discharge rate m²/sec	Discharge rate Rank[a]
Nile	6,670	1	3,349	3	2,830	NA
Congo	4,630	10	3,822	2	39,000	2
Niger	4,100	14	2,092	9	5,700	37
Zambezi	2,650	33	1,331	15	7,070	29
Ubangi	2,460	40	773	33	7,500	28
Orange	2,250	50	855	26	215	NA

[a]Rank among world rivers.

Data source: V. Showers. *World Facts and Figures*, 3rd ed. New York: Wiley, 1989.

vegetation characteristics favor runoff instead of infiltration, heavy rains often lead to flash flooding (see Figure 23.1, Tiga) and the erosion of soil from farmland. On the other hand, savanna rivers with low-lying floodplains are likely to be inundated annually. These floods have traditionally brought many benefits, including silt to fertilize the soil, and water to grow flood-tolerant crops such as rice. The inundation of floodplains also delays and spreads out flood peaks, in the process sparing places downstream from destructive floods (see Figure 23.1, Gashua).

Where there is a great seasonal variation in discharge, the utility of even the largest rivers for navigation and for hydroelectrical genera-

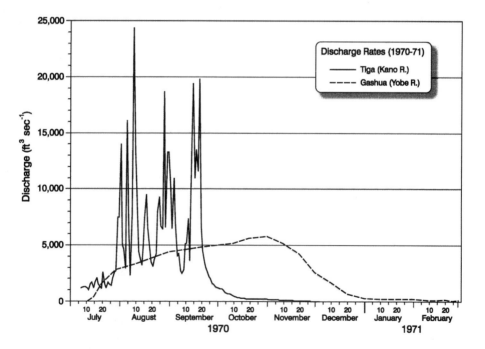

FIGURE 23.1. Flooding regimes in a dry savanna region. The hydrographs show runoff in 1970 along the Yobe River at Gashua and along its tributary, the Kano River, at Tiga. They illustrate two distinct flow regimes. At Tiga, an area with steep slopes and heavy soils, runoff fluctuates wildly and flash floods are commonplace after large rainstorms. At Gashua, 480 km downstream from Tiga, the floods arrive months later. The floodwaters increase and recede gradually, without the extremes found at Tiga. Data source: G. Kerekes and B. Alma'ssy. *Hydrological Yearbook*. Kano, Nigeria: Kano State Ministry of Works and Surveys, 1974.

tion may be limited. Sites with very high generating potential during times of peak flow may produce little or no power late in the dry season. Many rivers in southern Africa have steep gradients and deep gorges that would be ideal for dam construction, but also have insufficient or strictly seasonal runoff.

Natural and Human-Constructed Lakes

Africa's lakes, like its major rivers, include some of the largest in the world (Table 23.2). Lake Tanganyika is the world's second deepest and third largest lake by volume. Like Lake Malawi, another very deep and large-volume lake, Lake Tanganyika owes its depth to its position within the rift valley. Lake Victoria is considerably larger in its surface area—it is ranked third in the world—but is relatively shallow, and its water volume is much less than those of either Lakes Tanganyika or Malawi. Lake Chad is extremely shallow, and it contains little more than 0.1% of the water volume of Lake Tanganyika. Lake Chad is seriously threatened by reduced inflow due to desiccation and irrigation projects; the lake has shrunk significantly since the 1960s.

In addition to these world-class lakes, there are many other lakes of substantial size. Many of these lakes are located within the rift valley system. The construction of several large dams starting in the 1950s has created a new class of large human-created lakes. Lakes Volta, Kariba, and Cabora Bassa are among the largest artificial lakes in the world, but compared to the largest of Africa's natural lakes, they are rather small.

Africa's lakes provide many benefits for humans. These include very substantial catches of fish, as well as water for urban, industrial, and agricultural purposes. Kenya's Lake Magadi is one of several closed-basin, saline lakes in East Africa. The lake contains large deposits of sodium carbonate that are mined commercially.

Wetlands

Although freshwater wetlands occupy just 1% of Africa's area, they are of great hydrological, ecological, and economic significance (see Vignette 23.1 and Figure 23.2). These features are very diverse. Several distinct subtypes are commonly recognized as occurring in Africa: river floodplains, freshwater marshes, swamp forests, and peatlands. Estuaries and coastal lagoons are wetlands as well, but they contain a mixture of freshwater and saltwater.

River floodplains bordering rivers with large seasonal variations in flow are typically flooded each year. Very large floodplain wetlands adjoin the Senegal, Logone, Chari, Tana, and other rivers in savanna regions. For example, the Logone and Chari Rivers that flow north into Lake Chad inundate 63,000 km^2 of floodplain. The floodplains are typically a maze of interconnected channels, oxbow lakes, and shallow depressions, with scattered islands of higher ground that remain above the flood. Freshwater marshes or swamps occur adjacent to rivers and lakes, and typically contain dense growth of papyrus and other aquatic species. The largest of these features is the Sudd, a huge marshland located along the White Nile in central Sudan.

Wetlands tend to trap and slow the passage of inflowing water, and thus to reduce the magnitude of floods downstream. The slowing and

TABLE 23.2. Largest Lakes in Africa South of the Sahara

Natural lakes	Area (km²)	Max. depth (m)	Volume (km³)	Human-constructed lakes	Area (km²)	Max. depth (m)[a]	Volume (km³)
Victoria	66,400	92	2,656	Volta	8,480	70	148
Tanganyika	32,890	1,435	18,940	Kariba	5,250	100	160
Malawi	30,800	706	7,000	Cabora Bassa	5,120	100	66
Chad	18,000	12	27				

[a]Dam height.

Data source: J. Balek, *Hydrology and Water Resources in Tropical Africa*. Amsterdam, 1983. (Cited in W. M. Adams, A. S. Goudie, and A. R. Orme, eds. *The Physical Geography of Africa.*, Oxford: Oxford University Press, 1996, p. 123.)

VIGNETTE 23.1. The Okavango Delta

The Okavango Delta of northern Botswana is truly an environmental jewel. This Massachusetts-sized wetland is fed by the Okavango River, which has its source in the Bié Plateau of central Angola. After following the Angola–Namibia border for some 400 km, the river turns southward into Botswana and enters a flat, arid plain. Here the river divides into a maze of meandering channels, choked with papyrus and other aquatic weeds and with deposited silt. In addition to the channel network, which changes constantly, there are large, shallow basins filled with water and higher "islands" that remain unflooded. Virtually all of the water that flows into the delta is lost due to evaporation, transpiration, and infiltration. Three channels lead southward from the delta, but only about 3% of the water that enters the delta flows out. However, this outflow is extremely important, because it keeps the delta from becoming saline.

The delta goes through an annual cycle of expansion and contraction, reflecting the seasonality of the climate and the flow regime of the river system that feeds it. Rains swell the Okavango River's Angolan tributaries during the wet season, but it takes several months for the water to reach the delta. Peak water levels that occur between March and July. The arrival of 11 billion m^3 of floodwaters causes the delta to expand up to three times its dry season area, depending on the size of the floods in a particular year.

The Okavango supports a rich diversity of wildlife that includes large crocodile, hippopotamus, elephant, and Cape buffalo populations. It is an important dry season refuge for large herds of wildebeest, impala, and other herbivores. There are over 500 species of birds. The Moremi Game Reserve was established in the eastern delta to protect the biodiversity of the region.

Until quite recently, the Okavango experienced relatively little human pressure; the small population of indigenous BaTawana lived by hunting and gathering. That situation has changed dramatically since the 1960s. With cattle ranching expanding northward, the Botswana government decided to construct barrier fences south and west of the delta to protect livestock from diseases carried by wildlife. Unfortunately, these fences disrupted annual migrations of wildlife, resulting in a massive death rate. Tourism has grown exponentially, and the regional town of Maun has expanded rapidly. There are major diamond mines at Orapa, southeast of the delta.

Because water in northern Botswana is such a precious and scarce commodity, the delta is attracting much attention as a potential resource for development. Proposals have been devel-

(cont.)

a

b

FIGURE 23.2. Okavango Delta, Botswana. (a) Areal view of the Okavango River meandering through the Moremi Game Reserve. (b) Reeds in the Okavango Delta. Photos: R. Maconachie.

VIGNETTE 23.1. (*cont.*)

---

oped that would see Okavango water used for irrigation projects, cattle ranching, and diamond mining, and for the needs of the town of Maun. The government of Botswana has considered using dredges to create channels that would allow more of the water to pass through the delta, where it would be used for irrigation and other purposes. Namibia is proposing to construct a dam upstream on the Okavango to supply water to Windhoek.

Environmentalists argue that these schemes to divert "just a bit" of the Okavango's water for development projects would have a disastrous impact on the fragile ecosystem of the delta. Moreover, the precedent created by approving one project based on Okavango water would increase pressure to allow further water withdrawals and modifications to the natural system.

Tourism as a mode of development may help to protect what has been called "Africa's last Eden." Tourism is relatively sustainable; it does not extract resources on nearly the same scale as irrigation, for example. For this comparatively lucrative industry to have a healthy future, however, the ecological integrity of the delta must be preserved.

---

spreading of the water allows increased infiltration, thus helping to replenish groundwater supplies. Much of the "output" of water from freshwater wetlands occurs through evaporation and transpiration. In the case of the Okavango, 97% of inflowing water "disappears" within its inland delta.

Africa's wetlands support diverse and rich flora and fauna. Species that develop in these environments are adapted to living with floodwaters, whether year-round or seasonally. Several wetlands located near the southern margin of the Sahara, including the Senegal Valley and the inland delta of the Niger River, provide critical staging grounds for species of birds that migrate seasonally between Europe and tropical Africa.

In recent decades, African wetlands, especially those in areas with seasonal rainfall, have been coveted as sites for modern irrigation development. As a result of irrigation development, the annual floods have been eliminated or modified, and large areas of wetland vegetation have been cleared. These developments have changed not only the ecology but also the economy of affected wetland areas. Large areas formerly used as dry season pastureland by pastoralists have been appropriated, and the diverse economies of nearby settled communities have been undermined.

Groundwater

Africa's groundwater resources are linked closely to bedrock geology. In general, where the bedrock consists of ancient *basement complex* rocks or volcanic deposits, the capacity to hold groundwater is low, except where the bedrock is overlain with a thick layer of weathered sediments. In contrast, large sedimentary basins, such as the Lake Chad basin, have a high storage capacity. The largest reserves of groundwater are found beneath the Sahara Desert in sedimentary formations. These are actually fossil water deposits, formed thousands of years ago when the regional climate was much moister than today. Libya is developing irrigation schemes that will tap into these sub-Saharan aquifers; it is expected that the water will be depleted quite rapidly.

Paradigm Shift: From Living with and from Nature, to Harnessing Nature

Water is life. This simple truism is the single most important key to understanding where people live and do not live in Africa. For the earliest humans, rivers and lakes provided not only drinking water, but also places to fish and hunt for food, and perhaps corridors for travel and communication with other groups.

As humans developed techniques of agriculture and domesticated animals as a source of food, river valleys became even more important. Their relatively fertile soils and water helped to grow crops, and provided dry season pasturelands for pastoralists. Riverine areas provided a wealth of other valuable products—fish and game for food, and often distinct species of trees for construction materials, medicines, and other goods. These production systems were sustainable, in that they generally did relatively little to alter the environment. Even in densely populated regions, such as along the Nile, riverborne silt provided sufficient nutrients to support dense populations.

Under European colonialism, the relationship between people and freshwater resources began to shift. The first irrigation megaproject, the Gezira scheme, was developed in Sudan during the 1920s. The Gezira scheme was followed by the 50,000-ha *Office du Niger* irrigation project, developed during the 1930s to 1950s in present-day Mali. The Owens Falls Dam, constructed on the Nile near Lake Victoria during the 1950s, created the first large-scale hydroelectric project outside of South Africa.

The period from the late 1950s to the mid-1970s has been called Africa's era of the megadams. Several major dam projects—the Akosombo in Ghana, the Kariba on the Zambia–Zimbabwe border, the Kainji in Nigeria, the Inga in the present-day Democratic Republic of the Congo, and the Cabora Bassa in Mozambique—were completed. Their electrical generating capacity was large, ranging from 700 megawatts (MW) to 2,000 MW. Despite their very high costs, these dams were coveted as keystone development projects that were expected to produce many positive spinoffs, including irrigated agriculture, flood control, fish production, navigation, and reduced dependence on imported hydrocarbon energy.

Each of these dam megaprojects was undertaken in a unique political context. Akosombo, constructed in the early 1960s on the Volta River, was to be the cornerstone of Ghana's drive for modernization and industrialization (Figure 23.3). A large aluminum smelter was established, but otherwise there were few tangible industrial spinoffs. Cabora Bassa on the Zambezi was constructed during the last years of Portuguese colonial rule. The mutual self-interest of South Africa and Portugal lay behind the project—electricity for South Africa, and assured South African support for Portuguese colonialism. Independent Mozambique continued to sell Cabora Bassa power to South Africa, even in the midst of the protracted, bitter con-

FIGURE 23.3. Akosombo Dam, Ghana. This dam, constructed in the 1960s, was expected to become the cornerstone of Ghana's drive for modernization. The development that occurred fell far short of expectations. Photo: CIDA (B. Paton).

flicts between the two countries during the apartheid era, which repeatedly disrupted power transmission.

Anticipated benefits from these massive investments generally failed to materialize, and negative effects such as the loss of rich farmland and increased disease transmission were frequently reported. The repayment of dam-related debt became a major financial drain in countries such as Ghana, and international financing for construction of large dams became extremely scarce. No similarly huge projects have been completed in Africa since the 1970s.

The end of the megadam era did not signal a slowing of interest in water projects. On the contrary, World Bank funding for irrigation schemes strengthened demand for the construction of medium-sized barrages and reservoirs, especially in comparatively wealthy countries with semiarid and arid climates. Of the approximately 980 "large" dams in Africa south of the Sahara, according to the World Commission on Dams (see "Internet Sources"), 539 are in South Africa, 213 in Zimbabwe, and 45 in Nigeria. Fewer than 10 dams have been constructed in 36 of the 46 countries of Africa south of the Sahara.

Human Exploitation of Water Resources

Irrigation

Large-scale irrigation schemes have long been seen as a means of increasing the productivity of semiarid regions. Colonial initiatives such as the Gezira scheme in eastern Sudan were lauded as great achievements that sought to replicate the success of irrigation agriculture in many parts of the world.

The Sahelian drought of 1970–1974 increased interest in irrigation. Irrigation seemed to promise increased production—two annual crops instead of one—and greater security from drought. Various economic arguments were advanced to justify investment in irrigation. Domestic food production, including wheat and rice cultivation to reduce imports, was stressed in some schemes. Elsewhere, the possibilities for export earnings from selling vegetables,

flowers, and sugar have been emphasized. The World Bank helped to finance many of the schemes.

Large-scale irrigation projects brought many negative downstream effects not considered in project planning. Seasonal flooding of rivers traditionally formed the basis for active floodplain economies based on fishing and farming throughout West Africa. Dam construction, which reduced or eliminated these floods, severely damaged the economies of farming communities that relied on floodwaters to grow rice, vegetables, and other crops, and of pastoral communities that depended on floodplain pastures to feed their herds during the dry season. In a northern Nigerian study, the value of production from a wetland under an indigenous production regime was compared to the net value of production with modern irrigation. The study found that net economic benefits from the indigenous economy were more than five times as great per hectare as those obtained with modern irrigation.

Large-scale irrigation has also been associated with ecological effects that range from the salinization of soils in poorly designed, poorly maintained irrigation schemes, to the elimination of flood-adapted species downstream from dams. Reduced water quality, including herbicide and fertilizer pollution, adversely affects communities downstream from irrigation projects. Communities in the vicinity of reservoirs and irrigation projects are at increased risk from waterborne diseases such as schistosomiasis.

Major irrigation schemes, such as the Senegal River and Kano River projects, have not been worth the price—whether this is measured according to monetary, social, or ecological criteria. Nigeria's three largest schemes had a total development cost of approximately $1.8 billion, equivalent to $25,000 per hectare of land developed. Maintenance and operating costs are very high. This heavy investment cannot be justified as drought security; there is seldom enough water for irrigation during droughts.

During the 1980s, the World Bank began to promote the use of small irrigation pumps as an alternative to large-scale irrigation projects. Sometimes water was obtained from rivers, and sometimes groundwater resources were tapped with tube wells. Irrigation from tube wells has been developed in several countries, including Nigeria, Senegal, and Gambia (see Figure 23.4). This technology is much cheaper and more flexible than dams and channel networks for irrigating gardens and other small-scale farms. The key to making these projects sustainable is to ensure that the rate of removal of water does not exceed annual recharge of groundwater during the rainy season.

Hydroelectricity

Africa has vast potential for hydroelectric development. An estimated 40% of the world's potential hydroelectric resources are located in Africa; only 6% of this capacity has been developed. The continent, particularly High Africa (see Chapter 4, Figure 4.1), is bordered by major escarpments, over which or through which rivers must pass to reach the ocean. Where large rivers cut through axes of uplift, or where they cross erosion-resistant rock formations, there are deep gorges and waterfalls or rapids that provide ideal sites for dam construction. Africa south of the Sahara has some 20 substantial hydroelectric facilities with capacities of 100–2,000 MW (see Figure 23.5), as well as numerous smaller installations.

The greatest potential is in the Congo basin.

In the lower reaches between Kinshasa and Matadi, the Congo River has an average flow of 40,000 m^3 per second, exceeded only by that of the Amazon. In this 350-km stretch, the river drops 270 m through a series of 30 rapids and waterfalls. Only 1.3% of the total capacity of 103 million kW has been harnessed. In all, the Democratic Republic of the Congo has about 16% of the world's total potential hydroelectric resources.

Many newer generating facilities are producing thermal electricity from fossil fuels. These operations are most attractive in countries with reserves of fossil fuels, especially natural gas that is produced as a by-product of petroleum extraction. The development of smaller-scale projects for generating energy—from local, small-scale hydroelectric facilities, as well as from the sun and wind—eventually may provide Africa with a more cost-efficient and less environmentally damaging alternative to large dam projects.

In the longer run, if energy costs continue to rise as known fossil fuel reserves begin to decline, the massive potential of the Congo and other rivers may attract international investment to Africa. Such investment would be not only in hydroelectric facilities, but also in energy-intensive industries. The vision of industrialization based on hydroelectric energy that led Kwame Nkrumah to commission the

FIGURE 23.4. Smaller-scale irrigation scheme, West Pokot, Kenya. Photo: R. Maconachie.

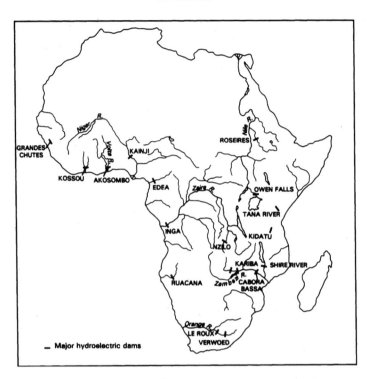

FIGURE 23.5. Major hydroelectric projects. Africa has several very large hydroelectric dams, but only a fraction of Africa's potential has been harnessed.

Akosombo Dam in the 1950s may yet come true, decades later, in a few parts of Africa that are rich in energy potential.

Fisheries

Fish are consumed in large quantities in much of Africa, especially along the coast and near large rivers and lakes with substantial freshwater fisheries. The nutritional value of these fisheries is great because many of these areas have a scarcity of domestic livestock. The importance of inland fisheries is often underestimated; the landlocked countries of Chad and Uganda have larger fish catches than the majority of coastal states.

According to the Food and Agriculture Organization (FAO), the fish catch from African inland waters amounted to 2,930,000 metric tons in 2000. Uganda was the largest producer with 355,831 metric tons, followed by Tanzania (280,000 metric tons), Kenya (210,000 metric tons), and the Democratic Republic of the Congo (205,000 metric tons). The large catches of these countries emphasize the great importance of Africa's inland lakes, especially Lakes Victoria and Tanganyika, as sources of fish. Nigeria, Mali, Cameroon, Chad, Ghana, and Zambia each had catches ranging from 50,000 to 132,000 metric tons in 2000.

The traditional fishing industry is large and very labor-intensive. Various fishing techniques are used, including fish traps, gill nets, and hooks and lines. Fishing is often done cooperatively, with several members of a family or community working together to drive fish into traps or to handle larger nets. Societies where fishing is important often practice resource conservation by observing traditions that regulate fishing seasons and that control access to key fishing locations.

Development agencies have attempted to improve the efficiency of traditional fisheries. For example, larger and stronger nets enable fishers to increase their catches, while improved preservation techniques may greatly reduce spoilage and enhance the quality of marketed fish. Fish farming, using fast-growing species like tilapia, has been promoted as a strategy for increasing protein supplies. Fish farming has often been introduced as a secondary enterprise in conjunction with irrigation projects.

Not all of these interventions have been problem-free. The introduction of the Nile perch and Nile tilapia into Lake Victoria had the desired effect of increasing the total catch from Lake Victoria, but the introduced species have caused ecological havoc, eliminating most endemic fish species from the lake (Vignette 23.2). As well, increased socioeconomic disparities were created, because most of the local fishers could not afford the large fishing gear needed to catch the Nile perch.

Navigation

Africa's major rivers and lakes are a mixed blessing in their impact on transportation.

VIGNETTE 23.2. Lake Victoria: A Case Study of Ecosystem Destabilization

Lake Victoria, the second largest freshwater lake in the world, is a major source of fish that supply protein for the people of Uganda, Tanzania, and Kenya. Like several other major lakes in Africa, Lake Victoria has been renowned for its biological diversity. More recently, it has become a lesson in the potential for massive ecological destabilization through ill-advised human interventions.

The fish fauna of the lake were, until recently, dominated by over 300 species of small fish known as *cichlids* that accounted for some 80% of the fish biomass. Virtually all of the cichlid species were unique to Lake Victoria. Catfishes, lungfish, and two species of tilapia were among the major indigenous species that preyed on the cichlids.

From the 1930s to 1960s, overfishing caused a gradual decline in the fish catch. In an effort to revive the fishery, new species were introduced to the lake in the 1950s. These included the Nile tilapia and the Nile perch. The Nile perch is a huge fish that commonly exceeds 40 kg in size, and is much valued by commercial fishers. However, it is also a voracious predator.

Initially, the new species had little effect. However, by the early 1980s, the indigenous tilapia had been displaced by the introduced tilapia species. There was also a massive increase in the Nile perch catch. However, the growth of the Nile perch population came at the expense of the cichlids that were its main food source. Not only did their population collapse, but also almost two-thirds of all cichlid species became extinct within a decade. Other ecosystem changes ensued, including considerable growth of phytoplankton and freshwater prawn populations because of reduced numbers of cichlids.

In addition to overfishing and predation by introduced species, other changes to the Lake Victoria ecosystem are contributing to the ecological crisis. Increased inflows of nutrients from fertilizers and urban effluent have resulted in a fivefold increase in algae growth and a decline in oxygen levels in the lake. Some of the algae blooms are toxic and have resulted in fish kills. The increased nutrient load has contributed to the two- to threefold increase in phytoplankton in the lake.

The water hyacinth is another recently introduced species causing great havoc. The hyacinth forms huge tangled masses of vegetation that may be virtually impenetrable for smaller boats, thus adversely affecting fishers on the lake. These masses of hyacinth also offer new habitat that favors some species but not others. As well, when large masses of water hyacinth die, the rotting vegetation releases toxic methyl mercury into the water and increases the level of toxins in fish, especially larger species such as the Nile perch.

Lake Victoria's extinct cichlids cannot be recovered; nor can the problem species, especially the Nile perch and water hyacinth, be readily eliminated. Lake Victoria offers a powerful warning to those who would tamper with nature with the objective of "improving" it.

Based on F. Witte et al. "Species extinction and concomitant ecological changes in Lake Victoria." *Netherlands Journal of Zoology*, vol. 42 (1992), pp. 214–232.

More often than not, rivers and lakes pose major barriers to road transportation. Bridges across major rivers are usually far apart, forcing traffic to take long detours or use ferries. Creating artificial lakes such as Lake Volta has further reduced transport efficiency.

In theory, Africa's rivers should offer an alternative to road or rail transportation. In practice, commercial navigation is quite unimportant, with few exceptions. One of the first investments of King Leopold's colonial regime was the establishment of a fleet of riverboats to carry goods and people along the Congo River and its major tributaries (see Chapter 8, Figure 8.7). Upstream from Kinshasa, the Congo system is navigable for thousands of kilometers, permitting relatively cheap access to much of the country throughout the year. In contrast, the seasonal climate and discharge regimes of savanna West Africa mean that the Niger and Benue rivers are navigable for only limited periods each year.

Very few African rivers offer the possibility of navigation from the coast into the interior. Rapids and waterfalls are common near the mouths of the major rivers, especially in southern Africa, where rivers drop precipitously from the high plateau to the coast over the fringing escarpments. This barrier, which helped to restrict Europeans to the coastal plain for centuries, continues to limit possibilities for the development of navigation.

Domestic and Industrial Use

Potable water is a basic necessity for all households. The great majority of people in the industrialized world take the availability of safe water for granted, but the same cannot be said for most Africans. In several countries, including Ethiopia, Madagascar, and Zambia, less than 30% of the population has access to safe water. In poorer rural areas, it is not uncommon for women and children to have to walk several kilometers daily to obtain water of any sort. Even in cities, where piped water systems exist, system breakdowns and user fees act as barriers, often forcing the poor to pay the high prices charged by water vendors or to make use of highly polluted water from streams and dugouts.

Water is a necessary input for most industrial production. The unreliability of water supply, and its high cost in areas where there is a water shortage, is a significant constraint on industrial expansion. In arid countries such as Botswana, the scarcity of water creates difficulty for mining companies, since mining and ore processing require substantial amounts of water. Governments are pressured to consider nonsustainable solutions, such as taking water from the Okavango Delta, to gain near-term benefits from new investment, employment, and profits.

Three Fundamental Challenges
Water Scarcity

As previously noted, Africa is characterized by gross regional and temporal variability in water supply. The problem of water scarcity in arid and semiarid regions has been exacerbated by abnormally low precipitation since the late 1960s. In semiarid regions with large and growing populations, and substantial economic activity, the scarcity of water has become a major concern.

At a broad regional scale, the greatest pressures are being felt in much of East Africa and the Horn, the West African Sahel, and large parts of southern Africa. There are large variations within these regions, with major problems in rapidly growing cities lacking a reliable local water supply. Nouakchott and Windhoek are examples of urban areas that have become increasingly desperate for water, forcing them to consider expensive solutions such as desalinization and long-distance water transfers. On the slopes of Mount Kenya, a serious water crisis has resulted from the recent boom in horticultural production. In South Africa, where water resources are already used very intensively for irrigated agriculture, mining, and other purposes, the increasing scarcity of water is likely to be a major limiting factor for future economic growth.

Africa's international politics of water pale in comparison to those in other regions with water deficits, such as the Middle East. Nevertheless, there have been several disputes over the allocation and development of contested freshwater resources. For example, Somalia and Kenya have been in dispute over the Tana River, while Mauritania and Senegal continue to argue about irrigation developments along the Senegal River. Egypt has been very watchful and determined to prevent projects upstream in Sudan and Ethiopia that could adversely affect the quantity or quality of Nile River water flowing into Egypt.

Water Quality: Pollution

Among the consequences of population growth and economic development has been an increase in water pollution. Pollution by chemical agents, excess nutrients, and alien species threatens to contaminate water supplies, making them less potable and in extreme cases unusable for many purposes. The greatest pollution often occurs in and around major urban areas, where there is the greatest density of population.

With the expansion of pesticide use, agriculture is becoming an increasingly widespread and serious source of chemical pollutants. Mines, especially poorly designed tailings ponds, may allow chemicals that have been used to extract minerals from the ore, or leached from tailings dumps, to enter streams or contaminate groundwater. In the Niger Delta, pollution from oil wells has resulted in widespread contamination. Too many manufacturers take advantage of the poor enforcement of environmental legislation in most African countries to dump their chemical wastes. Perhaps the most troubling sources of chemical contamination are dumps of toxic waste exported from the industrialized world to Africa for (unregulated) disposal.

The growing use of chemical fertilizers in agriculture has resulted in increased nutrient loading of lakes and rivers. The discharge of untreated sewage from large cities into nearby bodies of water is another major source of phosphates and other nutrients. This human-generated overnourishment, known as *cultural eutrophication*, promotes excess plant growth, lowers oxygen levels, and reduces water quality.

Biological pollution occurs with the intro-duction of new species that have a significant impact on water quality. For example, the introduction of the water hyacinth into Lake Victoria has resulted in the release of mercury and other toxins into the lake as the plant decays.

In the Cape region of South Africa, there is concern about how alien species are affecting not only the quality but also the quantity of water. Introduced species such as the black wattle and pine grow faster than indigenous trees, and thus have been valued by forest product companies. However, the introduced species consume more water, and thus reduce stream discharge by an average of about 7%. This reduced runoff has significant consequences for ecosystems downstream and for human water use. In an effort to protect indigenous species and conserve scarce water, the government has initiated a program to remove large stands of nonindigenous trees.

Ecosystem Degradation

Freshwater ecosystems are in a state of decline or crisis in many African rivers, lakes, and wetlands. The implosion of the cichlid populations of Lake Victoria, and with it the loss of hundreds of species found nowhere else, is among the most dramatic examples (see Vignette 23.2). Lakes Tanganyika and Malawi are also under significant ecological pressure due to human activities. In Kenya's Lake Nakuru, a large proportion of the flamingo population has been lost, due to the effects of heavy nutrient pollution.

Development projects have changed forever the ecology of many African river valleys. Some valleys have been drowned by the reservoirs created by dam construction, or cleared of their natural vegetation to establish new irrigation projects. Engineered changes to the runoff regime (e.g., trapping floodwaters behind a dam) or to the river channel have consequences for ecosystems located downstream.

The Okavango Delta and Lake Victoria case studies remind us of the fragility of ecosystems, as well as the potential for poorly conceived human intervention to have disastrous consequences. They point to the need to value and preserve the integrity of Africa's freshwater resource base and the biodiversity it supports.

Further Reading

For a review of the hydrology of African rivers, lakes, and wetlands, see the following sources, especially Adams et al.:

Adams, W. M., A. S. Goudie, and A. R. Orme, eds. *The Physical Geography of Africa.* Oxford: Oxford University Press, 1996. (See especially Chapters 6, 7, and 15.)

Christie, F., and J. Hanlon. *Mozambique and the Great Flood of 2000.* Oxford: James Currey, 2000.

Thompson, J. R. "Africa's floodplains: A hydrological review." In M. C. Acreman and G. E. Hollis, eds. *Water Management and Wetlands in Sub-Saharan Africa.* Geneva, Switzerland: World Conservation Union, 1996, pp. 5–20.

The following sources offer broad discussions of issues related to water management in Africa:

Adams, W. M. *Wasting the Rain: Rivers, People, and Planning in Africa.* London: Earthscan, 1992.

Godana, B. *Africa's Shared Water Resources: Legal and Institutional Aspects of the Nile, Niger, and Senegal River Systems.* Boulder, CO: Lynne Rienner, 1985.

Rachid, E. *Water Management in Africa.* Ottawa: International Development Research Centre, 1996.

Sharma, N. *African Water Resources: Challenges and Opportunities for Sustainable Development.* Washington, DC: World Bank, 1996.

Environmental and social impacts of dams and reservoirs are discussed in the following sources:

Adams, W. M. "Downstream impacts of dam construction: A case study from Nigeria." *Transactions of the Institute of British Geographers,* vol. 10 (1985), pp. 292–302.

McCully, P. *Silenced Rivers: The Ecology and Politics of Large Dams.* London: Zed Books, 1996.

Middlemas, K. *Cabora Bassa: Engineering and Politics in Southern Africa.* London: Weidenfeld and Nicolson, 1987.

Roder, W. *Human Adjustment to Kainji Reservoir in Nigeria: An Assessment of the Economic and Environmental Consequences of a Major Man-Made Lake in Nigeria.* Lanham, MD: University Press of America, 1994.

There is a large literature on irrigation in Africa. See, for example, the following studies:

Barrett, A. *The Gezira Scheme: An Illusion of Development.* London: Frank Cass, 1977.

Kimmage, K., and W. M. Adams. "Small-scale farmer-managed irrigation in northern Nigeria." *Geoforum,* vol. 21 (1990), pp. 435–443.

Van Beusekom, M. *Negotiating Development: African Farmers and Colonial Experts in the Office du Niger, 1920–1960.* Portsmouth, NH: Heinemann, 2002.

If you'd like to learn more about the Okavango Delta, see the following books:

Bailey, A. *Okavango: Africa's Wetland Wilderness.* Cape Town, South Africa: Struik, 1999.

Lanting, F., and C. Eckstrom. *Okavango: Africa's Last Eden.* San Francisco: Chronicle Books, 1993.

Internet Sources

The following websites provide access to a wealth of resources on issues related to water development and conservation:

Global Water Partnership—Southern Africa. www.gwpsatac.org.zw

International Rivers Network: Linking Human Rights and Environmental Protection. www.irn.org

National State of the Environment Report—South Africa. Freshwater Systems and Resources. www.ngo.grida.no/soesa/nsoer/issues/water

Southern African Development Community Water Sector Coordinating Unit SADCWSCU. www.sadc-wetlands.org

Waternet. *Building Capacity for Water Resource Management in Southern Africa.* www.waternetonline.ihe.nl

Water Policy International. *The Water Page: Incorporating the African Water Page.* www.thewaterpage.com

World Commission on Dams. *The WCD Knowledge Base.* www.dams.org/kbase

World Conservation Union (ICUN). *Wetlands and Water Resources Program.* www.iucn.org/themes/wetlands

24

Flora and Fauna
as Economic Resources

When non-Africans think of Africa, chances are that their first image will be of vast herds of animals grazing on the savannas. Yet this popular image is typical of only of certain game reserves and parks, mostly in East Africa. Moreover, the large mammals that command so much attention represent only a tiny portion of Africa's biotic heritage.

For Africans, flora and fauna are critical economic resources. In rural areas, vegetation provides food, fuel, and a variety of products that are utilized on a daily basis. Wildlife is used widely as a food source, in urban as well as in rural areas. These resources are also important for the national economies of several countries, as the basis for the forest industry and much of the continent's tourist industry.

The Economic Importance
of Vegetation

Statistics on production in the forest industry do not distinguish clearly between products produced and sold in the formal economy and those produced and sold within the informal sector. However, only 10% of the total production of industrial roundwood is exported; the other 90% is destined for local use as building materials, furniture, and other products made of wood (see Figure 24.1).

What is clear from Table 24.1 is that forestry is extremely important in both forested and savanna areas. Only 5 of the largest 10 producers of industrial roundwood have substantial areas of tropical forests. Ethiopia is a major wood producer, although it is primarily arid. Thus the future availability of adequate forestry resources will depend on the careful management of all ecosystems, not just the tropical forests.

Processed wood products, such as sawn wood, panels (plywood and similar products), and pulp and paper, account for only a small proportion of the industrial roundwood production in most African countries. Only in South Africa, Ghana, Côte d'Ivoire, and Cameroon is a substantial proportion of harvested wood processed. Indeed, wood panel production fell substantially from 1986–1988 to 1996–1998 in several countries, including Nigeria. In South Africa, the production of pulp and paper from

a   b

FIGURE 24.1. Tropical timber in export and local economies. (a) Hardwood for export, southern Côte d'Ivoire. Timber is Côte d'Ivoire's third largest export, but Ivoirian forests are being depleted rapidly. (b) Local sawmill, eastern Democratic Republic of the Congo. Enterprises processing timber for local use make an important contribution to local and regional economies. Photos: (a) author; (b) A. Harder.

softwood species is the most important sector in the forest industry.

The Timber Industry

Some of the strongest, hardest, and most beautiful woods available to humankind are harvested from tropical forests. Mahogany, widely used to construct fine furniture and paneling, is the best known of the tropical woods. Tropical forests also contain dozens of other species that are exploited commercially for diverse uses. The leading exporters of tropical hardwood, which together account for 75% of supplies, are Indonesia, Malaysia, and the Philippines. Africa produces fewer than 15% of tropical hardwoods entering international trade, most of them in the form of unprocessed logs.

The responsibility of the timber industry for the destruction of forest ecosystems is a matter of some debate. The timber companies stress, correctly, that considerably more damage is done by agricultural clearance than by logging. Ecologists, however, point to the disproportionate damage caused by a timber industry that is only interested in a few select species

TABLE 24.1. Major Producers of Roundwood and Wood Products, 1996–1998

Country	Industrial roundwood			Processed wood panels		
	Rank	(1,000 m^3)	% change since 1986–1988	Rank	(1,000 m^3)	% change since 1986–1988
South Africa	1	18,439	+54.6	1	653	+64
Nigeria	2	8,992	+14.3	8	38	−80
Congo (Dem. Rep.)	3	3,557	+27.9	10	21	−47
Cameroon	4	3,323	+20.2	4	124	+42
Côte d'Ivoire	5	3,013	−6.4	2	306	+42
Gabon	6	2,653	+103.8	9	36	−80
Ethiopia	7	2,468	?	13	13	?
Tanzania	8	2,246	+27.9	18	4	−67
Uganda	9	2,156	+32.8	16	5	+47
Sudan	10	2,082	+22.5	20	2	0

Data source: World Resources Institute. *World Resources 2000–2001*. Amsterdam: Elsevier, 2000.

scattered at extremely low densities throughout a forest. When these desired species are extracted, other vegetation is severely damaged; one study found that 55% of large trees were irreparably damaged by logging activities in which only 10% of large trees were harvested. The construction of logging trails may also result in indirect harm to the forest ecosystem by providing settler-farmers and hunters with easier access to previously unexploited areas.

Not only are the ecological costs of logging very high, but also the economic returns are surprisingly low. The relatively small economic contribution of the timber industry is evident in Table 24.2; only one African country, Cameroon, derives more than 10% of its total export earnings from the sale of timber.

Studies in Ghana by Owusu (see "Further Reading") showed how policies associated with structural adjustment, especially currency devaluation, stimulated much-increased production of timber for export. However, because international prices had fallen, export earnings from the sale of timber stagnated. Structural adjustment policies had contributed to more rapid deforestation—an environmental cost hardly justified by the modest economic returns.

The most desired species of tropical timber are increasingly scarce in accessible areas. They also take decades to grow and are difficult to propagate artificially. Once the commercially valuable timber is gone, the export timber trade will end. The Democratic Republic of the Congo is the only country with very large resources of yet unexploited timber. In fact, it has the third largest area of timber in the world, but it produces fewer than 1% of Third World timber exports. The country's political and economic instability has inhibited the development of the timber trade, and the relative inac-cessibility of the regions where these forests are found has also protected them. However, as Amazonian and Southeast Asian forests are depleted, Congolese forests inevitably will be opened for business.

Vegetation as a Resource in Domestic Economies

Timber for export represents only a small proportion of the economic contribution of the forests. Natural vegetation is a source of countless useful products, ranging from fever remedies and fishing nets to farming tools and fencing. Forest resources are particularly important in indigenous subsistence economies, where the forest takes the place of the building supply store, the specialty food shop, and the pharmacy (see Vignette 24.1, with Figure 24.2 and Table 24.3).

The forest provides timbers and sawn wood for the construction of houses and other buildings. The thatched roofs of traditionally constructed houses are made from tall grass, or in the forest from very large tree leaves. Furniture is constructed out of wood by local carpenters. Various kitchen utensils, garden tools, and decorative objects are made from wood, gourds, seeds, and other objects obtained from vegetation. Ceremonial items such as masks, statues, and walking sticks are most frequently fashioned out of wood.

Fruits, nuts, tubers, and leaves from uncultivated plants are an important source of vitamins, minerals, fiber, and carbohydrates in African diets. Where natural foodstuffs are extensively used, diets are quite varied and are likely to be more nutritionally balanced than where such foods are seldom consumed. The natural environment is also the source of many

TABLE 24.2. Major African Exporters of Tropical Timber, 1996–1998

Country	Annual exports (U.S.$, million)	Timber exports as % of all exports
Cameroon	426.6	17.5
Gabon	287.6	8.7
Côte d'Ivoire	280.1	5.7
Congo	147.4	8.2
Ghana	144.7	8.7

Data source: World Resources Institute. World Resources 2000–2001. Amsterdam: Elsevier, 2000.

VIGNETTE 24.1. Farmland Trees

African farmed landscapes, whether in the tropical forest or the savanna, very commonly include a variety of randomly spaced trees that have been planted or permitted to grow naturally (see Figure 24.2). These trees and shrubs provide a tremendous variety of foodstuffs, medicines, and other useful products (see Table 24.3). Products obtained from farmland trees may also provide substantial income; for instance, the locust bean trees commonly found on savanna farmlands yield fruits that are used as condiments and can provide several hundred dollars of income per year. A number of species are maintained because they contribute to soil fertility and improved crop yields. Research has shown that yields of sorghum and millet are doubled in the presence of white acacia, which fixes nitrogen in the soil, provides humus through leaf fall, and attracts grazing animals that fertilize the land with manure.

Only recently have agricultural economists begun to fully appreciate the importance of farmland trees for rural societies and economies. From their research has come a series of policies designed to encourage the development of agroforestry in Africa. *Agroforestry* is, in essence, a strategy for sustainable development encouraging the systematic integration of trees into farming systems. It is an attempt to understand, promote, and extend the use that African farmers and pastoralists have made of trees for thousands of years. As such, agroforestry builds upon the strengths of indigenous knowledge and methods of utilizing the environment, and provides additional scientific justification for the protection of vegetation.

(cont.)

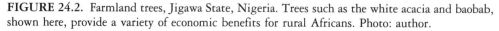

FIGURE 24.2. Farmland trees, Jigawa State, Nigeria. Trees such as the white acacia and baobab, shown here, provide a variety of economic benefits for rural Africans. Photo: author.

"famine foods"—emergency edibles that are seldom consumed except when preferred foodstuffs are unavailable—that have always been crucial for survival in drought-prone areas. Although the consumption of natural foods remains an important part of diets in many areas, there has been an attendant decline in the di-

etary role of foods from uncultivated sources as the use of modern, processed foodstuffs increases and their availability decreases.

The Hausa people of Nigeria and Niger refer to the forest as "God's medicine cabinet," reasoning that cures for any human ailment are to be found there. In all African cultures, exten-

---

VIGNETTE 24.1. *(cont.)*

TABLE 24.3. Uses of Common Farmland Trees of the West African Savanna

Silk cotton	• Kapok (silky fibers from seed pods) is used to make mattresses. • Wood is preferred for carving into utensils and tools. • Branches are "harvested" as firewood.
Baobab	• Young leaves are used as a vegetable. • Fruits are eaten. • Inner bark is used to make rope.
Tamarind	• Fruits are used to flavor food. • Bark and leaves are used to treat stomach ailments. • It is preferred as a shade tree.
White acacia	• It is valued highly because it increases crop yields by fixing nitrogen in the soil. • Bark and roots are medicinal. • Leaves are dry season fodder for livestock.
Shea butter	• Fruits provide an oil used in cooking and soap making, and as a medicine. • Dead wood is harvested as fuel.
Locust bean	• Fruits are used as a condiment. • It is leguminous, and thus it improves the soil.

---

sive use is made of herbal remedies. These traditional medicines continue to be extremely important, not only because of the inadequate development of modern health care, but also because of patients' confidence in them. Modern science, knowing that many herbal medicines have verifiable medical effects, is increasing its research on tropical plants as a source of new drugs. Although traditional healers are the true guardians of herbal knowledge, most rural Africans have some knowledge of how to prepare common herbal remedies.

Fuelwood

Africa is heavily dependent on fuelwood as a source of energy. Wood accounts for only 10% of Asia's and 20% of Latin America's total energy consumption, but constitutes three-quarters of the energy used in Africa south of the Sahara, excluding South Africa. Biomass (wood, charcoal, and crop refuse) accounts for over half of the total energy consumption in every country except South Africa, Zimbabwe,

Botswana, Mauritania, and Gabon. In Tanzania, Burundi, Ethiopia, Liberia, and Mozambique, more than 90% of the total national energy supply is from biomass.

The size of a country's population is closely related to the amount of fuelwood and charcoal that it produces and consumes (see Table 24.4). Nigeria, with a population of over 120 million, accounts for about one-fifth of the fuelwood and charcoal production of Africa south of the Sahara. Ethiopia, Sudan, Tanzania, and Kenya produce much more fuelwood and charcoal than countries such as Cameroon and Côte d'Ivoire, which, although better known for their forest resources, are also considerably less populous.

A major reason for the dominance of biomass energy is that wood was available until quite recently as virtually a free commodity. Even in urban areas, wood and charcoal have remained the cheapest sources of domestic energy. Moreover, the increased cost of petroleum since the 1970s severely limited the growth of fossil fuel consumption. Thus urban households and small industries that might have been expected to

TABLE 24.4. Leading Producers of Fuelwood and Charcoal, 1996–1998

	Production (1,000 m³)	% change since 1986–1988
Africa south of Sahara	445,783	+28.4
Nigeria	87,001	+18.0
Ethiopia	46,522	?
Congo (Dem. Rep.)	44,814	+42.0
Tanzania	35,947	+35.4
Kenya	26,879	+33.6
Ghana	20,678	+64.5
Mozambique	16,724	+17.0
South Africa	14,467	+35.6

Data source: World Resources Institute. *World Resources 2000–2001*. Amsterdam: Elsevier, 2000.

convert to more modern fuels have continued to rely on biomass energy. Wood fuel consumption in Africa south of the Sahara increased by 28.4% between 1986–1988 and 1996–1998—more than twice as much as the growth of commercial energy.

Although most biomass energy is used as a fuel for cooking, it is also an important source of energy for light industrial production. Smaller commercial enterprises—bakeries, brickworks, potteries, and metal forges—commonly use wood or charcoal as their primary source of energy. A study in Tanzania found that biomass accounted for four-fifths of the energy used by manufacturers.

As the hinterlands of major cities have been progressively deforested, particularly in drier savanna regions, wood has to be brought to the cities from farther and farther away. The consequent increases in transportation costs have raised the cost of fuelwood to urban consumers and have encouraged the substitution of charcoal for unprocessed wood. Charcoal is much less bulky, and thus is less costly to transport. Unfortunately, perhaps 60% of the energy contained in wood is lost when it is converted to charcoal via traditional African techniques.

As wood has become increasingly expensive and scarce, several approaches have been used to alleviate the fuelwood crisis. In the short run, the development of more energy-efficient stoves helps to conserve energy. These stoves are up to five times as efficient as open fires for cooking. Nevertheless, the popularity of stoves has been limited, because they are much less effective than open fires in providing warmth and light.

Programs of tree planting for fuelwood use have been widely implemented to increase wood supplies (Vignette 24.2). Tree planting is a useful medium-term solution, especially when it is integrated into peasant production systems. Large-scale tree planting in government or commercial plantations may help to improve fuelwood supply in urban areas. However, consumers tend to prefer specific forest species to those commonly grown in plantations because of their superior burning characteristics. In the longer term, other types of energy, such as hydroelectric, coal, solar, and wind, may increasingly replace fuelwood in higher-income households and commercial enterprises.

Fauna as an Economic Resource

Africa's wildlife is an important natural resource that contributes significantly to national, regional, and local economies. It is a vital source of food and of other valuable products, as well as a key resource for the tourist industry.

Fauna as Food: Meat from Wild Animals

Wild-animal meat is the most important source of dietary animal protein in many parts of Africa, particularly where heavy infestations of tsetse preclude livestock production. In the northern Congo, a study found that 85% of the population consumed bush meat on a daily basis. In areas where considerable wild-animal meat is consumed, it forms an important component of rural–urban, interregional, and even

VIGNETTE 24.2. Women's Tree-Planting Groups in Uganda

For rural women in Uganda, as elsewhere in Africa south of the Sahara, the environment is a vital source of resources to sustain themselves and their families. Because women are responsible for obtaining fuelwood, deforestation and the resulting scarcity of wood are of particular concern. In some parts of Uganda, rural women and children spend up to three hours per day gathering fuelwood.

The government of Uganda has given considerable attention to environmental protection and has stepped up efforts to increase public awareness of these issues. The formation of the Uganda Women Tree Planting Movement (UWTPM) occurred in response to the government's public-education campaign. The objectives of the UWTPM include encouraging women to plant trees as a source of energy and other useful products, and increasing public awareness of the importance of environmental protection.

In Mpigi District near Kampala, women responded to the messages of the UWTPM and the government by coming together to form 34 tree-planting groups. These groups are small in size (from 10 to 35 members each) and have only women as members. The women have done this to ensure that their projects, and the expected benefits from tree planting, remain under their own control. One woman explained it this way: "When you work with men, they make you do all the work, and they will have all the benefits."

The tree-planting groups in Mpigi District experienced a number of difficulties. Some of these difficulties were technical in nature, reflecting a lack of previous experience in growing and transplanting seedlings. A number of the groups were able to address these problems by getting help from the district forestry officer. Another problem for many groups was the severe shortage of land suitable for tree planting in this densely populated region. Finally, since tree planting is a very long-term investment, the members have had to find ways to sustain their projects for several years until they have an opportunity for returns from the sale of fuelwood or other products. Some groups have done so by selling seedlings, but this option has only been viable in communities located near the main roads where potential customers can be found.

Despite these problems, the organization of the women's tree-planting groups has been a very positive development. The international media have helped to create the perception that Africans are the passive victims of environmental crises beyond their control, and that the only potential solutions to these problems are coming from international aid agencies. This case study shows that Africans are very much involved in addressing issues of environmental degradation, mobilizing their own resources to address local needs.

Based on D. H. Kasente. "Performance of Uganda Women Tree Planting Movement." Unpublished research report, Makerere University, Kampala, Uganda, 1991.

international trade. Thus, pressures on wildlife may not necessarily be a reflection of hunting for local food needs. Because of hunting to supply the trade in wild meat, many species of animals, birds, and other kinds of fauna are suffering from hunting stress. Such stress is especially evident in western Africa because of very large human populations and shrinking wildlife habitats.

In west central Africa, the explosive growth of the bush meat trade is having a serious impact on wildlife. What was once a sustainable harvest supporting local populations is now the focus of large-scale trade that threatens the survival of many species, including some that are protected. In the food markets of Pointe Noire, Congo, some 150,000 metric tons of bush meat are sold annually. Newly constructed roads pushed into virgin forest by loggers or mining interests serve as conduits for commercial hunters to find fresh sources of supply and to ship meat to urban markets. Large cities such as Brazzaville are supplied on a regular basis with truckloads and planeloads of bush meat from the interior.

Game cropping and game ranching have been explored as alternatives to traditional hunting. Game cropping involves a regulated harvest of selected species of wildlife in designated areas, whereas game ranching involves the semidomestication of wildlife. Game cropping and ranching schemes have focused on several species of ungulates, such as the eland, wildebeest, and zebra. Although experimental game ranches established in Zimbabwe, Kenya, and Tanzania have had promising results, this strategy has not been widespread. Game cropping has been quite widely practiced in East and south central Africa, but is much less important than unregulated hunting as a source of food.

There has been interest in improving wildlife management to achieve higher productivity, especially since wild animals have several advantages over domestic livestock. Wildlife adapt more readily to varying ecological conditions, especially in marginal areas. They are more tolerant of heat stress, utilize a wider range of fodder resources, and are better able to survive when fodder and water are scarce. Consequently, the ungulate carrying capacity, as measured by the total weight of animal populations that can be sustained in a given savanna environment, is several times greater than the domestic livestock carrying capacity in the same environment. Resistance to certain diseases such as trypanosomiasis enables wild animals to thrive in areas where domestic livestock could not survive.

Fauna as a Source of Valuable Products

Africans have obtained hides and skins, ivory, and other items from fauna since time immemorial. Hides and skins were made into clothing, utensils, weapons, and musical instruments, and various animal parts were used in medicinal preparations. Ivory was also carved to make ceremonial objects and was also, for many centuries, one of Africa's most valued exports. It was an important component of the trans-Saharan trade and the Atlantic coast trade with Europe. From East and south central Africa, ivory was exported to the Middle East, India, and China.

During the colonial era, there was a flourishing trade in animal products, especially ivory and the hides and skins of animals as the leopard, crocodile, and zebra. Wealthy foreign hunters came in search of exotic trophies, and fauna was also trapped for live export to zoos and circuses. This extensive trade in African animals and animal parts posed a threat to the survival of some species, and it certainly resulted in reduced ranges for animals in areas where wildlife numbers were low. For example, the greatly reduced ranges of the cheetah and leopard reflect the intensity of hunting pressures on these valued species.

The trade in Africa's endangered wildlife and wildlife products continues on a massive scale, despite the prohibition of such trade by the Convention on International Trade in Endangered Species of Wild Fauna and Flora (CITES; see Vignette 24.3, with Figures 24.3 and 24.4). Many countries, both in Africa and in other parts of the world, have not ratified the treaty. Even where CITES is in effect, the flourishing underground trade persists, catering to the insatiable appetites of those willing to pay the price.

During the 1980s and 1990s, the elephant population of Africa fell sharply, due to illegal hunting and the loss of habitat. In Angola alone, 100,000 were killed by the National Union for the Total Independence of Angola (UNITA) guerrillas to fund their insurrection against the government, virtually wiping out the entire population. The once-huge herds of Sudan and the Democratic Republic of the Congo have also been decimated during civil wars. The majority of elephants in Tanzania, Zambia, Kenya, and Uganda have also been lost. According to a 1998 survey, Botswana, Zimbabwe, and Tanzania have between 70,000 and 100,000 animals, while seven other countries have populations of 10,000–25,000. Smaller numbers, generally under 1,000, are found in some 25 other nations. The remaining populations in West Africa are small and di-

VIGNETTE 24.3. How to Protect Africa's Elephants: The Debate

In October 1989, the Convention on International Trade in Endangered Species of Wild Fauna and Flora (CITES) voted to enforce a complete ban on international trade in ivory, in an attempt to reduce the poaching that had dramatically reduced the population of African elephants (see Figure 24.3). Subsequently, all of the major importers of ivory agreed to suspend ivory imports. This decision to ban the ivory trade came as a result of the failure of attempts to use a quota system to limit the slaughter of elephants. The quota system was not applied in a consistent and ecologically rational way. Each country set its own quota without due consideration to its elephant population; the CITES-approved quotas of 1986 totaled 108,000 tusks, up to ten times the sustainable level of harvest. Moreover, the quotas did not apply to worked ivory. The extremely high market prices for ivory ensured that poachers were willing to take ever-greater risks in order to cash in.

Kenya was among the strongest supporters of the CITES ban on ivory exports. Its government argued that only a total ban would allow effective policing of the ivory trade and would bring about a drop in prices sufficient to dissuade poachers. As a symbol of its determination, Kenya's government staged a public burning of its entire, huge stock of ivory seized from poachers.

The ban on the ivory trade was strongly opposed by several countries in southern Africa, including Zimbabwe, Botswana, Namibia, and South Africa. These countries all have stable (if not growing) herds of elephants, which the governments of these countries attribute to careful management of the resource. They argue that selective killing is necessary to maintain the health of the herds and of the ecosystems supporting them. Moreover, the sale of ivory can be used as a means of financing rigorous wildlife conservation and antipoaching programs; the wildlife services of most African countries are notoriously underfunded and underequipped.

The debate on how to save the elephant resumed at the March 1992 meeting of CITES. The supporters of the ban pointed to the collapse of the price of ivory on the world market since 1990, and argued that a continuation of the ban was needed to allow for the elephant herds in East Africa to recover. The southern African countries with more stable elephant populations claimed that they were being penalized for their good management, and again argued that a controlled harvest of elephants and a limited legal sale of ivory made sense. CITES voted to extend the ban—a decision that the southern African states called "political"

(cont.)

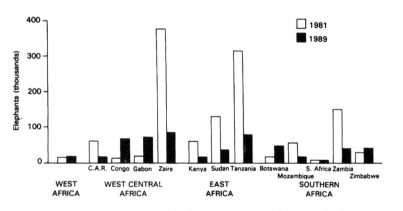

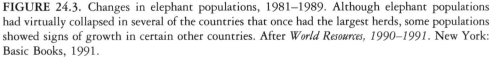

FIGURE 24.3. Changes in elephant populations, 1981–1989. Although elephant populations had virtually collapsed in several of the countries that once had the largest herds, some populations showed signs of growth in certain other countries. After *World Resources, 1990–1991*. New York: Basic Books, 1991.

VIGNETTE 24.3. *(cont.)*

(i.e., more influenced by a conservationist climate in Europe and North America than by good scientific data).

During the 1997 meeting of CITES, it was agreed to allow a one-time, tightly controlled sale of stockpiled ivory from three southern African states. New programs were initiated in order to monitor elephant populations to obtain information on poaching, and to monitor the illegal trade in ivory. Two years later, several countries argued that their poaching problems had increased following the 1997 decision. With the two sides remaining firmly committed to their respective positions, it was decided to retain the existing agreement and not to sanction additional ivory sales.

This debate shows that the conservation of African wildlife is a complex issue that needs to be addressed at the local, national, and international levels. Policies that seem to work in one jurisdiction and for one particular species may not be appropriate for others. Our understanding of the science, economics, and politics of conserving Africa's wildlife is still rudimentary (see Figure 24.4).

FIGURE 24.4. Elephant herd, northern Cameroon. The debate continues about what strategies provide the most effective protection for Africa's wildlife, and thus for Africa's important wildlife-based tourism industry. Photo: CIDA (R. Lemoyne).

vided into many small groups for which longer-term prospects are bleak.

The implications of the slaughter of elephants extend beyond the diminishing opportunities to see elephants in the wild. When elephant herds uproot trees and otherwise "damage" natural vegetation, they are helping to diversify the environment, and thus are enhancing its capacity to support many species. Fewer elephants may lead to a decrease in wildlife numbers and species diversity.

The devastation of Uganda's wildlife provides a case study of the impacts of poaching. The two largest parks in Uganda lost over 95% of their elephants between the early 1970s and mid-1980s, at a time of civil unrest in the country. Despite a total ban on elephant hunting since 1980, Uganda exported 280 metric tons of ivory in 1987. The quest for rhinoceros horn, used for dagger sheaths in Yemen, resulted in the total elimination of the rhinoceros from Uganda. Endangered species like the Nile crocodile and leopard continue to be hunted illegally for their skins. The country's exceptionally rich bird life has also come under increasing pressure from poachers.

Wildlife is in crisis, even in countries fully committed to conservation. Conservation officers must patrol vast territories and guard highly mobile wildlife. The poachers are very determined and often armed with high-powered automatic weapons. Although poachers receive only a small fraction of the final market value of the wildlife products, the profits are attractive in desperately poor countries with few job prospects. Zimbabwe has tried to protect its once large rhinoceros population in the Zambezi Valley from poachers, many of whom are based in Zambia. Despite extensive and sophisticated surveillance, and a policy of shooting to kill when poachers are encountered, Zimbabwe's wild rhinoceros population has declined by more than 90% since 1970.

Fauna and Tourism

For more than a century, African wildlife has served to attract tourists to the continent. As described in Chapter 2, the initial visitors were aristocrats and tycoons from Europe and North America who came to view, and often to shoot, wildlife. With advances in air transportation and reasonably priced package vacations, trips to Africa to view its fauna in the wild have become much more accessible to a mass clientele.

East Africa and southern Africa are the main destinations for travelers wishing to see African wildlife. Most head for the large national parks and wildlife reserves that serve the dual purposes of conservation and tourism. The parks and reserves have provided an environment in which wildlife populations have had some protection from poachers and expanding human occupancy. Nevertheless, the effect of poaching has increased as wildlife numbers beyond park boundaries have declined. Many of the sites chosen for parks are also places of unique scenic beauty. In Kenya, national parks and reserves (Figure 24.5) are located beside scenic lakes, on the slopes of huge volcanic peaks, and within rift valleys, providing spectacular settings for wildlife observation.

African communities have frequently had an uneasy relationship with parks and reserves set aside to protect wildlife and cater to tourists. During the colonial era, people and their livestock were often forced to move when these sanctuaries were established. The control of fires and limitations on livestock in the reserves allowed vegetation to flourish and wildlife to proliferate, creating ideal conditions for tsetse. Virulent epidemics of sleeping sickness subsequently appeared in human populations in East Africa and other parts of the continent. Livestock were decimated by trypanosomiasis in tsetse-infested areas. Colonial officials generally saw these epidemics as further evidence of the pervasive unhealthiness of Africa. Seldom was there any recognition of the ways in which colonial policies had contributed to the disaster.

People versus Wildlife?

From afar, the question of protecting African wildlife and the environments that sustain it may seem like a simple matter of recognizing the importance of wildlife and having a strong commitment to protecting the resource. It is seldom so simple. Wildlife and humans often compete for the same scarce environmental and financial resources.

The utilization of wetland habitats provides an excellent example of the people-versus-wildlife dilemma. Africa's wetlands are crucial ecological resources. For example, a few relatively small wetlands located south of the Sahara are absolutely essential staging points for many species of migratory birds from Europe. However, these areas are also extremely important for primary producers, since they have fertile soils and water for irrigation. Farmers seldom have viable alternatives to using these areas, even though wildlife habitat is likely to be damaged in the process. "Debt-for-conservation" swaps, in theory, may provide a means of protecting vulnerable wetlands. The problem, however, is that the rural populations whose livelihoods are affected by such conservation efforts seldom receive just and adequate compensation.

In a similar way, the crisis of declining wildlife numbers from illegal and inadequately reg-

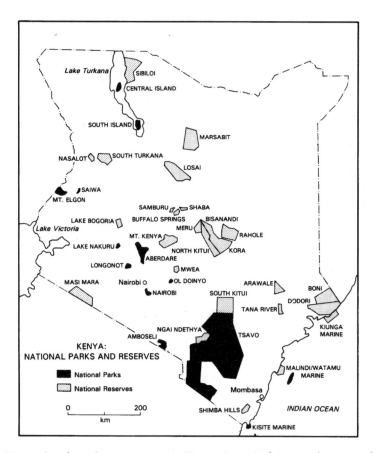

FIGURE 24.5. National parks and game reserves in Kenya. Several of these parks are the focal points of the country's thriving tourist industry. After *Msindi*, vol. 4, no. 1 (1993), p. 1.

ulated hunting reflects a wider societal malaise. The traditional control over hunting formerly exercised by African societies as a means of resource management has been effectively eroded. Moreover, the declining economic situation provides a strong incentive for wildlife to be slaughtered for sale or for food. In short, the fate of Africa's wildlife cannot be separated from the political and economic condition of Africans.

Are there opportunities for "win–win" solutions that protect both wildlife and the humans living in their midst? *Wildlife comanagement* is an increasingly popular response to this challenge. Whereas parks and reserves have often been established at the *expense* of local communities—depriving them of access to traditional hunting, grazing, and farming lands, and providing few if any benefits in return—comanagement seeks to involve local communities as *partners*. Community representatives join conservation officers on management committees that discuss issues of mutual concern. Agreements may be reached to allow regulated hunting of certain species for food, and may provide local communities with a share of profits from the hunting of trophy animals by tourists. Community members have priority for employment as guides and on poaching patrols.

The best of the comanagement schemes, such as Zimbabwe's successful CAMPFIRE program, succeed by giving local communities a tangible stake in the success of wildlife tourism and conservation efforts. The legitimacy of local concerns and economic needs are acknowledged. Indigenous environmental knowledge is valued and is often applied when making management decisions. In reality, comanagement is seldom an unqualified success, due to factors that range from power dynamics on the committee to differences between community and official priorities that cannot always be reconciled. However, without a strong commitment to conservation that is compatible with local communities and their needs, future prospects for "protected" environments remain bleak.

Further Reading

The following sources are useful sources on the tropical timber industry:

Guppy, N. "Tropical deforestation: A global view." *Foreign Affairs*, vol. 62 (1984), pp. 928–965.

Owusu, J. H. "Current convenience, desperate deforestation: Ghana's adjustment program and the forestry sector." *Professional Geographer*, vol. 50 (1998), pp. 418–436.

World Conservation Union (IUCN). *The Conservation Atlas of Tropical Forests in Africa*. Singapore: Simon and Schuster, 1992.

The nature and significance of agroforestry are discussed in the following sources:

Chambers, R., and M. Leach. "Trees as savings and security for the rural poor." *Unasylva*, no. 161 (1990), pp. 39–52.

Cook, C. C., and M. Grut. *Agroforestry in Sub-Saharan Africa: A Farmer's Perspective*. Washington, DC: World Bank, 1989.

For discussions of Africa's fuelwood crisis, see these sources:

Cline-Cole, R. *Contesting Forestry in West Africa*. Aldershot, UK: Ashgate, 2000.

Cline-Cole, R., H. A. C. Main, and J. Nichol. "On fuelwood consumption, population dynamics, and deforestation in Africa." *World Development*, vol. 18 (1990), pp. 513–518.

Leach, G., and R. Mearns. *Beyond the Fuelwood Crisis: People, Land, and Trees in Africa*. London: Earthscan, 1988.

The impact of colonial policy on wildlife is considered in the following volumes:

Kjekshus, H. *Ecology, Control, and Economic Devel-

opment in East African History*. London: Heinemann, 1977.

Matzke, G. *Wildlife in Tanzanian Settlement Policy*. Syracuse, NY: Maxwell School of Citizenship and Public Affairs, Syracuse University, 1977.

The significance of wild game as food and experiences with game ranching are discussed in these sources:

Jewell, P. A. "Ecology and management of game animals and domestic livestock in African savannas." In D. R. Harris, ed. *Human Economies in Savanna Environments*, pp. 353–381. London: Academic Press, 1982.

Walker, B. H. "Game ranching in Africa." In B. H. Walker, ed. *Management of Semi-Arid Ecosystems*, pp. 55–81. Amsterdam: Elsevier, 1979.

Issues concerning the implications of conservation for local communities are addressed in the following sources:

Enghoff, M. "Wildlife conservation, ecological strategies, and pastoral communities: A contribution to the understanding of parks and people in East Africa." *Nomadic Peoples*, nos. 25–27 (1990), pp. 93–107.

Hulme, D., and M. Murphree, eds. *African Wildlife and Livelihoods*. Harare, Zimbabwe: Weaver Press, 2001.

Kiss, A., ed. *Living with Wildlife: Wildlife Resources Management with Local Participation in Africa*. Washington, DC: World Bank, 1990.

Reid, D. G., ed. *Ecotourism Development in Eastern and Southern Africa*. Harare, Zimbabwe: Weaver Press, 1999.

Internet Sources

The FAO website is a rich resource. Look for databases, the annual *State of the World's Forests*, the journal *Unasylva*, and web pages on such subjects as wood energy, international trade, nonwood forest products, and biodiversity:

Food and Agriculture Association of the United Nations (FAO). **www.fao.org/forestry**

Controversies about the tropical timber industry may be explored in the following sites:

International Tropical Timber Organization. www.itto.or.jp

Tropical Forest Foundation. www.tropicalforestfoundation.org

World Rainforest Information Portal. www.rainforestweb.org

There are numerous superb websites that address wildlife conservation issues. Several of the sites listed below are comprehensive, while others focus on elephant conservation:

Convention on International Trade in Endangered Species of Wild Fauna and Flora CITES. www.cites.org

Convention on the Conservation of Migratory Species of Wild Animals. www.wcmc.org.uk/cms

International Union for the Conservation of Nature. *African Elephant Specialist Group (AfESG).* www.iucn.org/themes/ssc/sgs/afesg

Kenya Wildlife Service. www.kws.org

WildNet Africa. *The Elephant Culling Issue* [discussion forum]. http://wildnetafrica.co.za/cites

World Wildlife Federation. *Conserving Africa's Elephants.* www.panda.org/resources/publications/species/elephant

The Central African bush meat trade is discussed in the following source:

CITES. *Technical Meeting of the CITES Bushmeat Working Group in Central Africa, July 2001.* www.cites.org/eng/prog/BWG/0107_wg_report.shtml

African Economies

The chapters in this section survey several aspects of African economic development at three levels of spatial resolution: Africa within the world economy, national economic development, and community/local development.

Chapter 25 examines why Africa occupies such a marginal and vulnerable position in the global economy. The dependence on a small number of primary-product exports, which are subject to uncertain demand and prices in the global marketplace, has remained a source of economic weakness since the colonial era. More recently, Africa's position has become even more tenuous, owing to cutbacks in aid and investment and a growing debt burden. The economic weakness of African countries helps to account for their difficulty in achieving the sustained progress for which they have struggled.

Chapter 26 looks at changes in national development strategies. During the first three decades of the postcolonial era, African countries experimented with three broad approaches to development, each characterized by its own set of priorities, strategies, and outcomes. In recent years, structural adjustment has emerged as a new "unifying" ideology, supplanting the old ideologies of development. The chapter also looks at some of the strategies countries are using in attempts to diversify their economies.

In Chapter 27, the focus shifts to development at the local level, and points to some of the limitations of grassroots development initiatives. The chapter begins with a historical overview of locally based development, and continues with several brief case studies of civil society in action. Although communities can achieve much when they mobilize *their own resources to address their own, self-identified needs, civil society initiative is neither a panacea nor a substitute for appropriate macro-scale initiatives involving the state and international development organizations.*

25

Africa in the World Economy

Globalization, which refers to the increased pace of global change as societies and economies in different parts of the world become increasingly interconnected and interdependent, has been identified as the dominant trend of the late 20th amd early 21st centuries. Globalization has challenged many taken-for-granted assumptions about the world; change is constant, rapid, and often fundamental. In the economic sphere, globalization has involved major shifts in manufacturing from high-income industrial countries to emerging industrial powers in Asia and Latin America. International trade—the flow of goods and services connecting the far-flung points of production and consumption to each other—has boomed under globalization.

Where does Africa fit into the new-millennium world of globalization? To date, Africa has been a bystander. While rapid change was occurring elsewhere in the world during the last two decades of the 20th century, most African economies stagnated. Even when growth occurred, it was mostly in traditional areas of activity (i.e., primary production of agricultural, mineral, and other commodities).

In a 1976 article, "The Three Stages of African Involvement in the World-Economy," Immanuel Wallerstein (see "Further Reading") predicted an African future shaped by declining demand for nonessential exports, increasingly serious food crises, acute suffering, and social and economic disintegration. This would set the stage for the future growth of mechanized primary production, controlled by multinational and state corporations, ready to supply raw materials to global markets. He anticipated that only the African states with the greatest industrial and strategic raw-material capacity—notably South Africa and Nigeria—would prosper.

Almost three decades later, much of what Wallerstein predicted seems to be taking place. The global organization of industry has increasingly been shifting from the old core regions to newly industrializing countries, and the African periphery has experienced a dramatic decline in economic fortunes. What have not transpired are the predicted growth and relative prosperity of the stronger economies. Nigeria, for example, has experienced increased indebtedness, reduced industrial production, and a growing range of social problems.

This chapter examines Africa's role in the world economy in the era of globalization, focusing especially on trade, aid, debt, and the role of multinational corporations. It shows not only that Africa is peripheral to the world economy, but also that its position has become progressively weaker. As countries have become less and less able to exert meaningful control over their own destinies, new forms of neocolonial control have emerged. Vignette 25.1 (with Figure 25.1) provides a unique perspective on these developments by comparing the wealth of African nations to that of large U.S. companies.

The Role in World Trade of Africa South of the Sahara

Africa's role in world trade as a supplier of raw materials and a consumer of imported industrial goods was established in the precolonial era and solidified after the colonial conquest. Much of colonial policy was preoccupied with increasing raw-material exports by whatever means necessary. The development of industry, except for the primary processing of raw materials for export, was actively discouraged; African-based industry would provide unwelcome competition for the colonialists' own industrial enterprises. In short, each colony was established as a fully dependent satellite of its colonial metropole in matters of trade and economic development.

Despite more than four decades of efforts by African countries to escape the straightjacket of dependent trade relations, their position in the world economy has declined rather than grown stronger. During the 1980–1989 period, the value of exports from Africa south of the Sahara (excluding South Africa) declined by an average of 4.5% per year. Its share of total world primary commodity exports fell between 1980 and 1990 from 10.4 to 4.1%. Merchandise exports remained stable at approximately $70 billion per year during the 1990s; during the same period, exports from all low- and middle-income countries doubled to $1,450 billion. By 2000, Africa south of the Sahara accounted for only 2% of total world trade, down from about 5% in the 1980s.

As of 2000, 11 countries obtained at least 90% of their export earnings from the three leading export commodities. In Nigeria, petroleum alone accounted for 99% of export earnings. There were only 11 countries in which the three leading exports accounted for less than two-thirds of the total (Figure 25.2). The number of countries with diversified export economies decreased slightly during the 1990s. In the case of Mauritania and Liberia, an apparent trend toward diversification actually reflected economic contraction, resulting from the closure of iron mines that had formerly provided the bulk of exports.

Countries that are highly dependent on one or a small number of export commodities are vulnerable to demand and price fluctuations on the world market. An increase in the price may bring temporary prosperity, but also may bring higher inflation and debt that magnify the effect of subsequent price declines. In the case of agricultural economies, countries depending on one or two export commodities may be severely affected by droughts, crop diseases, or other natural disasters that reduce the quantity or quality of the harvest.

Several countries, most notably Côte d'Ivoire and Kenya, have been very successful in diversifying their farm exports. However, their recent economic difficulties show that diversification provides very imperfect protection from swings in the world market, in large part because demand and prices for broad categories of commodities tend to rise and fall in unison. Meanwhile, a number of other countries have become progressively less, rather than more, diversified over time. Nigeria is a good example: Although its export economy was quite small but diversified prior to the mid-1960s, its exports now consist almost entirely of one commodity, petroleum.

Diversification has frequently involved the production of a wider range of "traditional" agricultural and mineral commodities. As such, diversification may succeed primarily in increasing market supply and lowering the world price for a particular commodity. The more marginal producers of the commodity become the main losers, no longer able to compete effectively. Market shifts that result in minor gains for commodity *consumers* may spell disaster for *producers* who have "all of their eggs in one basket."

VIGNETTE 25.1. If Countries Were Companies: The Wealth of African Nations

The State of the World Atlas, now in its sixth edition, contains many thought-provoking maps. Figure 30 in the 1984 edition compared the sales incomes of the world's largest companies to the gross domestic products (GDPs) of countries for the year 1981. The map underscored the relative poverty, and hence the economic and political weakness, of many states—particularly in the Third World.

Figure 25.1 provides a similar comparison of countries and companies, in this case relating the GNIs of African nations to the sales incomes of the 1,000 largest U.S. companies in 2001. According to *Fortune* magazine, Wal-Mart, with sales of about $220 billion, was the largest U.S. company in 2001. PC Connection ranked 1,000th, with annual sales of $1,181 million.

If GNIs were equated with sales incomes, only 2 countries in Africa south of the Sahara would have ranked among the top 100 U.S. corporations. South Africa had a GNI of about $133 billion in 2001, equivalent to 6th place among U.S. corporations (between Enron and General Electric). Nigeria's GNI of about $38 billion was equivalent to the sales of the 39th-ranked corporation. Only 3 more countries, Kenya, Côte d'Ivoire, and Sudan, had GNIs as large as the almost $9.6 billion sales of Entergy, which was ranked 200th among U.S. corporations. A total of 21 African countries had GNIs greater than the approximate $3 billion in

(cont.)

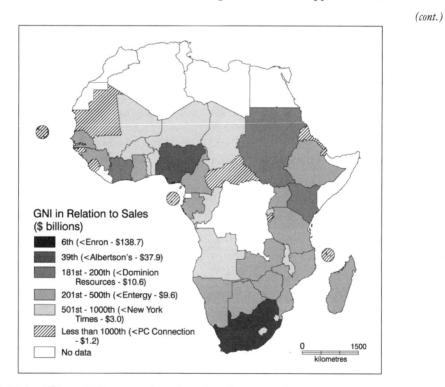

GNI in Relation to Sales
($ billions)

- 6th (<Enron - $138.7)
- 39th (<Albertson's - $37.9)
- 181st - 200th (<Dominion Resources - $10.6)
- 201st - 500th (<Entergy - $9.6)
- 501st - 1000th (<New York Times - $3.0)
- Less than 1000th (<PC Connection - $1.2)
- No data

0   1500
kilometres

FIGURE 25.1. African GNIs equated to the sales of major U.S. corporations, 2001. Even the largest African economies pale in comparison to the sales of major corporations, underscoring Africa's relative weakness in the global economy. Data sources: World Bank. *World Development Report 2002.* New York: Oxford University Press, 2002. "The 2002 Fortune 500." *Fortune,* April 15, 2002.

VIGNETTE 25.1. *(cont.)*

---

sales of *The New York Times*, *Fortune's* 500th-ranked U.S. company. Eight countries, including Eritrea, Mauritania, and Sierra Leone, would not have made the top 1,000 if they were U.S. corporations.

A comparison with the same data for 1989, contained in the first edition of this book, shows how much Africa's standing in the global economy slipped in just 12 years. In 1989, 4 countries had GDPs equivalent to top-50 status among U.S. companies. Whereas 22 countries would have ranked in the top 200 in 1989, only 5 did so in 2001, and the 22nd largest African economy in 2001 would not have made the top 500 U.S. corporations.

Data such as these highlight several important facts. First, they show the immense economic power wielded by large corporations, and demonstrate why these corporations have been able to exercise such great political influence. Second, they provide dramatic evidence of the economic and hence political weakness of African states. It is little wonder, given the paucity of resources at their command, that African states have been unable to address more than a portion of the myriad demands of their citizens for social and economic development. Third, the data show a large gap between the resources available to the largest and best-developed economies—in particular, South Africa and Nigeria—and those of the smallest and least developed countries. These disparities help to account for the different levels of influence of various African countries, both on the continent and in the world.

Most African commodities tend to be price-inelastic; that is, declines in price do not usually bring large increases in demand. This inelasticity applies both to foodstuffs such as coffee and cocoa, and to industrial raw materials such as copper and iron ore (demand for which depends much more on the overall health of the world industrial economy than on the cost of raw materials). The development of lower-cost or more adaptable substitute materials has been a further source of trouble for some commodities. These downward pressures are evident in Table 25.1, which shows trends in the world market prices for selected commodities. The world prices in 2000 for these commodities ranged from 35% of the 1980 price for cocoa to 86% of the 1980 price for crude petroleum.

Whereas the market price of African primary commodities showed a strong downward trend, the same was not true for the manufactured products that account for most of Africa's exports. With few exceptions, however, African countries have fared poorly in their terms of trade (i.e., the price index for their exports relative to the price index for imports). The relative purchasing power of Uganda's exports in 1999 amounted to only 20% of the 1980 value. Other countries that suffered a large loss in their terms of trade between 1980 and 1999 include Nigeria (100 to 33%), Zambia (100 to 40%), and Malawi (100 to 65% of the 1980 value).

The primary problem is simple: The prices of both exports from Africa and imports to Africa are effectively established in London, New York, and other major centers of the world economy. African countries cannot predict (much less control) the value of their commodities on the world market, and so cannot undertake medium- to long-range development planning with any certainty. Producers of several commodities, including coffee, cocoa, copper, and petroleum, have attempted to establish cartels to regulate market supply and price. In the 1970s, the Organization of Petroleum Exporting Countries (OPEC) had some notable success; in general, however, the establishment of cartels has not brought the stability and prosperity that commodity producers had envisaged.

During the 1960s and 1970s, African countries attempted to reduce their dependence on

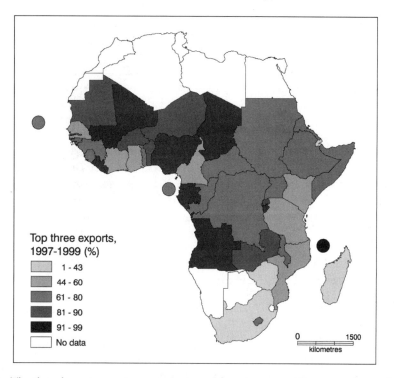

FIGURE 25.2. The three largest exports as a percentage of total exports. Most countries are highly dependent on earnings from three or fewer export products. Data source: United Nations Conference on Trade and Development (UNCTAD). *Commodity Yearbook 2000.* New York: United Nations, 2001.

imported goods through a strategy of import substitution industrialization. Industrial production increased, and many consumer goods and some products for the industrial sector are now made in Africa. However, many of these industries were unable to withstand competition from imported goods in liberalized trading environment after structural adjustment was implemented during the 1980s. Not only that, the industrialization process brought a new

TABLE 25.1. Export Price Index for Selected Commodities, 1980–2000 (1980 = 100)

Commodities	1980	1985	1990	1995	2000
Food products	100	66	69	90	69
Cocoa	100	86	49	55	35
Coffee	100	81	46	83	44
Agriculture nonfood	100	78	81	106	72
Cotton	100	75	95	103	62
Groundnuts	100	70	102	92	58
Minerals and fuels	100	93	71	58	87
Copper	100	64	123	135	86
Crude petroleum	100	91	68	54	86

Data source: United Nations Conference on Trade and Development (UNCTAD). *International Trade Statistics Yearbook* (1995 and 2000 editions). New York: United Nations, 1995, 2000.

form of dependency—namely, a dependency on imported technology and expertise, as well as spare parts and other production inputs. Most of the industrial capacity of numerous African countries is now idle. Africa accounts for only 0.4% of total world exports of manufactured goods.

Compared to the situation during the colonial era, African countries are much less closely tied to a single trading partner. A growing portion of African trade is now with the newer industrializing countries of Brazil, China, Korea, and India. Nevertheless, over three-quarters of African exports still go to the older industrialized countries. The long-held aspiration to reduce dependency on outsiders by increasing intra-African trade has not made great progress; less than 8% of trade is between countries within Africa south of the Sahara. Because Africa still relies so heavily on non-African trade partners, efficient transportation of its goods to these partners is essential—but is often lacking, as Figure 25.3 illustrates. Such transport bottlenecks constitute yet another impediment to African development.

Trade between the countries of Africa and the European Union has been governed since the mid-1970s by a series of treaties known as the Lomé Conventions. These agreements have been notable for recognizing some of the problems discussed above. Tariffs and other trade barriers restricting African access to European markets, particularly for some semiprocessed goods, have been progressively removed. A stabilization fund was created to compensate countries whose major exports are adversely affected by declining market prices. Lomé has enabled a few countries to develop new exports catering to European markets; Kenya's exports of fruits and vegetables are good examples. However, for most countries, the Lomé Conventions have not been sufficient to alleviate Africa's steadily deteriorating trading position in the world marketplace.

Aid, Debt, and Adjustment: In One Hand and out the Other

Since the late 1970s, it has become increasingly difficult to discuss meaningfully the role of foreign aid in African development without reference to the subjects of debt and structural adjustment. Most official development assistance (ODA) to Africa in recent years has been targeted toward the alleviation of debt-related crises, rather than toward such "development" objectives as improving social welfare and increasing the productive capacity of regional, national, and local economies. This critique would be disputed by many development planners, especially those working for the International Monetary Fund (IMF) and World Bank. They argue that current aid and structural

a   b

FIGURE 25.3. Linking Africa to the outside world. Several countries' export economies have been damaged by transport bottlenecks, such as those caused by overburdened facilities (e.g., Tanzania's ports), and by sabotage (e.g., during Angola's civil war). (a) Port of Dar es Salaam, Tanzania. (b) Benguela Railroad, Angola. Photos: CIDA (B. Paton).

adjustment policies are creating the preconditions for economic growth and progress, and that without appropriate adjustment Africa's economic crises will only deepen.

In 1989, Africa south of the Sahara received $16.5 billion in ODA, amounting to $33.10 per capita, or 10.1% of the gross national income (GNI) of the continent. A decade later, total ODA had declined by 29% to $11.7 billion, equivalent to $19.40 per capita, or 6.2% of the GNI (Table 25.2). This significant reduction in ODA speaks volumes about the declining commitment of the world's wealthiest countries to global development and social justice. The G8 countries pledged in 2002 to increase their aid to Africa, to support the New Economic Partnership for Africa's Development (NEPAD). However, with other competing priorities, such as the war on terrorism following the attacks of September 11, 2001, it seems unlikely that aid to Africa is about to increase substantially.

ODA to Africa south of the Sahara is very unevenly distributed, ranging from less than $2 per capita in Nigeria to $254 per capita in São Tomé e Principe (see Figure 25.4). The decline in ODA between 1989 and 2000 was especially evident in the front-line states of southern Africa after the demise of apartheid. Ghana and Zambia were among a handful of countries that received more ODA in 2000 than in 1989. ODA is equivalent to less than 10% of the GNIs of most countries, but it exceeds 25% of the GNI in Zambia, Eritrea, Guinea–Bissau, and Sierra Leone—all smaller economies affected by multiple challenges.

Despite the often-repeated commitment of aid donors to give priority to the world's least developed countries in the allocation of assistance, these least developed countries received only 27.5% of all assistance given to developing countries in 2000. Botswana, with a thriving mining sector and GNI per capita of $3,300, received about almost twice as much ODA per capita as Ethiopia, a country coping with extreme economic, social, and environmental crises and having a per capita GNI of only $100.

Much has been written about the dilemmas of aid—for example, about the costs to recipients of tied-aid programs that have been designed as much to provide a market for donor nations' products as to bring development to the Third World. The dilemmas of aid also extend to the question of who benefits from aid; large-scale projects such as irrigation development and mechanized farming have been criticized because they tend to confer major benefits on indigenous elites and multinational companies, and at the same time to displace or disrupt the lives of ordinary people and threaten the environment. The Tanzania–Canada Wheat Program (Vignette 25.2) illustrates several of these dilemmas of aid.

In the mid-1980s, following the emergence of the Third World debt crisis, the international banking system was seriously threatened by the possibility of a mass default on loan repayment commitments by several major borrowers. This crisis had developed in the 1970s and early 1980s, when international banks had lent huge amounts to the Third World, often

TABLE 25.2. Official Development Assistance to Selected Countries, 1990 and 2000

Country	U.S.$ million		U.S.$ per capita		As % of GNI	
	1990	2000	1990	2000	1990	2000
Nigeria	234	185	2.0	1.6	0.7	0.4
Ethiopia	888	693	17.4	11.0	14.8	10.6
Ghana	456	609	31.2	31.6	7.4	11.6
Zambia	438	795	54.0	76.3	14.0	27.3
Mozambique	946	876	60.2	47.9	65.7	23.3
Lesotho	138	42	78.0	20.4	24.5	4.6
Botswana	148	31	118.2	19.9	5.5	0.6
All of Africa south of Sahara	16,538	11,792	33.1	19.4	10.0	6.2

Data sources: World Bank. *World Development Report, 1992.* New York: Oxford University Press, 1992. United Nations Development Programme (UNDP). *Human Development Report, 2002.* New York: Oxford University Press, 2002.

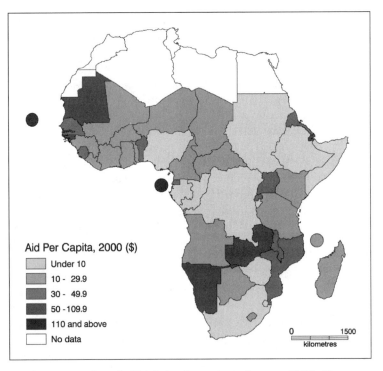

FIGURE 25.4. Receipts per capita of official development assistance, 2000. Data source: World Bank. *World Development Indicators 2002*. Washington, DC: World Bank, 2002.

for very risky ventures. Mexico's threatened default on its loan repayments in 1985 was patched over by means of the Baker Plan, whereby an additional $40 billion was to be lent to the most heavily indebted nations—thus, in effect, buying time for the banks. U.S. financial support for this scheme was made contingent upon the World Bank's and IMF's forcing recipient countries to implement harsh structural adjustment policies. Failure to adhere to the prescribed policy regimen would result in a total withdrawal of financial support from all donors. Thus heavily indebted nations had little option but to accept the neocolonial intrusion of the IMF and World Bank into their affairs. An increasing proportion of ODA consisted, in effect, of new loans to meet repayment and interest obligations on old loans.

Since the mid-1980s, the IMF and World Bank have been net *recipients* of funds from Africa south of the Sahara. In 2000, they collected $500 million more in debt repayment and interest than they provided in new loans. The decline in net transfers from the IMF and World

Bank has been offset by increased lending from the International Development Association (IDA). Support from the IDA is mostly targeted at the poorest countries and is on better terms, but it is still in the form of *loans* that must be repaid. Meanwhile, loans from bilateral sources (country to country) and multilateral sources (e.g., United Nations agencies and the European Union) increased only slightly during the 1990s. Although grants appear to have grown, much of this aid was in forms that did not contribute directly to African development—emergency food aid and military assistance, for example.

Debt repayment is a major constraint on the economies of poorer countries that constrains their attempts to develop (Figure 25.5). In the 28 least developed countries of Africa south of the Sahara for which data are available, the debt load ranges from 28% of the annual GNI in Equatorial Guinea to 538% of the GNI in São Tomé e Principe. The debt exceeds annual GNI in 15 of the 28 states. The annual cost of servicing this debt load is under 10% of export earn-

VIGNETTE 25.2. Wheat at What Cost?: An Assessment of the Tanzania–Canada Wheat Program

There have been prolonged debates about the true benefits and costs of foreign aid projects. These debates have also focused on the extent to which donors should be held accountable for the negative social, economic, and environmental effects of the projects they have supported. The relative benefits of larger projects, versus smaller-scale initiatives such as those undertaken by development nongovernmental organizations (NGOs), have also been debated at length.

Between 1972 and 1994, some $200 million of aid money was invested by the Canadian International Development Agency (CIDA) in the development of seven large, mechanized wheat farms in Tanzania, operated by Tanzania's National Agriculture and Food Corporation (NAFCO). These farms produced some 40% of Tanzania's wheat requirements, and the project was judged a major success by Canadian and Tanzanian government officials.

Closer analysis raised important questions about the project's true benefits and costs. Four-fifths of Canadian aid to Tanzania was tied to the purchase of Canadian goods and services. The use of imported technology on the large wheat farms provided many benefits for Canadian manufacturers, but created an expensive, long-term dependency for Tanzania. When the true costs of energy and other inputs were considered, the project was shown to be uneconomic. Wheat represents only 1% of Tanzania's food consumption. By encouraging Tanzanians to consume more wheat, the project ultimately increased Tanzanians' dependency on imported food.

The Tanzania–Canada Wheat Program was also expensive as far as foregone economic opportunities were concerned. The focus on large-scale wheat production drew scarce resources away from other crucial needs, and from more appropriate and economic approaches to growing wheat—such as from long-established smallholder farms elsewhere in Tanzania that are economically viable producers of wheat. The Barabaig, a pastoralist minority group, protested that 100,000 acres of their land had been taken over without their consent to establish the wheat farms. They claimed that the project was destroying their way of life by reducing their access to land and increasing environmental degradation.

Acknowledging that the project had been a costly mistake, CIDA abandoned the wheat project in 1994. In doing so, it left behind the various economic and human rights problems the project had created.

The Tanzania–Canada Wheat Program reflected a "frontier development" mentality reminiscent of the European settlement of North America. Large-scale development initiatives of this type have become less prevalent in recent years. Development is now more likely to be undertaken in the form of smaller projects, often coordinated by NGOs. Emphasis is placed on community consultation; participatory approaches to development are meant to ensure that development will reflect local priorities and fit into the local society and economy without ill effects, and to ensure that local environmental knowledge is integrated into project design. Although the "real world" implementation of this new approach is often difficult, there is substantial agreement that it represents real progress over the large-scale projects that prevailed in the past.

Based primarily on C. Lane. "Wheat at what cost?: CIDA and the Tanzania–Canada Wheat Program." In J. Swift and B. Tomlinson, eds. *Conflicts of Interest: Canada and the Third World*, pp. 133–160. Toronto: Between the Lines, 1991.

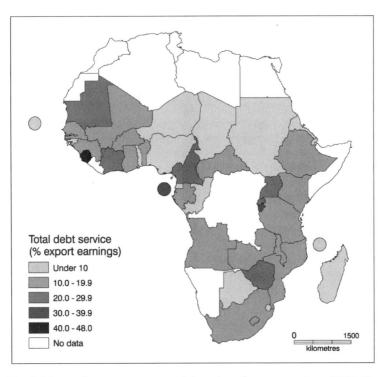

FIGURE 25.5. Total debt service as a percentage of the value of export earnings, 2000. Data source: World Bank. *World Development Indicators 2002*. Washington, DC: World Bank, 2002.

ings in only 3 countries, and exceeds 30% of exports in 6 of Africa's 28 least developed economies.

Much has been made of the role of IMF structural adjustment programs in controlling debt loads and creating opportunities for economic growth as a means of escaping the debt trap. When we compare the debt levels of 1985 and 1999 for Africa's 28 least developed nations, however, there is good reason to question the veracity of these claims. In only 4 of these 28 nations did the ratio of debt to annual GNI decline. The largest reduction occurred in Equatorial Guinea, which became a major oil exporter in the late 1990s. In 5 countries, the debt-to-GNI ratio remained constant (+/−10% change between 1985 and 1999). The proportional debt load increased in 19 countries, including 9 where the ratio of debt to GNI at least doubled. So much for the much-touted economic benefits of structural adjustment for Africa's poorest countries!

There have been widespread calls for debt alleviation programs, which would cancel much (if not all) of the debt owed by heavily indebted poor countries to multilateral agencies and other countries. Other programs have been developed to reschedule payments for the more than $50 billion of commercial bank loans and export credits owed primarily by a few countries with the largest economies. Without any relief, the debt load of several of Africa's poorest countries cannot be sustained. An initiative of the World Bank and IMF has provided significant relief to a small number of countries that have met the program requirements. Although the debt relief movement is gaining momentum—for example, as a basic component of the NEPAD program launched in 2002—the actual implementation of debt forgiveness has progressed slowly.

Transnational Corporations and African Underdevelopment

The growth in size, economic and political influence, and spatial reach of transnational corporations was one of the most important developments of the late 20th century. The major transnational corporations have more economic power than all but the largest countries. Decision making by these companies is based entirely on opportunities for profit; as such, transnational corporations have no inherent commitment to any particular country or enterprise.

Transnational corporations have been widely criticized for distorting the economic development of poorer nations. They represent a drain on resources, because most transnational corporations rely heavily on imported technology, production inputs, and expertise. Instead of supporting indigenous development through strong local economic linkages, transnational corporations often compete with domestic companies and squeeze them out of the market. These corporations tend to make very large profits, which are either repatriated or reinvested in order to acquire an ever-increasing share of the country's productive capacity. For example, data from the United Nations Centre on Transnational Corporations indicates that Nigeria received $3.8 billion in foreign investment between 1970 and 1980, but experienced a net loss of $2.7 billion during this period because of the repatriation of $6.5 billion in fees, royalties, and profits.

Compared to most other parts of the world, transnational corporations have an extremely small presence in Africa south of the Sahara. Net direct foreign investment amounted to $5.8 billion in 2000, a mere 0.4% of total world investment (Table 25.3). Although the amount of net direct foreign investment in Africa in 2000 was double the average for 1990–1995, its share of the world and Third World totals diminished significantly during the same period. The largest share of investment inflow goes to South Africa, Nigeria, and Angola. The flow of resources to these countries points to the importance of the mining and energy sectors for transnational investments.

Africa south of the Sahara, excluding South Africa, has attracted less than 2% of U.S. and 1% of Japanese foreign investment. It accounts for about one-tenth of British and one-quarter of French foreign investment, reflecting those countries' colonial histories. This is evident in the distribution of investment: That of France is concentrated in Côte d'Ivoire, Gabon, and Senegal, while British investment is heaviest in Nigeria, Kenya, and Zambia. In certain southern African countries, including Namibia, Botswana, and Lesotho, South Africa is the main source of foreign capital. Indeed, South Africa's De Beers Consolidated, the Anglo-American Corporation from the mining sector, South African Breweries, and the retailer Pepkor are listed among the 50 largest transnational corporations with headquarters in developing countries.

Royal Dutch/Shell, with a total of $3.8 billion invested in 33 African countries, is among the leading transnational investors in the continent. Others include Nestlé, which has 30 factories in 12 African countries, and Unilever, which has manufacturing operations in 13

TABLE 25.3. Trends in Net Direct Foreign Investment, 1980–1985, 1990–1995, and 2000

	Net direct investment (U.S.$, million/year)			% of Third World total investment		
	1980–1985	1990–1995	2000	1980–1985	1990–1995	2000
Developed countries	37,179	145,019	1,227,476			
Developing countries	12,634	74,288	237,894			
Africa south of Sahara	1,004	2,777	5,790	7.9	3.7	2.4
Oil producers	524	1,332	2,885			
South Africa	83	301	888			
Other countries	297	1,144	2,817			

Data sources: UNCTAD. *World Investment Report 1992: Transnational Corporations as Engines of Growth*. New York: United Nations, 1992.
UNCTAD. *World Investment Report 2002: Transnational Corporations and Export Competitiveness*. New York: United Nations, 2002.

countries and sales units in 40 countries of Africa south of the Sahara.

Lonrho Corporation, formerly the largest transnational presence in Africa, has declined steadily since the mid-1990s; indeed, most of its assets have been sold. In the late 1980s, Lonrho controlled more than 800 companies operating in 80 countries. Its 1988 sales income of $7.8 billion came from a wide range of businesses—agriculture, mining, tourism, motor vehicle distribution, general trade, publishing, and financial services. Lonrho was unusual among transnational corporations, in that its primary locus was in Africa. Although Africa accounted for only 18% of Lonrho's gross sales during 1985 to 1988, it produced half of its pretax profits. At the time, almost 70% of Lonrho's 98,000 employees were in Africa. Among the keys to Lonrho's success were the strong connections that its executive director maintained with ruling African elites.

In Africa, as elsewhere, transnational corporations have at times been criticized for their investment decisions and operating practices. Transnational corporations operating in South Africa under the apartheid regime were criticized for their implicit support of racial discrimination. In Nigeria, Shell (and, to a lesser extent, other petroleum producers) was accused of damaging the environment of the Niger Delta, and of complicity in the Nigerian state's execution of the leaders of the Movement for the Survival of the Ogoni People (MOSOP; see Chapter 28, Vignette 28.2). De Beers was accused of profiting from the illegal trade in blood diamonds originating in war zones (see Chapter 22, Vignette 22.2). Talisman Energy's investments in Sudan also provoked strong protests because of their implications for the country's civil war (Vignette 25.3).

The Changing World Economy: Effects on Development

The deterioration in recent years of African economic and social conditions cannot be understood without reference to the marginalization of the continent in the global economy. The sharp decline in world prices for Africa's pri-

mary products is tied to the growth of the debt crisis, which has in turn brought about a progressive loss of sovereignty to the IMF and World Bank. Transnational corporations continue to make substantial profits in countries where there is no money for schoolbooks and where hospitals function without medicines.

African countries have been given no choice but to focus on short-term objectives related to balance of payments and debt servicing. The social and economic costs of these policies have been very high for the African people. Moreover, Africa's longer-term recovery has been seriously jeopardized by the damage done to economic and social infrastructures through the diversion of scarce resources overseas.

The globalization processes that transformed the world economy, starting in the 1980s, have mostly bypassed Africa. Not only has Africa's role in the world economy—as a producer of some primary commodities—not changed, but its share of world trade and economic activity has continued to decline. The burden of debt, combined with other challenges, has limited Africa's ability to respond to the new realities of global economic organization. Despite optimistic forecasts from the United Nations Conference on Trade and Development (UNCTAD) at the beginning of the new millennium, which saw Africa as a prime target for international investment, the reality has remained quite different. If there is cause for optimism, it relates only to a few countries (e.g., South Africa and Senegal) and to certain industries (especially oil).

Further Reading

For overviews of Africa's role in the world economy, see the following sources:

Cooper, F. "Africa and the world economy." *African Studies Review,* vol. 24 (1981), pp. 1–86.

Wallace, L., ed. *Africa: Adjusting to the Challenges of Globalization.* Washington, DC: International Monetary Fund (IMF), 2000.

Wallerstein, I. "The three stages of African involvement in the world-economy." In P. C. Gutkind and I. Wallerstein, eds., *The Political*

VIGNETTE 25.3. Investment and Human Rights: Talisman Energy in Sudan

What is the role of foreign investment as a means of pressuring repressive governments to improve their human rights records? Does investment in countries whose governments have bad human rights records help to prop up these rogue regimes, or does it make them more accountable to international public opinion?

Many questions were raised about the role of foreign investment, and the responsibilities of transnational companies, during the 1980s and early 1990s—when companies operating in South Africa were subjected to sustained pressure by individual and institutional shareholders worldwide to withdraw from that country. It was argued that these investments both directly and indirectly supported the apartheid regime, and that these investments were immoral because the companies were profiting from the oppression of nonwhite South Africans. As a result, there was a massive flight of international capital out of South Africa; this divestment is considered to have been a major factor in causing the demise of the apartheid government.

During the late 1990s, questions were raised about the propriety of investments made by Talisman Energy, a Canadian multinational corporation, in Sudan. Significant oil reserves have been known to exist in the Sudd region of southern Sudan for some time. However, these reserves had remained untapped, largely because of the ongoing civil war in that part of Sudan. Talisman succeeded in developing Sudan's oil reserves, with the result that Sudan became a significant producer of oil for the first time.

Talisman's investments in the Sudan were subjected to harsh criticism at the United Nations and in the international press. Talisman's critics argued that oil revenue flowing to the Sudanese regime not only supported it, but also funded its military campaigns in southern Sudan. Sudan has been accused of many abuses against the minorities of southern Sudan, including indiscriminate bombing, enslavement, and forced conversion to Islam. Southern Sudanese liberation forces claimed that many people had been driven away from the oil-producing region by the Sudanese armed forces, in order to make it secure for Talisman. They also claimed that Sudanese government forces had used Talisman airstrips to launch attacks.

Talisman replied to these criticisms by denying any connection between its activities and the war in the southern Sudan, which, it pointed out, had been underway long before the company's arrival. Talisman's executives argued that their investments were a positive force for change in the Sudan—that oil profits would help to make the Sudanese government more accountable. They claimed that they were bringing benefits to the southern region, including improved transportation, services such as health clinics, and opportunities for employment. These changes, it was said, would help to end the marginalization of the southern region within the Sudan.

Responding to growing criticism of Talisman, the Canadian government commissioned a report on the company's role in Sudan. The Harker Report, released in 2000, supported the view that Talisman's investments were helping to prolong the war and to support repression by the Sudanese regime. However, it stopped short of recommending that the company face punitive sanctions because of its investments.

Talisman continued to protest its innocence, but the negative publicity and Sudan's uncertain investment environment depressed the value of the company's shares. Bowing to sustained pressure, Talisman sold its stake in Sudan to the Indian national petroleum company in 2002.

Economy of Contemporary Africa, pp. 30–57. Beverly Hills, CA: Sage, 1976.

On patterns of trade involving African countries, see the following sources:

Bevan, D., et al., eds. *Trade and Fiscal Adjustment in Africa.* Basingstoke, UK: Palgrave Macmillan, 2000.

Helleiner, G.K. *Non-Traditional Export Promotion in Africa.* Basingstoke, UK: Palgrave Macmillan, 2002.

Mshomba, R. *Africa in World Trade.* Boulder, CO: Lynne Rienner, 2000.

There is a large literature on issues related to international development aid for Africa. See the following:

Carlsson, J., et al., eds. *Foreign Aid in Africa: Learning from Country Experiences.* Uppsala, Sweden: Nordiska Africaninstitutet, 1998.

Crewe, E., and E. Harrison. *Whose Development?: An Ethnography of Aid.* London: Zed Books, 1999.

Cumming, G. *Aid to Africa: French and British Policies from the Cold War to the New Millenium.* Aldershot, UK: Ashgate, 2001.

Lancaster, C. *Aid to Africa: So Much to Do, So Little Done.* Chicago: University of Chicago Press, 1999.

Omoruyi, L. O. *Contending Theories on Development Aid: Post-Cold War Evidence from Africa.* Aldershot, UK: Ashgate, 2002.

Africa's debt and the role of international lending agencies are discussed in the following sources:

George, S. *A Fate Worse Than Debt.* London: Penguin, 1988.

Helleiner, G. K. "The IMF, the World Bank and Africa's adjustment and debt problems: An unofficial view." *World Development,* vol. 20 (1992), pp. 779–792.

Mistry, P. *African Debt: The Case for Relief for Sub-Saharan Africa.* Oxford: Oxford International Associates, 1989.

Simon, D., ed. *Structurally Adjusted Africa: Poverty, Debt, and Basic Needs.* London: Pluto Press, 1995.

World Bank. *Sub-Saharan Africa: From Crisis to Sustainable Growth.* Washington, DC: World Bank, 1989.

The role of multinational corporations is considered in the following sources:

Cronje, S., M. Ling, and G. Cronje. *Lonrho: Portrait of a Multinational.* Harmondsworth, UK: Penguin, 1976.

Langdon, S. *Multinational Corporations in the Political Economy of Kenya.* London: Macmillan, 1981.

Wells, L. T., Jr. *Third World Multinationals: The Rise of Foreign Investment from Developing Countries.* Cambridge, MA: MIT Press, 1983.

Internet Sources

The following are websites for major international financial institutions:

International Monetary Fund (IMF). www.imf.org/external

Organization for Economic Cooperation and Development (OECD). www.oecd.org

United Nations Conference on Trade and Development (UNCTAD). www.unctad.org

United Nations Development Programme (UNDP). www.undp.org

The World Bank Group–IMF. *Africa Club.* www.worldbank.org/afr

World Trade Organization (WTO). www.wto.org

Here are websites on Canadian, British, and American international aid initiatives:

Department of Foreign Affairs and International Trade, Canada. *Aid and Development—Africa.* www.dfait-maeci.gc.ca/africa/aid_development-en.asp

United Kingdom. *Department for International Development.* www.dfid.gov.uk

U.S. Agency for International Development (USAID) Bureau for Africa. *USAID in Africa.* www.usaid.gov/regions/afr

Africa Action is one of many NGOs that have lobbied about issues related to Africa's debt:

Africa Action. *Africa's Right to Health Campaign.* www.africaaction.org/action/debt.htm

26

National Economies: Strategies for Growth and Development

The first years of independence were character-ized by great optimism about Africa's future. The United Nations declared the 1960s to be the Development Decade and set lofty objec-tives for it, such as a minimum of 5% annual growth for national incomes. Four decades later, we can see that the expectations of immi-nent development proved sadly inappropriate. Fortunes have varied from strongly qualified success in some countries to decay and disinte-gration in many others.

This chapter provides a brief review of the approaches used by African nations since inde-pendence to order and shape national develop-ment, especially the following:

- The political ideologies that have informed African development, and the influence of ideology on development priorities and strat-egies
- The effect of International Monetary Fund (IMF) and World Bank structural adjustment packages on economic development perfor-mance
- Examples of economic growth and diversifi-cation strategies

Development Ideologies

During the first decades following independ-ence, external role models often inspired the economic development strategies of African countries. Some nations were influenced by the development of communist countries such as the Soviet Union and Cuba, while others looked to the United States and Japan for inspiration. Certain countries attempted to develop their own distinctly African approach to develop-ment. According to Crawford Young (see "Fur-ther Reading"), these political ideologies may be grouped into three categories: capitalism, populist socialism, and Afro-Marxism. Each is discussed in turn.

African Capitalism

Among the African states whose economies were organized according to capitalist princi-ples from the beginning were Côte d'Ivoire, Kenya, Nigeria, and Gabon. The private sector was encouraged to take a leading role in devel-opment, and foreign investment was welcomed, especially in economic sectors where indigenous companies had little expertise. At the same time, these African states tended to take a very active role in the development process. Governments not only attempted to create a climate favorable to investment by offering incentives to foreign investors, but they also become directly involved as partners or as major shareholders in new ventures. States also took steps to increase the role of indigenous firms in their economies by passing laws reserving some types of economic activity for local companies, requiring other companies to train and employ Africans in management positions, and creating incentives for the development of local supply linkages.

Capitalist countries, inspired by classical economic theory, emphasized economic growth as their primary objective; as such, there was less concern with issues of urban–rural, regional, and class-based equity in the distribution of benefits. Thus, although development by the private sector tended to be highly concentrated in both spatial and social terms, it was assumed that the benefits of growth (in the form of increased trade and employment opportunities, and more money for social expenditure) would eventually "trickle down" to benefit initially disadvantaged regions and social groups.

The distribution not only of wealth but also of social welfare—in particular, safe water, health care, adequate nutrition, and education—tended to be very uneven in African capitalist states. Whereas the small elite populations of top political figures, civil servants, and entrepreneurs have controlled a disproportionate share of wealth and have access to the very best in social amenities, almost everyone else has struggled to make ends meet and often has no access at all to even the most basic social services.

Côte d'Ivoire has been Africa's leading beacon of free enterprise. After independence, the country grew at the astonishing rate of 11% per year during the 1960s and by 7% per year in the 1970s. During the 1980s, the economy grew by only 1.2% annually, well below the 3.9% annual rate of population growth. Growth rebounded to 2.8% per year during the 1990s. The country has gone from one of the poorest countries prior to 1950 to one with the eighth-highest gross national income (GNI) per capita in Africa south of the Sahara.

Ivoirian growth has been strongly based in the agricultural sector, especially in coffee and cocoa production. The country is the world's leading producer of cocoa and fifth-ranked producer of coffee. These crops are grown primarily on some 600,000 family-owned plantations averaging 3–4 ha in size. Côte d'Ivoire is also one of the world's top producers of palm oil; Africa's largest rubber producer and third-largest cotton producer; and a major source of bananas and pineapples.

The Ivoirian economy is characterized by gross migrant–indigene, urban–rural, and regional disparities in wealth (Figure 26.1). With the exception of cotton, all of the major cash crops are produced in the southern forest regions. Modern industry, commerce, transportation, and government are also highly concentrated in the south, especially in Abidjan. Little has happened to change the north's original role—namely, as a source of cheap migrant labor for the southern region. For Africa's economic nationalists, Côte d'Ivoire is one of the ultimate examples of neocolonial underdevelopment, where growth has greatly benefited foreign capital and indigenous elites at the expense of the poor.

Populist Socialism

During the 1960s and 1970s, many regimes aimed to distance themselves from their colonial past by following a socialist path of development. One of the earliest and most influential attempts to implement populist socialism took place in Ghana under Kwame Nkrumah. In his own words, Nkrumah aimed to create "a welfare state based on African socialist principles, adapted to suit Ghanaian conditions, in which all citizens, regardless of class, tribe, color or creed, shall have equal opportunity." Nkrumah sought to combine the development experiences of other, primarily socialist countries with principles taken from African traditional society, which, he argued, was inherently socialist.

Ghana's experiment with populist socialism

a   b

FIGURE 26.1. Two faces of life in Côte d'Ivoire. (a) Modern office towers in Abidjan. (b) Rural women preparing food. Photos: CIDA (R. Lemoyne).

ended in failure with the overthrow of Nkrumah. (Vignette 26.1 gives an account of developments in Ghana after this event.) However, there were other countries, including Tanzania and Zambia, where the commitment to socialist development was more enduring. Each had a charismatic president who led the fight for independence and who clearly articulated a vision of egalitarian, self-reliant development. Although the policies that the leaders of these countries advocated often failed to achieve the intended results, they did provide a vision of a better African future that inspired many at the time.

Populist socialism was rooted in ideas about the importance of communities in traditional African society. The individualism and competitiveness of capitalism were criticized as being alien to African traditions of cooperative production, communal land ownership, and shared responsibility for welfare. Instead, community

mobilization for self-help projects and communal production was encouraged. These socialist regimes tended to be partial to rural development, perhaps because the values that they espoused were thought to be best preserved in rural communities. Because self-reliance was considered a precondition for real independence, there was heavy reliance on state-owned industrial and trading companies, as well as strict controls on foreign investment.

Although populist socialist regimes recognized the importance of economic growth, they tended to differ from capitalist nations in not treating growth as a preeminent objective. They also rejected the assurances of "trickle-down" economics that growth will ultimately benefit all, regardless of where it takes place or who controls it. Numerous parastatal companies were established to spur economic development in such sectors as agriculture, manufacturing, transportation, and trade. In many

VIGNETTE 26.1. Economic Policy Transitions in Ghana

In few countries was there more optimism about postindependence prospects than in Ghana. Unfortunately, much of Ghana's history since 1957 has been characterized by recurrent economic and political crises and by rapidly deepening poverty. Since the mid-1980s, however, Ghana has been seen widely as one of Africa's success stories—as having made a successful transition from military rule to democracy, and as achieving relative economic, political, and social stability.

Kwame Nkrumah led Ghana to independence in 1957 and retained power until 1966. He moved ambitiously to broaden the economic base beyond cocoa, of which Ghana was the world's top-ranked producer. The massive Volta River project was launched to provide power for industrial development; the new port city of Tema was built; and state farms were set up to modernize agriculture. However, the economic situation worsened as a result of too many overly ambitious and poorly implemented development projects, as well as a 75% decline in the world price of cocoa.

There was widespread public support when Nkrumah was toppled in a coup and replaced by the National Liberation Council in 1966, but what followed was a 15-year period of political instability and economic decline under a succession of weak military and civilian governments. The causes of the economic decline were many, including the weaknesses of cocoa prices, a growing debt burden, and poorly conceived state policies. Attempts to encourage a market-driven economic recovery under the civilian regime of Kofi Busia gave way to austerity measures (such as currency devaluation and drastic limitations on imports) under successive military regimes. The economic situation continued to worsen despite a temporary revival of cocoa prices, largely because so much of the crop was being smuggled into neighboring countries. Market prices for food rose sharply, causing much hardship for consumers and spurring a series of workers' strikes.

The political instability ended in 1981 when Lieutenant Jerry Rawlings seized power; his military regime ruled for a decade. At first, Rawlings initiated a series of strategies that suggested a return to a Nkrumah-like government. However, the regime's socialist rhetoric was eclipsed by economic policies that emphasized austerity and free-market principles. Because of these policies, Rawlings became a rather unlikely hero for the World Bank and IMF. Under the Economic Recovery Program, prices were increased and subsidies on imported food were removed to encourage increased production of export and food crops. Large cuts were made in the civil service and in social expenditure. User fees were introduced in the education and health sectors. The devaluation of the Ghanaian currency continued; by the late 1980s, the Ghanaian *cedi* retained less than one-thousandth of its initial value under Nkrumah.

Compared to the years before, economic performance since 1983 has been a major success. The gross domestic product (GDP) grew by 3.0% during the 1980s and 4.3% during the 1990s. Export revenues have grown considerably, and inflation has declined. However, most of the increased revenues have gone to service the debt, which had reached $3.9 billion in 1990 and continued to grow to $6.9 billion in 1998. The cost of debt servicing consumes over half of Ghana's export income, compared to only one-eighth in 1980.

A new democratic constitution was approved in 1992. After declaring himself a candidate for the presidency just days before the election, Rawlings won a landslide victory. He served for two terms, during which he continued the economic policies that had brought relative stability to Ghana's economy, but also had caused hardship for the poor. With Rawlings's term of office at an end, Ghana's voters elected the opposition leader, John Kufuor of the New Patriotic Party (NPP) in 2000. The NPP promised to maintain a market-oriented economic policy, but to accelerate economic growth through support for the private sector and by increasing foreign investment.

cases, however, these state-initiated projects were costly failures. Their difficulties were related both to internal problems (such as poor management and peasant resistance) and to external threats (principally, declining terms of trade on the world market). These difficulties forced the socialist states to adopt major revisions in their development strategies as the price for IMF support.

Afro-Marxism

Until the late 1980s, several African regimes maintained an explicit commitment to Marxism–Leninism. Some, including Mozambique and Angola, were forged through a process of armed struggle leading to independence. Other former Afro-Marxist regimes, including those of Burkina Faso, Benin, and Ethiopia, emerged as a result of military coups staged by disaffected lower-ranking soldiers.

Afro-Marxist regimes remained committed to gaining control of the "commanding heights" of the economy—commercial agriculture, industry, banking, trade, and commerce—through state direction. The Soviet Union and other communist states provided not only the model, but also the financial, technical, and military aid to make it possible. The heavy reliance on assistance from communist states was necessary, because these regimes could not attract substantial aid or investment from the West.

Most Afro-Marxist regimes sought to achieve social, as well as economic, progress. Mass literacy campaigns were organized; primary health care programs were implemented; and initiatives were taken to improve the status of women in society. Though the social and economic transformation occurred in a top-down fashion, most regimes were strongly committed to the mobilization of grassroots participation in the process of change.

As a general rule, Afro-Marxist regimes failed to achieve their objectives of rapid economic growth and comprehensive social and economic transformation. Nevertheless, their failures need to be put into perspective. Although in many cases the failures may be attributed to the naive application of inappropriate models and bad management, the importance of Western hostility to these experiments should not be underestimated. Very limited access to Western aid and investment, trade embargoes, and attempts at political destabilization created a severe disadvantage for these regimes. The withdrawal of support from the Soviet Union and its allies left the remaining Afro-Marxist states with no option but to attempt to reach out to Western aid agencies and investors.

In Mozambique, the liberation movement called Frelimo drew inspiration from a long history of resistance to Portuguese colonialism. The struggle was long and bloody, but Frelimo finally triumphed, gaining independence for Mozambique in 1975. The departing Portuguese left the country in shambles; the new state inherited very little in the way of infrastructure or of formal training and preparation for the task at hand.

Frelimo sought to achieve a total economic, social, and political transformation to erase the profound underdevelopment caused by colonial exploitation (see Figure 26.2). In agriculture, priority was given to the establishment of large, mechanized state farms. The program was a costly failure: The state lacked adequate management expertise; mechanized equipment was ill suited to the country's needs; and many peasants were not interested in collectivized farming. The development of social services, virtually ignored by the Portuguese, was another high priority of the Frelimo regime.

The gravest threat to Mozambique's development efforts came from the Mozambican National Resistance (Renamo), which received direction, funding, and material support from the South African government. Between 1978 and 1992, an estimated 1 million Mozambicans lost their lives as a result of Renamo's insurrection; 1.5 million others fled to other countries as refugees; and 2 million more were dislocated internally. Agricultural production declined by at least one-third, and interregional trade was severely restricted. The state was forced to allocate 40% of its budget to defense in an unsuccessful effort to achieve a military victory. Peace returned in 1992, following an end to South

FIGURE 26.2. Billboard advertising Frelimo's fifth party congress. Frelimo was the governing party that spearheaded Mozambique's Afro-Marxist development strategy during the 1970s and 1980s. Photo: CIDA (B. Paton).

African interference and the signing of a peace accord with Renamo.

The political and economic cost of the war, the effect of severe droughts, the failure of its economic policies, and the virtual absence of Western aid forced Frelimo to abandon its commitment to Afro-Marxism. The implementation of structural adjustment forced the dismantling of the state farm program, the introduction of free-market pricing, currency devaluation, and cutbacks in public-sector employment and social-service expenditure.

Structural Adjustment and the "End" of Ideology

During the 1980s and early 1990s there was an increasing convergence in the economic policies of African states. This convergence came about through the growing debt crisis and the insistence of creditors that strict conditionalities established by the IMF be met before countries could be considered for debt rescheduling or further loans. Virtually every country of Africa south of the Sahara has had to accept the intervention of the IMF. Through the implementation of these agencies' structural adjustment programs, almost all countries have now adopted neoliberal, free-market economic policies. As well, the demise of the Soviet Union and of communist regimes in Eastern Europe effectively removed the main source of both ideological inspiration and development aid for Afro-Marxist states, and thus narrowed the options of every country.

Structural adjustment programs reflect to some extent the political, economic, and social realities of the individual countries in which they are implemented. However, what is much more striking is the uniformity of policy direction established through structural adjustment packages. In general, adjustment programs attempt to increase the solvency of debtor states through policies designed to stimulate economic growth and investment, and to reduce the cost of government by cutting back on the public sector. The ultimate objective of these policies is to ensure that African debtor states will be able to meet their debt repayment commitments.

The specific policy directions established through structural adjustment programs generally include the following:

- Massive currency devaluation
- Reduction of internal and external government deficits by limiting the growth of credit and money supply
- Liberalization of the economy, guided by

"market forces" domestically and "comparative advantage" internationally

- Encouragement of foreign investment
- Removal of high tariffs and quotas to encourage a more efficient allocation of resources in the economy
- Elimination of price controls and subsidies to encourage increased productivity, especially in agriculture
- Increasing export producers' share of world market prices
- Cutbacks in the state sector, including privatization of state enterprises and reduction in public-sector employment
- Introduction of cost recovery (user fees) in health care, for education, and for other social amenities

Structural adjustment has imposed onerous social, political, environmental, and economic costs. Among others, the United Nations Children's Fund (UNICEF) has shown how the higher cost of food and reduced access to health care have increased malnutrition and infant mortality. The poor, women, and children suffer the most as a result of reduced social expenditure. Structural adjustment has also helped to accelerate the environmental crisis by forcing both governments and families to liquidate ecological capital—cutting trees, for example—to make ends meet. As for the economy, adjustment programs have significantly distorted economic planning by focusing on short-term, balance-of-payments objectives at the expense of longer-term planning for balanced and equitable development.

Academics, international social agencies, and African governments have criticized structural adjustment as development orthodoxy. How could programs that exact such a heavy price for such questionable gains be considered a success? The critics claim that IMF and World Bank policies have not addressed, but rather have diverted attention from, the causes of African underdevelopment. These policies have eroded sovereignty and undermined the ability of African states to plan and implement rational development policies. Structural adjustment has been externally dictated and designed to address the concerns of Africa's creditors, rather than the development needs of the people of Africa.

Table 26.1 lists selected macroeconomic indicators of performance for a number of African countries in 1990 and 2000. The use of 1990 as a base year eliminates the most difficult early years of structural adjustment, and focuses instead on a period when widespread recovery and growth were expected. Among the countries listed are Ghana and Uganda, frequently mentioned as countries where improved economic performance demonstrates the value of structural adjustment policies. The data indeed show that Ghana and Uganda experienced notable growth in per capita income during the 1990s, as well as a substantial growth in exports. In many other African countries, however, per

TABLE 26.1. Economic Indicators for Selected Countries under Structural Adjustment Programs, 1990–2000

Country	GDP[a] (U.S.$, billion)		GNI[b] per capita average increase		Exports (U.S.$, million)		Debt (U.S.$, million)	
	1990	2000	U.S.$	1988–2000	1990	2000	1990	2000
Cameroon	7.07	8.69	570	−2.5%	2,002	1,880	6,676	9,241
Chad	1.13	1.41	200	−1.0%	188	183	524	1,126
Ghana	3.50	5.42	350	1.6%	897	1,670	3,881	6,657
Tanzania	6.87	9.32	280	0.5%	415	663	6,457	7,445
Togo	0.96	1.28	300	−1.2%	268	340	1,281	1,485
Uganda	3.10	6.25	310	3.7%	147	380	2,583	3,408
Zambia	2.75	2.91	300	−2.2%	1,309	800	6,916	5,730

[a]GDP, gross domestic product.

[b]GNI, gross national income.

Data sources: World Bank. *World Development Indicators 2002*. Washington, DC: World Bank. World Bank, 2002. *World Development Report 2002*. New York: Oxford University Press, 2002.

capita income fell and export earnings stagnated or even fell during the 1990s. In Cameroon, for example, per capita income fell by an average of 2.5% per year during the 1990s. Debt loads also increased in most African countries during the 1990s, in some cases substantially. Debt increased in six of the seven countries listed in Table 26.1; in Ghana, for example, the total debt increased by 72% during the 1990s.

In response to the limitations of structural adjustment programs, the United Nations Economic Commission for Africa (UNECA) in 1989 set forth an alternative approach: the African Alternative Framework to Structural Adjustment Programs for Socio-Economic Recovery and Transformation (AAF-SAP). It was argued that the key to a resolution of this deepening crisis was for African governments to regain control of economic and social policy. The AAF-SAP proposal urged that adjustment be accompanied by measures to facilitate longer-term socioeconomic change. These measures included increasing the development of intra-African trade; strengthening productive capacity, especially to grow more food; and giving priority to equity considerations in growth and in the provision of basic needs.

During the 1990s, the IMF and World Bank supported the development in several countries of measures designed to ameliorate the impacts of structural adjustment on poor people. These programs, which included relief from health care user fees for the poorest segments of society, were generally too limited in their scope and too complex to administer to have any broad-based benefits for the majority who were poor. However, these programs did represent an acknowledgment that structural adjustment had had adverse impacts on society.

Strategies for Economic Diversification

Economic diversification has been one of the stated objectives of structural adjustment programs. Of course, economic diversification has been a goal of all states since they achieved their independence. Data on the concentration of export earnings (see Chapter 25, Figure 25.2) show that many countries have fared rather poorly in their attempts to diversify. Nevertheless, a number of countries have had some success in developing and expanding newer sectors of the economy.

Figure 26.3 classifies the countries of Africa south of the Sahara according to the source (structure) of export earnings. Although most countries fit into the well-known traditional categories of agricultural commodity exporters, mineral exporters, and petroleum exporters, several African countries now depend primarily on the service and industrial sectors for most export earnings. However, it is important to acknowledge that while production for export is an important part of economic activity, the majority of production is for the domestic market and for household consumption.

The discussion below looks briefly at three areas of diversification—in agriculture, manufacturing, and tourism. Of course, this is not an exhaustive list. Every government dreams of having the same good fortune as Botswana, Gabon, and Equatorial Guinea, all of which were transformed overnight from dire poverty into countries with among Africa's highest per capita incomes through discoveries of diamonds and petroleum.

Often forgotten, but of great importance in many parts of Africa, are the repatriated earnings of citizens working in other countries. For almost a century, the economies of South Africa's neighbors—Swaziland, Lesotho, Mozambique, Malawi, and others—have depended heavily on remittances from migrants working in South Africa's mines and on its farms. Remittances are extremely important elsewhere as well; remittances from migrants to West Africa's coastal states, especially Côte d'Ivoire, have long been of critical importance to the poor countries of the Sahel.

In recent decades, remittances to Africa have increasingly come from other parts of the world. The foreign exchange earnings of Somalia, for example, now consist almost entirely of funds sent home from Somali emigrants living in Europe, North America, and the Middle East. Interna-

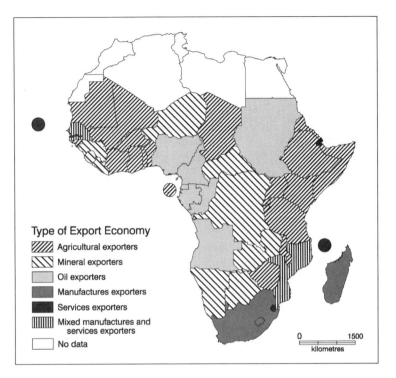

FIGURE 26.3. Structure of the export economies of African nations. Primary data source: United Nations Conference on Trade and Development (UNCTAD). *The Least Developed Countries Report 2002: Escaping the Poverty Trap*. New York: UNCTAD, 2002.

tional remittances are very important not only in poor, shattered nations such as Somalia, but also in more prosperous countries such as Senegal and Ghana, which happen to have very large, well-established expatriate communities in Europe and North America. These overseas communities have become an invaluable source of investment funds for a variety of business ventures in their home countries.

Agricultural Diversification

Agricultural diversification has long been touted as a development strategy for Africa's poorest countries. However, as noted in Chapter 14, diversification has often been achieved at a considerable cost. Land and water resources are diverted from production for local consumption, and the production schedules for local crops are distorted. Expanding the production of "traditional" export staples, such as cotton, coffee, and cocoa, has generally brought disap-

pointing results. Increased production of these price-inelastic commodities has only served to create large surpluses and to bring about a fall in market prices.

Countries that have sought to market new export crops have had mixed results, but some success. For example, Ghana has promoted the production of nontraditional crops such as cashews and pineapples. Fresh fruit and vegetables are produced for export to Europe in Kenya (now Europe's largest source of fresh produce) and other countries, including Senegal, Gambia, Tanzania, Zambia, and South Africa. Another major export of rapidly growing importance, especially from Kenya and South Africa, is fresh cut flowers. However, markets for these newer crops tend to be insecure, especially for smaller growers.

Manufacturing

Apart from South Africa, very few countries in Africa have a healthy manufacturing sector.

Nevertheless, manufacturing plays an important role in the economic development of certain countries. Senegal, once dependent almost entirely on groundnut exports, now exports more manufactured products and services than raw agricultural goods. Much of Senegal's industry involves the processing of raw materials for domestic use and exports—producing canned fish, fertilizers from phosphate ore, textiles from cotton, and cooking oil from groundnuts. Freer trade under the Economic Community of West African States (ECOWAS) has enabled Senegal to increase its sales of manufactured goods to neighboring countries that are less industrialized.

Very little of the global outsourcing of industrial production under globalization has reached Africa south of the Sahara. This is in marked contrast to many poorer countries in Asia and Latin America, where there has been considerable growth in labor-intensive industrial production. The U.S. African Growth Opportunity Act (AGOA) of 2000 has created new opportunities in the textile sector for African countries to increase exports to the United States, provided they meet specifications for the sourcing of raw materials (which must be of African or U.S. origin), social responsibility (e.g., labor practices), and quality assurance. Among the countries responding to these new opportunities are South Africa, Lesotho, Kenya, and Madagascar. Most African countries are unlikely to be able to meet AGOA requirements for some time to come.

Tourism

Between 1990 and 2000, both the number of tourists visiting Africa south of the Sahara and the receipts from tourism more than doubled. In 2000, the 17.6 million visitors visiting Africa south of the Sahara spent $6.6 billion, representing 8.4% of all export earnings. As Table 26.2 shows, tourism is concentrated in 10 countries, led by South Africa (number of tourists and receipts) and Tanzania (contribution to export earnings).

Tourism statistics are very difficult to interpret—among other reasons, because of the diversity of the visitors. Nigeria had 800,000 visitors in 2000, but had very low tourist receipts; this was probably because most arrivals were Nigerians living abroad, coming home to visit relatives. In contrast, most tourists in Kenya and Gambia are international visitors who patronize the formal "export" tourist sector.

Tourism is considered attractive because it not only earns foreign exchange, but also is an employment-intensive industry. Nevertheless, tourism has been criticized on a number of grounds. Tourism requires considerable upfront investment in hotels, roads, and other types of infrastructure. Many of the inputs, ranging from hotel furnishings to food and drink, may be imported. Much of the apparent earnings accrues to foreign-owned hotel chains, airlines, and travel companies. Most of the jobs in the tourist sector pay relatively poorly. Ille-

TABLE 26.2. Tourism in Africa: Leading Countries of Destination

Country	Inbound tourists, 2000 (thousands)	International tourist receipts, 2000	
		U.S.$, million	As % of exports
South Africa	6,001	2,526	7.5
Zimbabwe	1,868	202	8.6
Kenya	943	304	11.3
Botswana	843	234	7.7
Namibia	614	288	17.9
Zambia	574	81	9.7
Tanzania	459	739	57.7
Ghana	373	304	12.3
Senegal	369	166	11.5
Gambia	96	49	18.6

Data source: World Bank. *World Development Indicators 2002*. Washington, DC: World Bank, 2002.

gal activities such as prostitution and drug dealing thrive in tourist areas.

In spite of these limitations, tourism is likely to continue to expand in a minority of African countries that promise outstanding wildlife viewing opportunities and/or beautiful beaches, with international-quality facilities and service in a safe environment (see Figure 26.4). Unfortunately, few international tourists have much interest in learning more about African society and culture, or in visiting historical sites such as Great Zimbabwe.

NEPAD: Hope for Future Development?

The New Economic Partnership for Africa's Development (NEPAD) was established in 2002. This agreement between the African Union and the major economic powers in the G8 group of countries promised increased economic aid for African nations that met such criteria as economic reform and democratization. NEPAD represents recognition on the part of the world's richest and most powerful countries that African countries need considerably more external support if their economies are going to stabilize and grow. It also acknowledges that African countries need to undertake reforms

creating stable conditions conducive to economic development.

Africans are generally pessimistic about NEPAD, since promises of increased aid from the West and of reform by African leaders have been made many times before. However, the disasters that occurred in countries such as Somalia, Sierra Leone, Liberia, and the Democratic Republic of the Congo during the 1990s (Vignette 26.2) provide an ominous warning of the economic—not to mention the social—price to be paid for the prolonged lack of global interest in African underdevelopment.

Further Reading

The following are surveys of the current state of economies of African states:

Braga de Macedo, J., and O. Kabbaj, eds. *Reform and Growth in Africa*. Paris: Organization for Economic Cooperation and Development, 2000.

Brown, M. B. *Africa's Choices after Thirty Years of the World Bank*. Boulder, CO: Westview Press, 1997.

Leonard, D., and S. Straus. *Africa's Stalled Development: International Causes and Cures*. Boulder, CO: Lynne Rienner, 2003.

Lundahl, M., ed. *From Crisis to Growth in Africa*. London: Routledge, 2001.

United Nations Conference on Trade and Development (UNCTAD). *Economic Development in Africa: Performance, Prospects and Policy Issues*. New York: UNCTAD, 2001. (Available online at www.unctad.org/en/docs/pogdsafricad1.en.pdf)

Van de Walle, N. *African Economies and the Politics of Permanent Crisis, 1979–1999*. Cambridge, UK: Cambridge University Press, 2001.

Introductions to the role of ideology in African development are given in the following sources:

Isaacman, A., and B. Isaacman. *Mozambique: From Colonialism to Revolution, 1900–1982*. Boulder, CO: Westview Press, 1983.

Lubeck, P. M., ed. *The African Bourgeoisie: Capitalist Development in Nigeria, Kenya and the Ivory Coast*. Boulder, CO: Lynne Rienner, 1987.

Saul, J. "Ideology in Africa: Decomposition and recomposition." In G. M. Carter and P.

FIGURE 26.4. Tourism advertisement from Côte d'Ivoire. The traditional attractions of the African tourist industry—wildlife, sand and surf, and indigenous culture—are featured.

VIGNETTE 26.2. What Happens to the Economy When the State Collapses?

Contemporary Africa provides troubling examples of states in extreme crisis. More than a decade after the outbreak of civil war, Somalia still has no effective central government. Two regions, Somaliland and Puntland, have functioned as de facto independent states for several years, and regional warlords control most of the remaining territory and what remains of the economy. In Liberia and Sierra Leone, relative stability was restored after long civil wars that effectively destroyed each state as a functioning entity. Nevertheless, the postwar state is very weak and the peace is fragile in Sierra Leone, and war has broken out again in Liberia. The Democratic Republic of the Congo has been mired in a bitter civil war since the mid-1990s. Rebel groups have controlled much of the country, especially in the eastern region, since the mid-1990s. What has been most troubling about the Congolese conflict is the direct involvement of several other states as supporters of each side.

Civil wars have had devastating impacts on societies and economies in the affected countries. Liberia was formerly a major exporter of iron ore from mines located in three regions of the country. Iron ore accounted for half of the country's exports in 1989, just before conflict erupted. Liberia has ceased to be an iron ore producer. Not only has it been impossible to maintain production while fighting is in progress; the infrastructure to support the mining industry, notably equipment and transportation systems, has also been destroyed, damaged, or looted. The large investments that would be needed to reactivate the industry are unlikely to occur, given that Liberia's situation remains chaotic and that there are many alternate sources of iron ore.

In Sierra Leone, gross national income (GNI) per capita declined at an average rate of 6.1% per year during the 1990s when civil war raged. The country's exports fell from $138 million in 1990 to only $13 million in 2000. Gem-quality diamonds continued to be mined in Sierra Leone throughout the war. However, the diamond-producing areas were controlled by the rebel forces, which used profits from their sale to finance their purchase of weapons. Most of the gems were exported via neighboring countries; diamonds became Liberia's leading export for a time, even though Liberia produces few diamonds of its own. In an attempt to regain control of the diamond-producing regions, the government of Sierra Leone signed an agreement with a Canadian-based mining company that granted the company exclusive mining rights in return for its support in the civil war. The company hired expatriate mercenaries to fight for control of the mining areas. This troubling arrangement ended with the demise of the regime that had signed the agreement.

In the Democratic Republic of the Congo, war is estimated to have resulted directly or indirectly in 3.5 million deaths, and millions have been forced to abandon their homes and livelihoods. The country's 2000 GDP of $3.1 billion was only one-third the size of the economy a decade earlier. The most disturbing impact of the war has not been the shrinkage of the nation's economy, but rather the development in contested regions of a war economy based on the looting of natural resources. According to a United Nations report issued in 2002, gold, diamonds, platinum, and *coltan* (columbite–tantalite, an ore that is refined to make tantalum and used in various electronic devices) worth billions of dollars have been stolen by occupying military forces from Rwanda, Uganda, and Zimbabwe, and by Congolese soldiers and officials. Some 85 international companies were cited in the report for having participated directly or indirectly in the looting of Congolese natural resources. Indeed, the opportunity for immense profits was cited as a reason for the perpetuation of the war. Normal economic activities are unlikely to resume for many years, perhaps decades, after the war comes to an end.

O'Meara, eds., *African Independence: The First Twenty-Five Years,* pp. 300–329. Bloomington: Indiana University Press, 1985.

Young, C. *Ideology and Development in Africa.* New Haven, CT: Yale University Press, 1982.

There is a vast literature on the nature and effect of the structural adjustment packages offered by the IMF and World Bank. For example, see the following sources:

Cornia, G., and G. Helleiner, eds. *From Adjustment to Development in Africa: Conflict, Controversy, Convergence, Consensus?* London: Macmillan, 1994.

Mkandawire, T., and C. C. Soludo. *Our Continent, Our Future: African Perspectives on Structural Development.* Ottawa: International Development Research Centre, and Dakar, Senegal: Council for the Development of Social Science Research in Africa (CODESRIA), 1999.

Onimode, B. *The IMF, the World Bank and African Debt. vol. 1: The Economic Impact. vol. 2: The Social and Political Impact.* London: Zed Books, 1989.

Simon, D., et al. *Structurally-Adjusted Africa: Poverty, Debt and Basic Needs.* London: Pluto Press, 1995.

World Bank. *Africa's Adjustment and Growth in the 1980s.* Washington, DC: World Bank, 1989.

World Bank. *Adjustment in Africa: Reforms, Results and the Road Ahead.* Washington, DC: World Bank, 1994.

The following sources address the impacts of war on the economies of selected countries:

Hanlon, J. *Mozambique: Who Calls the Shots?* London: James Currey, 1991.

Richards, P. *Fighting for the Rain Forest: War, Youth and Resources in Sierra Leone.* Oxford: James Currey, 1996.

Internet Sources

Global Policy Forum. *Poverty and Development in Africa.* www.globalpolicy.org/socecon/develop/indexafr.htm

Third World Network (TWN). http://twnafrica.org

United Nations Economic Commission for Africa (UNECA). *Economic Report Card on Africa, 2002: Tracking Performance and Progress.* www.uneca.org/era2002

U.S. Department of State. *Africa—Trade and Economic Development.* http://usinfo.state.gov/regional/af/trade

27

Civil Society and Local Self-Reliance in Development

For too long, development theory and practice failed to pay sufficient attention to the strong tradition of self-help within local African communities. One of the basic premises of modernization theory, for example, was that the traditional economies had no real future, and that their present role was confined for the most part to contributing labor and savings in support of the emerging modern economy. Attempts to transform rural economies were usually based on introducing imported technologies that were too expensive for most people, often inappropriate for local conditions, and based on outsiders' assumptions about the needs and objectives of local residents.

The failure of conventional, top-down models of development brought calls for a new approach in which development is defined and implemented by the residents of local communities. Universal solutions are rejected in favor of approaches that are of necessity diverse—grounded in the unique socioeconomic and ecological conditions of particular localities. Several variations on community-focused, bottom-up development have been promoted since the 1970s; they include *development from below*, *development from within*, and *participatory development*. The currently popular civil society approach looks at the sociopolitical dynamics that underlie local development processes.

Local Initiative and Development in Historical Perspective

The support for local initiatives in development belatedly recognizes the importance of what African communities have done since time immemorial—namely, attempting to improve the quality of their own lives by using locally available resources. Unlike the predominant Western model, which puts emphasis on individual initiative, ownership, and economic gain, African societal traditions tend to emphasize collective security and cooperative forms of production. Land and other key resources are communally owned; making one's livelihood from these resources carries an inherent obligation to protect them for the use of future generations.

Collective security in precolonial African societies was achieved through working and celebrating together. Among the Langi of Uganda, for example, collective security was formalized through *wang-tic*—an institution in which community members came together at a member's request to undertake larger farming tasks such as bush clearing, weeding, or granary construction. Participants were rewarded with food and millet beer, and with the right to assume that the beneficiary would honor their own future requests for assistance. Cooperative farming involving the members of an extended family or a community not only spreads the workload in times of heavy labor demand, but also reaffirms the interdependence of the group. In times of crisis, those who are better off have a moral obligation to share their resources with relatives and neighbors who are in need.

The African tradition of sharing also extends to the products of specialized knowledge. Persons who know how to prepare medicines for particular illnesses, for example, carefully guard their secrets but willingly share their formulations with others in need. Thus the community benefits from the collective expertise of all its members.

Colonial and postcolonial policies often undermined traditional institutions and encouraged Western-style individualism, sometimes deliberately and at other times unwittingly. Among the Langi, the cooperative traditions of *wang-tic* were weakened by the increasing importance of cotton, a labor-intensive commercial crop. With the large-scale cultivation of cotton by merchants (many of whom were also money-lenders), *wang-tic* evolved into what was often essentially a forced-labor team—in which poorer farmers were obliged to provide labor on demand, but did not benefit from reciprocal assistance of equal value.

Several countries sought to capitalize on the tradition of community self-reliance as a means of addressing the high expectations of the early postindependence period. The Kenyan government encouraged communities, mostly rural, to undertake self-help development projects. These local *harambee* projects have accounted for the majority of Kenya's secondary schools, as well as a large proportion of its rural health facilities and water-supplying systems. Although individual *harambee* projects varied in terms of

their viability and longer-term success, collectively they contributed greatly to Kenyan rural development, at relatively low cost to the state. However, the dynamic self-reliance of the early phases of *harambee* was increasingly undermined by the Kenyan government's determination to direct local development initiatives to suit its own political purposes.

"Discovering" the Development Potential of Communities

The importance of encouraging communities to undertake their own self-help initiatives has emerged, particularly since the mid-1980s, as one of the most popular themes of the "development industry." At the most basic level, local development initiatives are seen as a necessary survival strategy, given the current state of crisis in Africa. Because governments have been unable and often unwilling to provide adequate development support to enable local people to improve the quality of their lives, communities have had to become as self-reliant as possible.

The shift in official thinking from macro-scale development projects to smaller-scale initiatives at the local level began with the 1973 speech by Robert McNamara as president of the World Bank, lauding the development potential of small farmers (see Chapter 14). However, the vision of locally based development promoted by the World Bank and other large agencies has been quite different from that envisioned by radical and populist supporters of local development, including such early proponents as E. F. Schumacher (*Small Is Beautiful*) and Paulo Freire (*Pedagogy of the Oppressed*) (see "Further Reading"). Freire's popular-education techniques, for example, were designed to help people to analyze their own situations as a basis for mobilization and change. As such, they were intensely political exercises, likely to result in the kinds of challenges to authority that occurred in conjunction with the organization of the Ada Salt Miners' Cooperative Society (Vignette 27.1). Conversely, the World Bank's vision of local development was apolitical; communities mobilized their own resources, worked together to solve local problems, and

VIGNETTE 27.1. Community Mobilization and the Struggle for Resources: The Salt Miners' Cooperative Society of Ada, Ghana

For centuries, the people of Ada on the coast of Ghana have exploited salt deposits produced by the evaporation of seawater. A French visitor in the late 18th century noted that attached to each household was a storage hut capable of holding at least 50 tons of salt. Salt was traded widely and at times was as valuable as gold. Chiefs and priests regulated the collection of salt, in the process ensuring that the entire community had fair access to the resource. In return, chiefs and priests received a substantial share of the salt that was collected.

In the early 1960s, ecological changes in the lagoons caused by the construction of the Akosombo Dam reduced natural salt production and severely damaged the traditional industry. A decade later, local chiefs granted exclusive mining rights to two commercial companies in return for promises of royalty payments. The people of Ada effectively lost access to their resource base. Conflict erupted in the early 1980s between the mining companies, which had fared poorly in their attempts to exploit the salt deposits on a commercial scale, and the local population. In order to fight for their rights to gather salt, 3,000 residents of Ada came together to form a Salt Miners' Cooperative Society. Their fight pitted the cooperative against local chiefs and elements in the government that favored commercial development of the resource.

The ensuing struggle was long and bitter. Many members of the cooperative were jailed, and one person was shot and killed. The government finally agreed in 1985 to grant open access to the salt lagoons. The cooperative's struggle continues to ensure that the hard-won right to mine salt is not lost. The cooperative's adversaries in this struggle have included not only the commercial companies, but also local and national governments. Following the early financial success of the salt cooperative, the district council and local council of chiefs both attempted to impose new taxes on salt mining. Factions within the Ghanaian government clearly continue to support the claims of a remaining commercial company to exclusive rights over the deposits.

The evolution of the salt cooperative illustrates that development from within involves much more than local communities' undertaking their own small projects. Genuine development from within is a political exercise that may well pit communities against outside interests, including those of the state. It may also divide communities, in this case along class lines. Still, the political dimension of the struggle at Ada may be seen as one of its greatest strengths. The people of Ada fought and won this battle for themselves, and in the process they learned much about the power of working together to achieve a common goal.

Based on T. Manuh. "Survival in rural Africa: The salt cooperatives in Ada District, Ghana." In D. R. F. Taylor and F. Mackenzie, eds. *Development from Within: Survival in Rural Africa,* pp. 102–124, London: Routledge, 1992.

did not really question or challenge the status quo.

Development from Within, a 1992 book by geographers Fraser Taylor and Fiona Mackenzie (again, see "Further Reading"), identified *territoriality* and *participation* as fundamental characteristics of genuine, progressive local development. Because of Africans' strong attachment to place, Taylor and Mackenzie argued

that development should be undertaken within distinct, locally recognized communities, not in spatial and social aggregates defined by outsiders. Within these local regions, participation involves community mobilization in an organized struggle over resources, and to improve the local quality of life.

Participatory approaches to development came into favor during the 1990s. Inspired by the work of scholars from the Institute of Development Studies at the University of Sussex, participatory development encourages local people to work together to analyze their situations and develop solutions in accordance with their own priorities and needs. Outsiders may help to facilitate the process and may offer professional advice, but ultimately are expected to take direction from the community. Participatory methods have been embraced by many development organizations operating in Africa (see Vignette 27.2, with Figure 27.1).

Political scientists with interests in governance and development have emphasized the importance of civil society for development, and called for the strengthening of civil society in Africa. *Civil society* refers to organized groups that operate in the "space" between the state and unorganized society. Business associations and local community development groups are examples of civil society organizations. These groups may exert political pressure on governments and thus make them more accountable. They help to provide voice for local needs, and seek support from the state, development agencies, or benefactors to achieve their objectives. They organize grassroots initiatives, such as community cleanups and tree-planting programs.

The growth of civil society as a force for development has been fostered by diverse interests. Frustrated by the failures of the African state to respond to their needs, local communities and interest groups (e.g., women's associations, trade unions) have mobilized to seek more resources and to do more for themselves. External donors have viewed these local organizations as a way to bypass governments and channel aid to local communities, and to encourage democratization through support for civil society. The increased opportunities for external funding in turn have brought about a proliferation of new nongovernmental organizations (NGOs), many of them relatively weak.

African governments have been ambivalent about the growth in civil society activity. Some countries, such as Zimbabwe and Cameroon, have regarded civil society movements as a potential challenge to their own authority, and have sought to limit their scope for political activism. Governments have also been weakened by the tendency of external donors to bypass them and direct funds directly to indigenous NGOs. On the other hand, many governments acknowledge the necessity and value of the work of NGOs at the grassroots, taking some pressure off the hard-pressed state.

Civil Society and Development

Contemporary local development initiatives come in all shapes and sizes. The following examples provide a sense of this diversity in the organization, objectives, and outcomes of development activities involving African civil society. The first two examples are organizations that operate at the local level but have grown to become major national organizations in their countries. The final two examples illustrate the operation of civil society on a smaller, community-based scale. (Figure 27.2 depicts another community-level project.)

The Naam Movement in Burkina Faso

The Naam Movement shows how community development may take place in a way that not only mobilizes local resources, but also educates and empowers those who become involved. This movement, first established in 1967 by a local teacher, had taken root in 1,350 Burkinabe villages by the mid-1980s. Naam was modeled on the traditional cooperatives that were organized around annual planting and harvesting activities. It is an approach that recognizes the creativity of ordinary Africans and emphasizes their right to be involved fully in decision making about their community's future.

The philosophy of the movement is to make each participating village responsible for its own development; to base development on what peasants know and want; and to rely as much as possible on locally available, low-cost materials and tools. An integral part of the development process is village-level "animation," through which communities may analyze their own situations and develop collective responses. The locally selected animators also help to educate villagers by setting up demonstration

VIGNETTE 27.2. Participatory Development in Theory and in Practice

Participatory research is a set of methods used to support bottom-up decision making through local people's organized analysis of their own needs. The inspiration for participatory methods came from diverse sources, including applied anthropology, farming systems research, and popular theatre. These sources contributed both general concepts (such as an emphasis on the creativity of the poor and the importance of empowerment) and specific methods (such as spatial analysis via transects and sketch maps, time analysis via work calendars, and social analyses via methods such as wealth ranking). Community members use these methods to make decisions about their situations and priorities, and to develop strategies for change. Figure 27.1 illustrates the evolution of participatory development from earlier assumptions about development over the past several decades.

The key to successful participatory development—according to Robert Chambers, who has written extensively about the approach—is a genuine reversal of roles, in which the poor have genuine control over the process and the outside experts listen, facilitate analysis, and wait for consensus to emerge from the process. It is an eclectic approach that will vary in relation to the character of individual communities and the nature of the problems being addressed. Sometimes there will be a clear consensus, but in other cases it will be difficult to overcome divisions in the community.

Critics of participatory development question the notion of community consensus building; all communities are divided in relation to such criteria as gender, age, wealth, and ethnicity. Not only does the community seldom speak with a single voice, but also local political and cultural dynamics make it difficult for disadvantaged groups (e.g., women and the poor) to be heard. The process itself may be very threatening for local leaders, since it invites citizens to question the status quo.

Other criticisms have been leveled at the supposed neutrality of the outside agents. Indeed, the approach has been widely adopted by the World Bank and other development agencies.

(cont.)

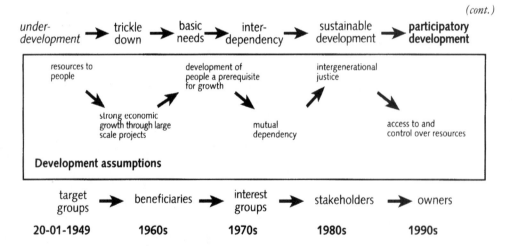

FIGURE 27.1. Participatory development in historical context. The diagram shows major changes in the dominant perspectives in development theory and practice, from the 1940s to the beginning of the new millennium. Source: A. Vainio-Mattila. "The seduction and significance of participation for development interventions," *Canadian Journal of Development Studies*, vol. 21 (2000), pp. 431–445. © 2000 by Arja Vainio-Mattila. Reprinted by permission of Canadian Journal of Development Studies.

VIGNETTE 27.2. *(cont.)*

They have often used a superficially participatory approach to conduct rapid reconnaissance surveys and to manipulate public opinion in support of predetermined objectives.

The application of participatory methods in community environmental management in Malawi illustrates both their limitations and possibilities. The supposedly participatory methods adopted by agencies such as the Department of Forestry were brief exercises that asked community members to rank their problems, but made no attempt to involve them in developing strategies for change. There was seldom a serious effort to ensure full participation by women in the deliberations. The agents frequently used the process to promote a predetermined "outsider" agenda. Visited communities frequently became cynical about participatory exercises that produced no benefits, often not even a follow-up visit.

An indigenous forest management project undertaken by the Wildlife Society of Malawi also used participatory methods, with greater success. Communities involved in the process engaged in a comprehensive analysis of indigenous environmental knowledge and their use of forest resources, as well as needs and projects they wished to pursue. Their decisions helped to shape the donor response. In some places, help was provided to individuals and community groups to establish woodlots. Other initiatives included the establishment of bee-keeping clubs and enterprises to make furniture from bamboo. Among the most important benefits of these projects were the change in attitude toward forest resources, and the replacement of destructive practices such as illegal charcoal production with sustainable income-generating activities.

Based primarily on R. Chambers. *Whose Reality Counts?: Putting the First Last.* London: Intermediate Technology, 1997. N. Moyo and F. Epulani. *Examples of CBNRM Best Practices in Malawi.* Blantyre: COMPASS–Malawi, 2001.

---

projects and informal schools. Although technical and material support from external sources is now available for Naam chapters, the key to the movement's success has been that projects continue to be locally conceived and locally controlled.

Naam communities have undertaken a wide variety of projects, including planting trees, controlling erosion, digging wells, establishing programs for primary health care, and organizing cereal banks that store grain as a protection against seasonal price increases and shortages. The construction of low stone barriers along slope contours to reduce soil erosion and trap runoff for crops and trees is of vital importance in areas with marginal amounts of precipitation and frequent drought. Building stone barriers is a traditional technique for collecting water used by the Mossi of Burkina Faso. Naam-sponsored animation informed villagers about how these structures could be built more effectively; individual Naam groups

made the decision themselves about whether this would be their immediate priority, and used their own resources to organize and do the work.

The AIDS Support Organization in Uganda

The devastating impact of HIV/AIDS has overwhelmed the resources of governments and international organizations, for the most part leaving those who are sick wholly dependent on their families for support. The AIDS Support Organization (TASO) has shown that there are alternative means of combating the epidemic. When 16 Ugandan volunteers formed TASO in 1987, it became the first African NGO to respond to the challenge of the HIV/AIDS epidemic. From these humble beginnings, TASO has grown into a large organization with many programs that has served as a model for community-based organizations in several other countries. Uganda has been lauded for achieving a significant reduction in HIV in-

FIGURE 27.2. Communal work on the construction of a channel to bring water to farmland, Niger. Photo: CIDA (R. Lemoyne).

fection rates. In this national campaign against HIV/AIDS, TASO has shown the difference that a determined, well-organized civil society organization can make.

From its inception, TASO's primary objectives were to fight the stigmatization of persons with AIDS and to break the silence about the disease. As of 2001, TASO had served over 65,000 people living with AIDS and their families, providing them with counseling, medical support, and social support. TASO has an extensive training program for volunteers to enable them to assist people living with AIDS in their communities. The organization also organizes public education campaigns to increase awareness about HIV and ways to prevent its spread.

Hometown Improvement Associations in Nigeria

Hometown improvement associations are found in many parts of Africa, but nowhere are they as widespread and instrumental in development efforts as in southern Nigeria. Many hundreds of these hometown associations exist and have existed for many decades. The primary purpose of these associations is to promote hometown development through self-help activities, such as the construction of a school or place of worship, or improvements to a road or drainage system. Donations are solicited from wealthier citizens, including both local residents and those living elsewhere. Citizens with limited financial means may contribute "sweat equity" as workers on a community development project.

Hometown improvement associations have a second role—namely, as a means of retaining ties to members of the community who have moved away and are residing elsewhere. Hometown societies based in larger towns have branches in large cities such as Lagos. These urban branch associations help to welcome and support newcomers to the city and to raise money for home. One measure of their success is the frequency of return migration of many urban-based Africans to their hometowns after they have retired from work. With increasing numbers of Nigerians living abroad, many overseas branch associations have been formed in Great Britain, the United States, and other countries.

The Mathare Youth Sports Association in Kenya

Mathare, a district in Nairobi, is widely known as one of Africa's largest and poorest slums. In response to the lack of recreational opportunities in the community, local activists organized the Mathare Youth Sports Association (MYSA) in the late 1980s to give youth a chance to play soccer. By 2000,

24,000 young people were playing on over 1,000 MYSA teams, and its senior club—against all odds—had won the Kenyan national championship.

But soccer is only the beginning. The MYSA also mobilizes its young members to engage in community development. It helps its members to organize environmental projects; MYSA soccer teams get league points not only for winning games, but also for collecting garbage! MYSA youth are also actively engaged in providing HIV/AIDS education for other youth and schoolchildren. The MYSA was honored at the 1992 Rio Earth Summit for its innovative programs. More significantly, Mathare has benefited not only from a cleaner environment, but also from happier youth and a sense of community pride.

The 1990s: The New Case for Civil Society Participation in Development

The proponents of civil society often refer to the failure of conventional, top-down approaches to development, and argue that universal solutions should be rejected in favor of strategies grounded in the ecological and socioeconomic realities of specific places. They contend that local communities have a vital interest in sustainability, since the communities have to live with the consequences of whatever development takes place. They see new opportunities for the involvement of women and disadvantaged minorities in expressing and acting upon their own aspirations. By encouraging broader participation, this approach may tap into previously ignored resources—for example, the varied kinds of indigenous knowledge found at the local scale.

This diversity of local communities results in a very complex and politically sensitive dynamic. Different classes, ethnic groups, genders, and age cohorts have distinct and often divergent aspirations. As for working together, actual communities are seldom perfect models of caring and sharing. It cannot be taken for granted that an entire community will benefit, and certainly not that its members will benefit

equally, from community development activities. Subordinate socioeconomic groups remain disadvantaged in comparison to local elites, and women continue to be dominated by men.

Primary health care is an example of a program that seems well suited for community development and empowerment: Communities can take responsibility for identifying and rectifying situations that threaten their collective health. However, development agencies have tended to emphasize the contribution of *funds* by communities, as opposed to community *empowerment*. The result has been a series of schemes to recover the costs of providing services. In the end, rural areas usually get inferior care for which they pay themselves, while most government resources continue to go into highly subsidized conventional hospitals in urban areas.

Development from within emphasizes what communities can do for themselves, but there are limits to what they can accomplish without active government support. For a number of years, youth groups in several wards of Kano organized community cleanup campaigns to remove accumulated garbage and to clear drainage channels. Frustrated by the municipal government's lack of support and failure to remove the collected refuse, community groups began to pile it in the middle of major roads, forcing the government to respond with better arrangements for refuse collection. They continue to resort to blocking roads with garbage whenever regular garbage collection is interrupted.

One of the assumptions implicit in much of the writing about local development initiatives is that African communities have substantial resources—whether of land, labor, or money—that can be tapped for local development. In reality, many African communities exist near the limits of survival and have few resources to spare. Thus, for example, the impositions of apparently modest user fees for health care and education have brought about notable reductions in utilization. Even the availability of free labor cannot be taken for granted; people may be able to participate in communal work only at certain times during the year, between farming seasons.

Civil Society and Local Development: What Role for the Future?

The recent wave of enthusiasm about mobilizing civil society for community self-help as a development strategy needs to be examined critically. After all, it is but the latest of several approaches to development that have promised to succeed where earlier models had failed. In each case, early enthusiasm gave way to more sober assessment and ultimately to widespread disillusionment. There is no reason to believe that the newly popular approach provides the last, best answer to the development question. What follows is a summary of some key strengths and limitations of this approach.

1. *In focusing on civil society, some of the most glaring weaknesses of earlier approaches to development are addressed.* Without doubt the most pervasive feature of past development models has been a shared arrogance about the assumed benefits and universal relevance of imported, packaged, designed-by-experts solutions that were conceived, driven, and controlled from above and outside, with little regard for local needs and realities. New approaches (especially participatory development) redirect control back to local communities, so that initiatives can be shaped in relation to local priorities, resources, and customs. The many potential benefits of this approach—including the political mobilization of marginalized groups, the development of a sense of community "ownership," and the application of indigenous knowledge—have been reviewed earlier in this chapter. Unlike top-down strategies that rely heavily on external inputs, local initiatives that are tailored to the availability of local resources have a better chance of long-term sustainability.

2. *Development initiated by civil society is very complex and unpredictable.* Genuine community development happens at its own pace and develops its own unique, local dynamic. It cannot be "willed" to happen. The process is complex because different groups in the community—defined in relation to gender, age, class, and ethnicity—have their own histories and agendas, as well as differing abilities to participate in and benefit from particular initiatives.

Difficulties also arise because many governments and external agencies are determined to control and direct local development efforts. There is a pervasive suspicion of genuine grassroots mobilization that could question or challenge the political status quo. The continuing arrogance of educated outsiders, who assume that the ignorant peasants need guidance, is often a problem. It may be possible to achieve temporary success through externally manipulated mobilization of local resources, but not a deeper, longer-term transformation.

3. *Local communities are not isolated, autonomous entities.* The linkages of local economies to the broader regional, national, and global economic spheres provide opportunities to strengthen local initiatives, but at the same time are key to understanding the marginalization of people and places in the African periphery. Local economies obtain some benefits from the urban–rural and interregional exchange of people, goods, money, and knowledge. Remittances by family members working away from home represent an important source of funds for local development, especially in poorer areas that have long served as labor reserves for more developed regions (Figure 27.3). The efforts of civil society groups to spur "development from within" depends partially or primarily on "resources from without."

One of the most important institutions linking local economies to the world beyond is the periodic market (Figure 27.4). At the level of the local economy, markets provide a venue for the exchange of a wide range of goods and services, as well as opportunities for people to exchange information and to socialize. However, as the degree of incorporation of rural Africa into regional, national, and international economies has increased, the role of rural markets has progressively shifted away from local exchange toward urban–rural and interregional exchange. Locally produced primary goods are purchased by urban-based traders for resale in the city, in other regions, or abroad. Manufactured goods of many kinds move in the opposite direction, destined for sale to rural consumers.

Although the terms of trade in the local marketplace represent only one factor determining the economic health of communities, it is an important one. Terms of trade influence

FIGURE 27.3. Modern house under construction, Jigawa State, Nigeria. This house, being built for an urban-based banker in his home village, symbolizes the importance of migrant remittances as a source of funds for local development. Photo: author.

decisions about what to grow—cash crops or food crops, for instance. When world prices for African primary products decline, the negative effects are felt at the local as well as the national level. When cheap imported manufactured goods flood the marketplace, the viability (and, ultimately, the survival) of local craft industries is at stake.

4. *Promoting the role of civil society in development must not be allowed to become merely "development on the cheap."* The new enthusiasm for locally initiated, locally funded development comes at a time when governments have few resources. Local self-reliance and the role of civil society are being endorsed at a time when foreign aid to Africa continues to be reduced, and when there is an increased emphasis on debt repayment and balancing national budgets. Critics of local development spearheaded by civil society have suggested that the enthusiasm for this approach has been driven by the desire of governments and development agencies to

FIGURE 27.4. Periodic market, Niger State, Nigeria. Photo: author.

absolve themselves of their responsibilities. In essence, "development at the grassroots" threatens to become "development on the cheap."

The possibility for greater self-help at the community level does not absolve the global community of its moral responsibility to support the quest for development and social justice in Africa. Nor does it in any sense absolve African governments of their responsibility to facilitate development at all levels, from the local to the national. The initiatives of individual communities are undertaken within the context of the state's social, economic, and political policies. Does the state take seriously its responsibility to bring development to all regions and all peoples, or does it serve only a portion of its citizens? Does the state strive to create an atmosphere of harmony, or does it resort to terror and to divide-and-rule tactics? Does the state encourage independent action by community groups, or is it threatened by such action?

Development or Postdevelopment?

The mobilization of civil society is not a panacea; it is not a substitute for a well-conceived, carefully implemented national strategy for development. It cannot by itself overcome the negative effects in peripheral regions of deteriorating terms of trade in the global economy. Without basic changes in the political–economic relations that underlie the current crisis in Africa, civil society is unlikely to achieve more than small, sporadic victories for the disadvantaged majority.

Civil society, out of both impatience and necessity, has taken up the development challenge. Successful local movements help to build a sense of empowerment among ordinary people, and with it the hope of further progress. However, governments and external donors remain ambivalent—generally welcoming, on the one hand, initiatives in which communities donate resources to do things for themselves while asking for little or no external assistance, but responding, on the other, with suspicion and often hostility to popular initiatives that present a real challenge to the status quo.

Writers from the postdevelopment school, such as Gustavo Esteva and Arturo Escobar (see ("Further Reading"), discuss the implications of the often uneasy relationship between communities and the state. Writing primarily about Latin America, they refer to development as being in a state of crisis: Promised benefits have seldom materialized, and indigenous social and economic institutions have been undermined by development. As a result, they state, an increasing number of marginalized, oppressed communities have rejected development as a goal, and instead work together to resurrect and strengthen local institutions and enhance local self-sufficiency.

Although the great majority of civil society organizations in Africa are striving, in one way or another, to promote development for their region or group, there are examples of civil society groups that function with little regard to the furthering of development goals, or that even organize to oppose modern development. Some African Christian churches, which have developed a distinct syncretic set of beliefs and rituals that have replaced Western Christian practice, would be an example of the first phenomenon. The rise of Islamic fundamentalism in Nigeria, such as the 'Yan Tatsine movement (see Chapter 28, Vignette 28.2), is an example of the second type of movement; its followers utterly rejected Western development and sought unsuccessfully to bring about its downfall.

Further Reading

Some of the precursors of the contemporary interest in local development initiatives include the following sources:

Freire, P. *Pedagogy of the Oppressed.* New York: Seabury, 1970.
Lipton, M. *Why Poor People Stay Poor: Urban Bias in World Development.* London: Temple Smith, 1977.
Schumacher, E. F. *Small Is Beautiful.* London: Abacus, 1978.

The case for community-based rural development is made in the following sources:

Ake, C. "Sustaining development on the indigenous." In *Background Papers: From Crisis to Sustainable Development.* Washington, DC: World Bank, 1989.

Brokensha, D. W., D. M. Warren, and O. Werner, eds. *Indigenous Systems of Knowledge and Development.* Lanham, MD: University Press of America, 1980.

Taylor, D. R. F., and F. Mackenzie, eds. *Development from Within: Survival in Rural Africa.* London: Routledge, 1992.

Participatory development, in theory and practice, is discussed in these sources:

Chambers, R. *Whose Reality Counts?: Putting the First Last.* London: Intermediate Technology, 1997.

Dow, H., and J. Barker. *Popular Participation and Development: A Bibliography on Africa and Latin America.* Toronto: University of Toronto Centre for Urban and Community Studies, 1992.

Gujit, I., and M. K. Shah, eds. *The Myth of Community: Gender Issues in Participatory Development.* London: Intermediate Technology, 1998.

Nelson, N., and S. Wright, eds. *Power and Participatory Development: Theory and Practice.* London: Intermediate Technology, 1995.

For studies that examine the problems and prospects of local development initiatives, see the following:

Chambers, R., A. Pacey, and L. A. Thrupp. *Farmer First: Farmer Innovation and Agricultural Research.* London: Intermediate Technology, 1989.

Harbeson, J. W., D. Rothchild, and N. Chazan. *Civil Society and the State in Africa.* Boulder, CO: Lynne Rienner, 1994.

Honey, R., and S. Okafor. *Hometown Associations: Indigenous Knowledge and Development in Nigeria.* London: Intermediate Technology, 1998.

Mackenzie, F. "Local initiatives and national policy: Gender and agricultural change in Murang'a District, Kenya." *Canadian Journal of African Studies,* vol. 20 (1986), pp. 377–401.

Tostensen, D., I. Tvedten, and M. Vaa. *Associational Life in African Cities: Popular Responses to the Urban Crisis.* Uppsala, Sweden: Nordiska Afrikainstitutet, 2001.

The following sources discuss a postdevelopment era, in which local civil cooperation is of critical importance:

Escobar, A. "Imagining a post-development era." In J. Crush, ed. *Power of Development,* pp. 211–227. London: Routledge, 1995.

Esteva, G., and M. S. Prakash. *Grassroots Postmodernism.* London: Zed Books, 1996.

Manzo, K. "Black consciousness and the quest for a counter-modernist development." In J. Crush, ed. *Power of Development,* pp. 228-252. London: Routledge, 1995.

Robinson, J., ed. *Development and Displacement.* Oxford: Oxford University Press, 2002.

Internet Sources

For information on participatory development, see the following websites:

Institute of Development Studies, University of Sussex. *Eldis: Participatory Monitoring and Evaluation.* www.eldis.org/participation/pme

World Bank Group. *Poverty Net: Participatory Methods.* www.worldbank.org/poverty/impact/methods/particip.htm

Websites for various types of African NGOs have been identified in previous chapters (e.g., see "Internet Sources" in Chapter 19 for a list of women's NGOs). The following are examples of the many African civil society organizations dealing with HIV/AIDS:

AIDS Foundation of South Africa. www.aids.org.za

AIDS NGOs Network in East Africa (ANNEA). www.annea.or.tz

The AIDS Support Organization (TASO). www.taso.co.ug

A growing number of community development organizations have websites. The second site is an umbrella association for Nigerian community groups, such as those from Idanre and Onitsha.

Idanre Development Association. www.idanre.org

Nigerian Community Associations in the United States. www.odili.net/community.html

Onitsha.org. www.onitsha.org

Political Geography: Regional Case Studies

In contrast to the thematic approach used in previous sections, the last three chapters are structured regionally. They explore aspects of the contemporary political geography of Africa south of the Sahara that have unfolded, and continue to unfold, in specific countries and regions. These processes have large implications for the future shape and "health" of African societies and economies.

Chapter 28 surveys the political geography of Nigeria since independence. The stability of Nigeria, with Africa's largest population and second largest economy, is important not only for Nigeria but also for Africa as a whole. Nigeria has survived a civil war, several coups d'état, and various other crises. Various policies to accommodate societal diversity have served Nigeria well, but with religious, ethnic, and regional tensions again rising, Nigeria's future remains in doubt.

Chapter 29 focuses on South Africa and its transition from apartheid to democratic rule by the majority. The chapter provides a historical review of the evolution and eventual demise of apartheid. South Africa today grapples with many social and economic challenges. Nevertheless, the political transition during the 1990s was far smoother and more peaceful than most outside observers had anticipated.

The final chapter is about political transitions in East Africa. Despite their many cultural and geographical similarities, Kenya, Uganda, and Tanzania have followed quite different political paths since they achieved independence. Now the three countries are moving gradually toward greater economic integration and political cooperation. The attempt to revive the East African Community after a quarter-century of inactivity is a tangible expression of this spirit of cooperation.

407

28

Nigeria: The Politics of Accommodating Diversity

The countries of Africa have all faced the challenge of nation building—the challenge of creating a sense of national identity and common resolve among their disparate peoples. African countries are characterized not only by ethnic, linguistic, cultural, and religious diversity, but also by colonial histories in which divide-and-rule tactics often increased the tensions caused by this diversity.

There are few nations in which the complexities of nation building are more obvious than in Nigeria. The political history of Nigeria has repeatedly revolved around such problems as the regional and ethnic divisions of power and wealth, and the search for balance between the demands of special interests and of national unity. The bloody civil war of 1967–1970 provides ample proof that the process has sometimes been very traumatic. Nevertheless, Nigeria has managed to survive, in large part because of a series of measures (e.g., the creation of new states) designed to reduce tensions by accommodating diversity.

Diversity

Nigeria was, and remains, a creation of British imperialism. The country was created in 1914,

when the protectorates of Northern and Southern Nigeria were amalgamated. Previously, the Colony of Lagos, the Niger Coast Protectorate, and the Royal Niger Company's territory had been joined together to form the Colony of Southern Nigeria. Nigeria's international borders were not "natural," in that they followed neither major physical features nor preexisting political borders. Moreover, the negotiations in Europe that drew Nigeria's borders brought about the separation of several groups between neighboring countries—for example, the Hausa–Fulani between Nigeria and Niger; the Kanuri between Nigeria, Niger, and Chad; and the Yoruba and Bussawa between Nigeria and Benin.

The rather artificial creation known as Nigeria brought together a great diversity of peoples. *Nigeria in Maps,* by Barbour et al. (see "Further Reading"), indicates that the country has 395 indigenous languages belonging to 10 different major language groupings. The diversity of peoples also extends to religion—including indigenous religions still widely practiced in some parts of the country, as well as the introduced religions of Christianity and Islam.

Nigeria is often characterized as an amalgam of three ethnic nations: the Hausa–Fulani in the north, the Yoruba in the southwest, and the Igbo in the southeast. According to the 1963 census, these three ethnic nations accounted for two-thirds of the Nigerian population; the Hausa–Fulani made up 29.5%, the Yoruba 20.3%, and the Igbo 16.6% of the population. Yet Nigeria also has several other large ethnic groups, including the Kanuri in northern Nigeria; the Tiv and Nupe in central Nigeria; and the Ibibio, Edo, and Ijo in southern Nigeria (Figure 28.1). Each of these groups makes up at least 1% of Nigeria's population, estimated in 2001 to be in excess of 125 million (Vignette 28.1). The remaining ethnic groups together represent less than 20% of the total population.

Each of these major ethnic nations evolved with its own cultural, religious, and political traditions. The Hausa culture developed in a series of autonomous city-states that were incorporated into the Fulani-ruled Sokoto Empire in the early 19th century. Although Islam had been practiced among the Hausa for hundreds of years, it was only after the ascendancy of the Sokoto Empire that strict adherence to Islamic teachings became the foundation upon which all aspects of Hausa–Fulani life were structured.

In southern Nigeria, the Yoruba and Edo were organized into a network of large and sophisticated semi-independent city-states. Conversely, the Igbo in precolonial times lived in spatially dispersed communities that functioned democratically without chiefs or other strong traditions of centralized government. Christianity made strong inroads during the colonial era in southern Nigeria, especially in Igboland. Among the Yoruba, there are more or less equal numbers of Muslims and Christians, as well as many adherents of indigenous religions.

The territories separating the major groups are occupied by a profusion of smaller groups with varied cultural and religious backgrounds. These groups represent a particularly important element in the *middle belt*, the zone located between the Hausa–Fulani heartland in the north and the Igbo and Yoruba homelands in the south.

The administrative structure under which the colony was divided into three regions (Northern, Eastern, and Western Nigeria) helped to intensify regional differences. This framework tacitly, if not explicitly, legitimated the domination of the various minority groups by the Hausa–Fulani, Igbo, and Yoruba. As a result, there was much discontent among the ethnic minorities in each region. This was particularly the case in Northern Nigeria, where the implementation of indirect rule meant that Muslim Fulani emirs continued to exercise considerable authority over many non-Muslim minority peoples.

The colonial administration of the three regions differed in a number of respects, the most important being the decision to limit the work of Christian missionaries and to give low priority to Western education in Islamic parts of northern Nigeria. These policies created a large north–south gap in education and development. At independence, southern Nigeria had 12 times as many students in primary schools and 10 times as many in secondary schools as equally populous northern Nigeria had. Interregional distrust intensified, and southerners, even in northern Nigeria, continued to hold most jobs in the public sector. The colony's economy was less diverse than its politics, however, being based primarily on the export of agricultural products (see Figure 28.2).

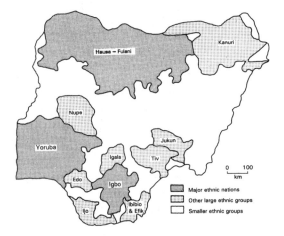

FIGURE 28.1. Major ethnic groups in Nigeria. The three largest groups dominate political discourse at the national level.

VIGNETTE 28.1. How Many Nigerians?

Few issues in Nigeria have evoked greater and more lasting controversy than that of the census; the question is not only how many Nigerians there are, but also where they live. Population figures are important, because they are the primary determinants of revenue allocations and parliamentary representation. The distribution of population also has a strong psychological effect, related to the concern of most southerners about "northern" control, and vice versa.

Conducting a census in this populous and diverse country has always been a difficult task. During colonial times, there was a strong incentive for individuals to avoid being counted because of the perceived linkage between the census and taxation. It has always been hard to get an accurate count in Islamic areas, where the seclusion of women in purdah is widely practiced. Since independence, concern about the regional distribution of political and economic benefits has provided a powerful incentive for localities and states to inflate their populations, not least because of the belief that there will be cheating in other parts of the country—"cheat to avoid being cheated."

The 1952–1953 colonial census recorded 30.4 million Nigerians, of whom 55% were in the Northern Region. It is usually considered to have been an undercount. The first postindependence census in 1962 proved to be a political bombshell: The population of the Eastern Region and Western Region was recorded as having increased by 70% in one decade, compared to only 30% growth in the Northern Region. With northern hegemony clearly at stake, the government rejected the 1962 figures and ordered a recount. The new 1963 census showed a population of 55.6 million, including nearly 30 million in the north. This time, even more people were found; the Northern Region's population was 67% higher than in 1952–1953, the Eastern Region was 65% higher, and the Western Region was almost 100% above the 1952–1953 figures.

Although the 1963 census results were widely rejected by demographers, the Nigerian government had little choice but to use them. Unfortunately, the next census to be accepted by the government did not take place until 1991. For three decades, estimates of Nigeria's national and regional populations used the 1963 figures as baselines for projections. Nigeria was forced, in effect, to "plan without figures."

Nigeria's next census, undertaken in November 1973, provided more surprises. The preliminary figures gave a national population of 79.8 million, more than 40% above the inflated 1963 total. Major regional differences in growth were recorded; whereas two of the southern states had lower populations, two of the northern states allegedly had almost doubled between 1963 and 1973. The release of preliminary results sparked a predictable response of outrage, especially in the southern states. The government had no option but to rule the 1973 census null and void.

A number of nationwide programs during the 1970s and early 1980s—national vaccination programs, the Universal Primary Education program, a national fertility survey, and voters' enumerations—provided partial population data suggesting that the 1963 totals might not be as inaccurate as once believed. Until 1992, there was general agreement that Nigeria's population was well over 100 million, and perhaps as high as 125 million. The World Bank's estimate for 1990 was 117.4 million. The release of preliminary figures from the 1991 census provided yet another shock: Only 88.45 million Nigerians were enumerated—at least 20–30 million less than anticipated. Large regional disparities in growth between 1963 and 1991 were found.

With an election campaign in progress at the time, it is hardly surprising that some claimed that the population had been undercounted in their parts of the country. However, the public response to the release of these figures was generally positive. The government was seen to have made every effort to depoliticize the process and to ensure as accurate a count as

(cont.)

VIGNETTE 28.1. *(cont.)*

possible. United Nations representatives who observed the census exercise expressed confidence in the results.

For Nigerian officials, having widely accepted population data helped to facilitate more effective development planning. It meant that one of the country's most divisive political issues could be set aside for at least a decade. However, no one familiar with the chaotic history of previous Nigerian censuses is likely to predict what surprises and controversies might emerge when the results of future censuses are released.

Based on R. M. Prothero. "Nigeria loses count." *The Geographical Magazine,* vol. 47 (October 1974), pp. 24–28. P. Idowu, N. Adio-Saka, and B. Olowo. "A game of numbers." *West Africa,* (March 30–April 5, 1992), pp. 539–541.

Nigeria since Independence

Nigeria became independent in October 1960, after several years of strenuous debate about the future shape of political institutions. Under the new constitution, considerable power was reserved for the regional governments of the federation. Much to the consternation of the southern politicians who had fought long and hard for independence, the federal government was dominated by conservative politicians from northern Nigeria who had been relatively inactive in the independence movement. Interregional—and, to a lesser extent, intraregional—tensions arose concerning, among other things, the allocation of funds for development, access to government jobs, the alleged rigging of census figures, and electoral abuse.

With Nigeria's fragile political institutions near collapse, several junior officers staged a bloody coup d'état in January 1966 and established an Igbo-dominated military regime. A northern countercoup occurred in July 1966, leading to the installation of Lieutenant Colonel Yakubu Gowon as leader. Disillusioned by this turn of events and the occurrence of widespread, bloody riots directed against Igbos living in northern Nigeria, the Eastern Region declared itself the independent Republic of Biafra in May 1967.

FIGURE 28.2. Groundnut pyramids, Jigawa State. Nigeria's colonial-era export economy, based on the export of agricultural products, has faded into insignificance since the rise of the petroleum economy in the 1970s. Photo: author.

For the next two and a half years, a bitter civil war was fought over the future shape of Nigeria. At stake was not only its territorial integrity, but also the question of who would benefit from the recently discovered petroleum deposits in the Niger Delta. Although there was widespread public sympathy in other countries for the Biafran cause, most governments supported the Nigerian campaign to reunite the country. To do otherwise, it was argued, would call into question the legitimacy of almost every African border and nation-state. However, the support that Biafra received from France, South Africa, and Portugal prolonged the war and increased the level of suffering. The Biafran resistance ended in January 1970, and the breakaway region was reincorporated into Nigeria with surprisingly few recriminations.

Military rule continued until 1979, when an elected civilian regime took power. Between 1970 and 1979, there were actually three military regimes. General Gowon retained power until July 1976, when he was overthrown. General Murtala Mohammed ruled for only seven months before he was assassinated in an attempted coup staged mostly by military personnel from Gowon's home base in the middle belt. Murtala Mohammed's former deputy, General Olusegun Obasanjo, then assumed power and presided over the return to civilian rule.

The 1970s brought hitherto unimagined growth to the Nigerian economy. Petroleum exports increased from 150 million barrels per year in 1966 before the war to 1.1 billion in 1980 (see Figures 28.3 and 28.4), and the value of oil in the world marketplace grew rapidly with the formation of the Organization of Petroleum Exporting Countries (OPEC). Both federal and state governments used the expanded income to undertake many social and economic projects throughout the country. Industry and commerce experienced rapid growth and diversification. The 1970s also saw the growth of conspicuous consumption among those able to gain access to oil wealth. The rapid growth of revenue, and the military government's policy of distributing these funds equitably to all states, led to a period of comparative political peace.

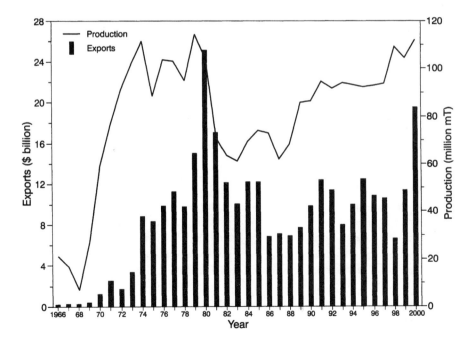

FIGURE 28.3. Petroleum in Nigeria's export economy, 1966–2000. Data source: United Nations Conference on Trade and Development (UNCTAD). *Commodity Yearbook*. New York: United Nations (various years).

FIGURE 28.4. Oil rig, Niger Delta. Photo: CIDA (B. Paton).

When civilian rule was reestablished in 1979, the conservative northern-based National Party of Nigeria (NPN) won, with support from the minority regions. The NPN was reelected in very controversial elections in 1983. It was felt by many that Nigeria's second democratic experiment had been grossly abused—especially by a clique of NPN insiders, popularly known as the Kaduna Mafia, who distributed wealth and privileges to themselves and their friends at an unprecedented rate. The government ignored a major economic crisis linked to the collapse of the global petroleum market, wasteful public spending, and large increases in the volume of nonessential imports.

The NPN government was overthrown in a northern-dominated coup on December 31, 1983. This coup was widely seen as a necessary response to the worsening economic and political situation. However, the new regime headed by Major General Muhammadu Buhari lost much of its public support as it revealed itself to be inflexible and autocratic. This regime was deposed in August 1985, in a coup led by Major General Ibrahim Babangida. Babangida's regime had its share of controversies and opposition, but it brought back a degree of stability to the country. Having inherited a massive foreign debt (most of it incurred during the 1979–1983 period), much-reduced export earnings, and a stagnating domestic economy, Babangida was forced to undertake a massive devaluation of the currency and to reduce public-sector spending greatly. Babangida displayed considerable political skill in managing to "sell" these highly unpopular measures to Nigerians.

Babangida took several initiatives to bring about a return to civilian rule in 1993. He imposed a rigid two-party system, arguing that defining the shape of political institutions would break the cycle of ethnic, religious, and regional rivalries that had caused Nigeria so much past grief. The mid-1993 presidential election was relatively orderly: The old regional voting patterns were less intense, and one of the candidates, Chief M. K. O. Abiola, won decisively. Just when Nigerian democracy seemed on the verge of a major triumph, Babangida annulled the election and declared that he would retain power. However, with political turmoil on the rise, Babangida soon resigned and put a hand-picked civilian president into office. Within weeks, the new president, who had absolutely no legitimacy, was overthrown in a bloodless coup staged by Major General Sani Abacha, the head of the armed forces throughout Babangida's years in power. The country was back to square one.

In mid-1994, the political climate of Nigeria became increasingly tense. Chief Abiola, winner of the annulled 1993 election, was imprisoned after declaring himself president. Prodemocracy forces launched a campaign of civil disobedience, including a general strike that crippled the vital petroleum industry and other sectors of the economy. The military government responded to these challenges by steadily increasing the level of intimidation and repression. Because the campaign against the northern-dominated military government was centered in the southern states, it served to revive long-standing regional, ethnic, and religious antagonisms.

For five years, Abacha ruled Nigeria with an iron fist. Many opponents of the regime, including former President Obasanjo, were jailed. Nigeria faced increasing internal tensions (which were repressed by force), as well as international alienation (Vignette 28.2). As the country's economic and political situation continued to deteriorate, there seemed no end in sight until Abacha died suddenly of natural causes.

Abacha's death set the stage for Nigeria's latest experience with civilian, parliamentary democracy. Multiparty elections were held in early 1999, with former President Obasanjo emerging victorious. Although Obasanjo was highly regarded internationally as a senior statesman, many Nigerians opposed his election. Indeed, because he was elected with only minority support both from his own Yoruba people, and from the Hausa-majority Muslim states of the far north, Obasanjo was unable to establish himself as a strong, take-charge president. He inherited an economy in rapid decline, with double-digit inflation and a $30 billion debt, despite oil production amounting to over 2 million barrels per day. In spite of his attempts to reduce introduce stronger fiscal management and to reduce the government's stake in inefficient parastatal enterprises, the economy continued to stagnate.

Obasanjo's major political challenge came from several northern states that introduced Islamic *sharia* law. Even during colonial times, *sharia* had continued to be applied in civil cases (e.g., divorces) involving Muslims. However, the current extension of a very strict form of *sharia* to all cases fueled a massive Christian–Muslim conflict. It became a political flashpoint that led to deadly riots in several cities, and caused many Nigerian migrants to abandon their businesses and return to their home regions. On a broader scale, the *sharia* controversy served as the locus for north–south political struggle. Obasanjo won reelection in 2003 with a high level of southern support, but with little support in the Muslim north.

Accommodating Diversity

Nigerian governments since independence, in their quest for political stability, have used several specific strategies and a stated commitment to equitable sharing of government resources, development, and power. Some of the strategies used by the Nigerian state have explicit spatial dimensions to them. Three of these—the creation of new states (see Figure 28.5); the decision to relocate the capital city; and the commitment to an equitable formula for sharing revenues and other benefits dispensed by the federal government—are discussed below. In every case, political imperatives have sometimes conflicted with economic rationality. It may be argued, however, that it has been a necessary price to pay for a united Nigeria.

The Formation of New States

As described earlier, Nigeria inherited a very unwieldy political structure at independence. The Northern, Eastern, and Western Regions were large and powerful enough to be semi-autonomous. Development tended to be highly concentrated, and large areas were virtually ignored. Minority groups felt excluded from decisions and believed that any sign of political insubordination would result in even greater penalties for their homelands.

The creation in 1963 of the Mid-Western Region in part of the Western Region was the first step in the dismantling of the old political structure, but long-standing appeals for the creation of regions in minority-dominated areas were not addressed. The government announced the formation of a new 12-state structure in 1967. This decision was taken hastily in response to the imminent threat of Biafran secession. The move was justified as a response to the aspirations of minorities, but in reality it was designed to deprive the rebellious Eastern Region of much of its territory, economic base, and population. For convenience, the boundaries of the new states followed existing provincial boundaries within the regions. The 12 newly created states varied greatly in area (from 3,577 to 272,726 km²), population (from 1.5 to 9.5 million), and resource endowment. For example, whereas some states had a surplus of qualified personnel, the development of other states was slowed because of an acute shortage of well-educated workers. Most of the states' revenue was derived from the federal government's transfer payments.

VIGNETTE 28.2. Uneven Development, Local Resistance, and State Violence

The politics and economics of oil have been basic defining features of Nigerian development since the 1960s. For many Nigerians with direct access to public resources, the oil boom created undreamed-of opportunities for accumulation. For others, the oil boom brought not only very few rewards, but also the destruction of their ways of life. Whereas most of those who had been marginalized simply struggled to survive, some responded with defiance and resistance. The *'Yan Tatsine* of Kano and the Ogoni of the Niger Delta are two such groups.

The *'Yan Tatsine* were followers of an unorthodox, outspoken religious leader, Alhaji Mohammed Marwa Maitatsine. The meaning of *Maitatsine* in Hausa is "the one who damns"; this alluded to his frequent, bitter public condemnation of the Nigerian state and the effects of modernization and corruption on northern Nigerian society. His greatest wrath was directed at the conspicuous consumption of the wealthy elites and the police, whom he accused of being the elites' agents. Maitatsine became increasingly influential, especially among Kano's unemployed migrants and among the *gardawa* (young men studying the Koran and working part-time to support themselves).

On December 18, 1980, Maitatsine and his followers staged an uprising, taking over a part of the old city of Kano and for 10 days fighting off the police. The uprising was finally quelled by the armed forces, following several hours of artillery bombardment. The official death toll was 4,177 *'Yan Tatsine,* several police and military personnel, and numerous civilians. Maitatsine himself was killed as he and his followers attempted to escape.

To the religious, political, and economic elites of Kano, Maitatsine was a fanatic and infidel who had espoused false Islamic beliefs and had falsely proclaimed himself a prophet. However, Maitatsine's analysis of Nigerian society, and especially his call for a return to basic religious values, had wide support among the common people of Kano and other northern cities. The Nigerian government blamed the uprisings on foreign Islamic fundamentalists intent on destabilizing Nigeria. However, the official inquiry noted that the vulnerability of unemployed and alienated migrant youth had contributed to the uprising.

The Ogoni are an ethnic group of some 500,000 people inhabiting part of the Niger Delta. In theory, living at the epicenter of Nigeria's oil region should have been a boon to the Ogoni. In reality, oil production devastated their home region. Much of their farmland was lost when pipelines and other facilities were constructed. Constantly recurring oil spills polluted creeks and destroyed marine life, and irreparably damaged farmland. Air pollution, especially from the flaring of natural gas, damaged crops and adversely affected people's health. The Nigerian state and the multinational oil companies operating in Ogoniland were oblivious to the protests of the Ogoni. Not only were they denied compensation for their losses, but they received almost no benefits in the form of employment, health care, or other services.

The Movement for the Survival of the Ogoni People (MOSOP) was formed to give voice to the community's outrage. Abacha's government responded to the protests by arresting several of the MOSOP leaders, including Ken Saro-Wiwa, a prominent writer, on dubious charges of planting explosives. Ignoring local and international protests, the Nigerian state executed Saro-Wiwa and eight others in November 1995. As a result, Saro-Wiwa became a martyr to the cause of social and environmental justice for the people of the Niger Delta, and attention was drawn to the complicity of the oil multinationals in the oppression that was taking place.

The Nigerian government responded to the challenges of *'Yan Tatsine* and MOSOP with brute force that quelled the dissents, but in doing so it revealed fundamental contradictions in the Nigerian state. At the very least, these movements exposed the fiction that the Nigerian leadership represented the interests of the people. More fundamentally, they may have revealed that the Nigerian petro-state, functioning as a coalition of elite interests, literally has no future.

Based on M. Watts. "Black gold, white heat: State violence, local resistance and the national question in Nigeria." In S. Pile and M. Keith, eds. *Geographies of Resistance*, pp. 33–67. London: Routledge, 1997.

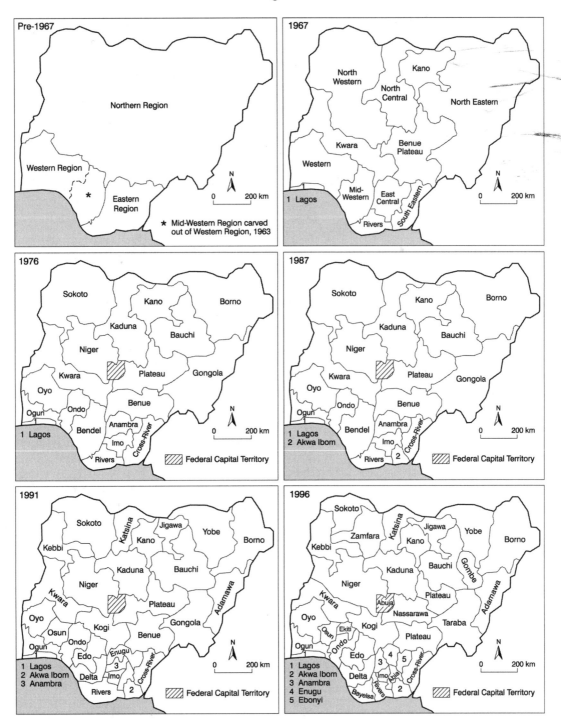

FIGURE 28.5. The evolution of Nigeria's political map, 1960–2002. Nigeria's political map has been in a state of flux since independence, but there is pressure from regional and ethnic groups to create even more states.

The 1967 reform of the political map did not end the clamor for additional states. Indeed, the end of the civil war and Gowon's promise of a return to civilian rule intensified demands for statehood in many areas. The basis of these demands varied; among the most common complaints were alleged discrimination or neglect by the existing state governments, long-standing distrust between ethnic groups, the righting of colonial wrongs, the need for balanced development, and the large size of certain states compared to others. Predictably, the creation of more states was one of the first promises made in 1975 following the Murtala Mohammed coup d'état. The number of states was increased subsequently to 19, 10 of which were located in the north.

With the return to civilian rule in 1979, and especially at the time of the 1983 election, the clamor for more new states reached a fever pitch. A senate committee recommended a total of 45. However, the growing economic crisis forced the government to postpone any decision. Following the 1983 coup, the new military government banned all efforts aimed at creating new states. Nevertheless, under Babangida's administration, 2 new states were added in 1987, and 9 more in 1991; an additional 6 were created under Abacha in 1996. Nigeria now has 36 states, varying considerably in size, wealth, and prospects for future development.

The proliferation of states may be seen as a valid response to political reality or as a massive waste of scarce resources. On the one hand, it addresses the aspirations of many minority groups, and it has brought about a more even distribution of development than would have been likely with the old regional structure. Moreover, the subdivision of large, wealthy states has contributed to political stability by ensuring that no state will be able to seriously challenge the authority of the federal government. On the other hand, the growing number of states is a financial burden that Nigeria can ill afford. The use of scarce resources to pay more civil servants' salaries and to build more government office towers should not be a high development priority. Among the strongest proponents of state creation have been entrepreneurs, civil servants, and politicians—groups whose self-interest in gaining more direct access to government-related opportunities is obvious.

Nevertheless, the proliferation of states has helped to diffuse regional, religious, and ethnic hostilities, and has contributed to the dispersion of oil wealth and economic growth in previously neglected areas. On balance, this expensive and often wasteful exercise has helped to make possible the survival of Nigeria.

Creating a New Capital

Lagos, Nigeria's capital city, was the subject of controversy for a long time. Some argued that Lagos was too strongly associated with the Yoruba and that the capital city should be a symbol of all Nigerians. Even in colonial times, it was a crowded, rather unattractive city. The oil boom of the 1970s brought rapid population growth and much new development—commercial, industrial, and institutional. Day-to-day activities and longer term planning became increasingly difficult in this already congested city. Nigerian government offices were situated in crowded downtown locations where orderly expansion was difficult.

In 1976, the government established a commission to consider the capital-city question. It recommended establishment of a new capital at Abuja, at the time a small village located north of the confluence of the Niger and Benue Rivers. The area around the new city was designated the Federal Capital Territory, to differentiate it from any one state. The choice of Abuja, centrally located within the country and away from the Hausa, Yoruba, and Igbo heartlands, symbolized a new beginning for the nation.

Making Abuja the new capital has been an expensive undertaking; by 1986, the costs had reached some $4.5 billion. Many civil servants were reluctant to move from the cosmopolitan city of Lagos to the still-developing Abuja. Nevertheless, the project was welcomed as an important symbol of progress. Within two decades of its establishment, Abuja had grown to about 1 million people, five times the number planned for in the original design. The utopian, planned inner core of the city contrasts markedly with the sprawling squatter settlements that have grown around its periphery. Although most Nigerians still agree with the decision to move the capital to Abuja, many are unhappy with the way it has developed.

Governing According to the "Federal Character" of Nigeria

For each Nigerian government, one of the most daunting challenges has been to allocate resources and rewards in a way that minimizes the inevitable charges of regional and ethnic bias. In a country where the federal government controls the vast majority of public wealth, revenue allocation is bound to be a controversial matter. The Nigerian government provides about 90% of the revenue needed by state governments, including a 70% share from a common pool of oil revenues. Half of this money is allocated equally to each state, and the rest is divided in proportion to each state's population.

There has been continuing debate about the allocation of other rewards—ranging from positions in the federal cabinet, civil service, and military; to the allocation of admissions to universities; to the location of such federally funded projects as new industries, colleges, and roads. It is seldom possible to treat all states equally in such undertakings. Even if equal distribution were possible, it would often be counterproductive, with efficiency and other reasonable criteria being sacrificed in the quest for equality.

The constitution of the Second Republic (1979–1983) was explicit about how the "federal character" of Nigeria was to be protected: It provided for the appointment of at least one cabinet member from each state, and specified that ambassadors and the armed forces should also reflect the nation. The principle of equity was also extended to scholarships and admission to federal training programs. However, even the inclusion of these provisions in the constitution did not stop the bickering and charges of favoritism.

Planning on the basis of political/equity principles has produced some dubious results. A glaring example concerns the location of rolling mills to process output from the steel com-

plex at Ajaokuta near the junction of the Niger and Benue Rivers. Over 25 years after it was started, Ajaokuta has yet to become operational. The three rolling mills were built at Oshogbo, Katsina, and Jos—cities that have very little other industry and are not well located to supply what the existing steel-utilizing industries need. All of these mills are badly underutilized.

As states continue to proliferate, the problems of adhering to established equity principles increase. In the late 1970s, there was a major expansion of the national university system to ensure that institutions of higher learning would be located in every state. It is doubtful that a further expansion of this system to open more universities in the most recently established states can be justified, given the unmet needs of existing universities and the scarcity of resources.

Continuing Challenges

Nigeria has survived four tumultuous decades of independence. The future promises to be, if anything, even more challenging. The discussion turns briefly to three key questions that will remain unresolved for a long time to come.

The Future Role of the Military

Nigeria has had a succession of military and civilian governments. Obasanjo's current regime is the third civilian government to take office. Nigerians have been ambivalent about the role of the military. Most aspire to a truly democratic system that would produce an effective and even-handed government. Their experience with civilian rule has left them skeptical and cynical. In general, military governments have been more successful than their civilian counterparts in quelling regional bickering. Nigerian civilian regimes have found equity concepts troublesome, owing to inevitable charges of political favoritism. However, under Abacha's rule, the tense political environment and suppression of dissent, especially in the south, created strong antimilitary sentiments.

Given the military's past history of repeatedly intervening in politics, Nigeria is unlikely to have seen its last coup. The nature of future military regimes cannot be taken for granted. Nigeria has been fortunate to have had several military rulers who were relatively competent and committed to national unity. The Abacha regime's divisive and repressive rule represented a new and disturbing trend. All military rule is ultimately based on the power of the sword; there is always a tangible risk that this power may be used as an instrument of terror and oppression.

Religious Fundamentalism and Intolerance

Religious tensions, which played a prominent role in events leading up to the civil war, have become Nigeria's most contentious issue. The most serious cause of these tensions has been the growth of religious fundamentalism among both Muslims and Christians. Fundamentalism has encouraged a much less tolerant attitude toward "unbelievers" and a more aggressive approach to proselytizing. Tensions have been especially great in communities where both faiths have been competing for converts, as in the middle belt.

There has been a long-standing religious controversy about the role of the Islamic legal code, *sharia*. In successive constitutional discussions, Muslims from the northern states demanded that the Islamic legal code be used in criminal cases involving Muslims. Non-Muslims strongly rejected the notion of adopting *sharia*, even in limited circumstances. The example of the Sudan, where attempts by an Islamic government to impose Islamic institutions in the south have led to civil war, served as a potent symbol for many Nigerians of why *sharia* was seen as a threat to religious freedom.

Following Obasanjo's election in 1999, 12 northern states (Figure 28.6) implemented the *sharia* code in defiance of the Nigerian federal government and minority groups within the states. The type of *sharia* introduced into Nigeria was a rigid version of the code, which permitted punishments such as amputation of limbs and death by stoning. A series of internationally publicized cases involving death sentences created a storm of protest both within Nigeria and abroad.

The imposition of *sharia* increased tensions between Christians and Muslims, culminating in a series of deadly religious riots across Nigeria. Although riots had occurred several times and in several places during the 1990s, the 2000–2002 clashes in Kaduna, Kano, Jos, and Lagos were much deadlier than the previous disturbances. A considerable number of lives have been lost, but in the longer run the most important casualty may have been Nigerians' trust in their ability to live together in harmony (which the placement of the principal religious buildings in Abuja was intended to symbolize; see Figure 28.7).

The Sustainability of Nigerian Development

Modern Nigeria has been built with petroleum revenue. Petroleum now accounts for 99% of exports by value; the old export economy based on agricultural products has all but disappeared. Petroleum has made a few Nigerians very wealthy, but it has not made Nigeria a wealthy nation. Nigeria's per capita income of $260 was the 15th lowest in the world. Its per capita income was the same in 2000 as it had been a decade before.

In spite of Nigeria's massive profits from oil, some 70% of the population lives below the international poverty standard of $1 per day. Poverty has doubled since 1980. Ordinary Nigerians in both rural and urban areas face major challenges in trying to make ends meet. Poverty is the root cause not only of hunger and ill health, but also of political and social unrest. The alienation and disentitlement of the poor seem especially intolerable in the context of the country's oil wealth. It is not coincidental that the most determined resistance to the state has come from the Niger Delta, the center of oil industry. The Ogoni and other delta peoples were reacting to the injustice of environmental damage from construction projects, oil spills, and air pollution that threatened their livelihood; these oil riches had all gone elsewhere, leaving them without even health care or other basic services (see Vignette 28.2).

Nigeria's macro-level economy is no health-

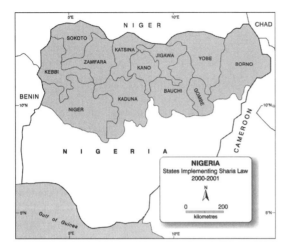

FIGURE 28.6. States implementing *sharia*, the Muslim legal code, 2000–2001.

developing a strong, coherent economic strategy for development. The World Bank substantially reduced its loans to Nigeria in 2002, citing the country's continuing macroeconomic instability and lack of overall growth.

The prospect of a future Nigeria with rapidly declining petroleum reserves, few other sources of wealth, and an economy addicted to large petroleum subsidies is very real. It will be a Nigeria with a much larger population than today, and inevitably will still be coping with the same problems of accommodating diversity. It will be a Nigeria with disparities of wealth and power even greater than those of today, in which the poor will be left even more to their own devices. The shape of a future Nigeria without substantial oil wealth is uncertain in the extreme.

ier. The creation of new states, and the initiatives sponsored by the states, have siphoned off a large proportion of the country's resources. The unstable political environment has both adversely affected the domestic economy and discouraged foreign investment in the country. Despite its huge population and resource wealth, and the devaluation of its currency to less than 1% of its 1980 value, Nigeria has continued to have difficulty attracting investment. Obasanjo's government has proven incapable of

Further Reading

Good, albeit dated, introductions to the geography of Nigeria are found in the following sources:

Barbour, K. M., J. S. Oguntoyinbo, J. O. C. Onyemelukwe, and J. C. Nwafor. *Nigeria in Maps.* New York: Africana, 1982.
Morgan, W. T. M. *Nigeria.* London: Longman, 1983.
Oguntoyinbo, J. S., O. O. Areola, and M. Filani,

FIGURE 28.7. View of Abuja, showing the Central Mosque and the Christian National Church. The proximity of these two buildings in the new capital city symbolizes the quest to reconcile religious differences as one of the keys to securing a peaceful future for the country. Photo: R. Maconachie.

eds. *A Geography of Nigerian Development.* Ibadan, Nigeria: Heinemann, 1978.

Udo, R. *Geographical Regions of Nigeria.* London: Heinemann, 1970.

On the history of Nigeria, see the following sources:

Crowder, M. *West Africa under Colonial Rule.* Evanston, IL: Northwestern University Press, 1968.

Falola, T. *The History of Nigeria.* Westport, CT: Greenwood Press, 1999.

Isichei, E. *History of Nigeria.* London: Longman, 1983.

Lugard, F. D. *The Dual Mandate in British Tropical Africa.* London: George Allen and Unwin, 1922.

On the development of the modern Nigerian state, see the following sources:

Ajayi, J. F. A., and B. Ikara. *Evolution of Political Culture in Nigeria.* Ibadan, Nigeria: Ibadan University Press, 1985.

Graf, W. D. *The Nigerian State.* Portsmouth, NH: Heinemann, 1988.

Kirk-Greene, A. M. H., and D. Rimmer. *Nigeria since 1970: A Political and Economic Outline.* London: Hodder and Stoughton, 1981.

The following are studies of Nigeria's political economy, with particular reference to oil:

Bevan, D. *The Political Economy of Poverty, Equity and Growth: Nigeria and Indonesia.* Oxford: Oxford University Press, 1999.

Biersteker, T. J. *Multinationals, the State, and Control of the Nigerian Economy.* Princeton, NJ: Princeton University Press, 1999.

Ihonvbere, J. O., and T. M. Shaw. *Towards a Political Economy of Nigeria.* Brookfield, WI: Avebury, 1988.

Ihonvbere, J. *Nigeria: The Politics of Adjustment and Democracy.* Somerset, NJ: Transaction, 1994.

Khan, S. A. *The Political Economy of Oil.* Oxford: Oxford University Press, 1995.

"Oil, debts and democracy in Nigeria." *Review of African Political Economy,* no. 37 (1986), pp. 1–105.

The following studies examine Nigeria's contemporary social and political crises:

Beckett, P., and C. Young, eds. *Dilemmas of Democracy in Nigeria.* Rochester, NY: University of Rochester Press, 1997.

Falola, T. *Violence in Nigeria: The Crisis of Religious Politics and Secular Ideologies.* Rochester, NY: University of Rochester Press, 1998.

Maier, K. *This House Has Fallen: Nigeria in Crisis.* Boulder, CO: Westview Press, 2003.

Manby, B. *The Price of Oil: Corporate Responsibility and Human Rights Violations in Nigeria's Oil Producing Economy.* New York: Human Rights Watch, 1999.

World Bank. *Poverty Profile for Nigeria: 1985–1996.* Washington, DC: World Bank, 1998. (Available online at http://www4.worldbank.org/afr/poverty/pdf/docnav/03007.pdf)

Internet Sources

Those interested in keeping up with Nigerian current affairs or obtaining further information about the country will find several comprehensive gateway websites:

AllAfrica.com. http://www.allafrica.com

Federal Government of Nigeria. *Nigeria.gov.ng: The Building of a Democratic Nation.* http://www.nigeria.gov.ng

Motherland Nigeria. http://www.motherlandnigeria.com

Nigeria.com: National News. http://www.nigeria.com

Nigeriaworld. http://odili.net/nigeria.html

Nigeria Business Info.com. http://www.nigeriabusinessinfo.com

The following sites provide information on current human rights issues in Nigeria:

Centre for Democratic Development Research and Training (CEDDERT) [Zaria, Nigeria]. http://www.ceddert.com

Human Rights Watch. *Africa: Nigeria.* http://www.hrw.org/nigeria

Sierra Club/Movement for the Survival of the Ogoni People. *Human Rights and the Environment. International Campaigns: Nigeria.* http://www.sierraclub.org/human-rights/nigeria/mosop

Women's Rights Watch, Nigeria. http://www.rufarm.kabissa.org

29

South Africa in the Postapartheid Era

The year 1994 will be remembered as a turning point of epic importance in the history of South Africa. The prolonged struggle of the majority of South Africans for freedom and an end to apartheid culminated in April 1994 in the country's first truly democratic election. The geography of South Africa developed over centuries, but especially from 1948 to 1994, as the applied geography of racism. Apartheid was a complex system that grew out of a long history of racial oppression, and that operated on many levels to shape South African society. In spite of the demise of apartheid, the country's geography continues to reflect in countless ways the legacy of 300 years of white domination.

Apartheid in Historical Perspective

Although the implementation of apartheid as official ideology began in 1948 following the electoral victory of the Afrikaner-based National Party, discrimination and separation by race had been integral to South African society for a very long time before that, legitimized through a series of increasingly discriminatory laws. To understand the nature of apartheid, it is important to examine the deep historical roots of racism in South African society.

The Precolonial Era

Conventional European histories of South Africa perpetuated the convenient fiction that the history of South Africa began in 1652, with the establishment of the first European settlement at what is now Cape Town. For apartheid-era governments, this version of history provided a key justification for the European occupation of South Africa and the pursuit of "separate development." If, as they argued, blacks and whites had arrived in South Africa at the same time, no group had a prior claim to the country.

Archeologists have shown this claim to be completely false. At the time when Europeans first arrived, there were four main groups of indigenous South Africans:

- Khoikhoi pastoralists living near the Cape
- San hunter-gatherers of the semiarid interior
- Nguni peoples, including the Zulu, Swazi, and Xhosa, living in the Natal coastal plain and the interior
- Sotho and related groups occupying the central interior

The Khoikhoi and San are closely related peoples descended from the earliest Stone Age inhabitants of southern Africa. South African sites have yielded early australopithecine remains and diverse Stone Age artifacts (Figure 29.1a). The Nguni and Sotho are peoples of Bantu origin. Iron Age sites dating from as early as A.D. 270 have been found in South Africa, providing clear evidence of the antiquity of Bantu settlement. Centuries before the arrival of Europeans, most of the fertile valleys in the eastern part of the country had been occupied by farmers and herders. Iron, copper, and tin were being mined, smelted, and traded in large quantities. Oral histories and linguistic evidence both point to the antiquity of trading contacts between the Nguni and Khoikhoi.

The Era of Conquest

In the two centuries following the arrival of Europeans, there was a gradual migration of settlers—especially poor whites of Dutch origin (the Boers)—into the interior. In 1795, the British annexed the Cape. The Boer migration out of the Cape quickened after the arrival of a large number of British settlers in 1820 and the abolition of slavery in 1833. The Great Trek of the late 1830s and 1840s involved the migration of about one-quarter of the Cape's white population north to lands beyond the Orange River. The independent Boer republics of Transvaal and the Orange Free State were founded there in 1854.

The Boer expansion into the interior was fiercely contested. Between 1779 and 1846, there were six major battles between the Xhosa and white settlers. The armed conquest of African kingdoms continued until the end of the 19th century.

For Africans, this was a time of great change, known as the *mfecane* ("the crushing") or among Sotho speakers as the *difaqane* ("the scattering"). It was characterized by a series of bitter wars, the emergence of formidable military states, economic crises, and large-scale migrations. Previously, these events were explained in relation to the assumed enmity of rival ethnic groups and the predatory militarism of the Zulu kingdom led by Shaka. More recent scholarship attributes the *mfecane* to severe ecological crises and profound social and economic

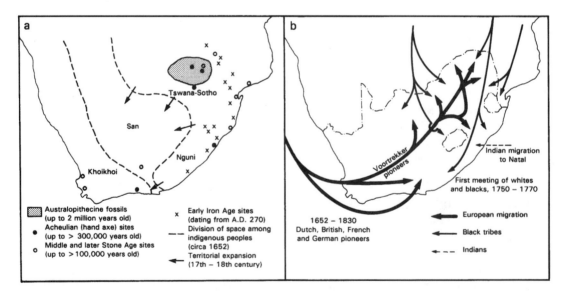

FIGURE 29.1. The settlement of South Africa: Two views. (a) The archeological and historical record. Human occupation of South Africa began hundreds of thousands of years ago. All parts of modern South Africa were settled prior to 1652. (b) The official view under apartheid. The configuration of arrows suggests falsely that whites found an unpopulated land when they arrived at the Cape and only made contact with Africans much later and far from Cape Town. After *Multi-National Development in South Africa: The Reality*, p. 21. Pretoria, South Africa: State Department of Information, 1974.

changes. Prolonged droughts gave rise to famine and, ultimately, struggles over scarce food and productive land. These ecological crises were manifested unevenly in space and within societies; the most vulnerable often had to seek help and protection from the powerful to survive. European demand for ivory, and later for slaves, helped to stimulate warfare and enslavement—and, increasingly, so did the white quest for land.

The northward expansion of the British also resulted in conflict with African kingdoms. The resistance of the powerful Zulu army was especially strong, but the British eventually prevailed in the battle of Ulundi in 1879, following a devastating loss to the Zulu at Isandhlwana.

The Era of Segregation

The political and economic transformation of South Africa began with the discovery of diamonds at Kimberley in 1874 and of gold at the site of Johannesburg in 1886 (see Figure 29.2). These discoveries brought a massive influx of capital and technology, as well as many white immigrants. The economic base of the region—hitherto exclusively in agriculture and trade—shifted increasingly to mining, with its promise of vast fortunes for those who were the luckiest, most astute, and most ruthless.

The conquest of African territory continued until the entire region had been carved up. Conquered peoples lost their land and livelihoods, and were increasingly confined to scattered pieces of marginal land known as *reserves*. Laws were passed that increasingly restricted blacks' freedom of movement and rights to employment. Many blacks sought work at the mines and in the mining towns. Although some were attracted by opportunities to accumulate wealth, increasingly they came because the alienation of their land left them with no viable alternative to labor migrancy.

The British seized control of South Africa following the Boer War of 1899–1902. In 1910, the Union of South Africa was formed through the amalgamation of the formerly separate colonies of Cape of Good Hope and Natal and the republics of Transvaal and the Orange Free State. Under the new government, discrimination against blacks continued to intensify. Legislation was passed that reserved skilled mining jobs for whites, declared illegal the occupation of land by blacks outside their reserves, and decreed the segregation of white and black residential areas in cities.

The African National Congress (ANC) spear-

FIGURE 29.2. Johannesburg. The pithead of a gold mine in the foreground and the impressive skyline in the background remind us of the importance of the mining industry in the development of modern South Africa. Photo: Chamber of Mines Library, Johannesburg (courtesy of J. Crush).

headed black resistance and pressed for the establishment of a democratic, nonracist South Africa, although racist laws excluded blacks from parliament. Blacks also pressed their claims for justice by engaging in strikes, protests, and innumerable forms of passive resistance. As a rule, however, their successes were partial and temporary.

The Era of Apartheid

Although the government of South Africa continued to extend the legislative basis for racial segregation, these measures were deemed insufficient by extreme nationalists among the Boers. Inspired largely by the nationalist ideology of Nazism, these Boers formed the "Purified" Nationalist Party in 1934 to fight for a South Africa fully organized according to apartheid principles. After winning the 1948 election, they moved to implement their vision.

The National Party government based its case for implementing apartheid on the premise that different groups needed to live and develop separately, each at its own pace and in accordance with its own cultural heritage, resources, and abilities. To do otherwise would defy the natural laws of peaceful coexistence between peoples. Through a policy of "creative self-withdrawal" (i.e., the creation of tribal homelands), the national aspirations of the various African groups would be resolved peacefully. The government, it was claimed, was sincerely committed to protecting the best interests of all residents of the country. In short, apartheid was said to exemplify ideals of fairness, justice, and freedom, and to have nothing to do with exploitation.

The reality of apartheid for the majority of South Africans was far removed from this picture of benevolence, harmony, and universal progress. Prime Minister Hendrik Verwoerd's dream became a nightmare for black South Africans.

The popular perception of the nature of apartheid has tended to focus on the strict racial segregation of beaches, buses, sports teams, and the like. Most such measures, often categorized as "petty" apartheid, disappeared during the 1980s as pressure for the abolition of apartheid

mounted. Much more fundamental, and much more resistant to change, was the ruthlessly enforced macro-scale division of space and resources between the country's officially designated racial groups. In his book *Endgame in South Africa?*, Robin Cohen (see "Further Reading") identified four major pillars upon which apartheid was constructed: the white monopoly of political power; the manipulation of space to achieve racial segregation; the control of black labor; and urban social control. These measures are outlined below.

Apartheid South Africa repeatedly contrasted its own democratic institutions with the undemocratic regimes governing many African countries. However, it was hardly a model democracy. Whites monopolized political power, effectively excluding the other 85% of the population. Asians and so-called "coloreds" were granted limited political rights in 1983, but the exclusion of blacks from the political system continued for a decade more. Blacks supposedly had political rights in the homelands, but most of these territories were ruled with an iron fist by puppets of the government.

When apartheid was proclaimed in 1948, the government set out to reorganize the sociopolitical map of the country. It commenced a program of forcibly removing blacks from their long-established homes in areas now designated for white use and dumping blacks in the remote, inhospitable relocation areas designated as reserves, with little or no provision for their welfare. Some 3.5 million people had been moved to the reserves (later combined into *homelands*), and a further 2 million lived under threat of displacement when removals ceased in the late 1980s (see Vignette 29.1 and Figure 29.3). Most removals were from "white" farming areas and areas of black-owned farmland in the "white" countryside that predated the passage of the Native Lands Act of 1913. Spatial reconstruction according to apartheid principles also occurred in urban areas—resulting in the removal of over 200,000 blacks, often from older communities within the city, to remote periurban ghettos.

Forced resettlement in remote and marginal locations created a vast pool of unemployed workers. This dependent, tightly controlled

VIGNETTE 29.1. The Homelands: "Separate (Under)Development" under Apartheid

The creation of identified, separate *reserves* for black South Africans dates back to 1913 and the Native Lands Act. A few more reserve lands were added during the 1930s. Under apartheid, many of these scattered reserves were consolidated. The Bantu Self-Government Act of 1959 amalgamated the various reserve areas to form 10 *homelands*, also called *Bantustans*. Each of these territories was designated as the territorial homeland for a separate ethnic group.

In each homeland, the South African government created a puppet government that was totally dependent on South African financial and military support. The homeland governments were notoriously repressive and corrupt. In 1976, Transkei became the first of four homelands to be declared independent. The fact that only South Africa recognized their independence did not deter the "independent" homelands from acquiring some of the symbolic trappings of nationhood, such as the issuing of their own stamps.

All black South Africans, regardless of where they lived, were to be designated homeland citizens and "repatriated" to the homelands. Between 1960 and 1983, an estimated 3.5 million people were forcibly removed, most into the already overcrowded homelands that represented only 13% of South Africa's area. The population of one homeland, QwaQwa, increased 20-fold from 23,000 to 453,000 between 1970 and 1991.

The homeland areas were often on barren, erosion-prone land that offered few opportunities for agriculture, much less for other types of employment. For most, migrant labor provided the only viable source of income.

The homelands were abolished in 1994, and the territories were amalgamated into the country's revamped provincial structure. Nevertheless, they are sure to remain areas of deep poverty and extremely high unemployment for many decades to come.

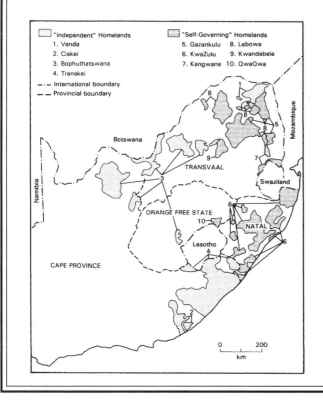

FIGURE 29.3. South Africa's homelands. Of the many effects of apartheid on South Africa's spatial economy, none rivals the creation of ethnic homelands as incipient "nations." Only South Africa recognized their "independence."

pool of migrant labor ensured that South African mines and farms would remain profitable. The state implemented a series of policies known as *influx controls* to severely limit the mobility of black South Africans. Blacks were forced to carry passbooks indicating where they were supposed to be. Anyone found without a passbook or found to be illegally present in an urban area was subject to prosecution.

Apartheid defined several categories of black workers. Urban workers were those who had been residents of one particular city since before 1945. From their homes in segregated, high-density townships, they commuted daily to jobs in the white industrial and commercial economy. Migrant workers, who left their families behind in the rural areas while employed in one-year contracts, were found in many sectors of the economy, particularly in the mining industry. Hundreds of thousands of homeland residents commuted on a daily basis to jobs in white areas of South Africa, spending up to six hours per day on the bus between home and workplace.

The employment of large numbers of migrant workers from countries other than South Africa predated the rise of apartheid but served apartheid well. The presence of foreign miners enabled the state and employers to use divide-and-rule tactics to keep all workers, foreign and South African alike, vulnerable and compliant. The migrant labor system was extremely profitable for employers, but it had devastating effects on family life in rural communities wherever this system operated, and these effects continue into the present day.

Social control, the fourth pillar of apartheid, was perpetuated through various political and social institutions. The police, secret police, and armed forces, whose role was allegedly to maintain law and order, repeatedly used their power to terrorize black citizens, murder key leaders of the resistance movement, and foment violence between various groups (especially in the urban townships). Much more subtle than the "stick" of police and military terror was the "carrot" of ideological manipulation of ideas, exercised through the schools, churches, and state-allied media. These channels were used to perpetuate and legitimate the official line that apartheid was rational and necessary for orderly development and peace.

The Struggle to End Apartheid

The fight to end racial discrimination in South Africa began long before the formal initiation of apartheid in 1948. The ANC, founded in 1912, intensified its campaign against racial discrimination. Its military wing, *Umkhonto we Sizwe*, launched sporadic attacks against strategic targets. The popular history of the struggle has focused for the most part on certain places where horrific events dispelled any illusions that the world might have had about the nature of apartheid—places such as Sharpeville, Soweto, and Boipatong. The popular history of the struggle also focused on the role of a few of its key leaders—among them Steve Biko, Desmond Tutu, and, above all, Nelson Mandela. Yet the heart of the struggle remained the daily resistance of millions of ordinary South Africans to the oppression of the system. For example, the abolition of the pass laws resulted from the mass disobedience of these laws, which ultimately made them impossible to enforce. Likewise, when the government attempted to co-opt Asian and "coloreds" by giving them limited political rights, they responded with a mass boycott of elections.

The workplace was one of the most important points of struggle against apartheid. In 1973 a series of strikes by black workers paralyzed mines and industries. These strikes paved the way for significant concessions, including higher wages, improved trade union rights, and the removal of some regulations that had strictly limited access to skilled jobs. Subsequently, there was a massive expansion in black trade unions and increasingly frequent industrial action (see Figure 29.4). The success of the ANC-sponsored general strike of 1992 again showed the importance of the workplace as a focus of antiapartheid struggle.

Worldwide campaigns against apartheid contributed greatly to apartheid's demise. International organizations such as the United Nations and its member agencies, the Commonwealth of Nations, and the Organization of African Unity

FIGURE 29.4. National Union of Mineworkers meeting. Strikes and protests by black workers, especially by members of the National Union of Mineworkers, played a crucial role in the struggle against apartheid. Photo: *The Star*, Johannesburg (courtesy of J. Crush).

(OAU) repeatedly passed resolutions condemning apartheid. However, as the nature of apartheid became more widely known, the most important source of international pressure proved to be the growing force of public opinion. Public opinion forced Western governments to speak out more clearly against apartheid and to implement increasingly comprehensive political and economic sanctions against South Africa. Public opinion sparked organized campaigns for shareholders to divest themselves of stock in companies profiting from apartheid, and the campaigns caused many international companies to reconsider their otherwise profitable South African investments.

As opposition to apartheid grew in the late 1970s, the South African state fought a losing battle for survival. Attempts were made to suppress black political activity by force. A state of emergency was declared. Divide-and-rule tactics were used among the black population by promoting anti-ANC groups such as Inkatha, and by buttressing the fragile powers of the homeland governments. A series of limited reforms were announced that ended specific types of discrimination. There were attempts to deflect international attention to other issues, through such tactics as the prolonged defiance of United Nations resolutions calling for

Namibian independence. Meanwhile, South Africa was promoting itself as a strategic bastion for the West in a supposedly Marxist-dominated southern Africa.

Internal and international pressures on the South African state continued to intensify. Increasing pressures for a political settlement that would end the country's growing isolation and prevent a bloodbath, while protecting white privilege as much as possible, came from white South African liberals and from the boardrooms of South Africa's largest companies.

The pace of change quickened with the accession to power of F. W. de Klerk in 1989. Unlike his rigid predecessors, de Klerk recognized that fundamental reform was necessary and showed a willingness to take bold initiatives. Several ANC leaders, most notably Mandela, were released from detention; the draconian state-of-emergency law was repealed; and informal negotiations with the ANC were begun. Formal negotiations concerning majority rule commenced in 1992, after de Klerk won a substantial victory in a referendum asking white voters to endorse this process.

At first, the prospects for reform seemed very much in doubt. The initial positions of the ANC and the government were far apart on virtually every substantive issue. Both Mandela and de Klerk enjoyed majority support from their constituencies but seemed to have limited room for compromise. The ANC leadership was under pressure from growing militancy within the movement, especially among its younger members. The government was under pressure not only from right-wing opposition groups, but also from many moderate whites who remained ambivalent about reform.

In spite of these obstacles, the negotiating parties made major concessions, setting the stage for the historic World Trade Center Agreement of November 18, 1993. This agreement mapped out a new constitution for South Africa, with a National Assembly to be chosen by proportional party representation, thus eliminating such contentious issues as constituency boundaries. The Senate was to be appointed by the legislatures of the nine newly created regions (Figure 29.5) that incorporated both the former provinces and all of the homelands. A stipulation that all parties gaining at least 5% of the vote would be represented in the Cabinet ensured that smaller political—and racial—groups would not be totally marginalized as a result of the reforms. Among the many other provisions in the new constitution were a com-

prehensive Bill of Rights, the naming of 11 official languages, and arrangements for restructuring the armed forces.

Although the new constitution was supported by most political groups and the majority of the population, right-wing Afrikaner groups were resolutely opposed to the transition to majority rule; they continued to insist on the establishment of an Afrikaner homeland. Opposition also came from elements within the black population. The Zulu-based Inkatha Freedom Party argued for much stronger guarantees of regional autonomy for KwaZulu–Natal. The radical Pan-African Congress rejected the principles of negotiation and compromise with the white minority, and it accused the ANC of selling out black interests.

The Geography of Postapartheid South Africa

The transition to a postapartheid era has brought profound changes, including a fundamental restructuring of the country's spatial economy. Change often occurred prior to, or in anticipation of, the formal abolition of apart-

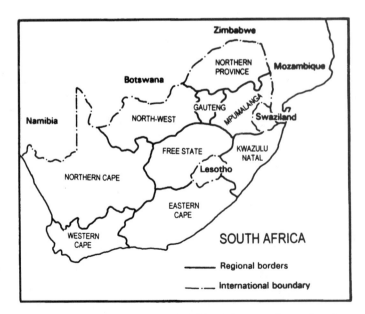

FIGURE 29.5. Political divisions in the new South Africa. The map shows the nine provinces created as part of the constitutional agreement of 1993, replacing the old provinces and homelands.

heid regulations. For example, some black families had begun to occupy dwellings in white-designated urban neighborhoods well before this practice became legal. Nevertheless, for many South Africans who had anticipated a new era of prosperity and security, the first postapartheid decade has been a disappointment. This situation reflects, among other factors, the limited financial resources available to the state, as well as its limited capacity to control the nature and direction of change in the less regulated environment of the new South Africa. It also reflects pragmatic decisions made by the state to implement policies designed to placate local business leaders and to build international confidence in the country and its future.

Political Transitions

The 1993 constitutional agreement abolished both the old provinces (each with its own history predating the formation of the country) and the homelands (including the four that had been designated independent). They were replaced by nine new provinces. This reconstruction of the political map was an attempt to dismantle the tribal identities of the former territories, especially the homelands, so carefully nurtured under apartheid.

The new provinces vary considerably in population and economic power. Northern Cape is the largest province by area, but it has only 2% of South Africa's population and 2% of the country's economic productivity. In contrast, the province of Gauteng (encompassing the Johannesburg area) has close to one-fifth of the country's population and almost 40% of its total economic production. These inequalities have contributed to the relative weakness of state governments, which tend to be limited by their weak resource bases.

The reorganization of political space did not occur only at the provincial level. In the Cape Town urban region, 25 racially divided townships were amalgamated to form a unified city with a single metropolitan council. The Cape Town government has moved to end race-based planning, and has undertaken to improve sig-

nificantly services in the poorest black neighborhoods. Although this long, costly process cannot be expected to eradicate the legacy of separate urban development, the recognition of the need to create an inclusive city was an important milestone. The perception of economic opportunity in the cities, in the midst of general economic stagnation, means that urban South Africa continues to grow rapidly as a result of migration.

The election of 1994 was a climatic moment in South Africa's political transition. The period before the election was very tense. The ultranationalist Afrikaner Resistance Movement, still calling for the creation of a separate white homeland, threatened to launch a civil war. In KwaZulu–Natal, where Inkatha was dominant, "black-on-black" violence between ANC and Inkatha supporters resulted in thousands of deaths. Inkatha vowed to boycott the election, and agreed to participate only after last-minute concessions were made.

The ANC, with 63% of the vote, scored a clear victory over the National Party and other opposition forces. The ANC also won power in six of the nine provinces. The election gave a strong mandate to govern to Nelson Mandela, who took office as the country's first black president. In keeping with his vision of the new South Africa as a "Rainbow Nation," Mandela brought members of the National and Inkatha Freedom Parties into his cabinet. The Truth and Reconciliation Commission (TRC) was launched as a strategy to help heal the painful emotional legacy of apartheid (Vignette 29.2). The difficult task of amalgamating the ANC's military wing with the South African armed forces was set into motion.

The events of the late 1990s demonstrated the wisdom of Mandela's strategies of working with his adversaries to achieve national reconciliation. Political violence in KwaZulu–Natal waned following the 1994 elections, and the extremist white groups faded into relative obscurity. In the 1999 election, voting patterns remained much the same, but levels of tension were considerably lower than in 1994. Mandela did not contest the 1999 election and went into retirement. The transition to the

VIGNETTE 29.2. The Truth and Reconciliation Commission

The Truth and Reconciliation Commission (TRC) was established under the Promotion of National Unity and Reconciliation Act of 1995, with Archbishop Desmond Tutu as its chair. The TRC offered the possibility of amnesty to individuals and groups for political crimes committed from 1960 to 1994, in return for public testimony. Some 7,000 people came forward to request amnesty; requests were granted in about 10% of the cases. Over 10,000 witnesses testified before the TRC.

The testimony before the TRC was often gripping, as witnesses testified about covert operations undertaken by the apartheid state and by the liberation forces that had never been made public. The testimony brought old wounds to the surface, especially for the relatives of those who had died. Many of those appearing were very remorseful, while others sought to minimize their own responsibility or evade the probing questions of TRC members.

The TRC hearings were not a complete success. There was some controversy over the granting of immunity for prosecution, given the horrible nature of many of the crimes. A number of prominent South Africans who were requested to testify refused to do so.

The TRC also offered ordinary citizens who had not committed a crime, but who desired to apologize for actions taken or not taken, the opportunity to express their remorse. The TRC's Register of Reconciliation (**www.doj.gov.za/trc/ror/page01.htm**) contains hundreds of poignant testimonies and calls for healing. The following is an example:

> As South Africa moves toward reconciliation, let us all bury our past and boldly stand up to be counted amongst other countries that we are leaders of Africa by example. This is the only way we can heal our past and reconcile with each other as a Rainbow Nation, which is prepared to rebuild the country as a whole! (Nyamben Thivhulawi, Cape Town and Venda)

The six-volume final report of the TRC not only provided a list of those seeking amnesty and a comprehensive overview of testimonies given, but also described the historical evolution of racism in South Africa and the institutional and social environment in which apartheid-era political offenses had taken place. The report dealt with human rights violations committed by the ANC and others involved in the liberation struggle, as well as by the apartheid regime. The ANC was angered by what it considered the TRC's failure to make a moral distinction between acts by the apartheid government and those committed by its opponents, but it did not succeed in its attempt to block publication of the report.

Whatever its limitations, the TRC helped South Africans to reconcile with their troubled histories, as well as with each other. It served as a model for a new, nonracist South Africa, dedicated to the principles of accountability, cooperation, and respect for human rights.

new presidency of Thabo Mbeki was routine and orderly.

Nonwhite South Africans have put their new political freedoms to good use. The new South Africa has countless civil society organizations, involved in all facets of the country's life. These organizations are committed to transforming the country through their work on democratization, human rights, gender and violence, HIV/AIDS education, and environmental protection, among many other causes.

Economic Transitions

Under apartheid, South Africa's prosperity depended heavily on foreign investment and the ability of businesses to make huge profits, mainly because of state-enforced control over labor. The ANC-led government made a concerted effort to reassure the business community. It downplayed its socialist roots, as set out in its Freedom Charter (1955) and Constitutional Guidelines for a Democratic South Africa (1988); instead, the government emphasized stability and respect for the rights of private businesses. The business community, both within South Africa and internationally, responded positively to these reassurances. It had excellent reasons for remaining and expanding: rich natural resources, Africa's wealthiest market, the prospect of increased economic activity following the end of sanctions, reopened markets in neighboring countries, and the increased purchasing power of 33 million black South Africans.

After taking office, Mandela's government began to implement its Reconstruction and Development Program (RDP), which was designed to spur socially just and sustainable development. The program sought to alleviate the worst manifestations of poverty through investment in social welfare (particularly in education, health care, and housing), and through economic development programs designed to spur job growth and reduce regional imbalances. The RDP's objectives were extremely ambitious—for example, the creation of 2.5 million new jobs and construction of 1 million new homes within ten years—in keeping with both the country's tremendous needs and the optimism of the moment.

Two years later, in 1996, the ANC government launched a new economic policy known as the Growth, Employment and Redistribution (GEAR) strategy. The GEAR program introduced a series of neoliberal, market-driven economic reforms, including major reductions in public-sector spending, large corporate tax cuts, and the privatization of some state functions and enterprises. Although the government continued to espouse the social welfare objectives of the RDP, in practice these programs were now allocated fewer resources and given less prominence. The implementation of GEAR was a bitter pill to swallow for many of the ANC's core supporters, who were strongly committed to the socialist agenda that the party had espoused for decades.

The South African economy continued to stagnate under GEAR; indeed, there was only a slight increase in the gross domestic product (GDP) and a decline in the per capita income during the late 1990s. The mining sector experienced considerable difficulty because of slumping world markets for gold and other minerals, the high cost of operating old mines with declining reserves of higher-quality ore, and increasing labor costs. Many mines were closed, resulting in major job losses. Large job losses also occurred in the agricultural sector as a result of increased mechanization. A significant number of factories that had survived in the protected market of apartheid South Africa were unable to compete with imported products and were forced out of business. Instead of the 2.5 million new jobs promised under the RPD, an estimated 500,000 jobs were lost between 1994 and 1999.

Economic change in the postapartheid era has accentuated prevailing social and spatial disparities. World Bank data show that the richest 20% of the population control some 63% of the country's total wealth, compared to only 3% accruing to the poorest 20% of the population. Unemployment and absolute poverty have actually increased in the former homelands, in part because of the collapse of many of the industries that had been situated nearby with state support. Homeland populations have also been burdened by the return of some who have lost their jobs in the agricultural and mining sectors, as well as of others with HIV/AIDS. Poverty levels in the former homelands range from some 50% (KwaNdebele) to over 90% (Transkei).

The ANC had long been committed to an affirmative policy of land reform to abolish racial restrictions on land ownership and provide redress for the victims of forced removals. The government pledged in 1994 to redistribute 30% of the arable land within five years. However, land reform has proceeded extremely slowly; only 1.3% of the agricultural land base had been transferred to black ownership by 2000. Land reform has been hampered by several factors, including the difficulty of resolving conflicting claims to specific parcels of land, compensation for white farmers displaced under land reform, and the inadequate provision of resources for the process by the government. Although forcible occupation of farmland has occurred in some areas (such as in the sugar-producing regions of KwaZulu–Natal), the slow pace of land reform has provoked surprisingly little backlash. Nevertheless, these gross disparities in land ownership, and the slow pace of land reform, pose a continuing serious threat to South Africa's peace and stability.

Social Transitions

Apartheid guaranteed that even the poorest of whites had privileges setting them apart from and above blacks. Now South African society is being transformed into one structured according to social class as well as race. The old racial divisions remain, but a new set of class divisions, both overall and within each of the racial groups, has been superimposed on them. The removal of employment barriers and restrictions on movement have helped to stimulate the growth of a black middle class with a lifestyle more closely resembling that of middle-class whites than that of most blacks. At the same time, there is evidence of increasing poverty and unemployment among some segments of white society—caused by decreased subsidies for small farmers, general economic decline, and increased opportunities for blacks. Despite the emergence of a new well-to-do black elite, white South Africans continue to control a vastly disproportionate share of wealth, while the majority of blacks remain a very large, poverty-stricken underclass with few prospects.

Separate and grossly uneven social services were inherent in the apartheid system. Progress toward the eradication of these inequalities has been a leading priority for the postapartheid state, and considerable success has been achieved in this area. Government housing programs provided improved shelter for 3 million people between 1994 and 1999. During the same period, close to 2 million homes were supplied with safe water, and the proportion of homes with electricity doubled to 63%. The imposition of user fees, however, has denied many of the poor access to these services.

The government has given highest priority to the reduction of black–white disparities in education (see Figure 29.6). There has been a massive investment in educational infrastruc-

ture and teacher training. Although black children's educational opportunities have been greatly increased, massive rich–poor and urban–rural disparities in educational quality remain, and these are likely to persist for decades to come. In order to insulate themselves from the deterioration in quality of public-sector schools, growing numbers of wealthy whites and blacks are seeking places in expensive private schools for their children.

Among the most troubling developments has been the recent increase in violent crime, especially in the Johannesburg and Cape Town urban regions. Crime has become a major impediment to development. The murder rate in the Johannesburg area was 105 per 100,000 people in 2000—some 10 times that of the United States. Violent crimes, including rape, armed robbery, and assault, increased by 30% in South Africa between 1994 and 2000. Crime is especially prevalent in the poverty-stricken suburbs created under apartheid, where deep poverty and a lack of employment opportunities have created a sense of despair. Street gangs are responsible for much of the criminal activity in these areas; others in the same communities are their most frequent victims.

No discussion of South Africa's contemporary challenges would be complete without a mention of the HIV/AIDS epidemic. In 2000, HIV had infected 20% of adult South Africans, and the epidemic was growing rapidly. Life expectancy had fallen by 10 years, and the challenge of caring for a growing number of "AIDS orphans" had become a major concern. The health system in many parts of South Africa has become overburdened by patients with AIDS. The direct economic costs associated with the care of such patients pales in comparison to the indirect costs, such as lost productivity in the economy. Poor rural areas, long denied basic social services under apartheid, can ill afford the loss of health care workers and teachers to the disease.

The fight against HIV/AIDS in South Africa has long been hampered by a climate of denial. For many years, migrant laborers from other African countries were blamed for introducing the infection. The government used the epi-

FIGURE 29.6. In the new "Rainbow Nation" of South Africa, this schoolgirl has access to opportunities unimagined for nonwhite residents just a few years ago. Education will be one of the keys to individual, as well as national, progress. Photo: CIDA (B. Paton).

demic as a pretext for banning migrant workers from Malawi. More recently, President Mbeki spoke often about his skepticism that sexually transmitted HIV is the cause of AIDS. Unfortunately, the climate of denial has hampered the development of strong government policies to counter the epidemic. Much of the health education effort for the treatment of the disease was left to civil society organizations. Fortunately, since the International AIDS Conference took place in Durban in 2000, the government has made greater efforts to counter the epidemic through health education, drug programs for HIV-infected pregnant women, and other measures.

The past decade has brought remarkable changes. There has been a genuine commitment by the government and most individual South Africans to work toward the creation of a new South Africa—the "Rainbow Nation" envisioned by Mandela. South Africa is far from a model society, as evidenced by the depth of poverty, the massive disparities between the rich and poor, the high rates of unemployment, the high rates of violent crime, and the devastation of HIV/AIDS. Nevertheless, the fears of white South Africans and many international observers that black majority rule would bring economic ruin, political chaos, and widespread revenge taking against white South Africans

have not materialized. There are good reasons to hope that the future will bring continuing reconciliation and progress.

Further Reading

On the history of South Africa, see the following sources:

Maylam, P. *A History of the African People of South Africa: From the Early Iron Age until the 1970s.* London: Croom Helm, 1986.

Worden, N. *The Making of Modern South Africa,* 3rd ed. Oxford: Blackwell, 2000.

Wright, H. M. *The Burden of the Present: Liberal–Radical Controversy over South African History.* London: Rex Collings, 1977.

Discussions of apartheid and its enforcement are found in the following sources:

Cohen, R. *Endgame in South Africa?* London: James Currey, 1986.

Freund, B. "Forced resettlement and the political economy of South Africa." *Review of African Political Economy,* no. 29 (1984), pp. 49-63.

Jeeves, A., J. Crush, and D. Yudelman. *South Africa's Labor Empire.* Boulder, CO: Westview Press, 1991.

Lemon, A. *Apartheid in Transition.* London: Gower, 1987.

Murray, M. *South Africa: Time of Agony, Time of Destiny.* London: Verso, 1987.

Smith, D. *Living under Apartheid: Aspects of Urbanization and Social Change in South Africa.* London: Allen and Unwin, 1986.

These sources, written late in the apartheid era, speculate about South Africa's future:

Collins, P., ed. *Thinking about South Africa: Reason, Morality and Politics.* New York: Harvester Wheatsheaf, 1990.

Smith, D. M., ed. *The Apartheid City and Beyond: Urbanization and Social Change in South Africa.* London: Routledge, 1992.

Wilson, F., and M. Ramphele. *Uprooting Poverty: The South African Challenge.* Cape Town, South Africa: David Phillip, 1989.

The following books provide insights on South Africa in the first years of the postapartheid era:

Bond, P. *The Elite Transition: From Apartheid to Neoliberalism in South Africa.* London: Pluto Press, 1999.

Deegan, H. *South Africa Reborn: Building a Democracy.* London: University College London Press, 1999.

Elbadawi, I. A., and T. Hartzenberg. *Development Issues in South Africa.* New York: St. Martin's Press, 2000.

Hart, G. *Disabling Globalization: Places of Power in Post-Apartheid South Africa.* Berkeley: University of California Press, 2002.

Lemon, A., ed. *The Geography of Change in South Africa.* Chichester, UK: John Wiley, 1995.

Lester, A., E. Nel, and T. Binns. *South Africa: Past, Present and Future: Gold at the End of the Rainbow?* Harlow, UK: Pearson Educational, 2000.

Koelbe, T. A. *The Global Economy and Democracy in South Africa.* New Brunswick, NJ: Rutgers University Press, 1999.

Shaw, M. *Crime and Policing in Post-Apartheid South Africa: Transforming under Fire.* Bloomington: Indiana University Press, 2002.

Internet Sources

For South African government information, see the following:

South Africa Government Online. www.gov.za

Truth and Reconciliation Commission. www.doj.gov.za/trc

The websites of the major political parties are identified below:

African National Congress. www.anc.org.za

New National Party. www.natweb.co.za

Welcome to the Democratic Alliance: South Africa's Consolidated Opposition Party. www.dp.org.za

The following websites are maintained by some of South Africa's leading civil society organizations for democratization:

Black Sash—Making Human Rights Real. www.blacksash.org.za

Institute for Democracy in South Africa (IDASA). www.idasa.org.za

Open Democracy Advice Centre (ODAC). http://opendemocracy.org.za

Polity.org.za—Policy and Law Online News. www.polity.org.za

South African Institute of Race Relations. www.sairr.org.za

Welcome to the South African Human Rights Commission. www.sahrc.org.za

To follow current events in South Africa, check the following media sites:

Mail and Guardian Online. www.mg.co.za

Sunday Times. www.sundaytimes.co.za

The Star. www.star.co.za

30

Kenya, Tanzania, and Uganda: Prospects for Integration

There are long-standing debates in Africa about national sovereignty in general, and specifically about the extent to which states should be willing to sacrifice aspects of national self-determination in the interests of some form of integrated development. On the one hand, most acknowledge the economic and political weakness of African nation-states, which could be addressed through the development of frameworks for cooperation and integration among these states. On the other hand, there is little enthusiasm for international agreements that would significantly limit state sovereignty without assured benefits for all participants.

Some of the founders of the Organization of African Unity (OAU) saw the OAU as a first step on the road to political unification. President Kwame Nkrumah of Ghana articulated a vision of a "United States of Africa," based on ideas espoused by Marcus Garvey and others in the Pan-African movement, and modeled broadly on the historical development of the United States of America. However, these calls for unification coincided with the termination of three major colonial experiments in unification—French West Africa, French Equatorial Africa, and the Central African Federation (Rhodesia and Nyasaland), which were replaced by a total of 14 independent states. Not surprisingly, many African leaders shied away from proposals that seemed to threaten their newly won national sovereignty.

Various groupings of African states have been formed to stimulate regional trade, among other objectives. The largest groupings are the Economic Union of West African States (ECOWAS), with 16 member countries; the Southern African Development Community (SADC), with 14 members; and the Common Market for Eastern and Southern Africa (COMESA) with 20 members.

This chapter examines the East African Community (EAC) as one example of a regional organization. The development of structures to integrate development of the three former British colonies of East Africa is noteworthy because of its longevity; such structures were first initiated in the 1920s. The fate of common organizations is linked significantly to what happens in individual member countries, so this chapter also surveys major political and economic developments in Kenya, Tanzania, and

437

Uganda. As background, Figure 30.1 depicts the major cities and transportation routes of these three countries.

East Africa during the Colonial Era

The European presence in eastern Africa commenced in the 16th century, when the Portuguese conquered the Swahili states. They ruled Mozambique for almost 500 years, but were evicted from the city-states to the north by Oman in a series of 17th-century battles. Zanzibar became the political and economic capital of an Omani colony that extended from Somalia to Mozambique and included substantial areas of the mainland. The value of trade in spices, ivory, slaves, and other goods from as far inland as the Congo basin was so great that Zanzibar was made the capital of the Sultanate of Oman in 1839.

Zanzibar's thriving trading economy attracted increasing European interest in the mid-19th century, and it was declared a British protectorate in 1890. Its mainland territories were ceded to the Germans, who took control of Tanganyika in 1885, and to the British, who established the Protectorate of British East Africa (later renamed Kenya) in 1895. Meanwhile, the British secured Uganda through a series of treaties it signed with the indigenous kingdoms of Buganda, Bunyoro, Toro, Ankole, and Bugosa between 1891 and 1894. Uganda was seen as a particularly valuable acquisition; the British referred to it as the "pearl of Africa." Following Germany's defeat in World War I, its East African possessions were divided and declared protectorates of Belgium (present-day Rwanda and Burundi) and Britain (Tanganyika).

Economy

The East African colonies were exploited in distinctive ways. Most of the Kenyan highland region, with its cool climate, was set aside for European settlement. By the late 1920s, some 2,000 white farmers had established farms pro-

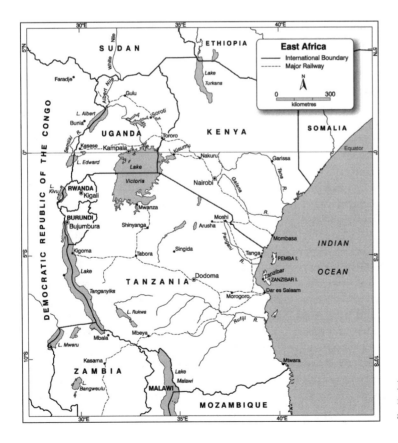

FIGURE 30.1. Kenya, Tanzania, and Uganda: Major cities and transportation routes.

ducing a variety of crops and livestock. European agriculture developed in a protected environment. Africans who were evicted to make way for European settlement became dependent on wage labor. Regulations limited opportunities for Africans to produce export crops, and ensured that they received lower prices than European farmers in the marketplace. Asian immigrants assumed a middle position in the economy as merchants and small-scale manufacturers. In areas away from the highlands, there was relatively little development during the colonial era.

The most fertile areas of Uganda had a relatively dense population and productive indigenous agriculture. There was little in the way of settler agriculture in Uganda; rather, its economy was based on smallholder agriculture, with coffee and cotton as its primary products. Uganda also had a substantial Asian immigrant population.

Germany envisaged that the economy of German East Africa would be based largely on plantation agriculture and enterprises owned by European settlers. Large-scale sisal plantations were established and became the mainstay of the economy, but European settlement did not take off as it had in Kenya, either under the Germans or after 1919 under the British. Thus the economy of Tanganyika evolved as a mixture of smallholder agriculture and large-scale settler and plantation enterprises.

Zanzibar experienced relatively little change as a British protectorate with an Omani sultan. Smallholder production of spices, especially cloves, remained the mainstay of the agricultural economy. Zanzibar's role as an important trading city gradually declined with the development of the mainland ports of Dar es Salaam and Mombasa as centers to serve the colonial export economy.

Political Dynamics

The settler dominance of Kenya's economy was mirrored in a political system that served white settlers. Labor legislation, agricultural regulations, and taxation policy reflected settlers' interests, protecting their markets and ensuring a steady supply of compliant labor. Although ultimate authority rested with the British colo-

nial office, each colony had its own legislative council that gave voice to local white residents. In Uganda, the presence of well-established African kingdoms enabled the development of a system of indirect rule, in which traditional rulers exercised local authority under the close control of British officials. Nevertheless, the status of these influential "states within the Ugandan state" was a complex issue that remained unresolved throughout the colonial period. Indirect rule was also implemented in Tanganyika, even though there had been no traditional structures of chieftaincy in many parts of the country.

Africans created their own political associations to fight for their rights. The Tanganyika African Association was founded in 1929; it became the Tanganyika African National Union (TANU) in 1954. In Uganda, the Uganda National Congress was formed in 1949 to fight for fair marketing regulations for African farmers. The Kenya African Union (KAU) was established in 1944 to fight for African land and political rights. The Mau Mau, a Kenyan secret society, initiated a campaign of terrorism in the early 1950s, directed at British settlers and African collaborators. Responding to increased pressure for independence within the region and throughout Africa, the British prepared to leave.

Independent Tanzania, Uganda, and Kenya

Tanzania

Tanganyika, under the leadership of TANU and Julius Nyerere, gained its independence in 1961. Two years later, Zanzibar became an independent sultanate. Within a matter of only four months, the sultan had been overthrown in an armed coup and replaced by a revolutionary government, which then entered into an Act of Union with Tanganyika. The new state was named Tanzania. Its constitution created a one-party state; Nyerere was appointed president, and Zanzibar's leader, Abeid Karume, became the first vice-president (Vignette 30.1).

Tanzania became a test case for the populist-socialist model of development. The turning point occurred in January 1967 with the adop-

VIGNETTE 30.1. Making Tanzania: The Union of Zanzibar and Tanganyika

During the past four decades, there have been a few attempts to unite sovereign states in Africa. Among the states that agreed in principle to unite were Senegal and Mali, Senegal and Gambia, and Ghana and Guinea. However, Tanganyika and Zanzibar are the only countries to have succeeded in overcoming the many hurdles to union. The very fact that the United Republic of Tanzania has existed for four decades is an inspiration to those committed to the cause of African unity.

Zanzibar gained its independence in December 1963, two years after Tanganyika. Within a month, the sultan was overthrown and the revolutionary new rulers, the Afro-Shirazi Party (ASP), declared a republic. Zanzibar was united with Tanganyika three months later; shortly thereafter, the new republic was renamed Tanzania. Under the new constitution of 1995, Tanzania was declared a one-party state, ruled by a TANU–ASP coalition.

The union was structured to ensure relative equity between the two original states, despite the fact that Zanzibar was far smaller than mainland Tanzania. Zanzibar's leader, Abeid Karume, served as vice-president under Nyerere. A Zanzibari, Ali Hassan Mwinyi, succeeded Nyerere as president, with a mainland Tanzanian serving as vice-president. The third political transition continued the tradition of sharing and rotating the country's top offices. The governments of Zanzibar (under the ASP) and the mainland (under TANU) retained considerable autonomy in their respective territories. Zanzibar developed its own distinctive political and social institutions. TANU and the ASP formed a single party, the Revolutionary Party (known as the CCM), in 1977.

By the mid-1980s, many Zanzibaris were demanding greater autonomy within Tanzania. As well, there were internal struggles between residents of the smaller island of Pemba and those of Zanzibar itself; Pembans felt excluded from power in Zanzibar. However, the ASP regime was ruthless in suppressing opposition to its rule.

Constitutional reforms in 1992 established multiparty democracy, with the proviso that all parties must have support in both parts of the union. The Civic United Front (CUF), representing those seeking greater autonomy for Zanzibar, won just under half of the seats and presidential votes on the islands in the 1995 elections. Since then, tensions have remained high. Members of the CUF protested their loss and refused to sit in the legislature for almost four years. The islands' CCM government was accused of many human rights abuses against CUF supporters, including the jailing of 18 members of the party for alleged treason. Troops from the Tanzanian army were sent to keep order on the islands during the election of 2000. The CCM won, but the CUF and its supporters rejected the result.

Representatives of the CUF and CCM met secretly in 2001 and reached an agreement to end the political stalemate. The agreement provided for a joint electoral commission, an improved voter registration system, and other changes that safeguarded the rights of the Zanzibari opposition.

The status of Zanzibar within the union, and the political rights of minority and opposition interests within Zanzibar and Pemba, remain difficult issues for Tanzania. The central government has generally taken a "hands-off" approach to Zanzibar's problems, seeking to support the development of responsible government through persuasion rather than force. Although this approach did little in the short run to safeguard the interests of the opposition in Zanzibar, in the longer run it appears to have contributed to an easing of political tensions. Indeed, it may be argued that the Tanzanian government's respect for Zanzibar's autonomy has been critical for keeping the union intact. In spite of the long-standing political tensions in Zanzibar, Tanzania is widely recognized as a country that has made great strides in strengthening democratic institutions since 1990.

tion of the Arusha Declaration by TANU. The declaration set forth a commitment to self-reliant, egalitarian socialist development, and it established a series of principles and policies for the implementation of the objective. It reserved a prominent role for the state and limited the role to be played by the private sector.

The cornerstone of Tanzania's rural development strategy was the *ujamaa* village. The scattered rural population was encouraged to settle in nucleated villages, where services could be provided more efficiently, communal production organized (see Figure 30.2), and political involvement encouraged. By 1974, 2.5 million people were occupying 5,000 *ujamaa* villages. However, problems began to multiply when other peasants were coerced into forming *ujamaa* villages. The communal farms failed in many villages, partly because of the peasants' lack of interest in communal work and partly because of inadequate marketing and technical support. Tacitly recognizing the limitations of the program, the state reduced pressures for communal production and began to encourage more production from farmers' private plots. What remained of the communal aspects of *ujamaa* disappeared under the influence of structural adjustment, starting in the early 1980s.

Tanzania established numerous parastatal companies to undertake and manage a wide range of enterprises in produce marketing, transportation, and manufacturing, among other things. By the latter part of the 1970s, four-fifths of all investment was by the state, and four-fifths of large and medium-sized enterprises were under state control. Although the overall performance of the parastatal sector was solid, most of these companies were privatized in the 1980s and 1990s.

The Arusha Declaration emphasized the importance of creating an egalitarian society. Investment in health care, education, and other social services received high priority. Tanzania played a pioneering role in developing primary health care, as a cost-effective means of addressing the basic health needs of the rural population. Disparities in wages between top officials and ordinary workers were reduced. Also justified on egalitarian grounds was the transfer of the capital from Dar es Salaam to the small regional center of Dodoma, where, it was argued, civil servants would be less able to isolate themselves from the everyday realities of ordinary Tanzanians' lives.

Nyerere retired in 1985 and was replaced by his vice-president, Ali Hassan Mwinyi, the leader of Zanzibar. Tanzania was facing a severe

FIGURE 30.2. Communal farm work, Tanzania. The encouragement of communal production was one of the main goals of Tanzania's populist-socialist government under Nyerere's leadership. Photo: CIDA (D. Barbour).

economic crisis that had begun in the late 1970s. The crisis has been attributed to a large decline in the price of coffee and other farm commodities, and to the $300 million cost of Tanzania's war with Uganda. Responding to a Ugandan attempt to annex some Tanzanian territory, Tanzania had invaded Uganda and brought about the collapse of the despotic regime of Idi Amin. The extent of the economic crisis forced Mwinyi to accept International Monetary Fund (IMF) conditions for financial support; Tanzania's socialist policies were progressively dismantled, and market-driven economic policies were implemented in their place. Unfortunately, many of the gains made in the social sector—especially in narrowing the disparities between urban and rural, and between rich and poor, Tanzanians—were eroded. For example, although 93% of the primary school-aged population was in school in 1980, the percentage had declined to only 64% in 1988.

Tanzania amended its constitution in 1992, and as a result it became a multiparty state. Following successful multiparty elections in 1995, Benjamin Mkapa was inaugurated as president. Four decades after achieving independence, Tanzania is widely regarded as an economic and political success story. New developments in the mining sector, as well as growth in agriculture and tourism, have fueled economic growth averaging 3.1% per year since 1990. It has successfully transformed itself from a one-party state into a multiparty democracy. The many challenges of merging two states with rather different social and political cultures have been addressed successfully to date.

Uganda

Uganda gained its independence in 1962, with Milton Obote of the Uganda People's Congress as the first prime minister. Under the constitution, Uganda was a federation of four regions based broadly on the traditional kingdoms. The largest region, Buganda, played a dominant role in Obote's initial government. However, tensions grew between Obote and the Bugandan elite, led by their Kabaka (king), who had been appointed the country's

nonexecutive president. Obote suspended the constitution and assumed all executive powers in 1966, in what was in effect a bloodless coup. A year later, a new constitution abolished the traditional kingdoms and established a one-party, unitary state. His government became increasingly repressive.

In 1971 Obote was deposed in a coup led by the army commander, Colonel Idi Amin Dada. During his eight years in power, Amin conducted a reign of terror against his opponents in the armed forces and in Ugandan society as a whole. The constitution was suspended, and Amin ruled by decree. The economy declined precipitously as a result of the lack of any coherent economic policy, the expulsion of noncitizen Asians in 1972, and the withdrawal of international aid. Seeking to divert attention from the country's internal problems, Amin attacked Tanzania in October 1978. Amin was deposed in early 1979 after Tanzanian armed forces, together with Ugandan dissidents, invaded Uganda.

The years 1979–1986 brought little in the way of respite. Two weak transitional regimes held power until elections were held in late 1980. Obote was returned to power, but there were widespread allegations of electoral fraud. The National Resistance Army, under the leadership of Yoweri Museveni, emerged as the largest of several guerrilla movements formed to challenge Obote. Years of guerrilla warfare, together with widespread repression by the government, ensured that Uganda remained in chaos. The economy showed very few signs of recovery. Hundreds of thousands were displaced from their homes, especially in areas where Obote had little support. Obote was deposed in a military coup in July 1986. Within six months, two other regime changes had occurred; the second of these brought Museveni to power.

Museveni formed a National Resistance Council (NRC), consisting of both civilian and military members. His government set about to restore peace and order, and to achieve national reconciliation. Although his regime made considerable progress toward the restoration of order in the south and central parts of the country, it was much less successful in the north, where a series of rebel movements emerged. Uprisings have continued to disrupt the societies and economies of northern Uganda, despite repeated attempts by the government to end the rebellion through military campaigns, promises of amnesty, and reconstruction programs.

The Museveni government has had considerable success in restoring Uganda's economy. Uganda has been praised by the IMF as one of Africa's most successful examples of economic recovery under structural adjustment. The regime has attracted substantial international aid and investment, some of it from Asian entrepreneurs who were permitted to return to Uganda some 20 years after they had been expelled by Amin. The country's gross national income (GNI) grew by an average of 6.8% between 1990 and 2001. However, Uganda's ill-advised intervention into the civil war in the eastern Democratic Republic of the Congo between 1997 and 2003 slowed economic growth and tarnished Museveni's image.

Museveni has resisted pressure, from both within Uganda and from abroad, to restore multiparty democracy. Although parties are allowed to organize, the government is elected on a "no-party" basis. In effect, Uganda has remained a one-party state ruled by the National Resistance Movement. Among the strategies used by Museveni to forestall political opposition was the restoration in 1993–94 of traditional monarchies in Buganda and the other kingdoms (Figure 30.3).

Uganda has made significant progress under Museveni; as a result, most of the country has recovered from the chaos and decay of the 1970s and 1980s. His regime, for example, has been perhaps the most proactive and successful in Africa in fighting the spread of HIV/AIDS. Nevertheless, Uganda remains a nation with significant difficulties—guerrilla warfare that has raged almost continuously in northern Uganda; often tense relations with neighboring states, such as Sudan and the Democratic Republic of the Congo; and uncertain economic prospects for the future. Moreover, there is widespread discontent with Museveni's long-standing reluctance to establish multiparty democracy.

FIGURE 30.3. Commemorative poster issued for the coronation in 1993 of the Kabaka of Buganda. The Ugandan government sought to win political support through the restoration of traditional monarchies that had been abolished by Obote.

Kenya

Kenya achieved independence in December 1963, and thus was the third of the East African colonies to do so. The legacy of the Mau Mau uprisings and complex issues related to white settlers and their land holdings were causes of delay. The Kenya African National Union (KANU), successor to the KAU, assumed power under the leadership of Prime Minister Jomo Kenyatta. Kenyatta remained in power until his death in 1978. During his term in office, his regime became increasingly autocratic and intolerant of political opposition.

Daniel arap Moi, Kenyatta's vice-president, assumed power in 1978 and remained in power until 2002. The Moi presidency was widely criticized for corruption and for its suppression of political opposition. Governing powers were increasingly concentrated in the presidential office; the independence of the judiciary was compromised; potential rivals were expelled from KANU; and others critical of the regime were detained without trial. The replacement of the secret ballot by a "queue-voting" system in 1986 further compromised political freedom in what was already a one-party state. Internal and international pressure for democratization culminated in constitutional changes in 1991 that established a multiparty system and restored the secret ballot.

Moi was exceptionally skilled at divide-and-rule politics, repeatedly forming and then breaking coalitions of convenience as a means of retaining power and splitting the opposition. Whereas KANU had originally been especially strong among the Kikuyu, Kenya's largest ethnic group, the Kikuyu were increasingly marginalized in Kenyan politics. Instead, Moi built a coalition of minority ethnic groups generally opposed to Kikuyu interests. When Kikuyu settlers in the Eastern Rift Valley were attacked and driven from their homes on several occasions during the 1990s, the government reacted weakly, creating the impression that the attackers had the tacit support of the state.

During the 1990s, opposition groups formed a coalition party known as the Forum for the Restoration of Democracy (FORD). However, FORD was soon destroyed by internal rivalries and policy differences, and split into a number of parties. Divisions in the opposition, together with intimidation and electoral fraud, ensured that Moi was able to hold power in the 1992 and 1997 elections.

Unlike both Tanzania and Uganda, Kenya has maintained a capitalist market economy continuously since independence. Economic development has been built upon the growth of smallholder agriculture, as well as investments in manufacturing, tourism, and other service industries by foreign and domestic investors. Kenya's large and wealthy Asian community has continued to play a key role in the economy. The country once had one of Africa's strongest and most diversified economies. However, the economy was increasingly weakened under Moi's presidency by poor economic planning, fiscal mismanagement, and massive corruption. The IMF suspended Kenya's access to loans in 1997, pending appropriate policy responses to address these issues. The country also lost the majority of its international aid, due to human rights abuses, economic mismanagement, and corruption. Concerns about crime and security caused a reduction in the number of tourists. By the year 2001, the growth of the gross national income (GNI) had slowed to −1.0% per year.

Moi was forced to retire in 2002, having completed the two terms allowed under the constitution of 1992. The National Rainbow Coalition (NARC), under the leadership of Mwai Kibaki, promised to restore human rights and good governance, as well as to stimulate economic growth. NARC gained a decisive victory over KANU, and a peaceful transfer of power took place. Kibaki's reformist regime assumed power in an atmosphere of high hopes and high expectations, both from Kenyans and from many in the international development community.

Coordination and Integration

As previously noted, East Africa is noteworthy because of the persistence of attempts to increase political and economic integration in the region. These attempts began quite early in the colonial era and were formulated in the creation in 1967 of the EAC. Despite the failure of the earlier attempts, Kenya, Uganda, and Tanzania have recently resumed their quest for a structure to facilitate effective regional cooperation.

Colonial Era

During the period of colonial rule, the British maintained separate governments in its three adjoining colonies. However, they moved gradually, especially during the 1920s, to integrate the economies of their East African possessions. A single, integrated East African Railway system was established. However, decisions about the location of new routes, as well as prices charged for different commodities, assisted the Kenyan settler economy at the expense of the

FIGURE 30.4. Postage stamps issued by the joint postal administration created for the three colonies. The establishment of a unified postal service was one of the first forms of integration. Common-issue Kenya, Uganda, and Tanganyika/Tanzania stamps were used from 1935 until 1977.

other colonies. Postal services were also integrated (Figure 30.4). Most importantly, a Customs Union was created in the early 1920s to enable a free flow of goods among the colonies. The Customs Union was of particular benefit to Kenya, facilitating the expansion of Kenya-based commercial enterprises into Uganda and Tanganyika. Customs duties were structured so as to protect Kenyan industries and to secure the entire East African market for them.

Despite their limitations, colonial efforts to integrate the three economies set the stage for postindependence cooperation. The common railway system was maintained, and cooperation continued in such fields as trade, communications, and higher education.

The EAC, 1967–1977

The Treaty for East African Cooperation, signed in 1967, created the three-country EAC. An East African Common Market was established; a common tariff policy was implemented; and a regional Development Bank was formed. Joint councils were created to develop policy in key areas, such as economic policy, social affairs, and justice. The EAC's activities and governance were dominated by the public sector; there was

little in the way of participation from civil society or the private sector.

From the outset, the EAC struggled with the differences in economic structure and policy in the member states: Kenya had a capitalist economy, while Tanzania and Uganda promoted large-scale state intervention within a socialist development framework. Kenya benefited disproportionately from intra-EAC trade, and it continued to attract most of the new investment. Personality differences existed among the leaders from the beginning, but they became worse when Amin seized power in Uganda. These growing problems led to the dissolution of the EAC in 1977. Most of the formal structures were dissolved, and the EAC's jointly owned assets were dispersed.

The Revived EAC

In 1984, Kenya, Tanzania, and Uganda commenced informal discussions about ways to facilitate regional cooperation. The process was formalized in 1991, when a permanent Tripartite Commission was established. The heads of state initiated an annual summit to discuss issues of common concern. Several areas of priority were identified, among them economic development policy, transportation and communications, agriculture, energy, natural resources, and justice. Coordinating committees were formed to initiate research and oversee policy discussions in these areas.

In 1999, the heads of state met at Arusha and signed a treaty for the establishment of a new EAC. The treaty made provision for the establishment of a Customs Union and Common Market to harmonize tariffs and trade regulations, as well as fiscal and monetary policies. Seven coordinating bodies were created to develop and implement programs in such sectors as transportation, agriculture, energy, and tourism. Members also agreed to work toward a harmonization of foreign policy and defense.

Kenya, Tanzania, and Uganda have proceeded deliberately but cautiously in their revival of the EAC. By initiating cooperation little by little over several years, the countries sought to create a sense of trust and common purpose. Priority has been given to the establishment of strong coordinating structures. Rwanda and Burundi have expressed an interest in joining. These countries have strong economic linkages to Kenya, Uganda, and Tanzania, but their rather different colonial experiences, as well as their serious political and social problems, would make it difficult to incorporate them into the EAC.

Opportunities and Challenges

Compared to Africa's other major common markets, the new EAC has some advantages. It is relatively small and geographically compact, with only three members. The members are quite comparable in size, culture, and political dynamics; for example, they share Swahili and English as lingua francas. In contrast, SADC, ECOWAS, and COMESA have between 14 and 20 members apiece. Each of these organizations has been challenged by major disparities in the size and political heritage of its member states.

The three member states are able to draw upon over 80 years of experience with various forms of integration for lessons, both positive and negative. With Kenya's economic decline in recent years, coinciding with the more rapid growth of Tanzania and Uganda, the economic imbalances that were a major challenge in previous decades may now be less of a problem. (Table 30.1 presents recent basic statistical indicators for the three countries.) Moreover, with the 2002 change of government, Kenya has begun to inaugurate economic and political reforms that will facilitate regional harmonization.

This is not to deny the major challenges that the EAC faces. Its economies are broadly similar in what they produce, so opportunities to increase trade are somewhat limited. Intra-EAC

TABLE 30.1. Kenya, Uganda, and Tanzania: Basic Statistical Indicators

	Kenya	Tanzania	Uganda
Population (million)	30.7	34.5	22.8
Urban (%)	33	28	14
GNI[a] (U.S.$, billion), 2001	10.3	9.2	6.3
GNI per capita	340	270	280
Annual growth of GDP,[b] 1990–2001	2.0	3.1	6.8
Value of exports (U.S.$, million), 2000	2,891	642	401
Intra-EAC exports (% of total exports)	12.3	7.4	14.5
Merchandise imports (U.S.$, million), 2000	1,571	1,588	935
Intra-EAC imports (% of all imports)	6.3	7.0	25.2
Development aid (U.S.$, per capita)	17	31	37
Debt (U.S.$, million), 2000	6,295	7,445	3,408
Debt as % of GNI	46	50	16
Foreign direct investment (U.S.$, million), 2000	111	193	220

[a]GNI, gross national income.

[b]GDP, gross domestic product.

Data sources: World Bank. *World Development Report 2003*. New York: Oxford University Press, 2003. United Nations. *International Trade Statistics 2000: Vol. 1. Trade by Country*. New York: United Nations, 2000.

trade currently accounts for only about one-tenth of the total trade of member countries. Each of the members also belongs to another common market—Kenya and Uganda to COMESA, and Tanzania to SADC. Harmonizing trade policies within the EAC, while also working within the frameworks created by these other organizations, will be a massive challenge. Uganda and Tanzania continue to be wary about the possible impacts on their own economic health of opening their borders to free trade. Energy and transportation costs for Ugandan industries tend to be somewhat higher than those for Kenya, in part because Uganda is a landlocked state. It is feared that these cost disadvantages could put some Ugandan manufacturers out of business in a free-trade environment.

Nevertheless, the establishment of a new EAC is an encouraging development. The creation of structures to bring about greater regional cooperation is one approach to overcoming the barriers to development that confront many smaller African states. The EAC has several advantages over other regional associations, and as a result may have the potential to achieve the greatest success. However, only time will tell whether the EAC will grow from strength to strength, or will repeat its failure of 1977.

Further Reading

For insights into British colonial rule in East Africa, see these sources:

Berman, B., and J. Lonsdale. *Unhappy Valley: Conflict in Kenya and Africa.* Athens: Ohio University Press, 1992.
Brett, E. A. *Colonialism and Underdevelopment in East Africa: The Politics of Economic Change 1919–1939.* New York: Nok, 1973.

There is a large literature on politics and development in Kenya. See, for example:

Haugerud, A. *The Culture of Politics in Modern Kenya.* Cambridge: Cambridge University Press, 1997.
Kimenyi, M. S., et al., eds. *Restarting and Sustaining Economic Growth and Development in*

Africa: The Case of Kenya. Aldershot, UK: Ashgate, 2003.
Leys, C. *Underdevelopment in Kenya: The Politics of Neo-Colonialism, 1964–1971.* New York: Random House, 1975.
Throup, D., and C. Hornsby. *Multi-Party Politics in Kenya: The Kenyatta and Moi States and the Triumph of the System in the 1992 Election.* Athens: Ohio University Press, 1998.

To study Tanzania's socialist experiment, and its experience with structural adjustment, see the following:

Hyden, G. *Beyond Ujamaa in Tanzania: Underdevelopment and an Uncaptured Peasantry.* London: Heinemann, 1980.
Nyerere, J. *Ujamaa: Essays on Socialism.* Oxford: Oxford University Press, 1968.
Tripp, A. M. *Changing the Rules: The Politics of Liberalization and the Urban Informal Sector in Tanzania.* Berkeley: University of California Press, 1997.
World Bank. *Tanzania at the Turn of the Century: From Reforms to Sustained Growth and Poverty Reduction.* Washington, DC: World Bank, 2001.

The following sources look at Uganda under the Museveni regime:

Behrend, H. *Alice Lakwena and the Holy Spirits: War in Northern Uganda, 1985–97.* Athens: Ohio University Press, 2000.
Bigsten, A., and S. Kayizzi-Mugerwa. *Crisis, Adjustment, and Growth in Uganda: A Study of Adaptation in a Changing Economy.* London: Palgrave Macmillan, 1999.
Hansen, H. B., and M. Twaddle, eds. *Developing Uganda.* Athens: Ohio University Press, 1998.
Kreimer, A., et al. *Uganda: Post-War Reconstruction.* Washington, DC: World Bank, 2000.
Mitra, J. D. *Uganda: Policy, Participation, People.* Washington, DC: World Bank, 2001.

The following books offer comparative studies of East African nations:

Barkan, J., ed. *Beyond Capitalism vs. Socialism in Kenya and Tanzania.* Boulder, CO: Lynne Rienner, 1994.
Berg-Schlosser, D., and R. Siegler. *Political Stability and Development: A Comparative Analysis of*

Kenya, Tanzania, and Uganda. Boulder, CO: Lynne Rienner, 1990.

See the following sources for details about past and current attempts at integration:

Hazlewood, A. *Economic Integration: The East African Experience.* London: Palgrave Macmillan, 1976.

Nangale, G. *Know the Reborn East African Community.* Arlington, VA: Pen, 2002.

Sircar, P. K. *Development through Integration: Lessons from East Africa.* New Delhi: South Asia Books, 1990.

Internet Sources

The websites for COMESA, EAC, and SADC are as follows:

Common Market of Eastern and Southern Africa (COMESA). www.comesa.int

East African Community On-Line. http://193.220.91.23/eac

Southern African Development Community (SADC). www.sadc.int

Government websites are an excellent source for information on current policy directions:

Government of Uganda. www.government.go.ug

Kenya Government Online. www.kenya.go.ke

The United Republic of Tanzania. www.tanzania.go.tz

The following East African periodicals provide insight into current political issues in the region:

The Arusha Times on the Web. www.arushatimes.co.tz

The East African Standard. www.eastandard.net

The New Vision: Uganda's Leading Daily. www.newvision.co.ug

Conclusion

Which Future?

Africa and its people struggle against imposing odds in their quest for survival and development. Four decades of disappointment have shown conclusively that African development is a much more complex undertaking than was once believed. The optimism of the 1960s has long since faded. For most African governments and for many communities and families, development has taken a back seat to survival.

The Hausa of Nigeria have a proverb: "If the world was a just place, then the spindle would be clothed in a gown and trousers" (i.e., the spindle never benefits from the products of its labor). For hundreds of years, Africa has been that spindle—the source of vast wealth that has benefited others, but brought Africans few and often fleeting rewards.

As we look into Africa's future, the foremost question is whether the current crisis can be reversed so that sustainable and balanced development can be anticipated. Although prospects for such change remain very much in doubt, the proven resilience of African societies and systems of production provide seeds for hope. So too does the growing commitment to the vision of an African Renaissance.

Will There Be an African Renaissance?

Since 1996, President Thabo Mbeki of South Africa has been proclaiming the dawn of an African Renaissance: "Yesterday is a foreign country—tomorrow belongs to us." In this vision of an African Renaissance, Africa will rise from decades—indeed, centuries—of suffering, exploitation, and underdevelopment. The future depends on achieving open, democratic societies in which leaders are fully accountable. It involves putting to use peaceful methods of conflict resolution. It means achieving sustainable, people-centered economic growth—within the framework of the world economy, but without surrendering to the idea of the market as a modern god. It is to be achieved by galvanizing all Africans to become engaged in the struggle for Africa's renewal.

The concept of an African Renaissance has caught the imagination of South Africans—no doubt because it coincides with the building of a new national identity that acknowledges and learns from the past, but is resolutely focused on building for the future. The concept has also stirred increasing interest elsewhere in Africa.

If the discussion of an African Renaissance is to move beyond mere rhetoric, major changes will have to take place, not just in South Africa but also across the continent. The move toward democratization that has occurred in many states over the past decade will need to continue to spread. The progress that has occurred in countries as diverse as Benin, Mali, Ghana, and Lesotho bodes well for the future.

The growing strength and engagement of African civil society over the past decade are causes for optimism. In countries where significant progress toward democratization has occurred, civil society has played a major role in development at the grassroots level, often working effectively with the government. In states with repressive regimes, civil society organizations have exerted increasing pressure for political change; the same forces that finally pushed Kenya's President Daniel arap Moi into retirement are increasing the pressure on other autocratic leaders, such as President Robert Mugabe of Zimbabwe.

At a continental level, the establishment of the African Union represents the opportunity for a new beginning, replacing an Organization of African Unity (OAU) that had ceased to be either effective or credible in forging African consensus on key issues. Through oganizations such as the Southern African Development Community (SADC) and the Economic Community of West African States (ECOWAS) African states have gradually moved toward greater cooperation at the regional level. These organizations are not panaceas, but they do offer opportunities for more intra-African trade and collaborative ventures.

Or Will There Be Spreading Decline?

The events of the past decade have also demonstrated the uphill battle that Africa faces if it is to achieve major progress. The course of events in Somalia, Liberia, Rwanda, Sierra Leone, and the Democratic Republic of the Congo seem to epitomize the descent into chaos predicted by Robert Kaplan in his controversial 1994 article,

"The Coming Anarchy." In each of these countries, the state effectively ceased to function, and power passed to competing factions headed by regional warlords. In Rwanda and Sierra Leone, a tense peace has been established, with some movement toward national reconciliation. In Somalia, Liberia, and the Democratic Republic of the Congo, however, there are few signs of hope early in the new millennium. A decade after international peacekeeping forces pulled out of Somalia, the country remains essentially ungoverned; it is divided into regional fiefdoms, and for all intents and purposes has been abandoned by the world community.

The scourge of HIV/AIDS is a huge calamity that does not distinguish between rich and poor, autocratic and democratic states. Indeed, the countries hardest hit include several, such as Botswana, Swaziland, and South Africa, with incomes that are much above average for the continent. The largely hidden crisis of AIDS is its impact upon economies and societies; essentially, it sucks away their productive capacity. Children cannot learn in schools robbed by the disease of their teachers, nor can health care systems robbed of their doctors and nurses function effectively. The famine of 2002–2003 has demonstrated the pervasive interconnections of agrarian decline, hunger, and HIV/AIDS. Stephen Lewis, the United Nations Special Envoy on HIV/AIDS in Africa, has speculated that some African states may collapse under the burden of the disease.

Poverty frequently frustrates the efforts of states seeking to overcome their underdevelopment. Ethiopia, for example, has made great efforts since the famines of the mid-1980s to achieve food security. Storage facilities were constructed in many parts of the country to hold emergency food; community labor was mobilized to create terraces to reduce soil erosion and reservoirs to hold water; and agricultural research and extension were given high priority. Yet in a country with a per capita income of $100 per year and a population exceeding 60 million, it is hardly a surprise that these efforts to become more self-reliant proved inadequate when the rains failed in 2002. Regardless of its development efforts and needs, Ethiopia receives among the very lowest rates of aid—only $11 per person in 2000.

There can be little hope of a true African Renaissance in a country such as Ethiopia without strong support for its development efforts from the international community. Indeed, because of high levels of indebtedness, several countries have a net outflow of wealth; that is, they pay more in interest and loan repayments than they receive in new development assistance. How can sustainable development be expected to occur in such circumstances? In the majority of African countries, the International Monetary Fund (IMF) plays a quasi-colonial role in dictating basic economic policies. In short, much as Africa needs to depend on its own resources to develop, a different dynamic between Africa and the world's wealthier nations is necessary if fundamental, sustainable change is to have a chance of occurring.

The World and Africa

In September 2000, the United Nations issued its Millennium Declaration, with this pledge: "We will spare no effort to free our fellow men, women, and children from the abject and dehumanizing conditions of extreme poverty, to which more than a billion of them are currently subjected." The declaration pledged to work toward the achievement of eight broad objectives:

• Eradicate extreme poverty and hunger.
• Achieve universal primary education.
• Promote gender equality and empower women.
• Reduce child mortality rates by two-thirds.
• Combat HIV/AIDS, malaria, and other diseases.
• Ensure environmental sustainability.
• Develop a global partnership for development.

Unfortunately, a few years later, (in)actions speak louder than words. The United Nations Development Programme's (UNDP's) *Human Development Report 2002* indicates that individual African states are as likely to be falling behind in meeting these objectives as they are to be on target for meeting them. Since the end of the Cold War, international development aid to Africa has continued to decline.

The endorsement of the New Economic Partnership for Africa's Development (NEPAD) in 2002 by the world's leading economic partners brought new hope. Had the industrialized world finally recognized that both humanitarianism and prudent self-interest dictated that African underdevelopment could no longer be ignored? Regrettably, in the months that followed the signing of the NEPAD agreement, it was clear that no real change had occurred. As world attention focused on the prospect of war in Iraq, food shipments to meet the urgent calls for emergency food relief for over 100 million people in 20 countries where crops had failed fell far sort of need. A U.S. pledge in February 2003 to devote $10 billion to fight HIV/AIDS in Africa was welcomed enthusiastically. However, within days, the U.S. government announced that general development assistance for Kenya, Uganda, and Tanzania would be cut drastically. In effect, what had been given with one hand was being taken away with the other.

The acronym NEPAD speaks of *partnership* to achieve sustainable development in Africa. Partnership implies mutual responsibilities— for Africans to work for genuine people-centered democratization in every country and to mobilize all available resources toward facilitating sustainable and socially just development; and for wealthy states to support this process through increased aid, debt forgiveness, and changes to global financial and trading regulations that have been so harmful to Africa's interest. With a true commitment to such a partnership, the dream of an African Renaissance may yet become a reality.

Glossary

AAF-SAP (African Alternative Framework to Structural Adjustment Programs for Socio-Economic Recovery and Transformation): A set of strategies to address the current economic crises of Africa, proposed by UNECA in 1989 as an alternative to the structural adjustment programs of the IMF and World Bank.

Afar Depression: A region adjoining the Red Sea, centered on Djibouti; it is located at the junction point of three tectonic plates and is known for its constant earthquakes and numerous volcanic features, which reflect the region's extreme geological instability.

African diaspora: The scattered peoples of African origin in other parts of the world; their dispersal was primarily the result of the trade in African slaves in centuries past.

Africanity: A term coined by Jacques Maquet that refers to the cultural unity of the peoples of black Africa.

African Union: The political organization linking all sovereign African countries, established in 2002 as the successor to the OAU.

Afrikaners: White South Africans primarily of Dutch origin. The Afrikaner identity is rooted in the use of Afrikaans as a language, in affiliation with the Dutch Reformed church, and in a shared history of isolation from (and often mistrust of) Europe.

Afrocentrism: An academic perspective in historical–cultural studies that emphasizes the cultural unity of Africa and celebrates African civilization and achievements.

AGOA (African Growth Opportunity Act): U.S. legislation passed in 2000 granting favored access to U.S. markets for apparel manufactures from African countries that have met eligibility criteria, especially the liberalization of trade and economic policies.

Agroforestry: The full integration of different species of useful trees into farming systems. These trees may be a source of valuable products such as foods and fuelwood, and serve other purposes such as the preservation of soil fertility.

Alfisols: Moderately weathered, moderately fertile brown to reddish soils associated with humid and semihumid climates; in Africa, they are found primarily in savanna areas.

Alley cropping: An agricultural system that involves planting crops between rows of leguminous shrubs. It was developed through agricultural research as a stable alternative to shifting cultivation, primarily for tropical forest environments.

ANC (African National Congress): A multiracial, but primarily black-supported, party founded in 1912. The ANC led the struggle against apartheid in South Africa and won the first postapartheid election in 1994.

Apartheid: The official doctrine that formally institutionalized the separate development of racial groups in white-ruled South Africa.

Aridisols: Soils found in dry environments, characterized by low organic matter and accumulations of soluble minerals (salts and carbonates).

Bantu: A term applied to a large family of African languages of the Niger–Congo group, as well as to the peoples who speak them. Over three millennia, Bantu peoples spread out from a point of origin in central Nigeria, eventually dominating most of Africa east and south of the Cameroon Highlands.

Bantustans: See **homelands**.

Basement complex: Ancient, geologically very stable rock formations dating from the Precambrian era. Most of the African continent consists of basement complex formations.

Basic needs: The economist Dudley Seers argued that providing basic needs—notably food, shelter, employment and health—is the first priority for development. These ideas form an integral part of the "bottom-up" approaches to development that have grown out of the rejection of modernization theory.

Berlin Conference: A gathering of European nations in 1884–1885 that established the ground rules for the colonial carving up of Africa, following moves by King Leopold II of Belgium to annex the Congo.

Bilateral aid: Aid agreements made between two countries, a recipient and a donor country. See also **multilateral aid.**

Biomass: The mass of living matter in an ecosystem. Biomass as a source of energy refers to fuelwood, charcoal, crop refuse, and dung—energy obtained from living (nonfossil) sources.

Biome: A large terrestrial ecosystem with a common type of vegetation, as well as broadly similar soils, climatic conditions, and animal life.

Boserupian: A perspective in debates about the impacts of population growth on the environment, derived from the work of Ester Boserup. It sees population growth as an incentive for societies to innovate and use natural resources more intensively.

Bush meat: Wild fauna of many types that are harvested for food. Bush meat is widely harvested for household subsistence. In certain areas, such as the Congo basin, a large-scale commercial trade in bush meat is a major threat to wildlife populations.

Cape to Cairo: The imperialist dream of Cecil Rhodes, who promoted a vision of continuous British-governed territory from the northern and southern extremities of the continent.

Cassava: A high-yielding but nutritionally poor starchy root crop that is a widely grown and consumed staple in tropical Africa.

Chemical weathering: Decomposition and decay of constituent minerals in rock as a result of the minerals' chemical alteration.

CIDA (Canadian International Development Agency): International development arm of the Canadian government.

CITES (Convention on International Trade in Endangered Species of Wild Fauna and Flora): An international agreement to strictly regulate or prohibit international trade in endangered species.

Civil society: The organized political and social activity that occurs at the interface between the state and individual/unorganized society. Development NGOs, trade unions, and political pressure groups are examples of civil society organizations.

Civilizing mission: The self-defined objective of many early colonial administrators and missionaries, who saw the conquest of Africa as a charitable act of Christian love and responsibility.

Close-settled zones: Areas of high-density rural settlement, especially in the Hausa heartland of northern Nigeria (e.g., the Kano close-settled zone).

Coloured: South Africans of mixed-race ancestry. Coloured was used as an official racial category under apartheid, but in the postapartheid era coloured is a general term for mixed-race people.

COMESA (Common Market for Eastern and Southern Africa): A common market linking 20 poorer nations in eastern and southern Africa; formed in 2000 to encourage greater trade and policy harmonization among members.

Commoditization: The incorporation of goods and services formerly produced on a subsistence basis and exchanged informally into the formal, monetarized economy.

Commonwealth of Nations: An international political organization, with membership consisting almost exclusively of countries formerly ruled by Great Britain.

Conditionalities: A term for the requirements laid down by the IMF as preconditions for the granting of loans. These requirements generally involve neoliberal economic reforms such as the removal of tariffs and subsidies, as well as currency devaluation.

Conflict diamonds: Sometimes called *blood diamonds*, they are diamonds from areas of conflict, such as Angola, which were sold illegally by rebel organizations as a means of financing their military operations. A tracking system for marketed gems was introduced in 2000 in an attempt to eliminate the trade in conflict diamonds.

Copper belt: A mineral-rich zone in the southeastern Democratic Republic of the Congo (Shaba Province) and north central Zambia, which produces some 10% of the world's copper as well as several other minerals.

Craton: A massive body of very stable, ancient rocks. Three large cratons form the core of the African continent.

Crude birth (death) rate: The annual number of births (deaths) per 1,000 persons in a population.

De Beers: A South Africa-based multinational corporation that, until 2000, exerted monopoly control over the global trade in diamonds.

Demographic transition: A model of historical population change, based on the pattern of change in late-19th- and early-20th-century Europe and North America. It predicts sequential change from an initial state of high birth and death rates, through an intermediate stage when first birth rates and then death rates decline significantly, to a final stage when birth and death rates reach a new lower-level equilibrium.

Dependency theory: A theory of underdevelopment associated with the writings of Andre Gunder Frank and other Latin American scholars, who argued that the poverty and underdevelopment of the colonial periphery was directly related to the prosperity of the colonizing power. Developed as a rebuttal to modernization theories, this approach was especially prominent in the 1970s.

Development from within: An approach to development, increasingly popular in the late 1980s and 1990s, that emphasizes the importance and untapped further potential of local self-reliance in development.

DES (dietary energy supply): Usually, the daily average number of calories consumed in a population.

Double workload: A term used primarily in feminist literature to refer to the dual responsibilities of women for reproduction and production (of goods and services traded in an economy).

Duricrusts: Hardened, erosion-resistant outcrops of iron and aluminum rich laterites in tropical regions. They often form cap rocks on flat-topped hills known as *mesas*.

EAC (East African Community): A common market union of Kenya, Tanzania, and Uganda that existed from 1967 to 1977, and that was revived in 1999. The EAC seeks to harmonize trade, tariff, and fiscal policies, and to enhance cooperation in areas such as tourism and transportation.

ECOWAS (Economic Community of West African States): An organization founded in 1975, which links 16 countries of West Africa. The founding Treaty of Lagos made provision for the elimination of barriers to trade and mobility, and aimed at promoting economic, social, and cultural cooperation.

Enset: A tree sometimes known as the false banana; its stem and bulbous root are prepared as a starchy food. It is a primary staple in parts of Ethiopia, and is consumed in certain other regions as a famine emergency food.

Entitlements: A concept first articulated by Amaryta Sen; it refers to the inequitable access to food of different members of communities and households, in relation to factors such as social and economic status and kinship networks.

Expanded Immunization Program: A joint UNICEF–WHO program to achieve full vaccination coverage of Third World children against common diseases of childhood.

Export processing zones: Small development areas designated by certain governments for export-oriented industrial development. Policies such as low rates of taxation and exemption from tariffs are used to attract foreign investment to these zones.

FAO (Food and Agriculture Organization of the United Nations): An agency of the UN, founded in 1945 and based in Rome, Italy, that seeks to enhance agricultural productivity, improve nutrition, and improve living conditions for rural peoples.

Food-for-work programs: Programs organized by the World Food Program and other agencies to distribute food in areas of need in return for work by community members on specified conservation and development projects.

Food security: The assured availability of sufficient quantities of varied, nutritious, culturally acceptable food to the entire community.

La Francophonie: An international organization consisting of nations that were formerly colonies of France, or that had other close historical ties to France.

Formal sector: The portions of the economy that are officially recognized, regulated, and remunerated. The formal sector includes larger-scale retailing and manufacturing, as well as the public sector.

Frelimo (Front for the National Liberation of Mozambique): The political–military front that organized armed resistance in Mozambique against Portuguese colonialism from 1962 to 1975, and that has remained the governing party since independence.

Game cropping: The practice of systematically harvesting wild animal populations (especially ungulates) as a source of food.

Gender and development: Analysis of the situation of women in the South that focused on the need to change power relationships that perpetuated women's oppression. This approach was a response to the perceived inadequacies of the earlier women-in-development approach.

Global apartheid: Gross international disparities in human rights, power, and wealth structured by race and place and perpetuated by global economic and political structures.

Globalization: The increased interconnectedness of different parts of the world, bringing about increasingly rapid and pervasive economic, cultural, and political change.

Gondwanaland: Ancient megacontinent composed of Africa, South America, Australia, Antarctica, India, Madagascar, and Arabia; it began to break up some 250 million years ago, with the constituent parts slowly moving apart via continental drift.

Great Escarpment: A massive escarpment, up to 3,000 m in height, paralleling the coast of southern Africa from Angola to Mozambique.

GDP (gross domestic product): Annual value of all goods and services produced in an economy, plus income earned by residents outside the country, minus incomes accruing to foreign residents of the country.

GNI (gross national income): Annual value of all goods and services produced in an economy. GNI has become the term of choice in place of the previously-used gross national product (GNP).

Green Revolution: The introduction of high-yielding varieties of rice, wheat, maize, and other crops into Third World farming systems, starting in the 1960s. Because of the demands of these crop varieties, the Green Revolution entailed adoption of a package of innovations, including increased use of chemical fertilizers and pesticides.

Groundnuts: Peanuts—an important crop produced in savanna regions for both export and domestic consumption.

Harambee: A community self-help movement that flourished in Kenya in the early years of independence, the objective being to increase the pace of development by mobilizing local resources.

Harmattan: Dry, dust-laden winds originating over the Sahara Desert that occur widely across Western Africa during the dry season.

Hidden curriculum: Values and attitudes transmitted informally through the educational system, which may have the effect of undermining indigenous cultures and economies.

High Africa: The southern and eastern part of the continent of Africa, characterized by elevations averaging 1,000–2,000 m above sea level (see also **Low Africa**).

Homeland: The 10 ethnic territories set aside under apartheid for the exclusive use of black Africans. Four of them became "independent" between 1976 and 1981. All were reabsorbed into South Africa in 1994 as part of the 1993 constitutional agreement.

HDI (human development index): A multidimensional scale developed by the UNDP to distinguish between levels of development, defined as the range and quality of options available to people for shaping their own destinies.

IADP (integrated agricultural development projects): A development strategy mostly associated with the World Bank, which uses intensive investment to improve rural infrastructures (roads, schools, etc.) and to promote new agricultural technology to spur development.

IITA (International Institute of Tropical Agriculture): One of a series of institutions doing research in tropical agriculture. It is located in Ibadan, Nigeria, and specializes in crops and farming systems of tropical forest environments.

ILO (International Labour Organization): An agency of the UN founded in 1919 to promote social justice for workers through the development of international labor standards and the provision of support services for workers' organizations, employers, and governments.

IMF (International Monetary Fund): An international agency originally founded in 1944, which oversees the "creditworthiness" of nations and establishes terms for loans. In this role, it has forced most African countries to adopt structural adjustment programs.

Indigenous knowledge: A cumulative body of knowledge and beliefs, passed from generation to generation, that is specific to a given culture and often to subgroups (e.g., gender, occupation) within the culture. Indigenous knowledge evolves over time as a reflection of experimentation, interaction with other societies, and cultural change.

Indirect rule: A strategy of colonial rule at the local level that involved retention of indigenous governments, closely supervised by colonial officials. It enabled the colonialists to rule with relatively few European personnel and to co-opt African rulers as collaborators.

Infant mortality rate: The annual number of infant deaths per 1,0000 live births in a population.

Informal sector: Diverse forms of economic activity undertaken outside formal systems of regulation and compensation. This sector includes small-scale manufacturing, petty trade, and other smaller-scale economic activities (see **formal sector**).

Inselberg: An "island mountain"—a solitary dome of rock, resistant to chemical weathering and erosion, that rises above the surrounding plains. It is most common in savanna regions.

Intercropping: The practice of growing two or more crops side by side in the same field. It is commonly used in indigenous farming systems to capitalize on symbiotic relationships between certain species, to reduce erosion, and to maximize yields.

Internally displaced populations: People forced to flee their home regions by civil unrest, environmental catastrophes, and similar circumstances, who seek refuge in another part of their own country (see also **refugee**).

ITCZ (intertropical convergence zone): The zone of low pressure located between the tropics, where air converges and ascends. The seasonal movement of the ITCZ governs the change of seasons in the tropics.

Labor migrancy: A form of labor mobility in which workers and their families are dependent on both wage earnings and subsistence production to make ends meet. The term is particularly common in studies of South African migrant labor, in which legal restrictions on where people could live ensured the continuation of labor migrancy.

Labor reserves: Areas that were deemed by colonial administrators to have little potential for development, and hence were exploited as source areas for migrant workers for mines, plantations, and so forth elsewhere.

Landrace seeds: Indigenous varieties of crops associated with a particular region, which evolved and were genetically improved through a prolonged process of selection by local farmers.

Laterite: Highly leached, infertile soils of the wet tropics, often with a brick-like hardpan crust.

Legitimate commerce: The 19th-century term for trade in such commodities as palm oil, which superseded the slave trade in West Africa.

Lingua franca: A language used as a medium of communication by people whose native tongues differ. Swahili and Hausa are important regional lingua francas in Africa.

Lomé Conventions: A series of agreements between the European Union and most African, Caribbean, and Pacific states to facilitate trade and development assistance.

Low Africa: The western and west-central portion of Africa, consisting of comparatively low elevation plains (see also **High Africa**).

Malthusian: A perspective in debates about population growth and its impacts on the economy that argues that excessive population growth inevitably brings about resource scarcity and environmental degradation. These ideas were proposed initially by Thomas Malthus, an English economist, at the end of the 18th century.

Mfecane: A period of turmoil during the early 19th century among the African population of South Africa. Violent competition for land led to the expansion of the Zulu kingdom and the subsequent migration of some other groups, such as the Sotho and Ndebele, to new territories.

Modernization theory: A theory of development, popular during the 1960s, which envisaged Third World countries as developing modern economies and societies like those of the West. Walter Rostow's stage theory is the most prominent statement of the modernization perspective.

Montane vegetation: Distinctive types of vegetation associated with certain high mountain regions, such as in eastern Africa.

MOSOP (Movement for the Survival of the Ogoni People): Organization of the Ogoni people of the Niger Delta, which gained international prominence for its struggle against the damage to their land, economy, and society caused by oil development. Ken Saro-Wiwa and other leaders of MOSOP were executed by the Nigerian government in 1995

Multilateral aid: Aid channeled through international agencies (e.g., UNICEF, UNDP) rather than from a specific donor country.

Native Lands Act: Legislation passed by the government of South Africa in 1913 that severely limited the rights of Africans to own land outside of designated reserve areas.

Neocolonialism: A term devised by Kwame Nkrumah of Ghana, to describe the continuing economic control of Europe over its politically independent former colonies.

NEPAD (New Economic Partnership for Africa's Development): A framework to stimulate African economic development in the new millennium, signed in 2002, involving a commitment by the world's major economic powers to increase development assistance to Africa in return for a commitment by African leaders to establish stronger democratic and economic institutions.

NGO (nongovernmental organization): International or indigenous organizations, operating at arm's length from government. Their scope may be very broad (e.g., Oxfam), or may be specific either to a particular type of activity or a single community.

Nitrogen fixing: The process in which certain species of plants (legumes) convert atmospheric nitrogen into organic nitrogen compounds that are accessible to all plants.

Ogaden: An arid region of southeastern Ethiopia, adjacent to the Somalia border, long claimed by Somalia because its population consists primarily of ethnic Somalis.

Onchocerciasis (river blindness): A chronic disease in which the proliferation of microfilariae (tiny worms) called *Onchocerca volvulus*, transmitted by the black fly called *Simulium damnosum*, cause the formation of skin nodules—and, in extreme cases, reduced life expectancy and blindness.

OAU (Organization of African Unity): An organization of 52 independent African states, founded in 1963, with the objective of promoting African solidarity. It was succeeded by the African Union in 2002.

ODA (official development assistance): A more formal term for development aid, particularly bilateral aid.

OPEC (Organization of Petroleum Exporting Countries): A cartel linking most of the world's major petroleum producing countries.

Orientalism: A term associated with the writings of Edward Said, used to characterize the West's pervasive and long-standing cultural "construction" of non-Western societies (the Orient) as mysterious, irrational, and backward, and thus fundamentally different from Western societies (the Occident).

Orographic precipitation: Precipitation that results when moisture-laden air is forced to rise by a mountain range or other topographic barrier.

Oxisols: Soils found in very moist tropical environments, which are old, highly weathered, and of low fertility.

Pan-African movement: An international political and intellectual movement of people of African descent, founded in 1900, that fought for black cultural and political liberation. Its slogan, "Africa for the Africans," is symbolic of the key role the movement played in setting the stage for African independence.

Parastatals: Companies or agencies established by governments and left to operate semiautonomously in a designated commercial or service field. Examples might include a state airline, a well-drilling company, or a state-established industry.

Participatory development: A community-based approach to development that reverses the usual power relationships between outside experts and the community. Participatory development envisages the empowerment of communities to analyze their own needs, develop their own proposals for change, and take the lead in program implementation.

PHC (primary health care): An approach inspired by the example of China and set out in the Alma Ata conference of WHO in 1978, calling for health strategies at the community level that addressed local needs in accordance with local values and resources.

Political ecology: The analysis of social, economic, and political forces impinging on people's use of their environment. This approach examines the ecology of production in relation to ideological, symbolic, and physical struggle over resources.

Polygyny: The cultural practice of permitting a man to have more than one wife.

Populist socialism: A development ideology that emphasized African communal and/or cooperative traditions as the source of principles for modern development. Julius Nyerere's Tanzania is the best-known example.

Postcolonialism: An approach to studying the history of oppressed peoples that reverses the usual official analysis, focusing instead on the perspectives of poor, marginalized peoples and paying particular attention to their resistance to imposed authority.

Poststructuralism: A theoretical approach that rejects broad, structural generalizations, and instead emphasizes the importance of the diversity of local situations and of local responses to them.

Primate city: A city that is much larger than other cities in the same country, and that typically has a very high proportion of the nation's modern economic development and political power.

Precambrian: The earliest geological time period, from the earth's formation 4.6 billion years ago until the beginning of the Cambrian period about 570 million years ago.

Purdah: The practice in orthodox Islamic societies of confining women to their homes, with opportunities to go out strictly regulated by husbands. The strictness of purdah varies in relation to religious orthodoxy, wealth, and class.

Quaternary: The last approximately 1.6 million years of the Earth's history, when there was a human presence on the planet. Earth scientists interested in the Quaternary period study glaciation and other manifestations of comparatively recent environmental change.

Rainbow Nation: A term used by President Nelson Mandela to characterize the diversity and spirit of cooperation of the diverse peoples of the new, postapartheid South Africa.

Refugee: People forced to flee from their home regions by civil unrest, environmental catastrophes, or similar circumstances, who seek refuge in another country and are enumerated by/recognized by UNHCR (see also **internally displaced populations**).

Remittances: Sums of money and other material aid that migrants send to their home communities, often to support their families or as investments in anticipation of their own return.

Removals: The South African policy under apartheid, whereby people (mostly black) were forced to move off land designated officially for other groups. In the case of blacks, people were removed to their assigned homeland.

Renamo (Mozambique National Resistance): A guerrilla organization, supported throughout the 1980s by South Africa, which fought the Frelimo government of Mozambique, with devastating effects on the country's economy and society.

River blindness: See onchocerciasis.

Roundwood: Unprocessed logs (as opposed to sawn lumber and manufactured wood products).

RUF (Revolutionary United Front): A rebel organization that waged a brutal campaign against the government and people of Sierra Leone during the 1990s.

Salinization: Processes in arid and semiarid regions with high rates of evapotranspiration that create high concentrations of mineral salts at or near the surface of the soil, thus affecting its ability to support the growth of plant species that are not highly salt-tolerant.

SADC (Southern African Development Community): The successor organization to SADCC (see below), established in 1992 with the signing of the Treaty of Windhoek. It envisaged the development of SADC into a common market with growing political and economic integration, and the eventual inclusion of South Africa as a member.

SADCC (Southern African Development Coordination Conference): An organization founded in 1980 to reduce the dependence of countries in southern Africa on South Africa through coordinated development planning (see **SADC**).

Sahel: The semidesert region at the southern fringe of the Sahara, as well as the countries that fall within this region (usually defined as Senegal to Chad). Droughts in the 1970s and early 1980s focused international attention on the Sahel.

Schistosomiasis (bilharzia): A disease caused by blood flukes *(Schistosoma)* that mature in the liver, causing anemia and urinary problems. Bilharzia is often related to water development projects, since its spread depends on the presence of certain freshwater snails.

Sharia: The Islamic legal code, the adoption of which in northern Nigeria and several other Muslim countries has been a matter of considerable controversy.

Shifting cultivation: Also called *swidden* or *slash and burn*, it is an agricultural system in which farms are abandoned for extended periods after a short cycle of cultivation. It is suited to tropical forest environments with poor soils and low population densities.

Site-and-service schemes: An approach to urban development that seeks to achieve cost-effective, orderly, and community-responsive growth. The government constructs the basic infrastructure (e.g., roads and water) for a new suburb, and sells low-cost lots to people who then build homes suited to their individual needs and resources.

Sleeping sickness: See **tsetse**.

SPHC (Selective primary health care): A health care strategy, much promoted by development agencies in the late 1980s and 1990s, that focuses entirely on delivering a few, most cost-effective interventions (e.g., safe water, immunization) to as many people as possible.

Squatter settlement: A settlement located on land neither owned nor rented by the residents—and, hence, illegal.

State bourgeoisie: The comparatively well-off social class that has benefited from its close ties to (or control of) the postcolonial state. Members include bureaucrats, soldiers, and entrepreneurs with government contracts.

Structural adjustment programs: A package of reforms designed by the IMF and World Bank to restore nations' economic health through reforms to devalue currency, reduce public-sector expenditure, and strengthen market forces. Structural adjustment programs have been implemented in virtually all African countries.

Swahili: The synthetic (African–Arab) culture of coastal East Africa, and especially the language that developed within this milieu. Although Swahili is the native tongue of relatively few, it is used as a lingua franca by the majority of East Africans.

SWAPO (South West African People's Organization): The leading nationalist organization in Namibia. SWAPO launched its armed struggle against South African occupation in 1966, ending in independence under a SWAPO-led government in 1989.

Teff: A grass-like cereal crop native to the Ethiopian highlands, the primary staple crop of the Ethiopian highland region.

Termitaria: Termite nests, which have various sizes and structures among different termite species.

Tied aid: Aid programs specifying that certain or all inputs (e.g., machinery) be purchased from the donor country. Tied aid limits the ability of recipient countries to choose the cheapest and best technology, and fosters long-term dependence (e.g., for spare parts).

Tillites: Ancient, fossilized sedimentary deposits laid down by glaciers.

Total fertility rate: The average number of births that a woman would experience in her lifetime if current age-specific fertility rates persist.

Transnational corporations: Firms, usually based in the major industrial nations, that control operations in more than one country.

TRC (Truth and Reconciliation Commission): A tribunal established by President Mandela and chaired by Archbishop Desmond Tutu as a device to encourage national healing in South Africa in the wake of the apartheid era. The TRC encouraged victims of violence under apartheid to tell their stories, and perpetrators of that violence to publicly acknowledge their responsibility.

Triangular trade: Trade linkages (controlled by European powers) between Europe, Africa, and the Americas during the slave-trading era; slaves from Africa were the key commodity.

Trickle-down: The assumed gradual spread of modernization and its benefits from large, national urban centers to the periphery. Trickle-down was a key assumption underlying most development during the 1960s, and contributed to an urban bias in development.

Tsetse: A fly that is the vector for a parasite (trypanosome) that causes trypanosomiasis, or sleeping sickness—a serious disease affecting humans, cattle, and some wildlife species.

Ultisols: A suborder of Alfisols, associated in Africa with moist savanna environments; soils are weathered, reddish-yellow in color, and low in fertility.

UNCTAD (United Nations Conference on Trade and Development): An agency of the UN, established in 1964 and with headquarters in Geneva, Switzerland, that promotes the incorporation of developing countries into the global economy through the integrated treatment of trade and development issues.

UNDP (United Nations Development Programme): The UN agency that spearheads an agenda of international development through activities in program areas such as poverty reduction, democratic governance, and the use of communications technology.

UNECA (United Nations Economic Commission for Africa): Established in 1958 as the regional arm of the UN in Africa, with the mandate of supporting economic and social development of African states and promoting international cooperation among these states.

UNEP (United Nations Environmental Programme): The UN agency, based in Nairobi, that was established to promote programs that protect the environment, and that assist nations and peoples to improve their quality of life without compromising that of future generations.

UNHCR (United Nations High Commissioner for Refugees): Established in 1950 and based in Geneva, Switzerland, the UNHCR is mandated to lead international action to protect refugees and resolve refugee problems worldwide.

UNICEF (United Nations Children's Fund): Founded in 1946, UNICEF is the UN agency charged with protecting the interests of the world's children through programs that address their human rights, as well as their education and health needs.

Ujamaa: Literally "familyhood" in Swahili, an approach to development stressing cooperative effort and an ethic of sharing. *Ujamaa* was the philosophical basis for African socialism in Julius Nyerere's Tanzania.

UNITA (National Union for the Total Independence of Angola): Originally, a guerrilla group fighting for independence. After independence (1975), UNITA continued to fight the Angolan government, with South African backing. Attempts to end the conflict failed until UNITA's leader, Jonas Savimbi, was killed in 2002.

UPE (Universal Primary Education): An ambitious scheme launched by the Nigerian government in 1976, which attempted to make primary education universally available and compulsory. UPE greatly increased enrollments, especially in the north, but fell short of its objectives.

White Highlands: The region of Kenya near Nairobi that developed in colonial times as a center of European commercial farming, and that was the focus of the Mau Mau struggle over land in the 1950s.

Wildlife comanagement: An arrangement in which local communities and governments agree to work together in managing wildlife resources, sharing authority, responsibility, and benefits.

Witwatersrand: South Africa's economic heartland, centered on the city of Johannesburg. The initial development of the region was based on gold mining; mining remains the keystone industry.

Women in development: The initial perspective on Third World women, introduced during the 1970s. It recognized the marginalization of women in the development process and advocated programs to improve their well-being, often through income-generating projects for women (see also **gender and development**).

World Bank: Established in 1944 at the same time as the IMF, the World Bank (initially the International Bank for Reconstruction and Development) aims at assisting member states in their development efforts with the provision of loans and technical assistance.

World Trade Center Agreement: An agreement signed in 1993 between the South African government and the political opposition, including the ANC; it established a political framework for the new South Africa, and defined mechanisms for the transition to majority rule in the country.

WHO (World Health Organization): An agency of the UN, founded in 1919 and based in Geneva, Switzerland, that works to achieve the highest possible level of health for all the world's peoples.

Xeralfs: A suborder of Alfisols, associated with Mediterranean climates.

Index

About the Author

Robert Stock is International Coordinator in the College of Arts and Science, University of Saskatchewan. He previously taught in the Department of Geography and in the Development Studies program at Queen's University (Canada). He holds a PhD from the University of Liverpool.